AF389391

Marine Toxins from Harmful Algae and Seafood Safety

Marine Toxins from Harmful Algae and Seafood Safety

Editor

Shauna Murray

MDPI • Basel • Beijing • Wuhan • Barcelona • Belgrade • Manchester • Tokyo • Cluj • Tianjin

Editor
Shauna Murray
School of Life Sciences
University of Technology Sydney
Sydney
Australia

Editorial Office
MDPI
St. Alban-Anlage 66
4052 Basel, Switzerland

This is a reprint of articles from the Special Issue published online in the open access journal *Toxins* (ISSN 2072-6651) (available at: www.mdpi.com/journal/toxins/special_issues/Seafood_Safety).

For citation purposes, cite each article independently as indicated on the article page online and as indicated below:

LastName, A.A.; LastName, B.B.; LastName, C.C. Article Title. *Journal Name* **Year**, *Volume Number*, Page Range.

ISBN 978-3-0365-6194-3 (Hbk)
ISBN 978-3-0365-6193-6 (PDF)

Contents

About the Editor

Shauna Murray

Professor Shauna Murray is a marine biologist specialising in the research of ecology and evolution using molecular genetic techniques. Marine biotoxins produced by phytoplankton, single-celled marine protists, are amongst the most toxic substances recorded to date. Her team at the University of Technology, Sydney, Australia, researches novel genetic tools for the monitoring of marine water quality, and the implemention of these tools across important aquaculture and fishing industries. They work closely with industries and governments, such as in shellfish aquaculture, fisheries, food safety regulators, and the environmental biotechnology sector.

Article

Toxicity of the Diatom Genus *Pseudo-nitzschia* (Bacillariophyceae): Insights from Toxicity Tests and Genetic Screening in the Northern Adriatic Sea

Timotej Turk Dermastia [1,2,*], Sonia Dall'Ara [3], Jožica Dolenc [4] and Patricija Mozetič [1]

1. Marine Biology Station Piran, National Institute of Biology, 6330 Piran, Slovenia; patricija.mozetic@nib.si
2. International Postgraduate School Jožef Stefan, 1000 Ljubljana, Slovenia
3. National Reference Laboratory for Marine Biotoxins, Centro Ricerche Marine, 47042 Cesenatico, Italy; sonia.dallara@centroricerchemarine.it
4. Institute of Food Safety, Feed and Environment, Veterinary Faculty, University of Ljubljana, 1000 Ljubljana, Slovenia; jozica.dolenc@vf.uni-lj.si
* Correspondence: timotej.turkdermastia@nib.si

Abstract: Diatoms of the genus *Pseudo-nitzschia* H.Peragallo are known to produce domoic acid (DA), a toxin involved in amnesic shellfish poisoning (ASP). Strains of the same species are often classified as both toxic and nontoxic, and it is largely unknown whether this difference is also genetic. In the Northern Adriatic Sea, there are virtually no cases of ASP, but DA occasionally occurs in shellfish samples. So far, three species—*P. delicatissima* (Cleve) Heiden, *P. multistriata* (H. Takano) H. Takano, and *P. calliantha* Lundholm, Moestrup, & Hasle—have been identified as producers of DA in the Adriatic Sea. By means of enzme-linked immunosorbent assay (ELISA), high-performance liquid chromatography with UV and visible spectrum detection (HPLC-UV/VIS), and liquid chromatography with tandem mass spectrometry (LC-MS/MS), we reconfirmed the presence of DA in *P. multistriata* and *P. delicatissima* and detect for the first time in the Adriatic Sea DA in *P. galaxiae* Lundholm, & Moestrup. Furthermore, we attempted to answer the question of the distribution of DA production among *Pseudo-nitzschia* species and strains by sequencing the internal transcribed spacer (ITS) phylogenetic marker and the *dabA* DA biosynthesis gene and coupling this with toxicity data. Results show that all subclades of the *Pseudo-nitzschia* genus contain toxic species and that toxicity appears to be strain dependent, often with geographic partitioning. Amplification of *dabA* was successful only in toxic strains of *P. multistriata* and the presence of the genetic architecture for DA production in non-toxic strains was thus not confirmed.

Keywords: Adriatic; *dabA*; domoic acid; *Pseudo-nitzschia galaxiae*; ITS

Key Contribution: Determination of toxic *Pseudo-nitzschia* species in the Adriatic Sea. Study on the phylogenetic distribution of the capability to produce DA.

Citation: Turk Dermastia, T.; Dall'Ara, S.; Dolenc, J.; Mozetič, P. Toxicity of the Diatom Genus *Pseudo-nitzschia* (Bacillariophyceae): Insights from Toxicity Tests and Genetic Screening in the Northern Adriatic Sea. *Toxins* **2022**, *14*, 60. https://doi.org/10.3390/toxins14010060

Received: 14 December 2021
Accepted: 12 January 2022
Published: 15 January 2022

Publisher's Note: MDPI stays neutral with regard to jurisdictional claims in published maps and institutional affiliations.

1. Introduction

Pseudo-nitzschia H. Peragallo is a genus consisting of 54 confirmed species of diatoms, about half of which have been confirmed as producers of the neurotoxin domoic acid (DA) [1,2]. There are several methods for detecting DA that have evolved over time. Shellfish-monitoring programs use the standard reference method—liquid chromatography with ultraviolet detection, which is sufficient because threshold concentrations are usually high (20 µg/kg shellfish tissue). Immunoassays are also readily available from commercial manufacturers and offer high sensitivity and throughput. Finally, liquid chromatography coupled with tandem mass spectrometry (LC-MS/MS) is the main method used in research today, as it offers high throughput and analytical precision. Several mechanisms for the induction and upregulation of domoic acid have been proposed, often with conflicting evidence. From the synthesis of factors affecting the production of DA presented in [3], it appears that the physiological state of the cell has a significant influence although the evolutionary purpose of the production of DA is not fully understood. Originally, it was proposed that DA is a chelating agent for iron and copper ions [4–6] although there is still conflicting evidence for this hypothesis [7,8]. Nonetheless, the understanding of toxin regulation and physiology has improved significantly recently, particularly with the discovery of the DA biosynthetic pathway, where four enzymes (DabA–D) coded by four genes (*dabA–D*) were discovered [9]. Furthermore, it was established that DA production is induced by copepod grazer cues [10,11] although an increase in DA concentration did not significantly affect grazing in these studies. Recently, however, new evidence has been presented for the deterrent function of DA against grazers [12]. Whatever evolutionary or ecological advantage the production of DA may provide, questions remain as to why some strains of the same species produce it and others do not and how the ability to produce DA is distributed along the phylogenetic tree of *Pseudo-nitzschia*. This is particularly interesting since some studies suggest that species that do not produce DA do not benefit from the addition of DA into their iron-limited growth media [8]. Conversely, the deterrent effect on grazers seems to be of great benefit to DA-producing strains, so it could be assumed that the loss of this ability would be detrimental to such strains. The discovery of nzyme-encoding genes involved in the biosynthetic pathway of DA production [9] provides an opportunity to trace gene distribution and structure within the genus *Pseudo-nitzschia* and beyond and to answer these questions.

The aim of this work was to determine the toxicity potential of several *Pseudo-nitzschia* species found in the Gulf of Trieste (GoT), the northern Adriatic [13], and to complement these results with molecular data to determine whether the production of DA is phylogenetically linked. In this context, we also investigated whether the *dabA* gene is present in strains that we found did not produce DA. The GoT is a nutrient-rich environment with occasional phosphorus limitation. The temperature rarely drops below 7 °C in winter and can exceed 28 °C at the surface in summer. Although potentially toxic *Pseudo-nitzschia* species occasionally bloom here [13,14], DA is rarely found in shellfish and is generally not harmful to the food industry [15]. The diversity of *Pseudo-nitzschia* in the Gulf of Trieste has only recently been elucidated [13], while numerous reports are available for other regions of the Adriatic Sea, e.g., [16–20]. Species richness is comparable to that of other coastal regions of the Mediterranean and other temperate zones, while seasonality and species occurrence seem to be localized to some extent. Reports on toxicity are much sparser, but the presence of DA in cultures of *P. delicatissima* [17], *P. multistriata* [21], and recently *P. calliantha* [22] has been reported. The latter was previously suspected based on circumstantial evidence derived from the analysis of toxin-positive mussel samples and the accompanying net trawls in which *P. calliantha* and *P. pseudodelicatissima* (Hasle) Hasle cells were found [19,23].

We tested six species using different methods and report toxin production in three species, namely *P. multistriata*, *P. delicatissima*, and—for the first time in this area—in *P. galaxiae*. Several strains of each species were studied, and three morphological types were also recognized in *P. galaxiae* [13,24]. We complement our results with a global perspective on the phylogeny of toxic and non-toxic *Pseudo-nitzschia* and as well provide preliminary insight into the distribution of the *dabA* gene.

2. Results

2.1. Toxicity of Individual Strains

Thirty-three strains belonging to six species of *Pseudo-nitzschia* were analyzed for the content of DA. Table 1 shows the toxicity results for each strain tested. The method with the lowest limit of detection (LOD) was the ELISA method, with a LOD of 0.17 ng/mL of DA and a limit of quantification (LOQ) of 0.5 ng/mL. HPLC-UV had a LOQ of 2 μg·mL^{-1}, and LC-MS/MS had a LOQ of 0.8 μg·mL^{-1}. As you can see, some strains were confirmed to be toxic only by the ELISA test, while they did not prove positive in HPLC-UV and in the LC-MC/MS method. This prompted us to retest these samples with ELISA at lower dilutions to confirm the presence or absence of DA. Most retests resulted in concentrations below LOQ, with the exception of *P. galaxiae* strain B3S, where the concentration was still above the limit of quantification. There were also many borderline strains that had concentrations below 0.5 ng/mL (LOQ), but we could not rule out their toxicity because the absorbance was lower than the negative standard values, indicating some competitive binding in the ELISA. *P. galaxiae* strains BAT2 and B2S also showed inconclusive results, as the dissolved DA (dDA) fraction gave higher measured concentrations than the total DA (tDA) fraction. On repeated analysis, both particulate (pDA) and dDA in B2S were below LOQ, while in BAT2, the concentration of pDA was higher than dDA although we could not quantify it again as the signal remained above quantification. We could not confirm toxin production in strains of *P. mannii* Amato & Montresor and *P. subfraudulenta* (Hasle) Hasle and *P. calliantha* although for one strain of *P. calliantha* and one strain of *P. mannii*, results indicated minute concentrations below the LOQ of the ELISA assay.

2.2. DA Production in Different Growth Phases

We observed a decreasing trend in particulate toxin content with increasing cell density in *P. multistriata* strain 119-A4 (Figure 1A), while dDA increased slightly only on the last day of measurements (Figure 1B). It can be seen that the initial screening yielded a similar cell number as on day 11 but a completely different concentration for both the pDA and dDA fractions. While the pDA fraction was lower in the initial screening compared to day 11, the dDA fraction was significantly higher. In our case, the toxin was already produced in the exponential growth phase, while measurements in the stationary phase were not performed. After the initial confirmation of toxicity in *P. delicatissima* strain 119-B3, we could not confirm DA in the experiment where DA was sampled during different growth stages. The concentration of DA was probably very low in this case since the HPLC-UV method could not detect DA even in the initial screening (Table 1). We see here that the measurements between the HPLC-UV and ELISA methods are comparable and give very similar concentrations, except in the case of the first screening with dDA, where the ELISA method gives much higher concentrations. Reliable concentrations of pDA and dDA for the other strains tested in different growth phases could not be determined and so are not reported here.

2.3. Phylogeny and Toxicity

Our comprehensive phylogenetic analysis based on the ITS2 sequences of all species of *Pseudo-nitzschia*—where both ITS2 sequence data and toxicity data were available—shows that all major lineages of the genus harbor strains and species that are both toxic and non-toxic (Figure 2). The group with the lowest number of toxic strains was Group III *sensu* [25] although we see that a toxic *P. kodamae* S.T. Teng, H.C. Lim, C.P. Leaw, & P.T. Lim strain is included in this group. We also know that *P. calliantha*, which is a member of Group III, can be toxic, but no ITS sequences of toxic strains were available in GenBank. In some cases, toxic and non-toxic strains were separated by high statistical support. We see this pattern in *P. australis* Franguelli; *P. bipertita* S.T. Teng, H.C. Lim, & C.P. Leaw; *P. galaxiae*; *P. kodamae*; *P. multiseries* (Hasle) Hasle; *P. subcurvata* (Hasle) Fryxell; *P. subfraudulenta*; and *P. pseudodelicatissima*. In some cases, such as with *P. bipertita*, *P. pseudodelicatissima*, and *P. subfraudulenta*, the differing strains were from different geographical areas, while in other species, strains were from the same area (Table S1). The differences between geographically separated as well as non-separated strains are most pronounced in *P. galaxiae*, which stands out when we take a closer look at the phylogeny (Figure 3B). *P. galaxiae* shows major genetic differences that are consistent with the morphological characterization. Therefore, *P. galaxiae* is hereafter referred to as the *P. galaxiae* species complex although the phylogenetic position of these different strains has not yet been clarified. We see that two major clades have emerged, one consisting mainly of non-toxic larger strains, while the other consisted of well-separated medium and small strains, including three from this study (Figure 4). Both clades harbored a strain that was an exception to this rule although the BAT2 strain was morphologically quite distinct from the other three strains, exhibiting a peculiar baseball bat-like morphology (Figure 5). Strain (10)4A3 from Greece was also a medium-sized strain.

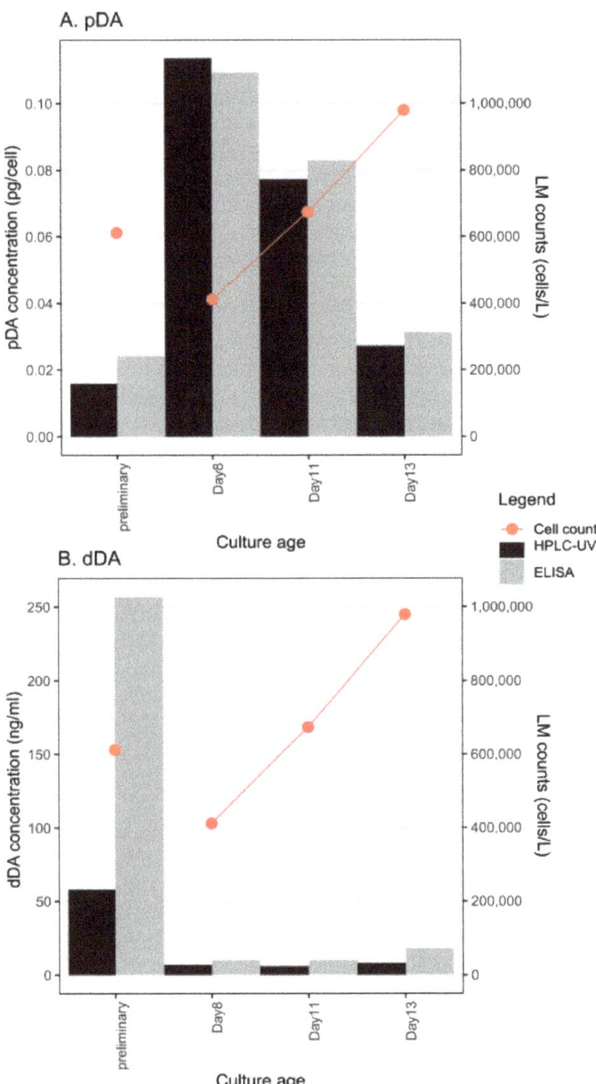

Figure 1. (**A**) Change in pDA content in *P. multistriata* strain 119-A4; (**B**) change in dDA content in P. multistriata strain 119-A4. The preliminary measurements were part of the screening experiment presented in Table 1.

Table 1. Summary of strains and tests performed on each strain and the concentrations of DA associated with them.

Species	Strain	Sampling Frequency [1]	Partition Tested	Test Method [2]	Result [3]	DA Concentration[4]	ELISA Retest [5]
P. delicatissima	219 A1	P	pDA	E	-		
	219 A2	P	pDA	E	-		
	219 A3	P	pDA	E	-		
	219 B1	P	pDA	E	-		
	219 B2	P	pDA	E	-		
	219 B3	P	pDA	E	-		
	219 B4	P	pDA	E	-		
	119 A2	P	pDA	E	-		
	119 A3	P	pDA	E	-		
	119 B3	P & C	pDA and dDA	E&H	+ (E)	1.5 fg/cell	<quant
	119 C1	P	pDA and dDA	E&H	+ (E)	NA	<quant
	119 C4	P	pDA	E	-		
P. multistriata	119 A4	P & C	pDA and dDA	E&H	+ (E&H)	pDA: 16–114 fg/cell; dDA: 6–257 ng/mL	
	119 C3	P	pDA and dDA	E&H	+ (E&H)	pDA: 121–160 fg/cell; dDA: 30–80	
	MS2	P & C	pDA and dDA	E; H; L	+ (E)	pDA: 0–32.8 fg/cell; dDA: 20 ng/mL	<quant
	MS3	P & C	pDA and dDA	E; H; L	+ (E)	pDA: 1.42–20.7 fg/cell; dDA: 14.6–38.5 ng/mL	<quant
	PN0DB2131216-A	P	pDA	L	-		
	PN0DB2131216-B	P	pDA	L	-		
	PN0DB2131216-C	P	pDA	L	+	0.217 fg/cell	
P. galaxiae—large	MA-919-C1	P	pDA and dDA	E&H	-		
	MA-919-A2	P	pDA and dDA	E&H	-		
	F919-C2L	P	pDA and dDA	E&H	-		
P. galaxiae—medium	F919-B1M	P	pDA and dDA	E&H	-		
P. galaxiae—bat-like	BAT2	P	pDA and dDA	E&H	o (E)	dDA: 13 ng/mL	<quant
P. galaxiae—small	B3S	P	pDA and dDA	E&H	+ (E)	dDA: 5–24.8 ng/mL; pDA > quant	>quant
	B2S	P	pDA and dDA	E&H	o (E)	dDA: 12.6 ng/mL	<quant
P. mannii	BF-819-B2	P	pDA and dDA	E&H	-		
	BF-819-A4	P	pDA and dDA	E&H	-		
	BD-919-A3	P	pDA and dDA	E&H	-		
	PNF_1020_5	P	pDA and dDA	E	o	<quant	
P. subfraudulenta	PNF_1020_1	P	pDA and dDA	E	-		
P. calliantha	PNF_1020_2	P	pDA and dDA	E	o	<quant	
	PN00BF281016-2A	P	pDA	L	-		

[1] P, point; C, continuous. [2] E, ELISA; H, HPLC-UV; L, LC-MS/MS. [3] o indicates an inconclusive result. [4] NA, not available; the toxicity range is obtained from different tests of the same culture. [5] </> quant, below/above quantification.

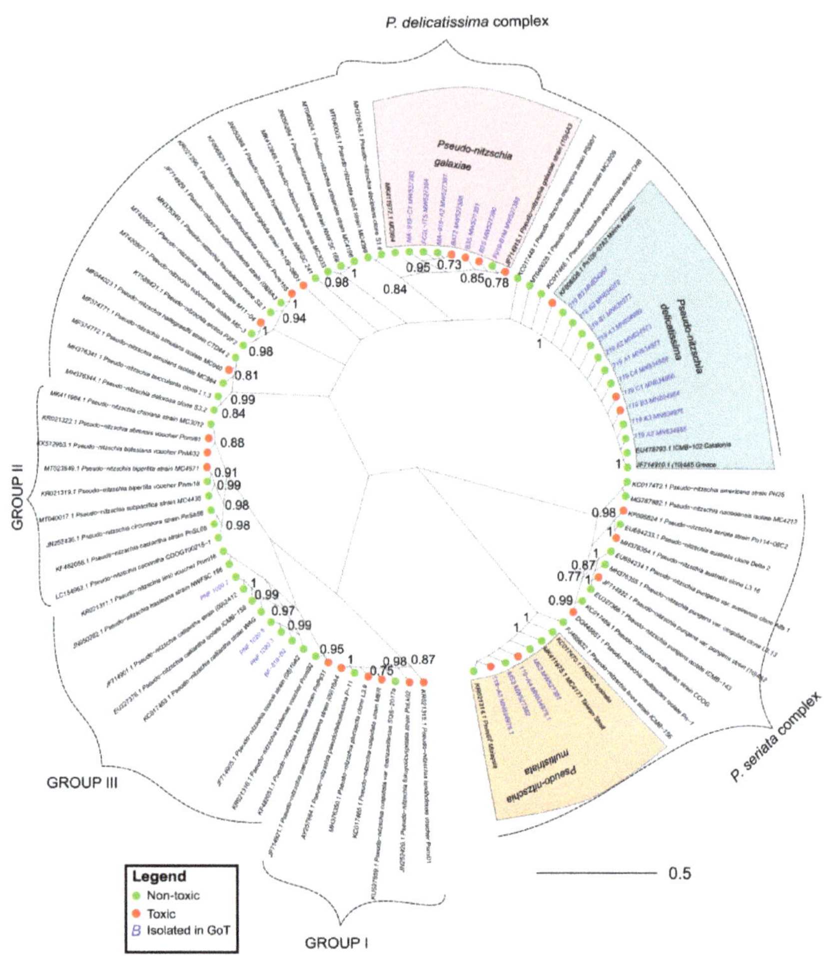

Figure 2. Maximum likelihood (ML) tree of the ITS2 marker constructed using 94 sequences gathered from GenBank (https://www.ncbi.nlm.nih.gov/genbank/; accessed on 19 November 2021), using the TVM + G + I evolutionary model and 10,000 bootstraps of the tree space. Only bootstrap supports higher than 0.7 are shown. >0.95 is considered high support. Note that not all strains presented in Table 1 are included because ITS sequences of some could not be obtained.

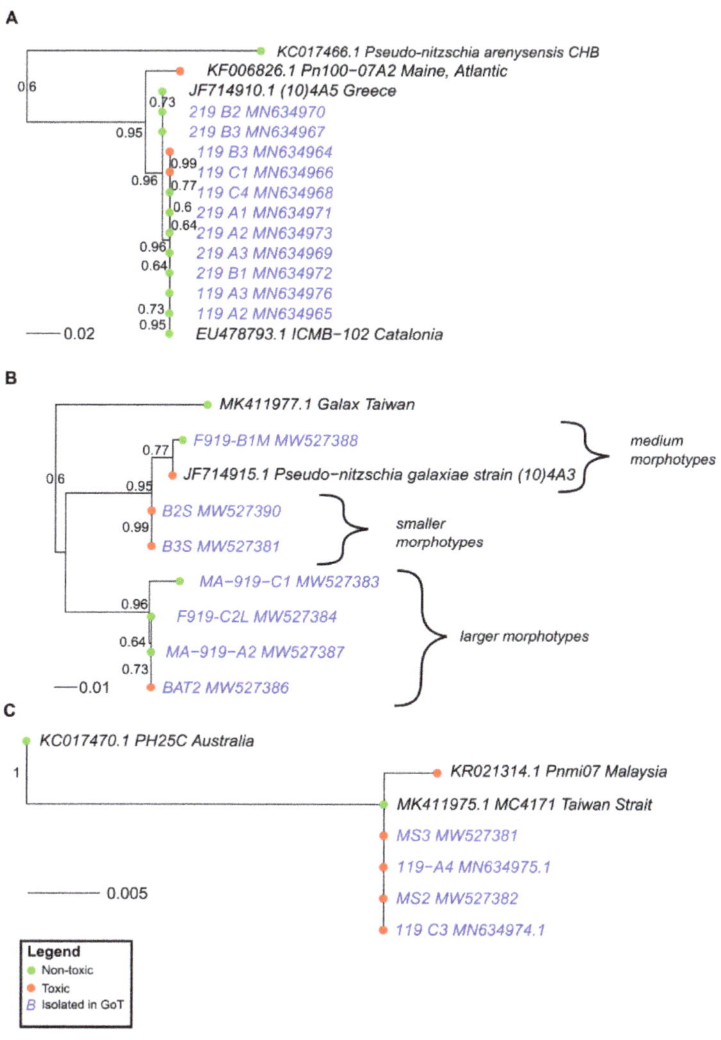

Figure 3. Pruned trees from Figure 2. (**A**) Tree of *P. delicatissima*; (**B**) tree of *P. galaxiae*; (**C**) tree of *P. multistriata*. The trees were drawn using the same conditions as the tree in Figure 2. Bootstrap supports higher than 0.5 are shown.

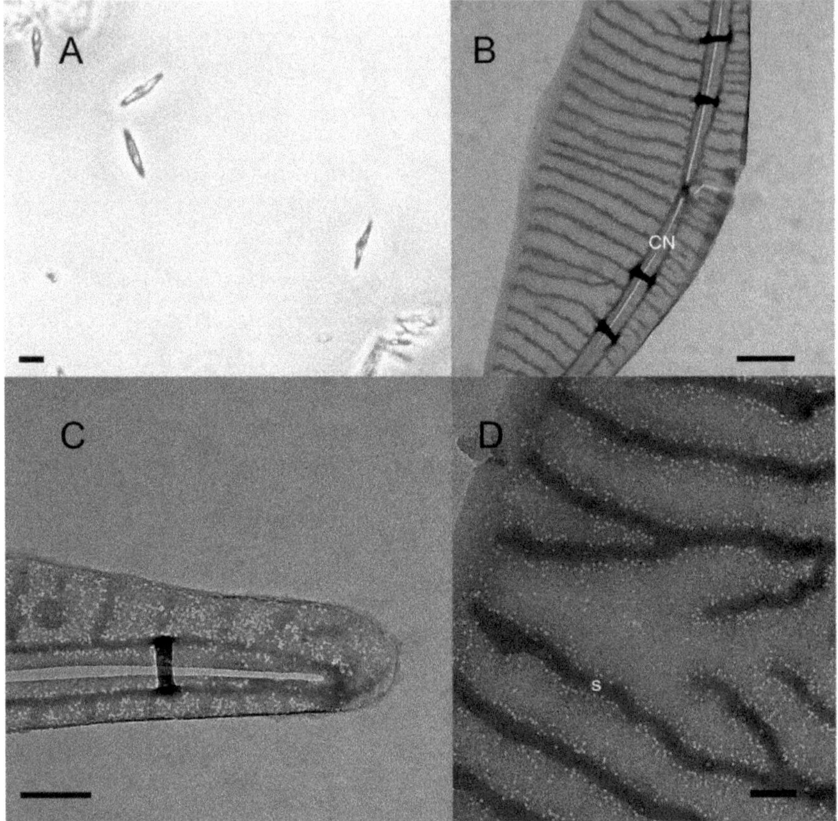

Figure 4. *P. galaxiae* small morphotype, strain B3S. (**A**) LM image of the cells in culture. Scale bar = 5 μm; (**B**) Transmission electron microscope (TEM)image of the cell valve at the position of the central nodule (CN). Scale bar = 500 nm; (**C**) TEM image of the cell tip with the random distribution of poroids. Scale bar = 200 nm; (**D**) TEM image showing the tightly packed poroids along the striae(s) with rare poroids in the interstrial space. Scale bar = 100 nm.

Species that we examined in more detail, since we had many strains available for toxicity testing, were also *P. delicatissima* (Figure 3A) and *P. multistriata* (Figure 3C). We see that in both cases, the genetic divergence of strains on the ITS2 marker was not as large as with the *P. galaxiae* species complex and that identical or slightly-divergent strains appeared to be both toxic and non-toxic. This is particularly evident in *P. multistriata*, which showed very little divergence. In *P. delicatissima*, the divergence was somewhat greater; in particular, the toxic strain PN100-07A2 from the western Atlantic was separated from ours and other Mediterranean strains, two of which were toxic. Unfortunately, we were unable to obtain ITS sequences of the *P. multistriata* strains that were found to be non-toxic in our analysis as well as the non-toxic *P. subfraudulenta* and *P. mannii* strains.

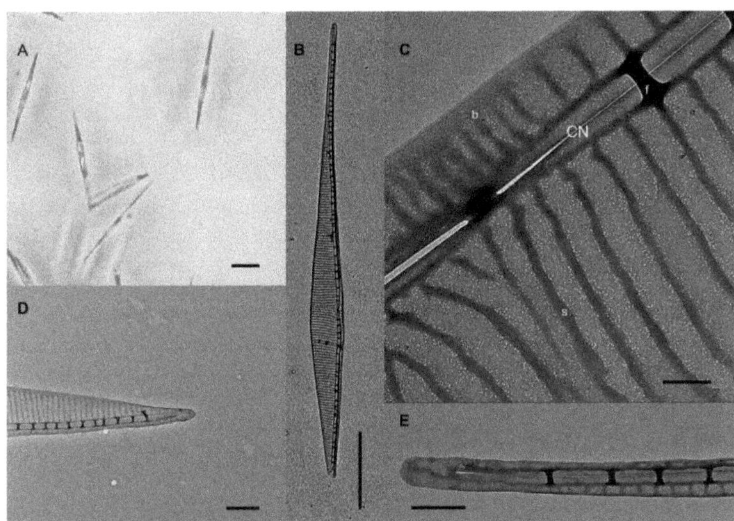

Figure 5. *P. galaxiae* strain BAT2 with a distinct bat-like morphology. Note that not all cells of the culture had this morphology and that the environmental isolate from which the culture was grown demonstrated this morphology. (**A**) LM image of the cells in culture. Scale bar = 10 μm; (**B**) TEM image of the entire valve. Scale bar = 5 μm; (**C**) Detailed TEM image of the valve and the central nodule (CN), with the visible random distribution of several poroids between striae(s), fibulae, and the band (**B**). Scale bar = 200 nm; (**D**) TEM image of cell tip at the unusually shortened end. Scale bar = 1 μm; (**E**) TEM image of the cell tip at the normally elongated end. Scale bar = 500 nm.

2.4. dabA Gene Screening in Toxic and Non-Toxic Strains

PCR with primers published in [9] did not yield specific products with any species, and so, we redesigned primers for the *dabA* gene using the available public sequences, which included a *P. multiseries* sequence from [9] and an incomplete *P. multistriata* sequence from [26], which came from genome sequencing. The designed primer amplified the gene in four of the sequenced *P. multistriata* strains (Table S1). The amplified product was approximately 1450 bp long and contained the intronic region, which was removed from the sequences after alignment with the only two published sequences. The product obtained was a partial gene sequence, as the primers designed in [9] did not yield a specific product, and so, the primers had to be designed internally on the available sequences. The *P. multistriata* sequences obtained were highly conserved and had only a few ambiguous sites, but these were all on degenerate codon positions and did not affect the amino acid sequence. The sequences all aligned well with the published sequence from the *P. multistriata* genome, which was incomplete, because it contained undefined positions, which resulted in a translated amino acid sequence with multiple stop codons. The published sequence of *P. multiseries* (MH202990) is 84% identical to the sequences of our *P. multistriata* strains, whereas the translated sequences show 89% similarity to the DabA protein of *P. multiseries*. There is significant structural similarity identified by homology modelling in Phyre2 (Figure S1), highlighting the functionality of the *P. multistriata* enzyme in producing DA. Our attempt to amplify the *dabA* gene in other species of *Pseudo-nitzschia* using the same primers did not yield specific products. After we purified, isolated, and sequenced the nonspecific bands from the other species, BLAST (https://blast.ncbi.nlm.nih.gov/Blast.cgi; accessed on 6 May 2021) did not yield any highly similar hits. Thus, we confirmed that all toxin-producing strains of *P. multistriata* harbored the *dabA* gene, while we were unable to resolve the dilemma of whether non-toxic strains have the genetic capacity to produce DA or whether they actually lack the required genes. In any case, based on the data we have from two closely related *Pseudo-nitzschia* species—*P. multiseries* and *P. multistriata*—the *dabA* gene is not highly conserved with respect to its nucleotide sequence, and it may well be that the primers we used did not work in *P. galaxiae* and *P. delicatissima* since these species are the furthest from the *P. seriata* (Cleve) H. Peragallo species complex in the phylogenetic tree (Figure 3).

3. Discussion

3.1. Toxicity

The results of the toxicity analyses complement the studies from the northwestern Adriatic in that mildly toxic strains of *P. delicatissima* and *P. multistriata* are present in the northeastern Adriatic Sea as well. However, the toxicity levels of *P. delicatissima are* 25-times higher (~1.5 fg cell^{-1}) than those reported in the northwestern Adriatic (0.063 fg cell^{-1}) [17]. During this study, we realized that many factors can affect pDA concentrations, including sample preparation, counting errors, culture condition, and other factors related to the production of DA that we did not control (reviewed in [3,27]). This was particularly evident in the experiment monitoring DA production of *P. multistriata* strain 119-A4 at different phases of growth, where the initial screens differed greatly from the concentration measured during the experiment. Regarding the threat of *P. multistriata* and *P. delicatissima* to the ecosystem and industry in our region, *P. delicatissima* from the northern Adriatic appears to be a mildly toxic species that can reach bloom abundances, especially in spring [13], although shellfish toxin monitoring programs did not detect any DA in shellfish in this period [15,28]. *P. multistriata*, in contrast, appears to have a higher cellular content of DA and also higher than strains from the northwestern Adriatic [21] and the recently identified Peruvian strains [29] but comparable to some strains from the Gulf of Naples [30,31]. However, this species rarely proliferates into high-abundance blooms and has only been detected in the winter months [13].

Furthermore, we can report toxic strains of *P. galaxiae*. These findings are preliminary, as no toxicity could be confirmed by chromatographic methods although repeated ELISA screens confirmed the presence of DA in at least one strain (B3S). DA was found only in small and medium morphotypes of the species [13,24] and even in these only in dDA during the first test, leading to some inconsistencies with pDA calculations. We attributed this to procedural errors and therefore repeated the ELISA with these samples, which resolved this discrepancy somewhat, at least for strain B3S. *P. galaxiae* also showed the greatest genetic divergence between different strains, which was also true for strains isolated on the same day and at the same location (see [13]). These results may additionally suggest that *P. galaxiae*, as now described, is indeed a species complex, which has been suggested previously [13,32] and is also supported by our morphological and toxicological observations. Our data contribute valuable information on the toxicological and phylogenetic position of this species. *P. galaxiae* was so far found to be toxic only in the Aegean Sea (Greece) and even here with trace levels of DA in culture [33]. All other studies investigating *P. galaxiae* toxicity found this species to be non-toxic ([1] and references therein). This is thus the first unequivocal report on *P. galaxiae* toxicity. This species is known to grow to high abundances both in culture—over 1 million cells per mL—and in the environment (unpublished data from Gulf of Trieste). However, due to the small size of the cells, similarly to *P. delicatissima*, it is probably not a threat.

Finally, the possible detection of DA in *P. calliantha*—even if below ELISA quantification—would need further investigation. However, it would not be surprising since DA has only recently been found in Adriatic *P. calliantha* strains [22] although it was previously suspected [19,23]. Conversely, the indication that even some *P. mannii* strains may be toxic is surprising—as this species has not been confirmed to be toxic before [1,34]—although we do not have enough evidence to confirm this.

The onset of production of DA in cultures depends on the species and strains [3] although in most species, DA production starts in the late exponential phase and increases during the stationary phase. We only had one stationary phase sample in our experiment. Our results corroborate those of [35], who found decreasing pDA levels with increasing cell number in *P.* cf. *pseudodelicatissima*, but contrast with the results of [31,36], who found increasing pDA concentrations with increasing cell number in *P. multistriata* and *P. multiseries*, respectively. In the study of [7], the pDA concentration remained constant with increasing abundance of *P. multiseries* although it decreased dramatically when DA was added externally to the growth medium, albeit during increased copper stress. The decrease in pDA concentration with increasing cell concentration may be due to an as-yet unknown quorum-sensing strategy of the cells to accommodate increasing DA in the environment by producing less DA [37]. Quorum-sensing responses could also depend on the resident bacterial community, so differences between strains and cultures would not be unexpected. However, we do not have sufficient data to conclude that this strategy is indeed responsible for our observed results. Our results also suggest that comparing DA concentrations of point samples between studies and basins is not useful because conditions and culture states can vary considerably [3].

Finally, we point to some of the inconsistencies in the measurements of pDA in MS2 and MS3, which prompted us to retest these samples with HPLC-UV and LC-MS/MS. However, the LC -MS/MS method used to screen mussel samples had a LOQ of 0.8 µg/mL and so was too high to detect DA in our culture samples with the exception of one strain (PN0DB2131216-C, GenBank 28S accession: MK682491.1), which was analyzed on another occasion. In addition, the repeated tests were performed on samples that had been stored in the original growth medium for six months to over a year after the initial ELISA screens, which may have resulted in some toxin degradation [38,39]. In contrast, the strain positive for DA in LC-MS/MS was specifically prepared for LC-MS/MS analysis and tested shortly after extraction.

3.2. Phylogenetic Relationships between Toxic and Non-Toxic Strains

To our knowledge, this is the first attempt to relate the toxicity of *Pseudo-nitzschia* strains to their phylogenetic relationship although speculation that ITS type is not related to the ability to produce DA has been made previously for *P. multistriata* [31]. This has also been shown for yessotoxin-producing dinoflagellates, where no correlations were found between toxicity and phylogenetic position [40]. Our study provides conflicting evidence for this question. Clearly, the ability to produce DA is widespread in the phylogenetic tree of *Pseudo-nitzschia* although it is possible that it was lost several times during the evolutionary history of the genus and its species. The genetic similarity of *P. delicatissima* and *P. multistriata* strains in relation to their toxicity suggests that toxicity in these two cases is irrelevant to the ITS phylogeny and may support the previously discussed idea that the physiological state of the culture determines whether or not DA is produced, where gene expression may also play a role [41].

In contrast, in *P. subfraudulenta*, *P. subcurvata*, *P. australis*, *P. pseudodelicatissima*, *P. multiseries*, and *P. galaxiae*, toxic and non-toxic strains were separated by high bootstrap support that was in certain cases also related to the geographic origin of the strains. This may contribute to the idea that genes required for DA production were lost in relatively recent evolutionary history or that the ability to produce DA is related to the environmental conditions to which strains are acclimatized. Conversely, the differential production of DA in strains from the same area cannot be explained following this logic. In any case, it appears that the ancestral state had the ability to produce DA, which also explains the occurrence of this toxin and its analogs in some other genera of diatoms and even red algae [42–44]. This is unless the biosynthetic apparatus was acquired during evolutionary history by horizontal gene transfer mechanisms perhaps on multiple occasions, as was suggested by [45] and which is the presumed pathway for the acquisition of saxitoxin production in dinoflagellates [46]. Horizontal gene transfer in protists is perhaps a neglected phenomenon and may be exacerbated by widespread viral infections [47]. For this hypothesis to be examined, a clear evolutionary role of DA should be established, which is at the moment not the case since proposed roles range from grazer deterrence [10,11,48] to metal chelation [4–6], while strains and species are all known to fare well in the absence of DA production. On the other hand, the competitive advantage of DA production may be clearer when local environmental conditions are considered. The third possibility is that the toxin was simply not produced at the time of sampling or that the methods were not sensitive enough, as we have also shown in some cases in this work. To prove this, we would need to trace the genes responsible for the production of DA in the genomes of several species and strains, which is what we attempted next.

3.3. DabA in Selected Strains

For the first time since the discovery of the DA biosynthetic pathway [13], we identified the *dabA* gene in a species other than *P. multiseries*. Although the *dabA* sequence of *P. multistriata* was deposited in GenBank (https://www.ncbi.nlm.nih.gov/genbank/, accessed on 14 December 2021) by genome sequencing [26], which allowed us to design primers, we filled in the missing sites in the genome sequence that made it untranslatable. The secondary structure of the protein appears to be conserved between *P. multistriata* and *P. multiseries*, although the nucleotide sequence is only 84% similar, which may contribute to the fact that the primers used did not amplify the gene in other species tested. *P. multistriata* and *P. multiseries* are closely related, whereas *P. calliantha*, *P. mannii*, *P. galaxiae*, and *P. delicatissima* are more distantly related ([25], Figure 2). For the strains for which good products were not obtained, the cause could be a missing gene or poor primers. If the latter is the case, this may not be a trivial task since our preliminary data show that the genetic as well as amino acid sequence differences between the closely related *P. multiseries* and *P. multistriata* are quite large. Therefore, further genomic and transcriptomic experiments with other *Pseudo-nitzschia* species—particularly those from Groups I–III and the *P. delicatissima* complex—need to be performed to populate the reference databases, which will facilitate the design of more universal *dabA* primers.

Recently, a transcriptomic study showed that only *dabA* and *dabD* of the *dabABCD* cluster were expressed in DA-producing *P. pungens*, and only *dabD* was found in *P. fraudulenta* transcripts. In contrast, *P. australis* expressed all four genes [41]. In the future, these transcriptomes could be mined to obtain sequence data and design new primers. However, there is also the possibility that the gene cluster is completely absent from the genome of some species or strains that are not actually toxic [1] although such an explanation is unlikely. This would indicate either multiple deletions or insertions of the gene cluster during the evolutionary history of the genus. Such a mechanism could imply either the horizontal gene transfer, discussed earlier, which is a plausible explanation for the occurrence of DabA analogs in red algae [45], or ongoing hybridization, which has, however, been demonstrated in the genus [48,49].

There is an idea that measuring the copy number of genes involved in the synthesis of DA—e.g., by qPCR—may be the key to accurately predict the threat of ASP. At the moment, however, the design of suitable probes is hampered by the lack of sequences from different species and could perhaps only be developed for *P. multiseries* and now with our data for *P. multistriata*. We acknowledge that there are other genes involved in the biosynthesis of DA [9,11] that we did not examine in this work and that may prove to be more conserved and thus better targets for such efforts. However, the focus of this work was on the toxicity profiles of NE Adriatic strains, and we hope that this work will open new opportunities for the study of gene expression and the discovery of genes involved in the synthesis of DA.

4. Materials and Methods

4.1. Species Cultures

The cultures used in toxicity tests were obtained from the National Institute of Biology—Marine Biology Station Piran, Slovenia culture collection and grown in 50 mL of L1 medium in 150-mL Erlenmeyer flasks, as described in [13]. Cultures of *P. calliantha*, *P. delicatissima*, *P. galaxiae*, *P. mannii*, *P. multistriata*, and *P. subfraudulenta* were used for this study.

4.2. Domoic Acid in Cultures

All cultures were tested at one point in the stationary growth phase. Fifty µL of the culture were taken in triplicate for counting under an Olympus BX51 microscope (Olympus, Tokyo, Japan) in a Fuchs–Rosenthal chamber (FRC). Then, depending on the number of cells, 10–20 mL of the culture were taken in duplicate. One of the replicates was sonicated on ice at 40 Hz for one minute to break up the cells and release the toxin. This sample was then filtered through a Millex 0.22-µm syringe filter (Merck-Milipore, Darmstadt, Germany) to remove debris, and the filtrate was stored at −20 °C. This represented the total DA (tDA). The other sample was centrifuged at 4500× g for 30 min, and the supernatant was stored at −20 °C. This was the dissolved DA (dDA) fraction. Since the centrifugation process is not perfect, the supernatant was recounted in the FRC to account for the remaining live cells. This was done as follows: the top layer of the supernatant was counted in duplicate, then a fraction of the supernatant was pipetted for toxin analysis. Then, the bottom layer of the supernatant was counted in duplicate. The cell count of both layers was then averaged, assuming a gradual increase in cell count from top to bottom in a centrifuged tube. To obtain the particulate DA concentration (pDA—toxin stored in cells), we subtracted the dDA fraction from the tDA and divided the resulting concentration by the number of cells in the lysate. A total of 33 strains from six species were tested.

For three strains of *P. multistriata* (119-A4, MS2, and MS3) and one strain of *P. delicatissima* (119-B3) that tested positive for DA in the first phase, an additional experiment was performed, namely the monitoring of DA at different stages of culture growth. These strains were cultured in 500 mL of L1 medium. Sampling was performed as described above at days 7, 11, and 13 post inoculation for strains 119-A4 and 119-B3 and at days 4, 6, and 11 and 4, 5, 6, 7, 11, and 13 post inoculation for strains MS2 and MS3, respectively.

4.3. Direct Competitive ELISA

The main method used to test most of the cultures examined was the competitive ELISA assay for the detection of domoic acid (Eurofins Abraxis, Warminster, MA, USA). The cultures were processed according to the manufacturer's guidelines. A constant temperature was maintained during plate preparation and pipetting. The reliability of the procedure was verified using internal and external controls and standards. The absorbance reader was turned on for an extended period of time prior to measurement to allow the light source to settle. This method is approved by the European Commission as a screening method for the determination of DA in shellfish (Commission Implementing Regulation (EU) 2019/627). Samples were diluted 1:25 with the dilution buffer provided and further diluted to 1:50 if the signal was still saturated. If the signal was borderline or unquantifiable, the samples were diluted less to 1:10. Some samples were not reanalyzed at higher or lower dilutions and are only reported as positive. Due to inconsistencies between ELISA and the analytical methods, some samples were measured multiple times by ELISA to confirm the presence or absence of toxin.

4.4. HPLC-UV Method for Selected Cultures

This method is generally intended for regulatory purposes, as it is sufficiently sensitive to detect toxicity when the toxin concentration approaches or exceeds the threshold. It is also capable of accurately measuring high concentrations of DA in cultures. Domoic acid content was determined after chromatographic separation on a reversed phase column (C18 reversed phase, 5 µm, 250 mm × 4.6 mm) under isocratic conditions (acetonitrile 5% with TFA 0.1% v/v). The analysis was performed with a UV-VIS detector set to 242 nm.

The amount of domoic acid was calculated using a certified DA standard from the National Research Council of Canada, which was used to prepare three dilutions that were used to calibrate the HPLC system.

4.5. LC-MS/MS for Selected Cultures

The LC-MS/MS analysis of domoic acid in extracts was performed using an Agilent UHPLC Infinity II (Agilent Technologies Inc., Santa Clara, CA, USA) equipped with an Agilent POROSHELL 120 EC C18, 2.1 × 100-mm 2.7-μm—LC column (Agilent Technologies Inc., Santa Clara, CA, USA) coupled to an Agilent 6460 Triple Quad Mass Spectrometer (Agilent Technologies Inc., Santa Clara, CA, USA). A certified DA standard from NRC-Canada was used to prepare five standard solutions that were used to calibrate the LC-MS/MS system in the multiple reaction monitoring (MRM) mode. The identification of the analyte was based on monitoring two ion products of DA (m/z 312 > 266, 312 > 161 from DA precursor ion (M + H) + m/z 312) in positive electrospray ionization (ESI+) mode. The most abundant fragment, 266, was selected for quantification, while the 161 ion was used for qualitative confirmation. A methanol/water solution of ammonium acetate/acetic acid was chosen the mobile phase for chromatographic separation. A total of 5 μL of the sample was injected into the LC-MS/MS system, and a 14 min gradient elution was used to separate the toxins.

4.6. ITS-2 Phylogeny Reconstruction with Tested Strains

The complete ITS region was sequenced as described in [13]. Sequences used for phylogenetic tree reconstruction were selected based on two factors: whether the publications under which they were published contained toxicity data and whether they were geographically representative. Phylogenetic trees based on the ITS2 region were constructed separately for each species with the phangron [50] and ape [51] packages implemented in R [52] using the implemented maximum likelihood algorithm with nearest neighbor interchange (NNI) optimization and a transversion model with estimated invariant sites and a gamma distribution of rates (TVM + G + I) that was established by the model-Test() command. Ten thousand bootstraps of the tree space were performed using the bootstrap.pml() command.

4.7. Amplification and Sequence Analysis of the dabA Gene

For the amplification of the *dabA* gene, we first unsuccessfully tried the primers published in [9]. Then, we designed new primers (Table 2) based on the alignment of sequences from shotgun sequencing of the *P. multistriata* genome [26], BioProject accession: PRJEB9419 (https://www.ncbi.nlm.nih.gov/bioproject/PRJEB9419/; accessed on 13 April 2021) and [9]. Internal sequencing primers were also designed to complete the gene region. Amplification was performed using Phusion HiFi Polymerase (New England Biolabs, Ipswich, MA, USA). For PCR reactions that resulted in nonspecific products, bands containing products of the appropriate size (~1500 bp) were excised from the agarose gels. To do this, products were run at 200 V for 30 s in pre-cut wells in the gel, followed by precipitation with 3M sodium acetate and isopropanol at −20 °C for one hour. The precipitates were then washed with 70% ethanol and dissolved in 1× Tris Low-EDTA (TLE). Products were also purified with exonuclease I and Alkaline Phosphatase—FastAP (Thermo Fisher Scientific, Waltham, MA, USA). Alternatively, bands were excised from the gel with a sterile scalpel when possible, and gel purification was performed with NucleoSpin Gel and PCR Clean-up (Machery-Nagel, Düren, Germany). The 3D structure of the predicted proteins was predicted using Phyre2 (http://www.sbg.bio.ic.ac.uk/phyre2/html/page.cgi?id=index; accessed on 22 January 2021) [53].

Table 2. Primers used for amplification and sequencing of the *dabA* gene in *P. multistriata*.

Primer Name	Primer-Sequence	Type
DabA_multF	ATGAAATTTGCAACGTCCATTGTC	PCR
DabA_degF	ATGAARTTTGCAACRTCCATYGTC	PCR
N1_R	TCCAAAAACGCTTTCATCAA	PCR
N2_R	AACGCTTTCATCAATGGTTTGTGG	PCR
Internal_multistriataF	CGATTGGATGAAGATCCCTTCA	Sequencing
Internal_multistriataR	GCAGAAGTCGACCATCCA	Sequencing

Supplementary Materials: The following supporting information can be downloaded at: https://www.mdpi.com/article/10.3390/toxins14010060/s1, Figure S1. Homology modelling of the translated sequence of the *dabA* gene from *Pseudo-nitzschia multistriata*, strain MS3. Most of the secondary structures are recovered and the protein resembles the published crystalline structure. Table S1. ITS and *dabA* accession numbers of strains used in the phylogeny reconstruction.

Author Contributions: Conceptualization, T.T.D. and P.M.; Methodology, T.T.D., S.D., J.D.; Software, T.T.D., S.D., J.D.; Validation, T.T.D. and P.M.; Formal Analysis, T.T.D.; Investigation, T.T.D.; Resources, T.TD., S.D., J.D.; Data Curation, T.TD.; Writing—Original Draft Preparation, T.TD.; Writing—Review & Editing, T.T.D. and P.M.; Visualization, T.T.D.; Supervision, P.M.; Project Administration, T.T.D.; Funding Acquisition, T.T.D. and P.M. All authors have read and agreed to the published version of the manuscript.

Funding: This work was conducted with the help of RI-SI-2 LifeWatch (Operational Programme for the Implementation of the EU Cohesion Policy in the period 2014–2020, Development of Research infrastructure for international competition of Slovene Development of Research infrastructure area— RI-SI, European Regional Development Fund, Republic of Slovenia Ministry of Education, Science and Sport). The authors acknowledge the financial support from the Slovenian Research Agency (research core funding No. P1-0237 and program for young researchers, in accordance with the agreement on (co) financing research activities.

Institutional Review Board Statement: Not applicable.

Informed Consent Statement: Not applicable.

Data Availability Statement: Sequence data is available in GenBank. Genomic resources are deposited at NIB MBP. Toxin analysis data are available from NIB MBP.

Conflicts of Interest: The authors declare no conflict of interest.

References

1. Bates, S.S.; Hubbard, K.A.; Lundholm, N.; Montresor, M.; Leaw, C.P. *Pseudo-nitzschia*, *Nitzschia*, and domoic acid: New research since 2011. *Harmful Algae* **2018**, 1–41. [CrossRef]
2. Huang, C.X.; Dong, H.C.; Lundholm, N.; Teng, S.T.; Zheng, G.C.; Tan, Z.J.; Lim, P.T.; Li, Y. Species composition and toxicity of the genus *Pseudo-nitzschia* in Taiwan Strait, including *P. chiniana* sp. nov. and *P. qiana* sp. nov. *Harmful Algae* **2019**, *84*, 195–209. [CrossRef]
3. Lelong, A.; Hégaret, H.; Soudant, P.; Bates, S.S. *Pseudo-nitzschia* (Bacillariophyceae) species, domoic acid and amnesic shellfish poisoning: Revisiting previous paradigms. *Phycologia* **2012**, *51*, 168–216. [CrossRef]
4. Rue, E.; Bruland, K. Domoic acid binds iron and copper: A possible role for the toxin produced by the marine diatom *Pseudo-nitzschia*. *Mar. Chem.* **2001**, *76*, 127–134. [CrossRef]
5. Maldonado, M.T.; Hughes, M.P.; Rue, E.L.; Wells, M.L. The effect of Fe and Cu on growth and domoic acid production by *Pseudo-nitzschia multiseries* and *Pseudo-nitzschia australis*. *Limnol. Oceanogr.* **2002**, *47*, 515–526. [CrossRef]
6. Wells, M.L.; Trick, C.G.; Cochlan, W.P.; Hughes, M.P.; Trainer, V.L. Domoic acid: The synergy of iron, copper, and the toxicity of diatoms. *Limnol. Oceanogr.* **2005**, *50*, 1908–1917. [CrossRef]
7. Liu, Y.; Gu, Y.; Lou, Y.; Wang, G. Response mechanisms of domoic acid in *Pseudo-nitzschia multiseries* under copper stress. *Environ. Pollut.* **2021**, *272*, 115578. [CrossRef] [PubMed]
8. Geuer, J.K.; Trimborn, S.; Koch, F.; Brenneis, T.; Krock, B.; Koch, B.P. Dissolved Domoic Acid Does Not Improve Growth Rates and Iron Content in Iron-Stressed *Pseudo-nitzschia subcurvata*. *Front. Mar. Sci.* **2020**, *7*, 478. [CrossRef]

9. Brunson, J.K.; McKinnie, S.M.K.; Chekan, J.R.; McCrow, J.P.; Miles, Z.D.; Bertrand, E.M.; Bielinski, V.A.; Luhavaya, H.; Oborník, M.; Smith, G.J.; et al. Biosynthesis of the neurotoxin domoic acid in a bloom-forming diatom. *Science* **2018**, *361*, 1356–1358. [CrossRef]

10. Tammilehto, A.; Nielsen, T.G.; Krock, B.; Møller, E.F.; Lundholm, N. Induction of domoic acid production in the toxic diatom *Pseudo-nitzschia seriata* by calanoid copepods. *Aquat. Toxicol.* **2015**, *159*, 52–61. [CrossRef]

11. Haroardóttir, S.; Wohlrab, S.; Hjort, D.M.; Krock, B.; Nielsen, T.G.; John, U.; Lundholm, N. Transcriptomic responses to grazing reveal the metabolic pathway leading to the biosynthesis of domoic acid and highlight different defense strategies in diatoms. *BMC Mol. Biol.* **2019**, *20*, 1–14. [CrossRef]

12. Zhang, S.; Zheng, T.; Lundholm, N.; Huang, X.; Jiang, X.; Li, A.; Li, Y. Chemical and morphological defenses of *Pseudo-nitzschia multiseries* in response to zooplankton grazing. *Harmful Algae* **2021**, *104*, 102033. [CrossRef]

13. Turk Dermastia, T.; Cerino, F.; Stanković, D.; Francé, J.; Ramšak, A.; Žnidarič Tušek, M.; Beran, A.; Natali, V.; Cabrini, M.; Mozetič, P. Ecological time series and integrative taxonomy unveil seasonality and diversity of the toxic diatom *Pseudo-nitzschia* H. Peragallo in the northern Adriatic Sea. *Harmful Algae* **2020**, *93*, 101773. [CrossRef] [PubMed]

14. Zingone, A.; Escalera, L.; Aligizaki, K.; Fernández-Tejedor, M.; Ismael, A.; Montresor, M.; Mozetič, P.; Taş, S.; Totti, C. Toxic marine microalgae and noxious blooms in the Mediterranean Sea: A contribution to the Global HAB Status Report. *Harmful Algae* **2021**, *102*, 101843. [CrossRef] [PubMed]

15. UVHVVR. *Yearly Report On Zoonosis and Their Causative Organisms*; 2019. Available online: https://www.gov.si/assets/organi-v-sestavi/UVHVVR/Varna-hrana/Porocila-bioloska-varnost/Nacionalno-porocilo-monitoringa-zoonoz-2019.pdf (accessed on 19 December 2021). (In Slovenian)

16. Arapov, J.; Skejić, S.; Bužančić, M.; Bakrač, A.; Vidjak, O.; Bojanić, N.; Ujević, I.; Gladan, Ž.N. Taxonomical diversity of *Pseudo-nitzschia* from the Central Adriatic Sea. *Phycol. Res.* **2017**, *65*, 280–290. [CrossRef]

17. Penna, A.; Casabianca, S.; Perini, F.; Bastianini, M.; Riccardi, E.; Pigozzi, S.; Scardi, M. Toxic *Pseudo-nitzschia* spp. in the northwestern Adriatic Sea: Characterization of species composition by genetic and molecular quantitative analyses. *J. Plankton Res.* **2013**, *35*, 352–366. [CrossRef]

18. Caroppo, C.; Congestri, R.; Bracchini, L.; Albertano, P. On the presence of *Pseudo-nitzschia calliantha* Lundholm, Moestrup et Hasle and *Pseudo-nitzschia delicatissima* (Cleve) Heiden in the Southern Adriatic Sea (Mediterranean Sea, Italy). *J. Plankton Res.* **2005**, *27*, 763–774. [CrossRef]

19. Marić, D.; Ljubešić, Z.; Godrijan, J.; Viličić, D.; Ujević, I.; Precali, R. Blooms of the potentially toxic diatom *Pseudo-nitzschia calliantha* Lundholm, Moestrup & Hasle in coastal waters of the northern Adriatic Sea (Croatia). *Estuar. Coast. Shelf Sci.* **2011**, *92*, 323–331. [CrossRef]

20. Giulietti, S.; Romagnoli, T.; Siracusa, M.; Bacchiocchi, S.; Totti, C.; Accoroni, S. Integrative taxonomy of the *Pseudo-nitzschia* (Bacillariophyceae) populations in the NW Adriatic Sea, with a focus on a novel cryptic species in the *P. delicatissima* species complex. *Phycologia* **2021**, *60*, 1–18. [CrossRef]

21. Pistocchi, R.; Guerrini, F.; Pezzolesi, L.; Riccardi, M.; Vanucci, S.; Ciminiello, P.; Dell'Aversano, C.; Forino, M.; Fattorusso, E.; Tartaglione, L.; et al. Toxin levels and profiles in microalgae from the North-Western Adriatic Sea - 15 Years of studies on cultured species. *Mar. Drugs* **2012**, *10*, 140–162. [CrossRef]

22. Arapov, J.; Ujević, I.; Straka, M.; Skejić, S.; Bužančić, M.; Bakrač, A.; Gladan, Ž.N. First evidence of domoic acid production in *Pseudo-nitzschia calliantha* cultures from the central Adriatic Sea. *Acta Adriat.* **2020**, *61*, 135–144. [CrossRef]

23. Arapov, J.; Ujevic, I.; Pfannkuchen, D.M.; Godrijan, J.; Bakrac, A.; Gladan, Z.N.; Marasovic, I. Domoic acid in phytoplankton net samples and shellfish from the Krka river estuary in the central Adriatic Sea. *Mediterr. Mar. Sci.* **2016**, *17*, 340–350. [CrossRef]

24. Cerino, F.; Orsini, L.; Sarno, D.; Dell'Aversano, C.; Tartaglione, L.; Zingone, A. The alternation of different morphotypes in the seasonal cycle of the toxic diatom *Pseudo-nitzschia galaxiae*. *Harmful Algae* **2005**, *4*, 33–48. [CrossRef]

25. Lim, H.C.; Tan, S.N.; Teng, S.T.; Lundholm, N.; Orive, E.; David, H.; Quijano-Scheggia, S.; Leong, S.C.Y.; Wolf, M.; Bates, S.S.; et al. Phylogeny and species delineation in the marine diatom *Pseudo-nitzschia* (Bacillariophyta) using cox1, LSU, and ITS2 rRNA genes: A perspective in character evolution. *J. Phycol.* **2018**, *54*, 234–248. [CrossRef]

26. Basu, S.; Patil, S.; Mapleson, D.; Russo, M.T.; Vitale, L.; Fevola, C.; Maumus, F.; Casotti, R.; Mock, T.; Caccamo, M.; et al. Finding a partner in the ocean: Molecular and evolutionary bases of the response to sexual cues in a planktonic diatom. *New Phytol.* **2017**, *215*, 140–156. [CrossRef]

27. Trainer, V.L.; Bates, S.S.; Lundholm, N.; Thessen, A.E.; Cochlan, W.P.; Adams, N.G.; Trick, C.G. *Pseudo-nitzschia* physiological ecology, phylogeny, toxicity, monitoring and impacts on ecosystem health. *Harmful Algae* **2012**, *14*, 271–300. [CrossRef]

28. HAEDAT Harmful Algae Event Database. Available online: http://haedat.iode.org/eventSearch.php?searchtext%5BcountryID%5D=58 (accessed on 8 June 2021).

29. Tenorio, C.; Álvarez, G.; Quijano-Scheggia, S.; Perez-Alania, M.; Arakaki, N.; Araya, M.; Álvarez, F.; Blanco, J.; Uribe, E. First report of domoic acid production from *Pseudo-nitzschia multistriata* in Paracas bay (Peru). *Toxins* **2021**, *13*, 408. [CrossRef]

30. Orsini, L.; Sarno, D.; Procaccini, G.; Poletti, R.; Dahlmann, J.; Montresor, M. Toxic *Pseudo-nitzschia multistriata* (Bacillariophyceae) from the Gulf of Naples: Morphology, toxin analysis and phylogenetic relationships with other *Pseudo-nitzschia* species. *Eur. J. Phycol.* **2002**, *37*, 247–257. [CrossRef]

31. Amato, A.; Lüdeking, A.; Kooistra, W.H.C.F. Intracellular domoic acid production in *Pseudo-nitzschia multistriata* isolated from the Gulf of Naples (Tyrrhenian Sea, Italy). *Toxicon* **2010**, *55*, 157–161. [CrossRef]

32. Ruggiero, M.V.; Sarno, D.; Barra, L.; Kooistra, W.H.C.F.; Montresor, M.; Zingone, A. Diversity and temporal pattern of *Pseudo-nitzschia* species (Bacillariophyceae) through the molecular lens. *Harmful Algae* **2015**, *42*, 15–24. [CrossRef]
33. Moschandreou, K.K.; Baxevanis, A.D.; Katikou, P.; Papaefthimiou, D.; Nikolaidis, G.; Abatzopoulos, T.J. Inter- and intra-specific diversity of *Pseudo-nitzschia* (Bacillariophyceae) in the northeastern Mediterranean. *Eur. J. Phycol.* **2012**, *47*, 321–339. [CrossRef]
34. Amato, A.; Montresor, M. Morphology, phylogeny, and sexual cycle of *Pseudo-nitzschia mannii* sp. nov. (Bacillariophyceae): A pseudo-cryptic species within the *P. pseudodelicatissima* complex. *Phycologia* **2008**, *47*, 487–497. [CrossRef]
35. Pan, Y.; Parsons, M.L.; Busman, M.; Moeller, P.D.R.; Dortch, Q.; Powell, C.L.; Doucette, G.J. *Pseudo-nitzschia* sp. cf. *pseudo-delicatissima*-A confirmed producer of domoic acid from the northern Gulf of Mexico. *Mar. Ecol. Prog. Ser.* **2001**, *220*, 83–92. [CrossRef]
36. Lundholm, N.; Hansen, P.J.; Kotaki, Y. Effect of pH on growth and domoic acid production by potentially toxic diatoms of the genera *Pseudo-nitzschia* and *Nitzschia*. *Mar. Ecol. Prog. Ser.* **2004**, 1–15. [CrossRef]
37. Amin, S.A.; Parker, M.S.; Armbrust, E.V. Interactions between Diatoms and Bacteria. *Microbiol. Mol. Biol. Rev.* **2012**, *76*, 667–684. [CrossRef] [PubMed]
38. Wang, Z.; Maucher-Fuquay, J.; Fire, S.E.; Mikulski, C.M.; Haynes, B.; Doucette, G.J.; Ramsdell, J.S. Optimization of solid-phase extraction and liquid chromatography-tandem mass spectrometry for the determination of domoic acid in seawater, phytoplankton, and mammalian fluids and tissues. *Anal. Chim. Acta* **2012**, *715*, 71–79. [CrossRef] [PubMed]
39. Smith, E.A.; Papapanagiotou, E.P.; Brown, N.A.; Stobo, L.A.; Gallacher, S.; Shanks, A.M. Effect of storage on amnesic shellfish poisoning (ASP) toxins in king scallops (*Pecten maximus*). *Harmful Algae* **2006**, *5*, 9–19. [CrossRef]
40. Howard, M.D.A.; Smith, G.J.; Kudela, R.M. Phylogenetic relationships of yessotoxin-producing dinoflagellates, based on the large subunit and internal transcribed spacer ribosomal DNA domains. *Appl. Environ. Microbiol.* **2009**, *75*, 54–63. [CrossRef]
41. Lema, K.A.; Metegnier, G.; Quéré, J.; Latimier, M.; Youenou, A.; Lambert, C.; Fauchot, J.; Le Gac, M.; Costantini, M. Inter- and Intra-Specific Transcriptional and Phenotypic Responses of *Pseudo-nitzschia* under Different Nutrient Conditions. *Genome Biol. Evol.* **2019**, *11*, 731–747. [CrossRef]
42. Pulido, O.M. Domoic acid toxicologic pathology: A review. *Mar. Drugs* **2008**, *6*, 180–219. [CrossRef]
43. Tan, S.N.; Teng, S.T.; Lim, H.C.; Kotaki, Y.; Bates, S.S.; Leaw, C.P.; Lim, P.T. Diatom *Nitzschia navis-varingica* (Bacillariophyceae) and its domoic acid production from the mangrove environments of Malaysia. *Harmful Algae* **2016**, *60*, 139–149. [CrossRef] [PubMed]
44. Smida, D.B.; Lundholm, N.; Kooistra, W.H.C.F.; Sahraoui, I.; Ruggiero, M.V.; Kotaki, Y.; Ellegaard, M.; Lambert, C.; Mabrouk, H.H.; Hlaili, A.S. Morphology and molecular phylogeny of *Nitzschia bizertensis* sp. nov.-A new domoic acid-producer. *Harmful Algae* **2014**, *32*, 49–63. [CrossRef]
45. Chekan, J.R.; McKinnie, S.M.K.; Noel, J.P.; Moore, B.S. Algal neurotoxin biosynthesis repurposes the terpene cyclase structural fold into an N-prenyltransferase. *Proc. Natl. Acad. Sci. USA* **2020**, *117*, 12799–12805. [CrossRef]
46. Orr, R.J.S.; Stüken, A.; Murray, S.A.; Jakobsen, K.S. Evolution and distribution of saxitoxin biosynthesis in dinoflagellates. *Mar. Drugs* **2013**, *11*, 2814–2828. [CrossRef] [PubMed]
47. Fuhrman, J.A. Marine viruses and their biogeochemical and ecological effects. *Nature* **1999**, *399*, 541–548. [CrossRef]
48. Casteleyn, G.; Adams, N.G.; Vanormelingen, P.; Debeer, A.-E.; Sabbe, K.; Vyverman, W. Natural Hybrids in the Marine Diatom *Pseudo-nitzschia pungens* (Bacillariophyceae): Genetic and Morphological Evidence. *Protist* **2009**, *160*, 343–354. [CrossRef]
49. Tesson, S.V.M.; Legrand, C.; van Oosterhout, C.; Montresor, M.; Kooistra, W.H.C.F.; Procaccini, G. Mendelian Inheritance Pattern and High Mutation Rates of Microsatellite Alleles in the Diatom *Pseudo-nitzschia multistriata*. *Protist* **2013**, *164*, 89–100. [CrossRef]
50. Schliep, K.P. Phangorn: Phylogenetic analysis in R. *Bioinformatics* **2011**, *27*, 592–593. [CrossRef]
51. Paradis, E.; Schliep, K. ape 5.0: An environment for modern phylogenetics and evolutionary analyses in R. *Bioinformatics* **2019**, *35*, 526–528. [CrossRef]
52. Team, R.C. *R: A Language and Environment for Statistical Computing*; R Foundation for Statistical Computing: Vienna, Austria, 2019.
53. Kelley, L.A.; Mezulis, S.; Yates, C.M.; Wass, M.N.; Sternberg, M.J.E. The Phyre2 web portal for protein modeling, prediction and analysis. *Nat. Protoc.* **2015**, *10*, 845–858. [CrossRef]

Article

Update of the Planktonic Diatom Genus *Pseudo-nitzschia* in Aotearoa New Zealand Coastal Waters: Genetic Diversity and Toxin Production

Tomohiro Nishimura [1,*], J. Sam Murray [1], Michael J. Boundy [1], Muharrem Balci [2], Holly A. Bowers [3], Kirsty F. Smith [1,4], D. Tim Harwood [1] and Lesley L. Rhodes [1,*]

1   Cawthron Institute, Nelson 7010, New Zealand; sam.murray@cawthron.org.nz (J.S.M.); michael.boundy@cawthron.org.nz (M.J.B.); kirsty.smith@cawthron.org.nz (K.F.S.); tim.harwood@cawthron.org.nz (D.T.H.)
2   Biology Department, Faculty of Science, Istanbul University, Istanbul 34134, Turkey; muharrem.balci@istanbul.edu.tr
3   Moss Landing Marine Laboratories, Moss Landing, CA 95039, USA; hbowers@mlml.calstate.edu
4   School of Biological Sciences, University of Auckland, Auckland 1142, New Zealand
*   Correspondence: tomohiro.nishimura@cawthron.org.nz (T.N.); lesley.rhodes@cawthron.org.nz (L.L.R.)

Abstract: Domoic acid (DA) is produced by almost half of the species belonging to the diatom genus *Pseudo-nitzschia* and causes amnesic shellfish poisoning (ASP). It is, therefore, important to investigate the diversity and toxin production of *Pseudo-nitzschia* species for ASP risk assessments. Between 2018 and 2020, seawater samples were collected from various sites around Aotearoa New Zealand, and 130 clonal isolates of *Pseudo-nitzschia* were established. Molecular phylogenetic analysis of partial large subunit ribosomal DNA and/or internal transcribed spacer regions revealed that the isolates were divided into 14 species (*Pseudo-nitzschia americana*, *Pseudo-nitzschia arenysensis*, *Pseudo-nitzschia australis*, *Pseudo-nitzschia calliantha*, *Pseudo-nitzschia cuspidata*, *Pseudo-nitzschia delicatissima*, *Pseudo-nitzschia fraudulenta*, *Pseudo-nitzschia galaxiae*, *Pseudo-nitzschia hasleana*, *Pseudo-nitzschia multiseries*, *Pseudo-nitzschia multistriata*, *Pseudo-nitzschia plurisecta*, *Pseudo-nitzschia pungens*, and *Pseudo-nitzschia* cf. *subpacifica*). The *P. delicatissima* and *P. hasleana* strains were further divided into two clades/subclades (I and II). Liquid chromatography-tandem mass spectrometry was used to assess the production of DA and DA isomers by 73 representative strains. The analyses revealed that two (*P. australis* and *P. multiseries*) of the 14 species produced DA as a primary analogue, along with several DA isomers. This study is the first geographical distribution record of *P. arenysensis*, *P. cuspidata*, *P. galaxiae*, and *P. hasleana* in New Zealand coastal waters.

Keywords: amnesic shellfish poisoning; domoic acid; geographical distribution; liquid chromatography-tandem mass spectrometry; molecular phylogenetic analysis; Pacific Ocean; phytoplankton

Key Contribution: This study updates information on the species diversity and toxin production of the genus *Pseudo-nitzschia* in New Zealand coastal waters. The latest classification of *Pseudo-nitzschia* species/clades/subclades was applied in the molecular phylogenetic analyses.

Citation: Nishimura, T.; Murray, J.S.; Boundy, M.J.; Balci, M.; Bowers, H.A.; Smith, K.F.; Harwood, D.T.; Rhodes, L.L. Update of the Planktonic Diatom Genus *Pseudo-nitzschia* in Aotearoa New Zealand Coastal Waters: Genetic Diversity and Toxin Production. *Toxins* **2021**, *13*, 637. https://doi.org/10.3390/toxins13090637

Received: 30 July 2021
Accepted: 1 September 2021
Published: 10 September 2021

Publisher's Note: MDPI stays neutral with regard to jurisdictional claims in published maps and institutional affiliations.

1. Introduction

The first amnesic shellfish poisoning (ASP) outbreak, caused by consumption of cultivated blue mussels, occurred in Canada in 1987, and several deaths were recorded [1–3]. At the same time, *Pseudo-nitzschia multiseries* (formerly reported as *Nitzschia pungens* f. *multiseries*) was identified as the causative species [1–4]. It produced a biotoxin, domoic acid (DA) [2], which has been monitored in Aotearoa New Zealand shellfish since 1993 [5], and is currently the only DA analogue regulated in New Zealand [6,7]. Additional DA isomers have been reported to date [8], however there is insufficient toxicity information to establish Toxicity Equivalency Factors (TEF) for the isomers [9]. As well as shellfish being tested, seawater samples are analysed to determine the presence and quantity of *Pseudo-nitzschia* cells, which bloom regularly in the coastal waters, to allow for risk assessments for the shellfish industry and regulators [10]. These assessments aid managers in their harvesting decisions and help public health officials determine whether to post warnings for recreational harvesters of potential toxins in seafood.

The occurrence of *Pseudo-nitzschia* blooms in New Zealand was summarised in 2000 [5] and again in 2011 [10]. An overview of harmful algal blooms (HABs) in Australasia, a region comprising Australia, New Zealand, and some neighbouring islands, then reviewed the monitoring data for *Pseudo-nitzschia* and its toxins up to 2018 [11]. Despite intensive monitoring of shellfish since 1993, only one critically high DA concentration has been reported, in scallop digestive glands in 1994, following the collapse of a *P. australis* bloom in Bream Bay, Northland, New Zealand [12]. More recently, a high DA concentration above regulatory level has been reported in Queen scallop flesh collected in deep water off the coast of Otago, New Zealand in December 2020 (Harwood et al. unpublished data). No human illnesses due to DA poisoning have occurred by consumption of shellfish harvested in New Zealand to date, due to the monitoring efforts.

By 2011, there were 35 *Pseudo-nitzschia* species known globally (14 of which were DA producers) and, of those species, 12 were recorded in New Zealand at that time (seven of which were known DA producers). Those species were *P. americana*, *P. australis*, *P. caciantha*, *P. calliantha*, *P. delicatissima* (including a genotype formerly reported as '*Pseudo-nitzschia turgidula*'), *P. fraudulenta*, *P. multiseries*, *P. multistriata*, *P. plurisecta* (formerly reported as *P. cuspidata* Hobart-5 type), *Pseudo-nitzschia pseudodelicatissima*, *P. pungens*, and *P. subpacifica* [formerly reported as *Pseudo-nitzschia* (cf.) *heimii*] [10]. Over the last decade, 23 additional species have been described globally (12 of those being known DA producers), resulting in 58 species as of 2021 [13], with 26 of those being DA producers.

The current study aims to refresh the data on the diversity of and toxin production by *Pseudo-nitzschia*, by presenting the results following the isolation of more than 100 *Pseudo-nitzschia* uni-cells/chains from the coastal waters of New Zealand's three main islands. Analysis of molecular phylogenetic data from these isolates enables reporting of newly recorded species of *Pseudo-nitzschia* in New Zealand. This new information, which includes toxin profiles, will allow risk assessments to be refined in New Zealand.

2. Results

2.1. Molecular Phylogenetic Characteristics

2.1.1. Molecular Phylogeny

Genotype screening using molecular phylogenetic analysis based on the nuclear-encoded ribosomal RNA gene (rDNA), specifically, internal transcribed spacer (ITS) 1–5.8S rDNA–ITS 2 region (ITS region) and large subunit (LSU) rDNA D1–D3 region (LSU rDNA D1–D3), was conducted for a total of 130 isolates of *Pseudo-nitzschia* established from 22 sites in subtropical and temperate coastal waters in New Zealand (Table S1). The screening revealed that the strains could be separated into 14 species: *P. americana* [number of strains (*n*) = 6], *P. arenysensis* (*n* = 7), *P. australis* (*n* = 14), *P. calliantha* (*n* = 3), *P. cuspidata* (*n* = 1), *P. delicatissima* (*n* = 28), *P. fraudulenta* (*n* = 13), *P. galaxiae* (*n* = 2), *P. hasleana* (*n* = 6), *P. multiseries* (*n* = 4), *P. multistriata* (*n* = 13), *P. plurisecta* (*n* = 1), *P. pungens* (*n* = 27), and *P.* cf. *subpacifica* (*n* = 5) (Table S2). Furthermore, the strains of five species (*P. arenysensis*, *P. cuspidata*, *P. delicatissima*, *P. galaxiae*, *P. hasleana*, and *P. pungens*) belonged to one or two clades/subclades as explained below.

As the sequences of the strains of each species/clade/subclade had high levels of identity, selected representative sequences from each species/clade/subclade were used for the molecular phylogenetic analysis using the maximum likelihood (ML) and Bayesian inference (BI) methods. The number of the representative sequences for the ITS region and LSU rDNA D1–D3 were 61 and 38, respectively (Table S2). The ITS region phylogeny comprising 52 *Pseudo-nitzschia* species is shown in Figures 1 and 2. The LSU rDNA D1–D3 tree comprising 47 *Pseudo-nitzschia* species is shown in Figure 3.

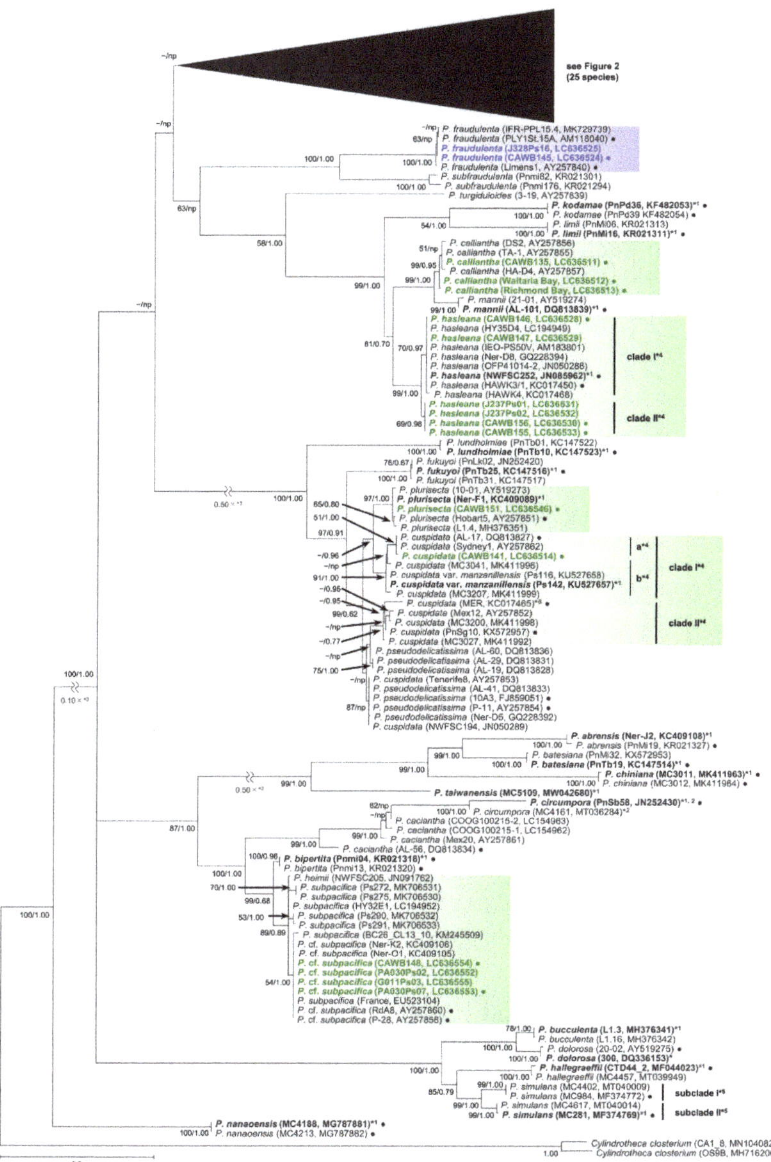

Figure 1. Molecular phylogenetic tree of 52 *Pseudo-nitzschia* species based on the ITS region sequences (253 sequences, 809 positions) using maximum likelihood (ML) analysis. See Figure 2 for details of 25 species. Strains from New Zealand are shown in colour fonts. Blue and green fonts indicate potential 'low and no ASP risk' species in New Zealand, respectively. A Black or coloured circle indicates a strain used in the LSU rDNA D1–D3 tree shown in Figure 3. Nodal support represents ML bootstrap value/Bayesian inference (BI) posterior probability. Nodal support under 50 in ML or 0.50 in BI is shown as a minus sign (−). A node that was not present in the BI tree is labelled as np. A scale bar indicates the number of nucleotide substitutions per site. [1]: A sequence obtained from holotype material is shown in bold font. [2]: A sequence having only ITS 2 region. [3]: A reduction ratio of reduced nodal length calculated from original nodal length. [4]: Clade separation reported by the present study. [5]: Subclade separation reported by Ajani et al. (2020) [17]. [6]: *Pseudo-nitzschia cuspidata* strain MER was re-assigned as *P.* cf. *cuspidata* by Ajani et al. (2021) [18].

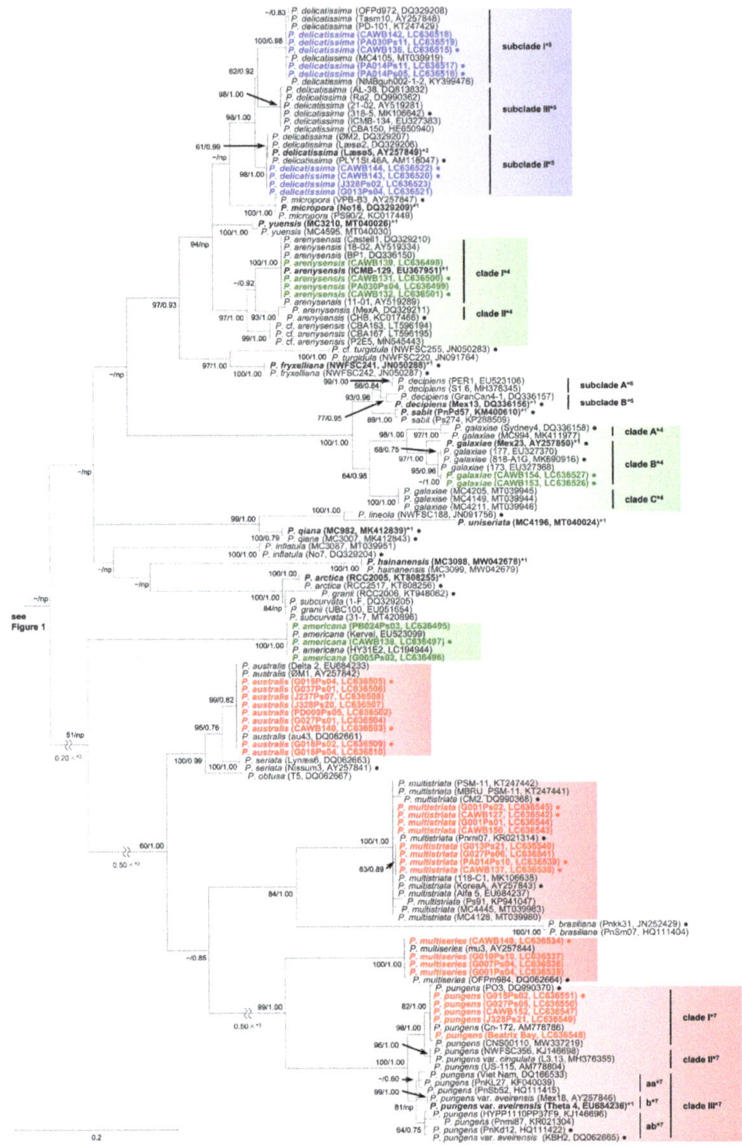

Figure 2. Molecular phylogenetic tree of 25 *Pseudo-nitzschia* species based on the ITS region sequences (144 sequences, 809 positions) using maximum likelihood (ML) analysis. Strains from New Zealand are shown in colour fonts. Red, blue, and green fonts indicate potential 'high, low, and no ASP risk' species in New Zealand, respectively. A Black or coloured circle indicates a strain used in the LSU rDNA D1–D3 tree shown in Figure 3. Nodal support represents ML bootstrap value/Bayesian inference (BI) posterior probability. Nodal support under 50 in ML or 0.50 in BI is shown as a minus sign (−). A node that was not present in the BI tree is labelled as np. A scale bar indicates the number of nucleotide substitutions per site. *1: A sequence obtained from holotype material is shown in bold font. *2: A sequence obtained from epitype material is shown in bold font. *3: A reduction ratio of reduced nodal length calculated from original nodal length. *4: Clade separation reported by the present study. *5: Subclade separation reported by Stonik et al. (2018) [14]. *6: Subclade separation reported by Gai et al. (2018) [15]. *7: Clade separation reported by Kim et al. (2015) [16].

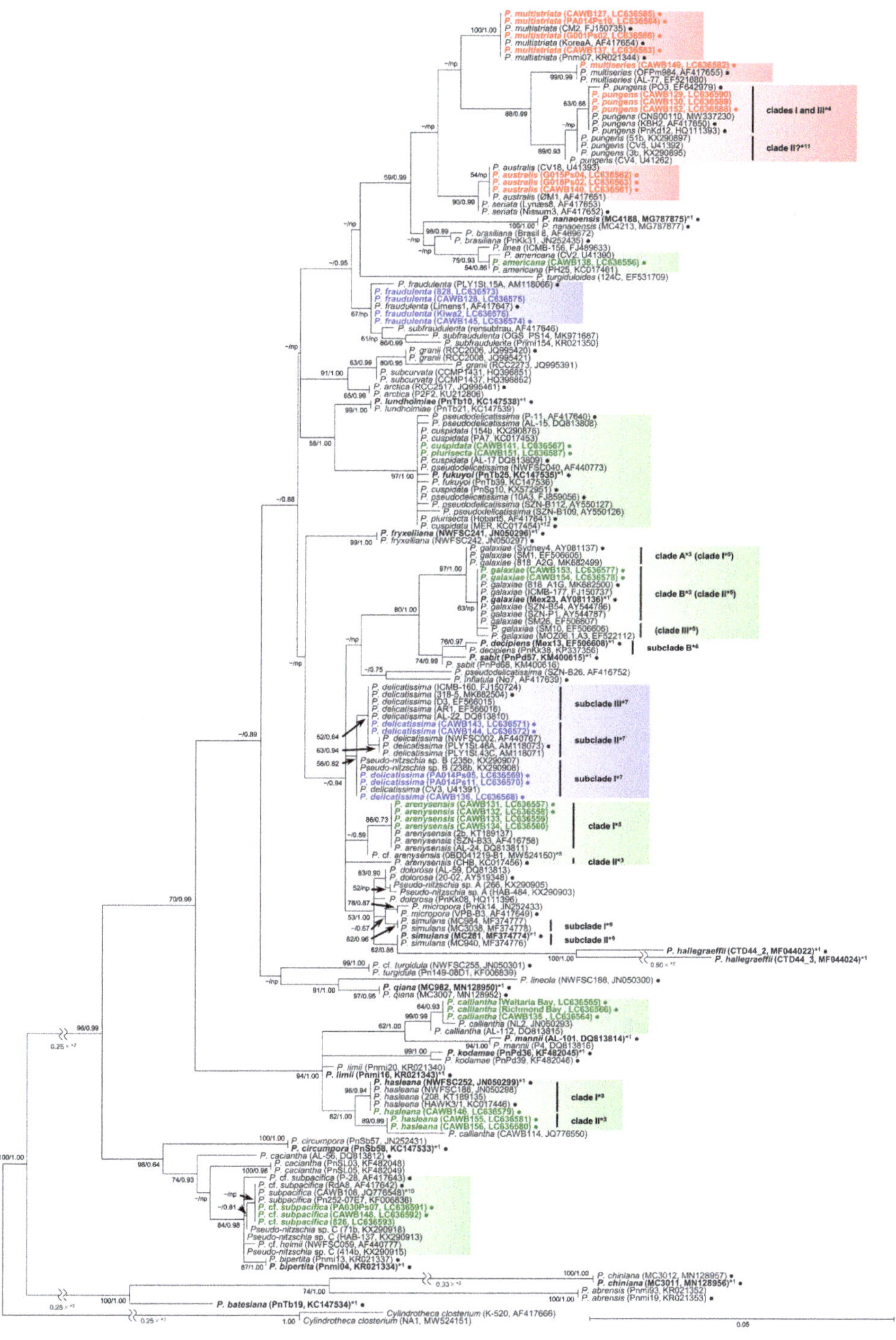

Figure 3. Molecular phylogenetic tree of 47 *Pseudo-nitzschia* species based on the LSU rDNA D1–D3 sequences (179 sequences, 660 positions) using maximum likelihood (ML) analysis. Strains from New Zealand are shown in colour fonts. Red, blue,

and green fonts indicate potential 'high, low, and no ASP risk' species in New Zealand, respectively. A Black or coloured circle indicates a strain used in the ITS region trees shown in Figures 1 and 2. Nodal support represents ML bootstrap value/Bayesian inference (BI) posterior probability. Nodal support under 50 in ML or 0.50 in BI is shown as a minus sign (−). A node that was not present in the BI tree is labelled as np. A scale bar indicates the number of nucleotide substitutions per site. [1]: A sequence obtained from holotype material is shown in bold font. [2]: A reduction ratio of reduced nodal length calculated from original nodal length. [3]: Clade separation reported by the present study. [4]: Clade separation reported by Kim et al. (2015) [16]. [5]: Clade separation reported by McDonald et al. (2007) [19]. [6]: Subclade separation reported by Gai et al. (2018) [15]. [7]: Subclade separation reported by Stonik et al. (2018) [14]. [8]: Strain 0BD041219-B1, originally reported as *P. delicatissima* by Dermastia et al. (unpublished data), is assigned as *P.* cf. *arenysensis* as its sequence was almost identical to those of *P.* cf. *arenysensis* reported recently by Giulietti et al. (2021) [20]. [9]: Subclade separation reported by Ajani et al. (2020) [17]. [10]: *Pseudo-nitzschia* cf. *heimii* strain CAWB106 is assigned as *P. subpacifica* based on morphological characters following discussions in Rhodes et al. (2013) [10]. [11]: *Pseudo-nitzschia pungens* strains CV4, CV5, 3b, and 51b [21,22] were isolated from California, USA where only *P. pungens* clade II was reported to date. These strains were tentatively assigned as putative clade II of *P. pungens* in the present study. [12]: *Pseudo-nitzschia cuspidata* strain MER was re-assigned as *P.* cf. *cuspidata* by Ajani et al. (2021) [18].

In the phylogenetic trees, eight *Pseudo-nitzschia* species (*P. arenysensis*/*P.* cf. *arenysensis*, *P. cuspidata*, *Pseudo-nitzschia decipiens*, *P. delicatissima*, *P. galaxiae*, *P. hasleana*, *P. pungens*, and *Pseudo-nitzschia simulans*) could be separated into various clades or subclades (Figures 1–3). Among these clades/subclades, the New Zealand strains belonged to *P. arenysensis* clade I, *P. cuspidata* clade Ia, *P. delicatissima* subclades I and II, *P. galaxiae* clade B, *P. hasleana* clades I and II, and *P. pungens* clade I (Figures 1–3). The clade/subclade separation previously reported for *P. delicatissima* [14], *P. decipiens* [15], *P. pungens* [16], and *P. simulans* [17] were applied in the present study. The clade/subclade separation for *P. arenysensis*, *P. cuspidata*, *P. galaxiae*, and *P. hasleana* were proposed in the present study. Among them, *P. hasleana* clade II was unreported so far and firstly reported by the present study. Although all species/clades/subclades, except for combinations of *Pseudo-nitzschia granii*/*Pseudo-nitzschia subcurvata* and *P.* (cf.) *heimii*/*P.* (cf.) *subpacifica*, were separated in the ITS trees (Figures 1 and 2), the sequences belonging to two combinations (*P. pungens* clades I/III and *P. cuspidata*/*Pseudo-nitzschia fukuyoi*/*P. plurisecta*/*P. pseudodelicatissima*) were clustered together respectively in the LSU rDNA D1–D3 tree (Figure 3).

2.1.2. Sequence Analysis

To reveal genetic divergence, especially for the newly reported *P. hasleana* clade II, the uncorrected *p* distances of the ITS region sequences of selected combinations between *Pseudo-nitzschia* species/clades/subclades were calculated. The selected species were *P. arenysensis*/*P.* cf. *arenysensis*, *P. cuspidata*/*P. pseudodelicatissima*/*P. plurisecta*, *P. decipiens*, *P. delicatissima*, *P. galaxiae*, *P. hasleana*, *P. pungens*, and *P. simulans* (Table S3). The comparisons revealed that the genetic divergence between *P. hasleana* clades I and II (the *p* distance: 0.012 ± 0.001, 0.011–0.015) were the same as that between *P. pungens* clades I and II (0.012 ± 0.001, 0.011–0.014) (Table S3). In contrast, that within *P. cuspidata* clade II (0.012 ± 0.008, 0.003–0.024) was similar to that between *P. cuspidata* clades Ia and Ib (0.013 ± 0.004, 0.010–0.021) (Table S3). Additionally, sequences of four New Zealand strains of *P. delicatissima* subclade II (strains CAWB143, CAWB144, G013Ps04, and J328Ps02) were slightly different (1–2 positions difference in 671 positions) from those of the others (strains ØM2, Læsø2, Læsø5, and PLY1St.46A) belonging to the same subclade (0.002 ± 0.001, 0.002–0.003) (Table S3, Figure 2).

The *p* distances of the LSU rDNA D1–D3 sequences of selected combinations between *Pseudo-nitzschia* species/clades, including *P. hasleana* clade II, were calculated. The selected species/clades were *P. hasleana* clades I and II, '*P. calliantha*', *P. calliantha*, *Pseudo-nitzschia mannii*, *Pseudo-nitzschia limii*, and *Pseudo-nitzschia kodamae* (Table S4). The comparisons revealed that the average *p* distance between *P. hasleana* clades I and II (0.006) were slightly lower than those of other combinations (0.007–0.035) (Table S4). The average *p* distances of '*P. calliantha*' strain CAWB114 towards *P. hasleana* clades I and II and *P. calliantha* were 0.014,

0.008, and 0.027, respectively (Table S4). Additionally, the values of '*P. calliantha*' strain CAWB114 towards *P. hasleana* clades I (0.014) and II (0.008) were higher than or similar to those between clades in the same species [*P. hasleana* clades I and II (0.006)] or between species [*P. hasleana* clade I and *P. limii* (0.007), or *P. hasleana* clade II and *P. limii* (0.008)] (Table S4).

2.2. Distribution

Among the 130 clonal *Pseudo-nitzschia* strains established and genetically identified in the present study, 29 strains originated from four subtropical sites and 101 strains originated from 18 temperate sites (Figure 4 and Table S2). In the subtropical zone, the number of species found from Northland and Bay of Plenty regions were nine and two, respectively. In the temperate zone, those from Golden Bay/Tasman Bay, Marlborough Sounds, and Big Glory Bay (Stewart Island) were seven, eleven, and five, respectively (Figure 4).

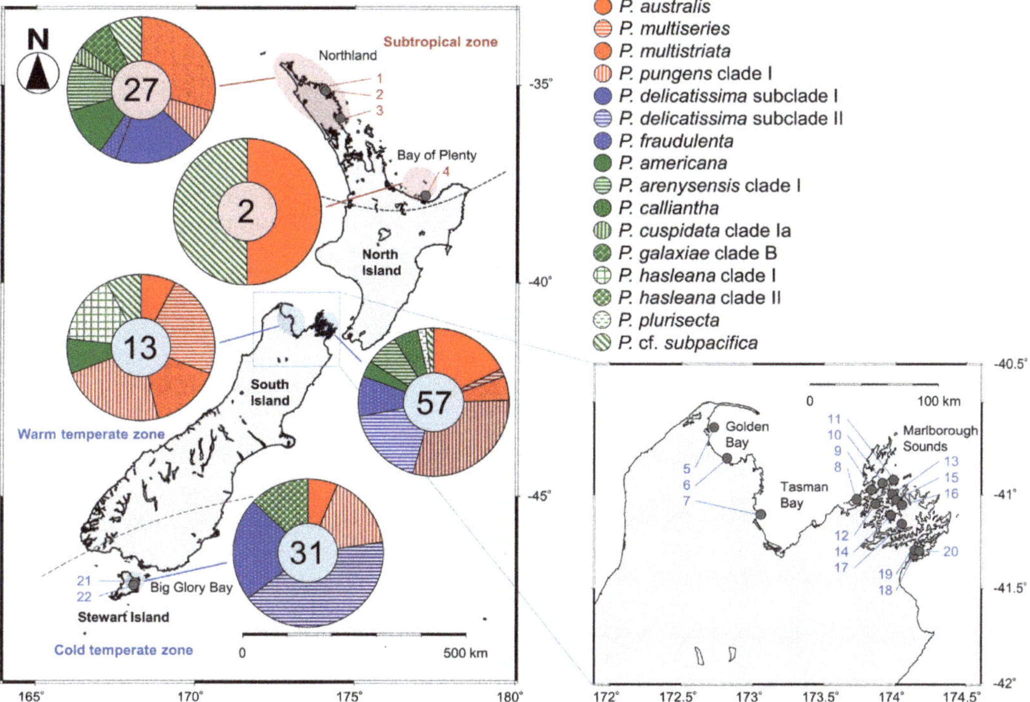

Figure 4. Sampling and distribution map of 14 *Pseudo-nitzschia* species in New Zealand coastal waters assessed between 2018 and 2020. The locations of 22 sampling sites (grey circles) and their site codes are shown. The numbers in each pie chart indicate the total number of clonal strains genetically identified by the molecular phylogenies. Red, blue, and green indicate potential 'high, low, and no ASP risk' species in New Zealand, respectively. Sampling details, including sampling site code, are shown in Table S1.

The species/clades/subclades composition plotted onto a map suggested that each of them showed a unique distribution pattern in New Zealand coastal waters (Figure 4). *Pseudo-nitzschia americana*, *P. arenysensis* clade I, *P. australis*, *P. fraudulenta*, *P. multistriata*, *P. pungens* clade I, and *P.* cf. *subpacifica* were widespread from subtropical to temperate zones (Figure 4). On the other hand, *P. cuspidata* clade Ia, *P. delicatissima* subclade I, *P. galaxiae* clade B were restricted in the subtropical zone, whereas *P. calliantha*, *P. delicatissima* subclade II, *P. hasleana* clades I and II, *P. multiseries*, and *P. plurisecta* were restricted in the temperate zone (Figure 4). *Pseudo-nitzschia hasleana* clades I and II were distributed in warm temperate and cold temperate zones, respectively (Figure 4).

2.3. Toxin Production

2.3.1. Production of Domoic Acid (DA) and Its Isomers

Production of DA and its isomers [c5′-epidomoic acid (epi-DA), isodomoic acids (iso-DAs) A, B, C, D, and E; Figure 5] was assessed in 73 representative strains, from 14 *Pseudo-nitzschia* species genetically identified above (Section 2.1), using liquid chromatography-tandem mass spectrometry (LC-MS/MS). Strain details, including cell quotas of DA and the monitored isomers, expressed as pg cell^{-1}, are shown in Table S5. Toxin cell quota detected between the limit of detection (LoD) (0.00005–0.0005 pg cell^{-1}) and lower limit of quantitation (LLoQ) (0.0005–0.003 pg cell^{-1}) was expressed as 'trace level' (Table S5). All the assessed strains of *P. australis* [number of strains (n) = 9] and *P. multiseries* (n = 4) produced DA with ranges of 0.004–2.02 pg cell^{-1} and 0.47–7.20 pg cell^{-1}, respectively (Figure 6A and Table S5). The DA cell quotas of *P. multiseries* strains were higher than those of *P. australis* strains (Mann–Whitney U test, $p < 0.05$). By contrast, the productivity of DA isomers of the strains was different between *P. australis* and *P. multiseries*. All *P. australis* strains produced low quantities of iso-DA A (trace level–0.060 pg cell^{-1}), iso-DA B (trace level), and iso-DA C (0.002–0.884 pg cell^{-1}), while some strains produced low quantities of epi-DA (trace–0.010 pg cell^{-1}), iso-DA D (trace level–0.003 pg cell^{-1}) and/or iso-DA E (trace level) (Figure 6A and Table S5). All *P. multiseries* strains produced low quantities of iso-DA A (0.014–0.206 pg cell^{-1}), iso-DA D (0.003–0.020 pg cell^{-1}), and iso-DA E (trace level–0.005 pg cell^{-1}), while some strains produced low quantities of epi-DA (0.002–0.012 pg cell^{-1}), iso-DA B (trace level) and/or iso-DA C (0.012–0.239 pg cell^{-1}) (Figure 6A and Table S5). Multiple reaction monitoring (MRM) chromatograms of DA and DA isomers standards and an extract of *P. multiseries* strain G001Ps04 are shown in Figure 7. Two out of the eight strains of *P. multistriata* assessed produced trace levels of DA and iso-DA A, and one strain of *P. cuspidata* clade Ia produced a trace level of DA (Table S5). None of the other 57 strains assessed showed detectable concentrations of the DA and DA isomers examined.

2.3.2. Relative Proportion of DA and Its Isomers

The relative proportion of DA and its isomers produced by the *P. australis* and *P. multiseries* strains assessed were compared (Figure 6B). Analogues detected with trace levels were ignored in the calculation. The comparison revealed that the primary analogue produced by the strains of both species was DA, whereas the proportions of DA isomers of the strains differed between *P. australis* and *P. multiseries*. The primary analogue produced by *P. australis* strains was DA (44.4–91.5% of total DA and its isomers quantified). The second primary analogue of the strains was iso-DA C (4.8–54.1%), and the other analogues were minor contributors (epi-DA, 0.4%; iso-DA A, 1.5–3.7%; and iso-DA D, 0.1%) (Figure 6B and Table S5). The primary analogue produced by *P. multiseries* strains was DA (93.9–96.5%), followed by iso-DA A (2.6–2.8%) and iso-DA C (2.3–3.1%), with the other analogues being minor contributors (epi-DA, 0.3–0.7%; iso-DA D, 0.3–1.2%; and iso-DA E, 0.1–0.2%) (Figure 6B and Table S5). The DA proportions of *P. multiseries* strains were higher than those of *P. australis* strains (Mann–Whitney U test, $p < 0.01$).

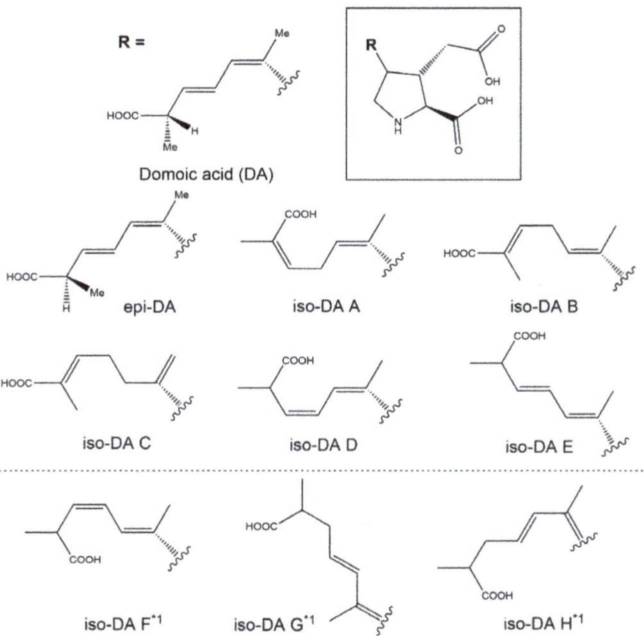

Figure 5. Structures of domoic acid (DA) and DA isomers. Adapted from Quilliam et al. (1995) [23]. *1 DA isomers not analysed in the present study.

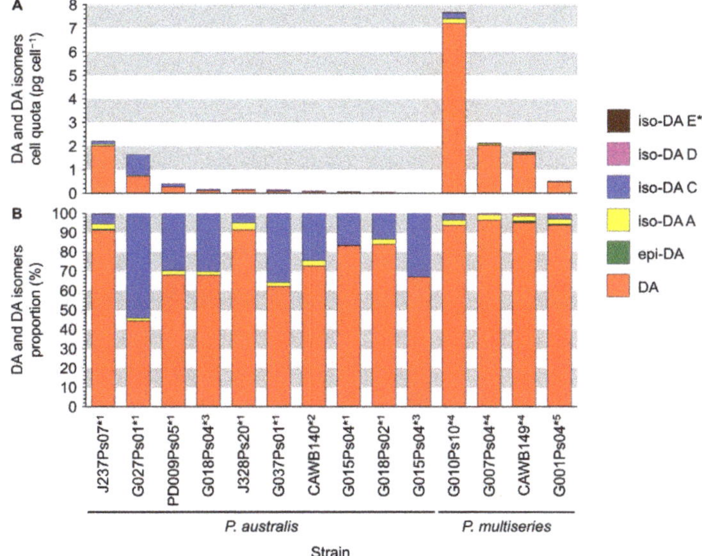

Figure 6. Domoic acid (DA) and DA isomers profiles of *Pseudo-nitzschia australis* and *P. multiseries* strains from New Zealand. (**A**) DA and DA isomers cell quota (pg cell^{-1}); (**B**) DA and DA isomers proportion (%). Asterisks: Strain was cultured using f/2−Si medium at *1 three, *2 four, *3 seven, *4 nine, or *5 ten months after the strain was established. *6: Three *P. multiseries* strains (G010Ps10, CAWB149, and G001Ps04) produced very low cell quotas of iso-DA E (0.002–0.005 pg cell^{-1}, 0.1–0.2%), resulting in difficulty in seeing these data in Figure 6 (**A**,**B**).

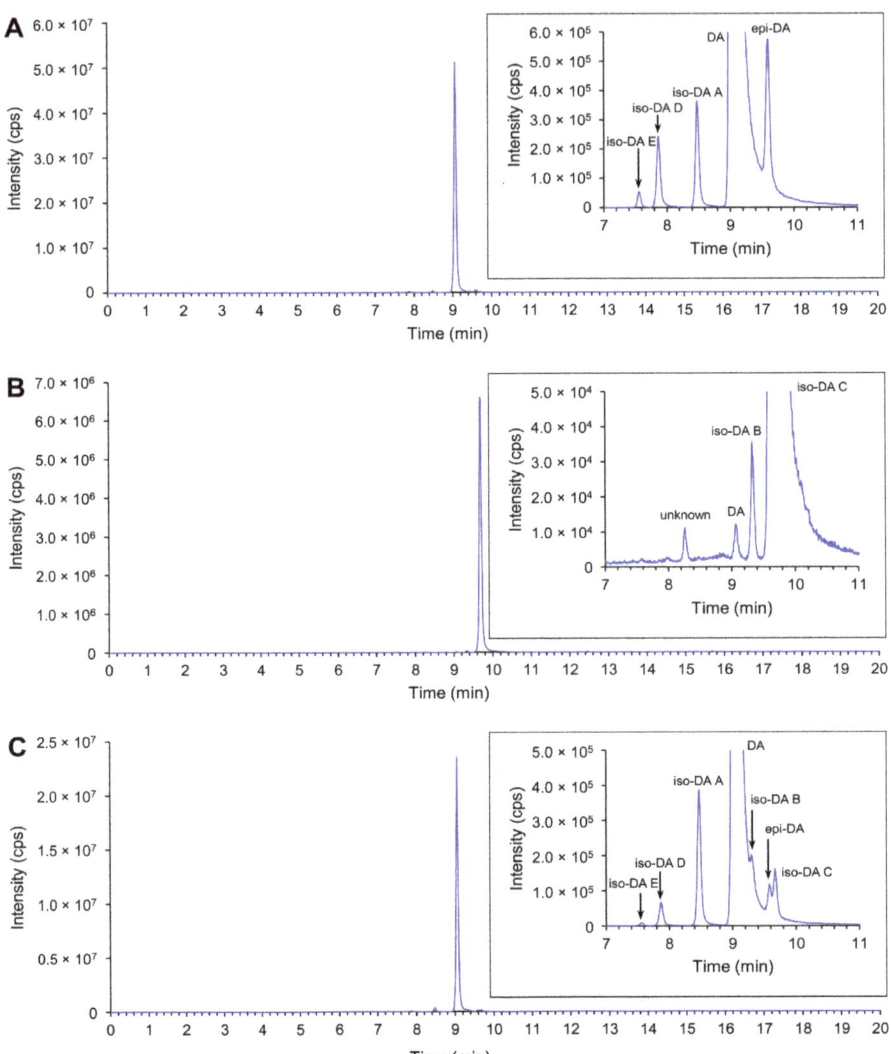

Figure 7. Total ion chromatogram of DA and DA isomers (combined 312 > 266 and 312 > 161) of DA and DA isomers standards and a *Pseudo-nitzschia* extract. (**A**) 1 µL injection of 1/50 dilution of NRCC DA-f DA and DA isomers standard; (**B**) 1 µL injection of iso-DA C standard (950 ng mL^{-1}); (**C**) 1 µL injection of an extract of *P. multiseries* strain G001Ps04.

2.4. Potential ASP Risk of Each Pseudo-nitzschia Species in New Zealand Coastal Waters

2.4.1. Classification of Potential ASP Risk for Each Species

The potential ASP risk for the 14 *Pseudo-nitzschia* species morphologically and/or genetically identified and two species morphologically identified previously was classified based on toxin production. As a result, four species (*P. australis, P. multiseries, P. multistriata,* and *P. pungens*) were classified as potential 'high ASP risk' species and three species (*P. delicatissima, P. fraudulenta,* and '*P. pseudodelicatissima*') as potential 'low ASP risk' species. The other nine species [*P. americana, P. arenysensis,* '*P. caciantha*', *P. calliantha, P. cuspidata, P. hasleana, P. galaxiae, P. plurisecta,* and *P.* (cf.) *subpacifica*] were classified as potential 'no ASP risk' species (Table 1).

Table 1. Summary of domoic acid (DA) and DA isomers production determined for isolates of 16 *Pseudo-nitzschia* species from New Zealand coastal waters between 1993 and 2020 (analysed by LC-MS/MS and/or LC-UV). Double circle: historical locality of the isolates recorded by 2011. Black circle: locality of the isolates recorded in the present study. This table was updated from previous review papers (Table 3 in Rhodes et al. 2012; Table 1 in Rhodes et al. 2013).

| | | | Range of DA and DA Isomers Cell Quota (pg cell^{-1})[b,c] | | | | | | | Locality of Clonal Isolates in New Zealand[d] | | | | | | | | | |
| | | | | | | | | | | Subtropical | | | Warm Temperate | | | | | Cold Temperate | |
Potential ASP Risk in New Zealand[a]	Species	First Record in New Zealand	DA	epi-DA	iso-DA A	iso-DA B[e]	iso-DA C	iso-DA D	iso-DA E	Northland[f]	Hauraki Gulf	Bay of Plenty[f]	Poverty Bay/Hawke Bay	Golden Bay[f]/Tasman Bay[f]	Marlborough Sounds[f]	Canterbury	Caroline Bay	Otago	Big Glory Bay[f]
High	*P. australis*	1993	0.004–2.20 (0–35.00)[g]	0–0.010	0–0.060	trace	0–1.700 (9.050[h])	0–0.003	0–trace	⊚	⊚	⊚[i] ●		●	⊚ ●	⊚[i]			⊚ ●
	P. multiseries	1997	0.47–7.20	0–0.012	0.014–0.206	0–trace	0–0.239	0.003–0.020	trace–0.005	⊚[i]				⊚ ●	●				
	P. multistriata	1997	0–1.60	0	0–trace	0	0–0.200	0	0	⊚ ●	⊚	⊚	⊚[i]	⊚ ●	⊚ ●				
	P. pungens	1993	0–0.47	0	0	0	0	0	0	●		⊚[i]	⊚	⊚ ●	⊚ ●	⊚			⊚ ●
Low	*P. delicatissima*[j]	1996	0–0.12	0	0	0	0	0	0	⊚[i] ●		⊚		⊚[i]	⊚ ●			⊚[i]	⊚ ●
	P. fraudulenta	1996	0–0.03	0	0	0	0	0	0	⊚ ●	⊚[i]	⊚[i]			⊚ ●	⊚			⊚ ●
	'P. pseudodeli-catissima'[k]	1996	0.12	NT	NT	NT	0	NT	NT						⊚		⊚	⊚[i]	⊚
No	*P. americana*	1994	0	0	0	0	0	0	0	●	⊚			●	⊚ ●	⊚			⊚
	P. arenysensis	2019	0	0	0	0	0	0	0	●					●				
	'P. caciantha'[k,l]	2005	0	NT	NT	NT	0	NT	NT						⊚				
	P. calliantha[l]	2005	0	0	0	0	0	0	0						⊚ ●	⊚	⊚		
	P. cuspidata[m]	2020[l]	0–trace	0	0	0	0	0	0						●				
	P. galaxiae	2020	0	0	0	0	0	0	0	●									
	P. hasleana	2020	0	0	0	0	0	0	0					●					●
	P. plurisecta[m]	2005[l]	0	0	0	0	0	0	0	●					⊚				
	P. (cf.) subpacifica[n]	1996	0	0	0	0	0	0	0	⊚ ●		●		●	⊚ ●	⊚[i]			⊚[i]

[a] Potential 'high, low, and no ASP risk' species represent potential high, low, and no toxin production species, respectively. [b] 0 indicates that no toxins were detected at the limit of detection (LoD). [c] NT indicates samples not tested. [d] The locality data include that of isolates both tested and not tested for toxins. [e] Iso-DA B was detected without quantitation due to lack of standard material. [f] Sampling site in the present study. [g] Rhodes et al. (1998b) reported a range of DA cell quota of wild *P. australis* cells with high concentration of up to 35.00 pg cell^{-1}. [h] Rhodes et al. (2004) reported high concentration of iso-DA C cell quota of 9.050 pg cell^{-1} from a *P. australis* isolate cultured in copper addition f/2−Si medium. [i] Additional localities of the isolates were reported by previous studies (Rhodes et al. 1998a, 1998b, 2000, 2004, 2012, and/or Casteleyn et al. 2008). [j] *'Pseudo-nitzschia turgidula'* reported previously in New Zealand is assigned as *P. delicatissima* subclade I in the present study based on an LSU rDNA D1–D2 sequence (DDBJ/EMBL/GenBank accession number: U92259), as discussed in Rhodes et al. (1998a, 1998b). [k] Species identification should be treated as tentative as discussed in Section 3.1. [l] Rhodes et al. (2013) reported *P. caciantha* strains by morphological identification from Marlborough Sounds, Canterbury, and Caroline Bay. Subsequently, the strains from the latter two sites were confirmed as *P. calliantha* based on sequences of the ITS region (Rhodes, unpublished data). Therefore, these *'P. caciantha'* strains from the two sites are assigned as *P. calliantha* in the present study. [m] Rhodes et al. (2013) reported a *P. cuspidata* strain as *Pseudo-nitzschia* sp. strain Hobart-5 type by morphological identification. Subsequently, strain Hobart-5 was assigned as *P. plurisecta* by Orive et al. (2013). Therefore, *P. cuspidata* reported previously in New Zealand is assigned as *P. plurisecta* in the present study. [n] *Pseudo-nitzschia* (cf.) *heimii* reported previously in New Zealand is assigned as *P. subpacifica* in the present study based on morphological characters as reported in Rhodes et al. (2013).

2.4.2. Distribution of Each Species

The distribution pattern of the 16 *Pseudo-nitzschia* species in New Zealand were summarised and mapped with their potential ASP risk information. The four potential 'high ASP risk' species (*P. australis*, *P. multiseries*, *P. multistriata*, and *P. pungens*) were distributed from the subtropical to temperate zones, with no records of *P. multiseries* and *P. multistriata* from the cold temperate zone (Figure 8 and Table 1). Regarding the three potential 'low ASP risk' species, *P. delicatissima* and *P. fraudulenta* were distributed from the subtropical to temperate zones, although '*P. pseudodelicatissima*' was restricted in the temperate zone (Figure 8 and Table 1). In terms of the nine potential 'no ASP risk' species, four species [*P. americana*, *P. arenysensis*, *P. plurisecta*, and *P.* (cf.) *subpacifica*] were distributed from the subtropical to temperate zones, with no records of *P. arenysensis* and *P. plurisecta* from the cold temperate zone. On the other hand, *P. galaxiae* was restricted to the subtropical zone, while the other four species ('*P. caciantha*', *P. calliantha*, *P. cuspidata*, and *P. hasleana*) were restricted to the temperate zone. Among the latter four species, *P. hasleana* was the only species recorded from the cold temperate zone (Figure 8 and Table 1). In detail, *P. hasleana* clade I was recorded from the warm temperate zone, whereas *P. hasleana* clade II was recorded from the cold temperate zone (Figure 4).

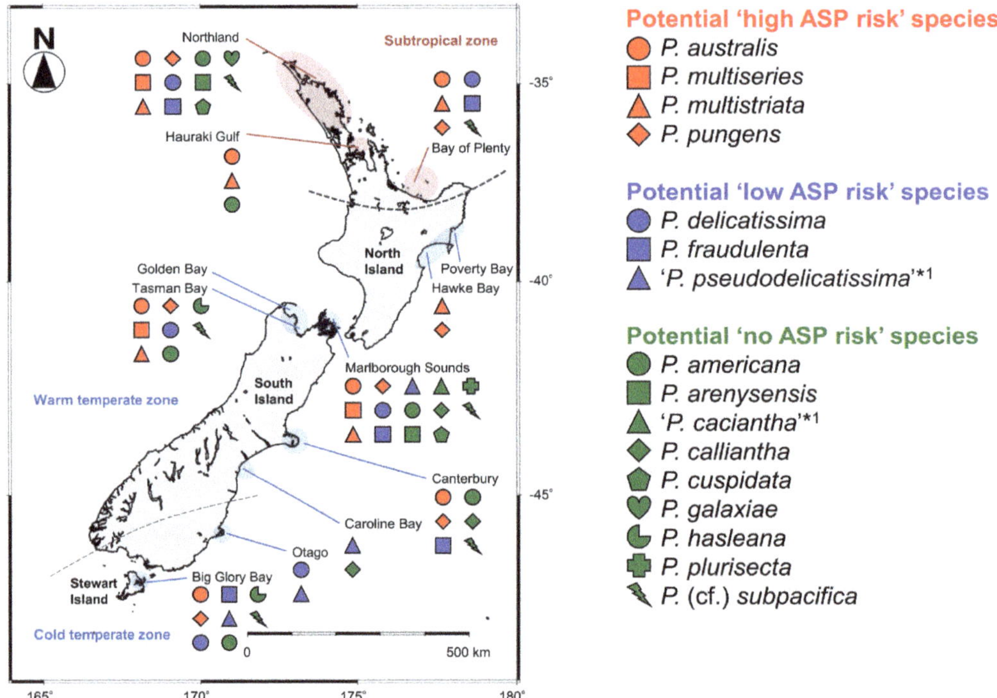

Figure 8. Distribution map of 16 *Pseudo-nitzschia* species in New Zealand coastal waters between 1993 and 2020. Red, blue, and green indicate potential 'high, low, and no ASP risk' species in New Zealand, respectively. *1: Species identification should be treated as tentative, as discussed in Section 3.1.

3. Discussion

3.1. Diversity of Pseudo-nitzschia Species in New Zealand

Among the current 58 *Pseudo-nitzschia* species in the literature, 57 species are accepted taxonomically. However, the remaining species, *Pseudo-nitzschia sinica*, was claimed to be an invalid name because its type was not indicated so far [24]. Historically, morphological characterisations were conducted, and all species were described with their specific morphological features. More recently, molecular phylogenetic characterisations have been added as an additional tool to classify *Pseudo-nitzschia* species. Ribosomal DNA [small subunit (SSU) rDNA; LSU rDNA; and ITS region], the mitochondrial encoded cytochrome *c* oxidase subunit 1 (*cox1*) gene, the chloroplast encoded Ribulose-1,5-bisphosphate carboxylase-oxygenase (RuBisCO) large subunit (*rbcL*) gene, and the RuBisCO small subunit (*rbcS*) gene have been used so far [25,26]. Among these gene regions, the rDNA regions, especially LSU rDNA D1–D3 and ITS region, are frequently used for molecular characterisation. The latter region, having a higher resolution than the former [27], can distinguish almost all *Pseudo-nitzschia* species whose sequences have been reported so far. Furthermore, sequences of several recently described species were reported with only the ITS region and not together with the LSU rDNA D1–D3 [13,15,28]. Additionally, among the 58 species, sequences of five species (*Pseudo-nitzschia antarctica*, *Pseudo-nitzschia prolongatoides*, *Pseudo-nitzschia pungiformis*, *Pseudo-nitzschia roundii*, and *P. sinica*) have not yet been determined. To date, sequences of 52 and 47 *Pseudo-nitzschia* species have been deposited in the International Nucleotide Sequence Database Collaboration [the DNA Data Bank of Japan (DDBJ)/the European Molecular Biology Laboratory Nucleotide Sequence Database (EMBL)/GenBank] for the ITS region and the LSU rDNA D1–D3, respectively. To complete the sequence data set of both regions, sequences of the five morphologically described species above, as well as the ITS region of *Pseudo-nitzschia linea* and the LSU rDNA D1–D3 of six species (*Pseudo-nitzschia bucculenta*, *Pseudo-nitzschia hainanensis*, *Pseudo-nitzschia obtusa*, *Pseudo-nitzschia taiwanensis*, *Pseudo-nitzschia uniseriata*, and *Pseudo-nitzschia yuensis*), still need to be determined.

The ITS region is considered a suitable marker for resolving *Pseudo-nitzschia* species classification. We used the ITS region for genotype screening of the clonal isolates established from New Zealand coastal waters. Previously, sequences from New Zealand isolates were reported for the ITS region of *P. pungens* clade I [29,30] and the LSU rDNA D1–D2 or D1–D3 of *P. australis* [31], '*P. calliantha*', *P. multistriata*, and *P. subpacifica* (registered as *P.* cf. *heimii* in the DDBJ/EMBL/GenBank) [10], and '*P. turgidula*' [32]. We also determined the LSU rDNA D1–D3 sequences of representative isolates to compare them with previous sequences from New Zealand. Genotype screening of the 130 isolates based on the ITS region and/or LSU rDNA D1–D3 reconfirmed the presence of ten previously reported species [*P. americana*, *P. australis*, *P. calliantha*, *P. delicatissima*, *P. fraudulenta*, *P. multiseries*, *P. multistriata*, *P. plurisecta*, *P. pungens*, and *P.* (cf.) *subpacifica*]. In addition, presence of four species (*P. arenysensis*, *P. cuspidata*, *P. galaxiae*, and *P. hasleana*), previously unreported from New Zealand's coastal waters, were confirmed with one to seven strains of each newly reported species. This result suggests that establishing a larger number of isolates for genotype screening would enhance research efforts for the identification of rare species. Therefore, in order to conduct comprehensive genotype screening globally, a large number of isolates are required to clarify *Pseudo-nitzschia* species diversity and geographical distribution, including the locations of rare species.

The clonal isolates of *P. delicatissima* belonged to two (I and II) of the three previously reported subclades (I, II, and III; [14]). Although previous studies in New Zealand [31–33] reported the existence of this species based on the results of morphological characterisation and/or *P. delicatissima*-specific Fluorescence In Situ Hybridization (FISH) assays, the sequences of wild cells or isolates of this species were not determined. Thus, the subclade information for the *P. delicatissima* specimens reported could not be confirmed. Regarding this species, Rhodes et al. (1998a, 1998b) [31,32] reported that strain CAWB12 from New Zealand showed morphological features of *P. turgidula*, while it was positively labelled with the *P. delicatissima*-specific FISH assay. Rhodes et al. (1998a) [32] also determined a sequence of the LSU rDNA D1–D2 for that strain and reported that the sequence was identical to that of *P. delicatissima* strain CV3, assigned as *P. delicatissima* subclade I in the present study. This molecular phylogenetic result suggests that the '*P. turgidula*' strain CAWB12 belongs to *P. delicatissima* subclade I. Considering this finding, along with the present study confirming *P. delicatissima* subclades I and II, there is a possibility that the species previously reported as '*P. delicatissima*' from New Zealand could correspond to *P. delicatissima* subclade II. Unfortunately, as the '*P. delicatissima*' strains reported previously no longer exist, this could not be confirmed. To resolve this issue, detailed morphological observations for the strains of *P. delicatissima* subclades I and II and comparisons of their morphological features with those previously reported for the strains of '*P. delicatissima*' and '*P. turgidula*' are needed in the future.

In the molecular phylogenetic trees, a discrepancy in subclade separation of *P. delicatissima* from New Zealand between the ITS region and LSU rDNA D1–D3 was found. The *P. delicatissima* subclade I strains were clustered in an identical position in both trees. However, the subclade II strains (CAWB143 and CAWB144), assigned based on the ITS region tree, did not cluster with another strain of subclade II (strain PLY1St.46A) but clustered with a strain of subclade III (strain 318-5) in the LSU rDNA D1–D3 tree. A similar discrepancy between the ITS region and LSU rDNA D1–D3 was also reported in *P. galaxiae* [34]. Despite that the ITS region sequence of *P. galaxiae* strain 818-A2G was similar to that of *P. galaxiae* clade II strains (e.g., 818-A1G and Mex23), the LSU rDNA D1–D3 sequence of strain 818-A2G clustered with those of clade I. Recently, Kim et al. (2020) [35] conducted a mating experiment using *P. pungens* clades I–III strains and reported that successful mating results were observed in two combinations of clades I/II and clades I/III. The authors also reported that offspring strains whose parents were clades I and III strains had ITS2 copies of clade I, clade III, and hybrid type. This finding revealed the possibility of mating between different genetic clades in the same species. More information on the possibility of mating between *P. delicatissima* subclades I, II, and III is required. Although the subclade II strains established in this study had the sequence of subclade III in the LSU rDNA D1–D3 as discussed above, no *P. delicatissima* subclade III strains have been established from New Zealand so far. There is a need to further investigate the diversity of *P. delicatissima* in New Zealand to reveal whether subclade III is widely distributed. It should also be noted that the ITS region sequences of four New Zealand strains of *P. delicatissima* subclade II were slightly different from those of the others belonging to the same subclade. This difference may relate to the discrepancy in the LSU rDNA D1–D3. Further genetic investigations of the New Zealand genotype of *P. delicatissima* subclade II need to be conducted in the future.

The present study successfully established clonal strains genetically assigned as closely related species, *P. cuspidata* clade Ia and *P. plurisecta*, from New Zealand. So far, Rhodes et al. (2013) [10] established a '*P. cuspidata*' strain from New Zealand and reported that the strain showed morphological features of *Pseudo-nitzschia* sp. strain Hobart5 [36]. Strain Hobart5 was then transferred to a newly described species, *P. plurisecta* in 2013 [37]. Based on this background, the present study updated the identification of the previous New Zealand '*P. cuspidata*' strain to *P. plurisecta*, making it the first record of *P. cuspidata* in New Zealand. The ITS phylogeny suggested that *P. cuspidata* strains were divided into at least two clades (I and II). A couple of other *P. cuspidata* strains (Tenerife8 and NWFSC194) were clustered together with *P. pseudodelicatissima* strains. Rivera-Vilarelle et al. (2018) [38] established *P. cuspidata* strains from Mexico (e.g., strains Ps116 and Ps142) and described them as a variety of *P. cuspidata*, *P. cuspidata* var. *manzanillensis*. These strains belonged to clade Ib and clustered with two Chinese *P. cuspidata* strains (MC3041 and MC3027) reported by Huang et al. (2019) [39]. However, in the publication from Huang et al. (2019) [39] there was no detailed morphological information on these Chinese strains, which is needed to determine if the strains morphologically resemble *P. cuspidata* var. *manzanillensis*. In the phylogeny, *P. cuspidata* clade II was a sister to a cluster comprised of *P. pseudodelicatissima* strains and two *P. cuspidata* strains (Tenerife8 and NWFSC194) rather than *P. cuspidata* clade I. Furthermore, the genetic diversity within *P. cuspidata* clade II was similar to that between *P. cuspidata* clades Ia and Ib, suggesting clade II was comprised of several genotypes. These results suggest the necessity to assess taxonomic status of *P. cuspidata* clade II in the future. Recently, Ajani et al. (2021) [18] re-assigned Australian *P. cuspidata* clade II strains as *P.* cf. *cuspidata*. It should be noted that the separation of the cryptic *P. cuspidata* and

P. pseudodelicatissima remains unresolved and should be clarified in the future, as discussed in previous studies [26].

This study presents the first record of *P. arenysensis* in New Zealand. *Pseudo-nitzschia arenysensis* was assigned previously as *P. delicatissima* clade B [40] or *P. delicatissima* del1 [27] and then described as a new species in 2009 [41]. Quijano-Scheggia et al. (2009) [41] reported that *P. arenysensis* was morphologically indistinct from *P. delicatissima* but differed in physiological parameters, such as the temperature, needed to trigger reproductive sexual, growth rates, and rate of cell size reduction. Therefore, the reason that *P. arenysensis* has not previously been recorded from New Zealand may be due to this species' morphological resemblance to *P. delicatissima* and/or that *P. arenysensis* is a rare species as discussed above.

This study is also the first record of *P. galaxiae* and *P. hasleana* in New Zealand. *P. galaxiae* was described from Mexico by Lundholm and Moestrup (2002) [42]. Later, McDonald et al. (2007) [19] conducted clone library sequencing of the LSU rDNA for field samples and isolates from Italy. The authors recognised five *P. galaxiae* clades (I–V) and revealed that clades I–III corresponded to those cell sizes based on observations using those clonal cultures [19]. Unfortunately, these sequences could not be included in our analyses as they were short reads (partial LSU rDNA D1–D2) but several sequences from clade I (strain Sydney4), clade II (strains Mex23 and SM26), and clade III (strain SM10) reported previously (Figure 2 in [19]) could be included. Although clades I and II (in the LSU rDNA) corresponded to clades A and B (in the ITS region), respectively, the correspondence of clade III was unclear in the ITS region phylogeny. Thus, the present study coded the *P. galaxiae* clades as A–C, instead of clades I–V, in the ITS region phylogeny. It would be desirable to determine the ITS region sequences of clades IV and V and the LSU rDNA sequences of clade C to clarify the clades correspondence and reveal the diversity of this species.

Pseudo-nitzschia hasleana was described from USA by Lundholm et al. (2012) [43]. Although the morphological features of this species were similar to other closely related species (e.g., *P. calliantha* and *P. mannii*), the molecular phylogenetic position of *P. hasleana* was separated from the other related species [43]. The present study found the previously reported genotype of *P. hasleana* (assigned as clade I) and discovered another genotype (assigned as clade II) closely relating to clade I. The comparison of the average *p* distance of the ITS region between these clades (I and II) and other species, having clades/subclades, revealed that the value between *P. hasleana* clades I and II was the same as that between *P. pungens* clades I and II. Moreover, the average value between *P. hasleana* clades I and II was smaller than the values calculated between other species combinations. This comparison suggests that *P. hasleana* clade II may belong to *P. hasleana* rather than the other species or new species.

Concerning *P. hasleana*, Rhodes et al. (2013) [10] reported a sequence of the LSU rDNA D1–D3 of '*P. calliantha*' strain CAWB114, whose morphology was similar to that of *P. calliantha*, from New Zealand. The LSU rDNA D1–D3 phylogeny in the present study demonstrated that '*P. calliantha*' strain CAWB114 was clustered together with *P. hasleana* clade II. The average *p* distance comparisons suggest that '*P. calliantha*' strain CAWB114 is closer to *P. hasleana* than to *P. calliantha*. The comparisons also suggest that '*P. calliantha*' would be a new clade of *P. hasleana* or an undescribed species. Unfortunately, '*P. calliantha*' strain CAWB114 died, and we could not compare the ITS region sequences between *P. hasleana* clades I and II and strain CAWB114. The establishment of new isolates of '*P. calliantha*' is, therefore, required. In the future, detailed morphological observations and comparisons of *P. hasleana* clades I and II and '*P. calliantha*' strains are needed. It is also necessary to assess their mating potential and ITS2 secondary structure to confirm their taxonomic relationship.

One of the remaining issues of *Pseudo-nitzschia* classification is between *P. heimii* and *P. subpacifica*. Although these two species closely resemble each other morphologically, both species can be distinguished by comparing some morphological characters (i.e., density of fibulae, striae, and poroids) [10,44]. In 1996, Rhodes et al. (1998a, 1998b) [31,32] reported *P. heimii* based on results of the *P. heimii*-specific FISH assay from New Zealand. The authors, then, re-evaluated this species using isolates collected in 1998 and suggested that the correct morphological identification of this species in New Zealand is *P. subpacifica* [10]. At the same time, the authors also determined an LSU rDNA D1–D3 sequence of *P. subpacifica* (reported as *P.* cf. *heimii*) strain CAWB106 and reported it to have 99% identity to *P. heimii* and *P. subpacifica* sequences deposited in DDBJ/EMBL/GenBank so far [10]. Molecular phylogeny of the LSU rDNA D1–D3 for some strains established in the present study suggested that they were clustered together with the previously reported *P. subpacifica* strain CAWB106. Although detailed morphological observations of these newly established strains have not been conducted yet, the present study assigned these strains as *P.* cf. *subpacifica* rather than *P.* cf. *heimii* as a temporary assignment. Detailed morphological observations should be made in the future to identify whether these strains are in fact *P. subpacifica*.

Other than the 14 species found in the present study, previous New Zealand studies reported '*P. pseudodelicatissima*' in 1996 [32] and '*P. caciantha*' in 2005 [10] based on the results of the species-specific FISH assay and/or morphological observations. There were however no rDNA sequences determined for these species from New Zealand. After the report of Rhodes et al. (1998a) [32], *P. pseudodelicatissima* classification became confused as the number of reports of *Pseudo-nitzschia* increased [36], and many morphologically similar species were molecular phylogenetically divided into three groups of the '*P. pseudodelicatissima* complex*' [26]. Rhodes et al. (2013) [10] reported '*P. caciantha*' strains by morphological identification from three sites in New Zealand. Subsequently, the strains from two of three sites were confirmed as *P. calliantha* based on sequences of the ITS region (Rhodes, unpublished data). Thus, the '*P. caciantha*' strains from the two sites should be re-assigned as *P. calliantha*. The assignment of the strains from the remaining one site needs to be re-evaluated. Therefore, the previous reports of '*P. pseudodelicatissima*' and '*P. caciantha*' in New Zealand should be treated as tentative until the clonal isolates, showing rDNA sequences of each species, are established in the future.

As mentioned above, although morphological features of each *Pseudo-nitzschia* species/clade/subclade in New Zealand were not investigated in the present study, these could be important for distinguishing them from each other. The features will be investigated in future research.

3.2. Toxin Production of Pseudo-nitzschia Species in New Zealand

There are so far ten known analogues of DA and DA isomers, namely DA, epi-DA, iso-DAs A, B, C, D, E, F, G, and H. From these ten analogues, only DA is currently regulated within the international Codex standard 292-2008 [6] and the New Zealand Animal Products Act 1999 [7] at a maximum permissible level of 20 mg kg^{-1} in shellfish. Additional isomers have however been reported [8], but there is insufficient toxicity information to establish Toxicity Equivalency Factors (TEF) for the DA isomers [9]. Because DA can convert to epi-DA during storage [45,46], the acute reference dose is currently applied to the sum of both DA and epi-DA, assuming equal relative toxicity. Therefore, DA and epi-DA are under surveillance in New Zealand shellfish. The European Food Safety Authority CONTAM panel concluded in their 2009 scientific opinion that setting TEF values for the other DA isomers was not required as the iso-DAs typically occur at lower concentrations and are less toxic than DA [47]. As the DA isomers do not have unique differentiating features by MS/MS analysis, only peaks which were identified as having consistent retention time with reference standards are able to be identified, hindering the research into DA isomer production. In the absence of iso-DAs F–H reference material, purified standards, or a QC sample, these analogues were not included in the present study.

Domoic acid is the primary ASP-causing toxin, with multiple studies assessing DA production by *Pseudo-nitzschia*. In contrast, there are only a few studies which assess DA isomer production by *Pseudo-nitzschia* strains: iso-DAs A and B from *P. delicatissima*, *P. multiseries* [48], and *Pseudo-nitzschia seriata* [49], iso-DA C from *P. australis* [50,51] and *P. subcurvata* [52], and epi-DA and iso-DAs A and D from *P. plurisecta* [53]. This is the first study to screen, and report, iso-DA E production by *Pseudo-nitzschia* strains. In addition, this is the first report of large scale screening of *Pseudo-nitzschia* strains (73 in total representing 14 species) for the production of DA, epi-DA, iso-DAs A, B, C, D, and E. The production of DA and five of the DA isomers (epi-DA and iso-DAs A–D) was subsequently detected in strains of *P. australis* and *P. multiseries*. In addition, a previously unreported DA isomer from *Pseudo-nitzschia*, iso-DA E, was detected in four strains of *P. multiseries* and two strains of *P. australis* for the first time. Details of DA and DA isomers production by New Zealand *Pseudo-nitzschia* strains are discussed below.

Rhodes et al. (2013) [10] reviewed the previous studies reporting the DA production by New Zealand strains assessed using an immunoassay, and liquid chromatography-ultraviolet detection (LC-UV) and/or LC-MS/MS analytical chemistry techniques. The authors reported that the strains belonging to seven of the 12 reported species produced DA. The authors also pointed out that there were differences in DA production and related potential ASP risk among these reported species: potential 'high ASP risk' species (four species with maximum DA production $\geq$0.5 pg cell^{-1}), potential 'low ASP risk' species (three species with maximum DA production $\leq$0.1 pg cell^{-1}), and potential 'no ASP risk' species (five species with DA production not detected) [10]. Of these, one of the potential 'high ASP risk' species, *P. australis*, was reported as the primary DA producing species in New Zealand, with all strains of this species analysed to date producing DA at concentrations ranging from 0.05 to 2.20 pg cell^{-1} [12,31–33,50,54]. Wild cells of this species were reported to produce DA ranging from 'not detected' to 35.00 pg cell^{-1} [31]. The present study cultured strains of 14 *Pseudo-nitzschia* species following the culturing method reported by previous studies in New Zealand, using an f/2−Si medium to make the cultures stressed to produce DA and DA isomers under silicate limitation [10,54]. As a result, all nine *P. australis* strains analysed produced DA, and their DA cell quotas ranged from 0.004 to 2.02 pg cell^{-1}. These results demonstrated that the DA cell quotas of the strains tested in the present study were similar to those of previous strains of the same species from New Zealand. On the other hand, the maximum DA cell quota of 35.00 pg cell^{-1} was reported so far from the wild *P. australis* cells as mentioned above, suggesting that the DA cell quota of the wild cells could be higher than those of cultured strains in some cases. In this regard, previous studies reported that abiotic (temperature, irradiance, salinity, pH/partial pressure of CO_2, trace metals, inorganic and organic nitrogen, limitation of phosphorus and silicon, and those interactions) and biotic (bacteria and grazing copepods) factors enhance DA production of various *Pseudo-nitzschia* species [25,55,56]. Rhodes et al. (2004) [54] also reported enhanced DA production of a *P. australis* strain from New Zealand under increased trace metal (i.e., copper) conditions in f/2−Si medium. Thus, the highest DA cell quota from the wild *P. australis* cells reported previously [31] might be enhanced by one or many of the factors above. In the future, the DA production of *P. australis* strains from New Zealand will be examined under a variety of culture conditions other than the factors reported previously by Rhodes et al. (2004) [54]. Another factor might be a decrease in DA production for the cultured strains resulting from long-term culturing, as reviewed in Lelong et al. (2012) [55]. The present study conducted culturing and toxin analysis for cultures of *P. australis* strain G015Ps04 at three and seven months after the strain was established. The results showed that the DA cell quota of the 'seven month old' culture (0.004 pg cell^{-1}) was almost ten times lower than that of the 'three month old' culture (0.05 pg cell^{-1}). Although this hypothesis needs to be evaluated using multiple strains in the future, this factor could potentially be one of the reasons why the DA cell quota from the wild cells appeared high in some cases.

The second potential 'high ASP risk' species in New Zealand is *P. multiseries*. Rhodes et al. (1998a) [32] reported that one *P. multiseries* strain produced 0.80 pg cell^{-1} of DA. The present study established four strains of this species and found that all strains produced DA ranging from 0.47 to 7.20 pg cell^{-1}. Although the lowest value was lower than that previously reported, the highest value was almost ten times higher than that previously reported. Considering this point, the difference in the DA cell quota might be due to the differences in DA productivity among the *P. multiseries* strains, similar to the results for *P. australis* strains above. It should be noted that all established strains of *P. australis* and *P. multiseries* tested so far produced DA, with the highest DA cell quota of the *P. multiseries* strains (7.20 pg cell^{-1}) being greater than that of the *P. australis* strains (2.20 pg cell^{-1}) in New Zealand.

The third and fourth potential 'high ASP risk' species in New Zealand are *P. multistriata* and *P. pungens*. Previous studies from New Zealand reported that some strains of these species produced DA, whereas others did not [12,32,33,50,54,57,58]. The present study revealed that all eight strains of *P. multistriata* and 13 strains of *P. pungens* tested did not produce quantifiable levels of DA, supporting the previous reports above. These results suggest that although *P. multistriata* ('not detected'–1.60 pg cell^{-1}) and *P. pungens* ('not detected'–0.47 pg cell^{-1}) were assigned as potential 'high ASP risk' species, their risk to ASP might be lower than the other two species, *P. australis* (0.004–2.20 pg cell^{-1}) and *P. multiseries* (0.47–7.20 pg cell^{-1}).

Rhodes et al. (2013) [10] assigned three species (*P. delicatissima*, *P. fraudulenta*, and '*P. pseudodelicatissima*') as potential 'low ASP risk' species in New Zealand. Previous New Zealand studies reported DA production from three *P. delicatissima* strains (0.03–0.12 pg cell^{-1}, two of them were reported as '*P. turgidula*') [12,32], only one of three *P. fraudulenta* strains (0.03 pg cell^{-1}) [32,50,54], and one '*P. pseudodelicatissima*' strain (0.12 pg cell^{-1}) [32]. The present study tested DA production in *P. delicatissima* subclades I (five strains) and II (four strains) and *P. fraudulenta* (five strains), and revealed these strains did not produce detectable levels of DA. This result supports the recommendation that these species should be kept as potential 'low ASP risk' species in New Zealand.

Rhodes et al. (2013) [10] assigned five species [*P. americana*, '*P. caciantha*', *P. calliantha*, *P. plurisecta* (formerly reported as *P. cuspidata* Hobart-5 type), and *P. subpacifica* (formerly reported as *P.* (cf.) *heimii*)] as potential 'no ASP risk' species in New Zealand as previous strains of these species analysed did not produce detectable levels of DA [58]. The present study tested DA production in *P. americana* (four strains), *P. calliantha* (three strains), *P. plurisecta* (one strain), and *P.* cf. *subpacifica* (four strains), confirming the previous report that these species in New Zealand do not produce detectable levels of DA. Regarding the four newly recorded species (*P. arenysensis*, *P. cuspidata*, *P. galaxiae*, and *P. hasleana*), previous studies from other countries reported that *P. arenysensis* strains tested did not produce DA, while some strains of *P. cuspidata*, *P. galaxiae*, and *P. hasleana* did [25,28,39,59]. The present study tested DA production in *P. arenysensis* (six strains), *P. cuspidata* (one strain), *P. galaxiae* (two strains), and *P. hasleana* (four strains), revealing none of the strains tested produced DA above the limit of detection. As an exception, one *P. cuspidata* strain, CAWB141, established in the present study did not produce DA on day nine, however there were trace levels (the value between LoD of 0.00008 and LLoQ of 0.0008 pg cell^{-1}) on day 43. Although this strain produced a trace level of DA, its cell quota was very low compared with those of potential 'high and low ASP risk' species. Thus, the present study tentatively assigned *P. cuspidata* as potential 'no ASP risk' species. It should be noted that more *P. cuspidata* strains need to be established and analysed for toxin production to determine if this species should remain assigned as a potential 'no ASP risk' species in New Zealand moving forward. In the case of *P. cuspidata* strains from New South Wales, Australia, the strains produced DA ranging from 'not detected' to 24.5 pg cell^{-1} [59]. In the ITS region tree, one of the Australian strains [strain MER, re-assigned as *P.* cf. *cuspidata* by Ajani et al. (2021) [18]] belonged to *P. cuspidata* clade II, which differs from clade I to which the New Zealand strain CAWB141 belonged. In addition, in the case of *P. cuspidata* strains from Barkley Sound, Canada and Washington state, USA, isolated from single-species blooms of *Pseudo-nitzschia* that produced DA (up to 63 pg cell^{-1}), the strains tested produced DA [60]. In the ITS region, one of these *P. cuspidata* strains (strain NWFSC194) belonged to a cluster comprised of *P. cuspidata* and *P. pseudodelicatissima* strains, which differs from clade I to which the New Zealand strain CAWB141 belonged. It is, therefore, suggested that the toxin production of each *P. cuspidata* clade and cluster may differ. The diversity and toxin production of additional *P. cuspidata* strains from various geographic locations need to be assessed in the future. As discussed above, some strains of the nine potential 'no ASP risk' species from other countries produced DA, except for *P. americana* and *P. arenysensis* [25]. However, none of the strains of these species tested from New Zealand produced DA. This difference might be caused by the clade/subclade/genotype/population-level differences of the tested strains among those localities, occurrence seasons, or bloom years. Although the result of the present study supports the recommendation that the nine species should be kept as potential 'no ASP risk' species, at least in New Zealand, continuous assessment for the toxin production of clonal isolates of these species is needed to reassess their assignations in the future.

Apart from DA, the present study detected all, or a combination of, the tested DA isomers (epi-DA and iso-DA A–E) from all *P. australis* and *P. multiseries* strains analysed. Recently, Olesen et al. (2021) [52] reported relative amounts of DA and iso-DA C in three *P. subcurvata* strains. The authors reported that the proportion of DA tended to be lower than that of iso-DA C (DA:iso-DA C = 34–49%:51–66%) [52]. Unlike the results for *P. subcurvata* strains [52], the present study found that the proportions of DA of *P. australis* and *P. multiseries* strains tended to be higher than those of iso-DA C [*P. australis*, DA:iso-DA C = 44.4–91.5%:4.8–54.1%; *P. multiseries*, 93.9–96.5%:2.3–3.1%], suggesting that the ratio of DA to DA isomers varies among species. In addition to this, Rhodes et al. (2004, 2006) [54,61] reported that increased trace metal conditions (e.g., zinc, cobalt, copper, or selenium) in f/2−Si medium enhanced the iso-DA C production in a New Zealand *P. australis* strain. The assessment on the enhanced DA and DA isomers production for *P. multiseries* strains has not been investigated yet. It is, therefore, necessary to examine the enhancements in the production of DA and DA isomers using other *P. australis* and *P. multiseries* strains in the future.

3.3. Distribution of Pseudo-nitzschia Species with Potential 'High, Low, or No ASP Risk' in New Zealand

Determining *Pseudo-nitzschia* species diversity and distribution, along with their respective DA and DA isomers cell quota, is essential in assessing the ASP risk in New Zealand. In 2011, Rhodes et al. (2013) [10] reviewed previous research on this genus in New Zealand, conducted over 20 years, and summarised the distribution of each species (Table 1 in [10]). As the status of *Pseudo-nitzschia* research has not been updated since this study, we aimed to refresh this information. The comparisons of distribution patterns of the 16 species, based on the analysis for the clonal isolates, recorded from New Zealand to date, suggest that most of the potential 'high and low ASP risk' species were widely distributed in the subtropical and temperate zones of New Zealand, compared to the potential 'no ASP risk' species which had restricted distributions. Additionally, the most critical species (*P. australis* and *P. multiseries*) for ASP risk assessment were recorded from both zones. Continuous monitoring of the potential 'high ASP risk' species, especially *P. australis* and *P. multiseries*, is therefore necessary to minimise the risk of ASP in New Zealand. As mentioned in previous study [56], it should be noted that bias can be introduced in these distribution records because the data were generated only from isolates that survived the culturing process employed in the previous and present studies in New Zealand. These isolates also might be biased by the collected field populations, which might have some factors influencing success in culture (e.g., seasonality and cell health).

Three of the four potential 'high ASP risk' species (*P. australis*, *P. multiseries*, and *P. pungens*) have larger cells compared with the other 13 recorded species. Another potential 'high ASP risk' species, *P. multistriata*, has smaller cells than the three species above, and it may be distinguishable due to its sigmoidal cell shape compared with the other 12 smaller recorded species from the coastal waters [62]. However, as Lelong et al. (2012) [55] discussed, precise determination of *Pseudo-nitzschia* species identity by light microscopy is difficult, if not impossible, because most of the frustule morphometrics needed for species determination are visible only by scanning or transmission electron microscopy. One of the solutions for this problem is the use of molecular tools, such as a real-time PCR or FISH assays, that enables morphologically similar species in field samples to be reliably differentiated for monitoring *Pseudo-nitzschia*. Real-time PCR assay uses a primer set or a primer/probe set, that reacts to each target species' specific DNA regions. This method can, therefore, specifically detect and enumerate each target species, and is expected to be a powerful tool for revealing detailed distribution patterns and for monitoring each target species' cell numbers and dynamics. These real-time PCR assays have been developed for several *Pseudo-nitzschia* species to date, specifically looking for: the genus *Pseudo-nitzschia* [63,64]; *Pseudo-nitzschia brasiliana*, *P. calliantha*, *P. delicatissima*, *P. arenysensis*, *P. fraudulenta*, *P. galaxiae*, *P. multistriata*, and *P. pungens* [65]; and *P. pungens* clades I/II and *P. pungens* clade III [30]. From the four potential 'high ASP risk' species in New Zealand, the assay has not been developed for *P. multiseries*. An assay targeting this species is currently under development (Bowers et al. unpublished data). It is expected that this method will be used to examine the detailed distribution and dynamics of each potential 'high ASP risk' species, and will be used to conduct a detailed ASP risk assessment for New Zealand in the future.

4. Materials and Methods

4.1. Sampling and Establishment of Clonal Isolates

As part of the New Zealand Marine Phytoplankton Monitoring Programme, 33 field seawater samples were collected using mainly a depth integrated hose sampling method, from 22 sampling sites (0–15 m depths) of New Zealand coastal waters, the south Pacific Ocean, between 2018 and 2020. Duplicate 100 mL samples were collected each time and Lugol's iodine solution was added to one of the two bottles in order to fix the microorganisms for future microscopic investigations, and the other bottle was used for cell isolations, and subsequent culturing. These bottles were immediately transported to the Micro-algae Laboratory at the Cawthron Institute, Nelson, New Zealand. The sampling locations comprised four sites from the subtropical zone (North part of the North Island) and 18 sites from the temperate zone (South part of North Island, South Island, and Stewart Island) (Figure 4). Details, including sampling site code, and the latitude and longitude coordinates of each site, are shown in Table S1.

At Cawthron Institute, the presence of *Pseudo-nitzschia* in the fixed samples was investigated using Utermöhl chambers and inverted light microscopes (CK-40 or CK-41; Olympus, Tokyo, Japan). Once the presence of *Pseudo-nitzschia* cells was confirmed, the replicate live seawater sample collected at the same time was used for cell isolation. The live sample was transferred to a 6-well flat-bottom cell culture plate (Costar 3516; Corning, NY, USA) and individual *Pseudo-nitzschia* cells or chains were isolated and washed with three drops of sterile f/2 medium [66], containing $Na_2SiO_3 \cdot 5H_2O$ at 59 μmol L^{-1} in the present study instead of $Na_2SiO_3 \cdot 9H_2O$ at 54–107 μmol L^{-1}, on a glass plate with a drawn-out Pasteur micropipette, using an inverted microscope (CK-2; Olympus). The sterile medium was prepared using autoclaved seawater (salinity adjusted to 33) and membrane filtration (GSWG047S6; 0.22 μm, 47 mm, Millipore, MA, USA). Each isolated cell or chain was then transferred to a separate well of a 24-well flat-bottom cell culture plate (Costar 3524; Corning) filled with 1 mL of the sterile f/2 medium. The 24-well plates containing the isolated cells were kept at 18 $\pm$ 1 °C and 40–90 μmol photons m^{-2} s^{-1} with a photoperiod of 12:12 h L:D. After a sufficient cell density was achieved, the culture media, including the cells in each well, were transferred into a clear polystyrene container (LBS32002NX; LabServ, Auckland, New Zealand) containing 20 mL of fresh sterile f/2 medium. The established clonal cultures were maintained under the culturing conditions described above and inoculated every ten days. Details of the clonal strains are shown in Table S2. A total of 30 representative strains established in the present study (CAWB127–CAWB156) are deposited and maintained in the Cawthron Institute Culture Collection of Microalgae (CICCM; Nelson, New Zealand).

4.2. Molecular Phylogenetic Characterisation

4.2.1. DNA Extraction, PCR, and Sequencing

Cultures of the clonal *Pseudo-nitzschia* isolates were harvested at stationary phase, around seven days after inoculation, in 1.5 mL screw-cap microcentrifuge tubes (Labcon, CA, USA) by centrifugation (at $9000\times g$ for 2 min, MiniSpin plus; Eppendorf, Hamburg, Germany). Genomic DNA of isolates was extracted using the DNeasy PowerSoil kit (Qiagen, Hilden, Germany) according to the manufacturer's protocol or by the chelex-based DNA isolation method [67]. Briefly, the cell pellets were dissolved in 200 µL of 10% ($w\,v^{-1}$) solution of Chelex 100 Resin (143-2832; Bio-Rad, CA, USA) in a 1.5 mL tube (Labcon) and incubated at 95 °C for 20 min with thorough mixing every 10 min. The tube was then centrifuged at $9000\times g$ for 2 min (Centrifuge 5430; Eppendorf). The supernatant containing genomic DNA was used as a template for PCR amplification below.

The rDNA, specifically, the ITS region and LSU rDNA D1–D3 was amplified by PCR. The PCR was performed using 25 µL reaction volumes as follows: nuclease-free water (AM9937; Ambion, CA, USA); 12.5 µL of MyTaq Red Mix, 2 × (Bioline, London, UK); each primer (0.4 µM as a final concentration) as below; non-acetylated bovine serum albumin (BSA, 16 ng; Sigma-Aldrich, Auckland, New Zealand), and template genomic DNA (approximately 5–100 ng) extracted as above. The ITS region was amplified using the primers: 4618F (forward, 5′-GTA GGT GAA CCT GCA GAA GGA TCA-3′) and LSU1R (reverse, 5′-ATA TGC TTA AAT TCA GCG GGT-3′) [68]. The LSU rDNA D1–D3 was amplified using the primers: D1R (forward, 5′-ACC CGC TGA ATT TAA GCA TA-3′) [21] and D3B (reverse, 5′-TCG GAG GGA ACC AGC TAC TA-3′) [69]. PCR was run using a thermocycler (Mastercycler nexus gradient; Eppendorf), and PCR cycling conditions were 98 °C for 4 min; 35 cycles of 95 °C for 30 s, 60 °C for 30 s, and 72 °C for 60 s; and 72 °C for 10 min. For identifying positive bands with a known standard, the PCR products were run on a 1.5% ($w\,v^{-1}$) agarose gel, stained with RedSafe Nucleic Acid Staining Solution (iNtRON Biotechnology, Seoul, Korea), and viewed under a UV light.

The PCR products were purified using NucleoSpin Gel and PCR Clean-up kit (Macherey-Nagel, Düren, Germany) according to the manufacturer's protocol and sequenced using the PCR primers by Genetic Analysis Services, University of Otago, Dunedin, New Zealand or Macrogen, Seoul, South Korea. Forward and/or reverse reads were edited and assembled using Geneious 8.0.5 (Biomatters, Auckland, New Zealand). Representative sequences were deposited in DDBJ (DDBJ accession numbers: LC636495–LC636593; the primer regions located at both ends of the sequences were excluded from the deposited sequences) (Table S2).

4.2.2. Molecular Phylogenetic Analyses

To estimate the molecular phylogenetic positions of New Zealand *Pseudo-nitzschia* based on the ITS region sequences, 190 publicly available sequences of 52 *Pseudo-nitzschia* species and two sequences of *Cylindrotheca closterium* as an outgroup were obtained from the DDBJ/EMBL/GenBank database. Compared with the LSU rDNA D1–D3 data set below, sequences of *P. linea* were not included in the ITS region data set because these sequences have not yet been determined. A total of 61 representative sequences of the ITS region of New Zealand *Pseudo-nitzschia* strains were aligned with the reference sequences above using ClustalW [70] in Geneious 8.0.5 (Biomatters). In this data set, the 5′ and 3′ ends were manually aligned to truncate, and both ends were refined. The final alignment consisted of 253 sequences of 1185 positions, including gaps [the alignment site corresponded to the 156–761 bp site of a sequence of *P. seriata* strain Lynæs6 (DDBJ/EMBL/GenBank accession number: DQ062663)]. Phylogenetic analysis was conducted using the maximum likelihood (ML) method by MEGA X [71]. To determine the best DNA model, all positions with less than 10% site coverage were eliminated (partial deletion option available in MEGA X), and the final data set was a total of 809 positions. The best-fit model of nucleotide substitution selected based on the Akaike information criterion (AIC) (find best DNA/protein models option available in MEGA X) was found to be the GTR + G + I model.

Bootstrap analysis of 1000 replicates was performed to examine the robustness of the clades. For the phylogenetic analysis, all positions were treated using the same parameters as those used for the DNA model analysis described above. Phylogenetic analysis was also conducted using the Bayesian inference (BI) method using MrBayes 3.1.2 [72] to estimate the posterior probability (pp) distribution using Metropolis-coupled Markov chain Monte Carlo (MCMCMC) [73]. For BI analysis, MrModeltest 2 [74] was used to determine the best-fit model of nucleotide substitution using PAUP 4.0b10 (Sinauer Associates, MA, USA). The best-fit model, according to the AIC, was the GTR + G + I model. The analysis was performed using four chains with temperature set 0.2. The analysis was performed with seven million generations, and the trees were sampled every 100 generations. For increasing the probability of chain convergence, 16,000 trees were sampled after average standard deviation of split frequencies (ASDSF) was below 0.01 to calculate the pp.

To construct the LSU rDNA D1–D3 molecular phylogeny, 139 publicly available sequences of 47 *Pseudo-nitzschia* species and two sequences of *Cylindrotheca closterium* as an outgroup were obtained from the DDBJ/EMBL/GenBank database. Compared with the ITS region data set above, sequences of six species (*P. bucculenta*, *P. hainanensis*, *P. obtusa*, *P. taiwanensis*, *P. uniseriata*, and *P. yuensis*) and three clade/subclades (*P. cuspidata* clade Ib, *P. decipiens* subclade A, and *P. galaxiae* clade C) were not included in the LSU rDNA D1–D3 data set because these sequences have not yet been determined. A total of 38 representative sequences of the LSU rDNA D1–D3 of New Zealand *Pseudo-nitzschia* strains were aligned with the reference sequences above using ClustalW, and both ends were refined following the method as described above. The final alignment consisted of 179 sequences of 710 positions for the LSU rDNA D1–D3 data set [the alignment site corresponded to the 74–729 bp site of a sequence of *P. seriata* strain Lynæs8 (DDBJ/EMBL/GenBank accession number: AF417653)]. Phylogenetic analysis was conducted using the ML method with the best-fit model (GTR + G + I model), following the procedures as above. The final data set was a total of 660 positions. Phylogenetic analysis was also conducted using the BI method with the best-fit model (GTR + G + I model), following the procedures as above. The analysis was performed with five million generations, and 19000 trees were sampled after ASDSF was below 0.01 to calculate the pp.

4.2.3. Sequence Analysis

Sequence analysis for the ITS region and the LSU rDNA D1–D3 was conducted by calculating the *p* distance of selected combinations within and between *Pseudo-nitzschia* species/clades/subclades using the uncorrected genetic distance (UGD) model (*p*-distance model available in MEGA 7; [75]). The alignments used for the phylogenetic analyses above were used. All positions with less than 10% site coverage were eliminated (partial deletion option available in MEGA 7). The number of positions used for the final data sets is shown in Tables S3 and S4.

4.3. Toxin Analysis

4.3.1. Culturing, Harvesting, and Toxin Extraction

To enable instrumental analysis (LC-MS/MS; Section 4.3.2) of DA and DA isomers (epi-DA, iso-DAs A, B, C, D, and E) production, 73 representative strains of the 14 *Pseudo-nitzschia* species (1–13 strains per species) genetically identified were cultured. These representative strains of each species were selected from each sampling site if multiple strains were established from multiple sites. Briefly, 7 mL of the strains cultured in f/2 medium for seven days (as described above) were inoculated into two clear polystyrene containers (LBS32002NX; LabServ), each containing 30 mL of f/2 minus $Na_2SiO_3 \cdot 5H_2O$ medium (f/2−Si medium) (approximately 20% inoculum) and cultured for four or five days. The use of f/2−Si medium stresses the cultures into producing DA and DA isomers by culturing under silicate limitation. Subsequently, approximately 7 mL of the cultures were inoculated into four containers containing 30 mL of f/2 or f/2−Si media, or K medium (not containing Si) [76] (approximately 20% inoculum) for an additional 8–13 and/or 43 d

(Table S5). The 10 mL of media containing the cultured cells, that had sunk to the bottom of the four containers, was removed using a micropipette and pooled into a 50 mL centrifuge tube (430829; Corning). A 100 µL aliquot of each pooled cultures was transferred into a 1.5 mL screw-cap microcentrifuge tube (Labcon) containing 890 µL of sterile seawater and 10 µL of Lugol's iodine solution [final concentration of Lugol's iodine solution was 1% (v/v)]. The cell counts for the fixed samples were conducted in triplicate drops on a boundary slide glass (S6113; Matsunami Glass, Osaka, Japan) using an inverted microscope (CK-2; Olympus). The 50 mL tube containing the pooled culture was centrifuged at $3214\times g$ for 40 min (Centrifuge 5810 R; Eppendorf), the supernatants were discarded, and resultant pellets were stored at $-20\ ^{\circ}$C until the toxins were extracted.

For the toxin extraction, 20% aq. acetonitrile (MeCN; $v\,v^{-1}$) (A955-4, Optima LC-MS Grade; Thermo Fisher Scientific, Waltham, MA, USA) was added to the 50 mL centrifuge tube containing the pellet, at a ratio of 1 mL per $\leq 2 \times 10^{6}$ cells, followed by sonication (10 min) using an ultrasonic bath (XUBA3, 35 W, 44 kHz; Grant Instruments, Cambridge, UK). The tubes containing the disrupted cells were centrifuged at $3214\times g$ for 5 min at 4 $^{\circ}$C (Centrifuge 5810 R; Eppendorf). The supernatants were transferred into 15 mL centrifuge tubes (339650; Thermo Fisher Scientific, MA, USA) and stored at $-20\ ^{\circ}$C overnight. The tubes were then centrifuged again at $3214\times g$ for 5 min at 4 $^{\circ}$C (Centrifuge 5810 R; Eppendorf) to afford a clear supernatant, and finally 1 mL was transferred into a 2 mL glass autosampler vial (AR0-3910-13; Phenomenex, CA, USA) and stored at $-20\ ^{\circ}$C until LC-MS/MS analysis.

4.3.2. Instrument Analysis

Calibration was performed using a dilution of CRM-DA-g, certified reference material from the National Research Council of Canada (NRC; Halifax, Canada), which contains certified concentration of DA + epi-DA, and uncertified concentrations of iso-DAs A, D and E. A well-characterised reference material of iso-DA C obtained from Cawthron Natural Compounds (CNC; Cawthron Institute) was used to quantitate iso-DA C. A sample of iso-DA B obtained from CNC (Cawthron Institute) was used to confirm retention times and ion ratios for qualitative presence/absence analysis. A mussel certified reference material ASP-Mus-d contaminated with DA and DA isomers (epi-DA and iso-DAs A, D, and E) was obtained from NRC for quality control and used to obtain reference spectroscopic data.

Screening of DA and DA isomers was performed on a Waters Xevo TQ-S triple quadrupole mass spectrometer coupled to an Acquity I-Class UPLC with flow-through needle sample manager (LC-MS/MS; Waters, Milford, MA, USA), and was based on the LC-MS/MS lipophilic toxins method published by McNabb et al. (2005) [77]. Chromatographic separation used a Waters Acquity BEH Shield RP18 column (1.7 µm, 130 Å, 50 × 2.1 mm) held at 40 $^{\circ}$C. The target analytes were eluted at 0.5 mL min^{-1} with (A) 5% aq. MeCN and (B) 95% aq. MeCN mobile phases, each containing 50 mM formic acid and 2.53 mM ammonium hydroxide. Initial conditions were 0% B and held for 0.2 min, then linearly increased to 15% B over 0.3 min, to 30% B over 0.5 min, to 80% B over 4 min then immediately increased to 100% B and held for 1.5 min before immediately returning to 0% B and holding for 0.5 min to re-equilibrate. The autosampler chamber was maintained at 10 $^{\circ}$C and the injection volume was 1 µL. The mass spectrometer used an electrospray ionisation source operated in positive ion mode. Other settings were capillary voltage 2 kV, cone voltage 50 V, source offset 50 V, source temperature 150 $^{\circ}$C, cone gas flow rate 150 L h^{-1}, desolvation temperature 600 $^{\circ}$C, desolvation gas flow rate 1000 L h^{-1}, nebuliser gas flow 7 bar and the collision cell was operated with 0.15 mL min^{-1} argon. Multiple Reaction Monitoring (MRM) transitions for DA and DA isomers were m/z 312.2 > 266.2 (quantitation) and m/z 312.2 > 161.1 (confirmation), with collision energies of 16 and 25 eV, respectively. Both MRM transitions had a dwell time of 100 ms. The typical limit of detection (LoD) and lower limit of quantitation (LLoQ) of total of DA + DA isomers were 0.0005 and 0.0025 pg cell^{-1} (1 and 5 ng mL^{-1}), respectively.

In addition, the screening of DA and DA isomers production was performed using LC-UV. This utilized a Waters Acquity I-Class UPLC with flow-through needle sample manager coupled to a photo diode array (PDA) detector (Waters, Milford, MA, USA) to confirm DA isomers identification. A Supelco Titan C18 column (1.9 μm, 80, 100 × 2.1 mm; Merck, Darmstadt, Germany), held at 40 °C, was used in conjunction with 17% aq. MeCN mobile phase, containing 0.1% trifluoroacetic acid. Isocratic elution was performed at a flow rate of 0.25 mL min^{-1}. PDA detection was performed with a range of 190–350 nm, at a resolution of 1.2 nm, a sampling rate of 20 points per second and a time constant of 0.1 s. Domoic acid and epi-DA were analysed at 242 nm, and iso-DA C was analysed at 220 nm. The autosampler chamber was maintained at 4 °C and the injection volume was 1 μL. The typical LoD and LLoQ for DA + epi-DA were 0.025 and 0.075 pg cell^{-1} (50 and 150 ng mL^{-1}), respectively.

Quantitative analysis of DA and DA isomers production, for the positive extracts screened by the LC-MS/MS or LC-UV techniques described above, was performed using LC-MS/MS on a Sciex 6500 + QTRAP tandem quadrupole mass spectrometer coupled to an ExionLC liquid chromatography system (Sciex, Framingham, MA, USA). A Biozen XB-C18 superficially porous column (1.7 μm, 100Å, 150 × 2.1 mm), held at 35 °C, (Phenomenex, Torrance, CA, USA) was used with (A) water and (B) MeCN mobile phases, each containing 0.1% formic acid. Initial conditions were 5% B at 0.3 mL min^{-1} linearly increasing to 15% B over 12.5 min, then increasing to 90% B over 2.5 min, held at 90% B for 1.5 min then returned to 5% B over 0.5 min and held to re-equilibrate for 3 min. Total run time was 20 min. The autosampler chamber was maintained at 4 °C and the injection volume was 1 μL. Pump compressibility compensation was enabled with pump A at 0.45/GPa and pump B at 1.20/GPa. Electrospray ionisation was performed in positive ion mode with curtain gas at 30, collision gas at high, ionisation voltage at 5500 V, temperature at 500 °C, ion source gas 1 at 50, and ion source gas 2 at 55. Declustering potential was set at 35, entrance potential at 9, and collision cell exit potential at 25. The quantitation MRM transition (m/z 312.2 > 266.2) was acquired with a collision energy of 21, and confirmation MRM transition (m/z 312.2 > 161.1) was acquired with a collision energy of 33 with a total 0.4 s cycle time. The typical LOD and LLoQ of each DA and DA isomers, except for iso-DA B, were 0.00005 and 0.0005 pg cell^{-1} (0.1 and 1 ng mL^{-1}), respectively.

To compare DA cell quota and DA proportion of the strains between *Pseudo-nitzschia* species assessed above, the data sets were assessed by statistical analysis using BellCurve for Excel (Social Survey Research Information, Tokyo, Japan). Nonparametric analysis (Mann–Whitney U test) was performed on the data sets since they did not show a normal distribution.

4.4. Classification of Potential ASP Risk and Distribution Mapping of Each Pseudo-nitzschia Species in New Zealand Coastal Waters

The potential ASP risk for the 14 *Pseudo-nitzschia* species morphologically and/or genetically identified and two species morphologically identified previously was classified based on toxin production assessed in the present and previous studies in New Zealand, following the criteria reported by Rhodes et al. (2013) [10]. Briefly, species with a maximum DA production of ≥0.5 pg cell^{-1} was classified as potential 'high ASP risk' species. Species with a maximum DA production of ≤0.1 pg cell^{-1} was classified as potential 'low ASP risk' species. Species with DA production not detected or with trace levels of DA cell quota were classified as potential 'no ASP risk' species.

The distribution pattern of each species, having different potential ASP risks in New Zealand, was assessed and plotted on the map, based on the results of the present and previous studies.

Supplementary Materials: The following are available online at https://www.mdpi.com/article/10.3390/toxins13090637/s1, Table S1: Details of 22 sampling sites in New Zealand between 2018 and 2020. Table S2: Details of 130 clonal strains of *Pseudo-nitzschia* established from New Zealand between 2018 and 2020. Table S3: Average, SD, minimum, and maximum uncorrected *p* distance of the ITS region of selected combinations between *Pseudo-nitzschia* species/clades/subclades. Table S4: Average, SD, minimum, and maximum uncorrected *p* distance of the LSU rDNA D1–D3 of selected combinations between *Pseudo-nitzschia hasleana* and several related species. Table S5: Details of 73 clonal strains of *Pseudo-nitzschia* from New Zealand tested for domoic acid (DA) and DA isomers analysis in the present study.

Author Contributions: Conceptualisation, T.N., K.F.S., and L.L.R.; methodology, T.N., J.S.M., M.J.B., and L.L.R.; software, T.N., J.S.M., and M.J.B.; validation, T.N., J.S.M., and M.J.B.; formal analysis, T.N., J.S.M., and M.J.B.; investigation, T.N., J.S.M., M.J.B., M.B., H.A.B., and L.L.R.; resources, K.F.S., D.T.H., and L.L.R.; data curation, T.N., J.S.M., and M.J.B.; writing—original draft preparation, T.N., J.S.M., M.J.B., and L.L.R.; writing—review and editing, T.N., J.S.M., M.J.B., M.B., H.A.B., K.F.S., D.T.H., and L.L.R.; visualisation, T.N.; supervision, K.F.S. and L.L.R.; project administration, K.F.S. and L.L.R.; funding acquisition, K.F.S., D.T.H., and L.L.R. All authors have read and agreed to the published version of the manuscript.

Funding: This research, including instrumental analysis, and APC were funded by the New Zealand Seafood Safety research platform, contract number CAWX1801 and the Cawthron Institute Internal Capability Investment Fund. This research was also funded by the Scientific and Technological Research Council of Turkey under a TUBITAK-BIDEB-2219-International Postdoctoral Research Fellowship, grant number 1059B191800400, which was awarded to M.B. for the present study at Cawthron Institute.

Institutional Review Board Statement: Not Applicable.

Informed Consent Statement: Not Applicable.

Data Availability Statement: The data presented in this study are in the article and supplementary materials.

Acknowledgments: We are grateful to New Zealand's Shellfish Industry, Ministry for Primary Industries, Marlborough Shellfish Quality Programme, and those sample collectors for their support and for allowing us to publish the data obtained from their field seawater samples. We thank Tony Bui and Catherine Moisan, Cawthron Institute, for their assistance in accessing the field samples. We thank Emillie Burger, Cawthron Institute, for the assistance with instrument operation and sample analysis. We also thank Lucy Thompson, Jacqui Stuart, and Janet Adamson, Cawthron Institute, for their technical support. We thank Sarah Challenger and Juliette Butler, Cawthron Institute Culture Collection of Microalgae, Cawthron Institute, for their assistance with culture management. We are grateful to Masao Adachi, Kochi University, for the software. We are also grateful to Nina Lundholm, University of Copenhagen, for helpful information. We acknowledge the anonymous reviewers and the editor for providing valuable comments to improve this manuscript's quality.

Conflicts of Interest: The authors declare no conflict of interest. The funders had no role in the design of the study; in the collection, analyses, or interpretation of data; in the writing of the manuscript, or in the decision to publish the results.

Abbreviations

AIC, Akaike information criterion; ASDSF, average standard deviation of split frequencies; ASP, amnesic shellfish poisoning; BI, Bayesian inference; CICCM, Cawthron Institute Culture Collection of Microalgae; CNC, Cawthron Natural Compounds; *cox1*, cytochrome *c* oxidase subunit 1; DA, domoic acid; DDBJ, DNA Data Bank of Japan; EMBL, European Molecular Biology Laboratory; epi-DA, c5′-epidomoic acid; FISH, Fluorescence In Situ Hybridization; iso-DAs, isodomoic acids; ITS region, internal transcribed spacer 1–5.8S rDNA-ITS 2 region; LC-MS/MS, liquid chromatography-tandem mass spectrometry; LLoQ, lower limit of quantitation; LoD, limit of detection; LSU rDNA D1–D3, large subunit rDNA D1–D3 region; MCMCMC, Metropolis-coupled Markov chain Monte Carlo; MeCN, acetonitrile; ML, maximum likelihood; MRM, multiple reaction monitoring; NRC, National Research Council of Canada; PDA, photo diode array; pp, posterior probability; *rbcL*, RuBisCO large subunit; *rbcS*, RuBisCO small subunit; rDNA, ribosomal RNA gene; RuBisCO, ribulose-1,5-bisphosphate carboxylase-oxygenase; SSU, small subunit; TEF, toxicity equivalency factors; UGD, uncorrected genetic distance.

References

1. Subba Rao, D.V.; Quilliam, M.A.; Pocklington, R. Domoic Acid—A Neurotoxic Amino Acid Produced by the Marine Diatom *Nitzschia pungens* in Culture. *Can. J. Fish. Aquat Sci.* **1988**, *45*, 2076–2079. [CrossRef]
2. Bates, S.S.; Bird, C.J.; de Freitas, A.S.W.; Foxall, R.; Gilgan, M.; Hanic, L.A.; Johnson, G.R.; McCulloch, A.W.; Odense, P.; Pocklington, R.; et al. Pennate Diatom *Nitzschia pungens* as the Primary Source of Domoic Acid, a Toxin in Shellfish from Eastern Prince Edward Island, Canada. *Can. J. Fish. Aquat Sci.* **1989**, *46*, 1203–1215. [CrossRef]
3. Quilliam, M.A.; Wright, J.L.C. The Amnesic Shellfish Poisoning Mystery. *Anal. Chem.* **1989**, *61*, 1053A–1060A. [CrossRef]
4. Hasle, G.R. *Pseudo-nitzschia pungens* and *P. multiseries* (Bacillariophyceae): Nomenclatural History, Morphology, and Distribution. *J. Phycol.* **1995**, *31*, 428–435. [CrossRef]
5. Hay, B.E.; Grant, C.M.; McCoubrey, D.-J. *A Review of the Marine Biotoxin Monitoring Programme for Non-Commercially Harvested Shellfish*; Part 1: Technical report, A report prepared for the Ministry of Health by AquaBio consultants Ltd.; NZ Ministry of Health: Wellington, New Zealand, 2000; ISBN 0-478-24349-9.
6. CODEX STAN 292-2008 Standard for Live and Raw Bivalve Molluscs. Available online: http://www.fao.org/fao-who-codexalimentarius/sh-proxy/en/?lnk=1&url=https%253A%252F%252Fworkspace.fao.org%252Fsites%252Fcodex%252FStandards%252FCXS%2B292-2008%252FCXS_292e_2015.pdf (accessed on 20 July 2021).
7. Ministry for Primary Industries Animal Products Notice: Regulated Control Scheme-Bivalve Molluscan Shellfish for Human Consumption. Available online: https://www.mpi.govt.nz/dmsdocument/30282/direct (accessed on 20 July 2021).
8. Clayden, J.; Read, B.; Hebditch, K.R. Chemistry of Domoic Acid, Isodomoic Acids, and Their Analogues. *Tetrahedron* **2005**, *61*, 5713–5724. [CrossRef]
9. FAO/WHO. *Toxicity Equivalence Factors for Marine Biotoxins Associated with Bivalve Molluscs*; World Health Organization: Geneva, Switzerland, 2016; ISBN 978-92-5-109345-0.
10. Rhodes, L.; Jiang, W.; Knight, B.; Adamson, J.; Smith, K.; Langi, V.; Edgar, M. The Genus *Pseudo-nitzschia* (Bacillariophyceae) in New Zealand: Analysis of the Last Decade's Monitoring Data. *N. Z. J. Mar. Freshw. Res.* **2013**, *47*, 490–503. [CrossRef]
11. Hallegraeff, G.M.; Schweibold, L.; Jaffrezic, E.; Rhodes, L.; MacKenzie, L.; Hay, B.; Farrell, H. Overview of Australian and New Zealand Harmful Algal Species Occurrences and Their Societal Impacts in the Period 1985 to 2018, Including a Compilation of Historic Records. *Harmful Algae* **2021**, *102*, 101848. [CrossRef] [PubMed]
12. Rhodes, L.; White, D.; Syhre, M.; Atkinson, M. *Pseudo-nitzschia* species isolated from New Zealand coastal waters: Domoic acid production in vitro and links with shellfish toxicity. In *Harmful and Toxic Algal Blooms: Proceedings of the Seventh International Conference on Toxic Phytoplankton, Sendai, Japan, 12–16 July 1995*; Yasumoto, T., Oshima, Y., Fukuyo, Y., Eds.; Intergovermental Oceanographic Commission of UNESCO: Paris, France, 1996; pp. 155–158.
13. Chen, X.M.; Pang, J.X.; Huang, C.X.; Lundholm, N.; Teng, S.T.; Li, A.; Li, Y. Two New and Nontoxigenic *Pseudo-nitzschia* Species (Bacillariophyceae) from Chinese Southeast Coastal Waters. *J. Phycol.* **2021**, *57*, 335–344. [CrossRef] [PubMed]
14. Stonik, I.V.; Isaeva, M.P.; Aizdaicher, N.A.; Balakirev, E.S.; Ayala, F.J. Morphological and Genetic Identification of *Pseudo-nitzschia* H. Peragallo, 1900 (Bacillariophyta) from the Sea of Japan. *Russ. J. Mar. Biol.* **2018**, *44*, 192–201. [CrossRef]
15. Gai, F.F.; Hedemand, C.K.; Louw, D.C.; Grobler, K.; Krock, B.; Moestrup, Ø.; Lundholm, N. Morphological, Molecular and Toxigenic Characteristics of Namibian *Pseudo-nitzschia* Species–Including *Pseudo-nitzschia bucculenta* Sp. Nov. *Harmful Algae* **2018**, *76*, 80–95. [CrossRef]
16. Kim, J.H.; Park, B.S.; Kim, J.-H.; Wang, P.; Han, M.-S. Intraspecific Diversity and Distribution of the Cosmopolitan Species *Pseudo-nitzschia pungens* (Bacillariophyceae): Morphology, Genetics, and Ecophysiology of the Three Clades. *J. Phycol.* **2015**, *51*, 159–172. [CrossRef]
17. Ajani, P.A.; Lim, H.C.; Verma, A.; Lassudrie, M.; McBean, K.; Doblin, M.A.; Murray, S.A. First Report of the Potentially Toxic Marine Diatom *Pseudo-nitzschia simulans* (Bacillariophyceae) from the East Australian Current. *Phycol. Res.* **2020**, *68*, 254–259. [CrossRef]
18. Ajani, P.A.; Verma, A.; Kim, J.H.; Woodcock, S.; Nishimura, T.; Farrell, H.; Zammit, A.; Brett, S.; Murray, S.A. Using QPCR and High-Resolution Sensor Data to Model a Multi-Species *Pseudo-nitzschia* (Bacillariophyceae) Bloom in Southeastern Australia. *Harmful Algae* **2021**, *108*, 102095. [CrossRef]
19. McDonald, S.M.; Sarno, D.; Zingone, A. Identifying *Pseudo-nitzschia* Species in Natural Samples Using Genus-Specific PCR Primers and Clone Libraries. *Harmful Algae* **2007**, *6*, 849–860. [CrossRef]
20. Giulietti, S.; Romagnoli, T.; Siracusa, M.; Bacchiocchi, S.; Totti, C.; Accoroni, S. Integrative Taxonomy of the *Pseudo-nitzschia* (Bacillariophyceae) Populations in the NW Adriatic Sea, with a Focus on a Novel Cryptic Species in the *P. delicatissima* Species Complex. *Phycologia* **2021**, *60*, 247–264. [CrossRef]
21. Scholin, C.A.; Villac, M.C.; Buck, K.R.; Krupp, J.M.; Powers, D.A.; Fryxell, G.A.; Chavez, F.P. Ribosomal DNA Sequences Discriminate among Toxic and Non-Toxic *Pseudo-nitzschia* Species. *Nat. Toxins* **1994**, *2*, 152–165. [CrossRef]
22. Bowers, H.A.; Marin, R.; Birch, J.M.; Scholin, C.A. Sandwich Hybridization Probes for the Detection of *Pseudo-nitzschia* (Bacillariophyceae) Species: An Update to Existing Probes and a Description of New Probes. *Harmful Algae* **2017**, *70*, 37–51. [CrossRef]
23. Quilliam, M.A.; Xie, M.; Hardstaff, W.R. Rapid Extraction and Cleanup for Liquid Chromatographic Determination of Domoic Acid in Unsalted Seafood. *J. AOAC Int.* **1995**, *78*, 543–554. [CrossRef]
24. Guiry, M.D.; Guiry, G.M. AlgaeBase. World-Wide Electronic Publication, National University of Ireland, Galway. Available online: https://www.algaebase.org/ (accessed on 20 July 2021).

25. Bates, S.S.; Hubbard, K.A.; Lundholm, N.; Montresor, M.; Leaw, C.P. *Pseudo-nitzschia, Nitzschia*, and Domoic Acid: New Research since 2011. *Harmful Algae* **2018**, *79*, 3–43. [CrossRef]
26. Lim, H.C.; Tan, S.N.; Teng, S.T.; Lundholm, N.; Orive, E.; David, H.; Quijano-Scheggia, S.; Leong, S.C.Y.; Wolf, M.; Bates, S.S.; et al. Phylogeny and Species Delineation in the Marine Diatom *Pseudo-nitzschia* (Bacillariophyta) Using Cox1, LSU, and ITS2 RRNA Genes: A Perspective in Character Evolution. *J. Phycol.* **2018**, *54*, 234–248. [CrossRef] [PubMed]
27. Amato, A.; Kooistra, W.H.C.F.; Levialdi Ghiron, J.H.; Mann, D.G.; Pröschold, T.; Montresor, M. Reproductive Isolation among Sympatric Cryptic Species in Marine Diatoms. *Protist* **2007**, *158*, 193–207. [CrossRef]
28. Dong, H.C.; Lundholm, N.; Teng, S.T.; Li, A.; Wang, C.; Hu, Y.; Li, Y. Occurrence of *Pseudo-nitzschia* Species and Associated Domoic Acid Production along the Guangdong Coast, South China Sea. *Harmful Algae* **2020**, *98*, 101899. [CrossRef] [PubMed]
29. Casteleyn, G.; Chepurnov, V.A.; Leliaert, F.; Mann, D.G.; Bates, S.S.; Lundholm, N.; Rhodes, L.; Sabbe, K.; Vyverman, W. *Pseudo-nitzschia pungens* (Bacillariophyceae): A Cosmopolitan Diatom Species? *Harmful Algae* **2008**, *7*, 241–257. [CrossRef]
30. Kim, J.H.; Kim, J.-H.; Park, B.S.; Wang, P.; Patidar, S.K.; Han, M.-S. Development of a QPCR Assay for Tracking the Ecological Niches of Genetic Sub-Populations within *Pseudo-nitzschia pungens* (Bacillariophyceae). *Harmful Algae* **2017**, *63*, 68–78. [CrossRef]
31. Rhodes, L.; Scholin, C.; Garthwaite, I.; Haywood, A.; Thomas, A. Domoic acid producing *Pseudo-nitzschia* species educed by whole cell DNA probe-based and immunochemical assay. In *Harmful algae = Algas nocivas: Proceedings of the VIII International Conference on Harmful Algae. Vigo, Spain, 25–29 June 1997*; Reguera, B., Blanco, J., Fernández, M.L., Wyatt, T., Eds.; Xunta de Galicia and Intergovernmental Oceanographic Commission of UNESCO: Santiago de Compostela, NY, USA, 1998; pp. 274–277. ISBN 84-453-2166-8.
32. Rhodes, L.; Scholin, C.; Garthwaite, I. *Pseudo-nitzschia* in New Zealand and the Role of DNA Probes and Immunoassays in Refining Marine Biotoxin Monitoring Programmes. *Nat. Toxins* **1998**, *6*, 105–111. [CrossRef]
33. Rhodes, L.L. Identification of Potentially Toxic *Pseudo-nitzschia* (Bacillariophyceae) in New Zealand Coastal Waters, Using Lectins. *N. Z. J. Mar. Freshw. Res.* **1998**, *32*, 537–544. [CrossRef]
34. Dermastia, T.T.; Cerino, F.; Stanković, D.; Francé, J.; Ramšak, A.; Žnidarič Tušek, M.; Beran, A.; Natali, V.; Cabrini, M.; Mozetič, P. Ecological Time Series and Integrative Taxonomy Unveil Seasonality and Diversity of the Toxic Diatom *Pseudo-nitzschia* H. Peragallo in the Northern Adriatic Sea. *Harmful Algae* **2020**, *93*, 101773. [CrossRef] [PubMed]
35. Kim, J.H.; Ajani, P.; Murray, S.A.; Kim, J.-H.; Lim, H.C.; Teng, S.T.; Lim, P.T.; Han, M.-S.; Park, B.S. Sexual Reproduction and Genetic Polymorphism within the Cosmopolitan Marine Diatom *Pseudo-nitzschia pungens*. *Sci. Rep.* **2020**, *10*, 10653. [CrossRef]
36. Lundholm, N.; Moestrup, Ø.; Hasle, G.R.; Hoef-Emden, K. A Study of the *Pseudo-nitzschia pseudodelicatissima/cuspidata* Complex (Bacillariophyceae): What Is *P. pseudodelicatissima*? *J. Phycol.* **2003**, *39*, 797–813. [CrossRef]
37. Orive, E.; Pérez-Aicua, L.; David, H.; García-Etxebarria, K.; Laza-Martínez, A.; Seoane, S.; Miguel, I. The Genus *Pseudo-nitzschia* (Bacillariophyceae) in a Temperate Estuary with Description of Two New Species: *Pseudo-nitzschia plurisecta* Sp. Nov. and *Pseudo-nitzschia abrensis* Sp. Nov. *J. Phycol.* **2013**, *49*, 1192–1206. [CrossRef]
38. Rivera-Vilarelle, M.; Valdez-Velázquez, L.L.; Quijano-Scheggia, S.I. Description of *Pseudo-nitzschia cuspidata* Var. *manzanillensis* Var. Nov. (Bacillariophyceae): Morphology and Molecular Characterization of a Variety from the Central Mexican Pacific. *Diatom Res.* **2018**, *33*, 55–68. [CrossRef]
39. Huang, C.X.; Dong, H.C.; Lundholm, N.; Teng, S.T.; Zheng, G.C.; Tan, Z.J.; Lim, P.T.; Li, Y. Species Composition and Toxicity of the Genus *Pseudo-nitzschia* in Taiwan Strait, Including *P. chiniana* Sp. Nov. and *P. qiana* Sp. Nov. *Harmful Algae* **2019**, *84*, 195–209. [CrossRef] [PubMed]
40. Lundholm, N.; Moestrup, Ø.; Kotaki, Y.; Hoef-Emden, K.; Scholin, C.; Miller, P. Inter- and Intraspecific Variation of the *Pseudo-nitzschia delicatissima* Complex (Bacillariophyceae) Illustrated by RRNA Probes, Morphological Data and Phylogenetic Analyses. *J. Phycol.* **2006**, *42*, 464–481. [CrossRef]
41. Quijano-Scheggia, S.I.; Garcés, E.; Lundholm, N.; Moestrup, Ø.; Andree, K.; Camp, J. Morphology, Physiology, Molecular Phylogeny and Sexual Compatibility of the Cryptic *Pseudo-nitzschia delicatissima* Complex (Bacillariophyta), Including the Description of *P. arenysensis* Sp. Nov. *Phycologia* **2009**, *48*, 492–509. [CrossRef]
42. Lundholm, N.; Moestrup, Ø. The Marine Diatom *Pseudo-nitzschia galaxiae* Sp. Nov. (Bacillariophyceae): Morphology and Phylogenetic Relationships. *Phycologia* **2002**, *41*, 594–605. [CrossRef]
43. Lundholm, N.; Bates, S.S.; Baugh, K.A.; Bill, B.D.; Connell, L.B.; Léger, C.; Trainer, V.L. Cryptic and Pseudo-Cryptic Diversity in Diatoms—with Descriptions of *Pseudo-nitzschia hasleana* Sp. Nov. and *P. fryxelliana* Sp. Nov. *J. Phycol.* **2012**, *48*, 436–454. [CrossRef]
44. Quijano-Scheggia, S.I.; Olivos-Ortiz, A.; Garcia-Mendoza, E.; Sánchez-Bravo, Y.; Sosa-Avalos, R.; Marias, N.S.; Lim, H.C. Phylogenetic Relationships of *Pseudo-nitzschia subpacifica* (Bacillariophyceae) from the Mexican Pacific, and Its Production of Domoic Acid in Culture. *PLoS ONE* **2020**, *15*, e0231902. [CrossRef] [PubMed]
45. Quilliam, M.A.; Sim, P.G.; McCulloch, A.W.; McInnes, A.G. High-Performance Liquid Chromatography of Domoic Acid, a Marine Neurotoxin, with Application to Shellfish and Plankton. *Int. J. Environ. Anal. Chem.* **1989**, *36*, 139–154. [CrossRef]
46. Quilliam, M.A. Chemical methods for domoic acid, the amnesic shellfish poisoning (ASP) toxin. In *Manual on Harmful Marine Microalgae, Monographs on Oceanographic Methodology*; Hallegraeff, G.M., Anderson, D.M., Cembella, A.D., Eds.; United Nations Educational, Scientific and Cultural Organization: Paris, France, 2003; pp. 247–266. ISBN 92-3-103948-2.
47. European Food Safety Authority Marine Biotoxins in Shellfish–Domoic Acid. *EFSA J.* **2009**, *7*, 1181. [CrossRef]

48. Kotaki, Y.; Lundholm, N.; Katayama, T.; Furio, E.F.; Romero, M.L.; Relox, J.R.; Yasumoto, T.; Naoki, H.; Hirose, M.Y.; Thanh, T.D.; et al. ASP toxins of pennate diatoms and bacterial effects on the variation in toxin composition. In Proceedings of the 12th International Conference on Harmful Algae, Copenhagen, Denmark, 4–8 September 2006; Moestrup, Ø., Doucette, G., Enevoldsen, H., Godhe, A., Hallegraeff, G., Luckas, B., Lundholm, N., Lewis, J., Rengefors, K., Sellner, K., et al., Eds.; International Society for the Study of Harmful Algae and Intergovernmental Oceanographic Commission of UNESCO: Copenhagen, Denmark, 2008; pp. 300–302.

49. Hansen, L.R.; Soylu, S.í.; Kotaki, Y.; Moestrup, Ø.; Lundholm, N. Toxin Production and Temperature-Induced Morphological Variation of the Diatom *Pseudo-nitzschia seriata* from the Arctic. *Harmful Algae* **2011**, *10*, 689–696. [CrossRef]

50. Rhodes, L.L.; Holland, P.T.; Adamson, J.E.; McNabb, P.; Selwood, A.I. Production of a new isomer of domoic acid by New Zealand isolates of the diatom *Pseudo-nitzschia australis*. In *Molluscan Shellfish Safety: Proceedings of the 4th International Conference on Molluscan Shellfish Safety, Santiago de Compostela, Spain, 4–8 June 2002*; Villalba, A., Reguera, B., Romalde, J.L., Beiras, R., Eds.; Consellería de Pesca e Asuntos Marítimos da Xunta de Galicia and Intergovernmental Oceanographic Commission of UNESCO: Santiago de Compostela, Spain, 2003; pp. 43–48. ISBN 84-453-3638-X.

51. Holland, P.T.; Selwood, A.I.; Mountfort, D.O.; Wilkins, A.L.; McNabb, P.; Rhodes, L.L.; Doucette, G.J.; Mikulski, C.M.; King, K.L. Isodomoic Acid C, an Unusual Amnesic Shellfish Poisoning Toxin from *Pseudo-nitzschia australis*. *Chem. Res. Toxicol.* **2005**, *18*, 814–816. [CrossRef] [PubMed]

52. Olesen, A.J.; Leithoff, A.; Altenburger, A.; Krock, B.; Beszteri, B.; Eggers, S.L.; Lundholm, N. First Evidence of the Toxin Domoic Acid in Antarctic Diatom Species. *Toxins* **2021**, *13*, 93. [CrossRef]

53. Caruana, A.M.N.; Ayache, N.; Raimbault, V.; Rétho, M.; Hervé, F.; Bilien, G.; Amzil, Z.; Chomérat, N. Direct Evidence for Toxin Production by *Pseudo-nitzschia plurisecta* (Bacillariophyceae) and Extension of Its Distribution Area. *Eur. J. Phycol.* **2019**, *54*, 585–594. [CrossRef]

54. Rhodes, L.; Holland, P.; Adamson, J.; Selwood, A.; McNabb, P. Mass culture of New Zealand isolates of *Pseudo-nitzschia australis* for production of a new isomer of domoic acid. In *Harmful Algae 2002. Proceedings of the Xth International Conference on Harmful Algae. St. Pete Beach, Florida, USA, 21–25 October 2002*; Steidinger, K.A., Landsberg, J.H., Tomas, C.R., Vargo, G.A., Eds.; Florida Fish and Wildlife Conservation Commission, Florida Institute of Oceanography, and Intergovernmental Oceanographic Commission of UNESCO: St. Petersburg, FL, USA, 2004; pp. 125–127.

55. Lelong, A.; Hégaret, H.; Soudant, P.; Bates, S.S. *Pseudo-nitzschia* (Bacillariophyceae) Species, Domoic Acid and Amnesic Shellfish Poisoning: Revisiting Previous Paradigms. *Phycologia* **2012**, *51*, 168–216. [CrossRef]

56. Trainer, V.L.; Bates, S.S.; Lundholm, N.; Thessen, A.E.; Cochlan, W.P.; Adams, N.G.; Trick, C.G. *Pseudo-nitzschia* Physiological Ecology, Phylogeny, Toxicity, Monitoring and Impacts on Ecosystem Health. *Harmful Algae* **2012**, *14*, 271–300. [CrossRef]

57. MacKenzie, A.L.; White, D.A.; Sim, P.G.; Holland, A.J. Domoic acid and the New Zealand Greenshell mussel (*Perna canaliculus*). In *Toxic Phytoplankton Blooms in the Sea: Proceedings of the Fifth International Conference on Toxic Marine Phytoplankton, Newport, RI, USA, 28 October–1 November 1991*; Smayda, T.J., Shimizu, Y., Eds.; Elsevier: Amsterdam, NY, USA, 1993; pp. 607–612. ISBN 978-0-444-89719-0.

58. Rhodes, L.; Jiang, W.; Knight, B.; Adamson, J.; Smith, K.; Langi, V.; Edgar, M.; Munday, R. The Genus *Pseudo-nitzschia* (Bacillariophyceae) in New Zealand: A Review of the Last Decade's Research Achievements and Monitoring Data. Prepared for Seafood Safety Programme Funded by Ministry of Science and Innovation. *Cawthron Rep.* **2012**, 2035B.

59. Ajani, P.; Murray, S.; Hallegraeff, G.; Lundholm, N.; Gillings, M.; Brett, S.; Armand, L. The Diatom Genus *Pseudo-nitzschia* (Bacillariophyceae) in New South Wales, Australia: Morphotaxonomy, Molecular Phylogeny, Toxicity, and Distribution. *J. Phycol.* **2013**, *49*, 765–785. [CrossRef] [PubMed]

60. Trainer, V.L.; Wells, M.L.; Cochlan, W.P.; Trick, C.G.; Bill, B.D.; Baugh, K.A.; Beall, B.F.; Herndon, J.; Lundholm, N. An Ecological Study of a Massive Bloom of Toxigenic *Pseudo-nitzschia cuspidata* off the Washington State Coast. *Limnol. Oceanogr.* **2009**, *54*, 1461–1474. [CrossRef]

61. Rhodes, L.; Selwood, A.; McNabb, P.; Briggs, L.; Adamson, J.; van Ginkel, R.; Laczka, O. Trace Metal Effects on the Production of Biotoxins by Microalgae. *Afr. J. Mar. Sci.* **2006**, *28*, 393–397. [CrossRef]

62. Rhodes, L.L.; Adamson, J.; Scholin, C. *Pseudo-nitzschia multistriata* (Bacillariophyceae) in New Zealand. *N. Z. J. Mar. Freshw. Res.* **2000**, *34*, 463–467. [CrossRef]

63. Penna, A.; Bertozzini, E.; Battocchi, C.; Galluzzi, L.; Giacobbe, M.G.; Vila, M.; Garces, E.; Lugliè, A.; Magnani, M. Monitoring of HAB Species in the Mediterranean Sea through Molecular Methods. *J. Plankton Res.* **2007**, *29*, 19–38. [CrossRef]

64. Fitzpatrick, E.; Caron, D.A.; Schnetzer, A. Development and Environmental Application of a Genus-Specific Quantitative PCR Approach for *Pseudo-nitzschia* Species. *Mar. Biol.* **2010**, *157*, 1161–1169. [CrossRef]

65. Andree, K.B.; Fernández-Tejedor, M.; Elandaloussi, L.M.; Quijano-Scheggia, S.; Sampedro, N.; Garcés, E.; Camp, J.; Diogène, J. Quantitative PCR Coupled with Melt Curve Analysis for Detection of Selected *Pseudo-nitzschia* Spp. (Bacillariophyceae) from the Northwestern Mediterranean Sea. *Appl. Environ. Microbiol.* **2011**, *77*, 1651–1659. [CrossRef] [PubMed]

66. Guillard, R.R.L. Culture of phytoplankton for feeding marine invertebrates. In *Culture of Marine Invertebrate Animals: Proceedings of the Conference on Culture of Marine Invertebrate Animals, Greenport, NY, USA, October 1972*; Smith, W.L., Chanley, M.H., Eds.; Springer: Boston, MA, USA, 1975; pp. 29–60. ISBN 978-1-4615-8714-9.

67. Richlen, M.L.; Barber, P.H. A Technique for the Rapid Extraction of Microalgal DNA from Single Live and Preserved Cells. *Mol. Ecol. Notes* **2005**, *5*, 688–691. [CrossRef]

68. Bowers, H.A.; Tomas, C.; Tengs, T.; Kempton, J.W.; Lewitus, A.J.; Oldach, D.W. Raphidophyceae [Chadefaud Ex Silva] Systematics and Rapid Identification: Sequence Analyses and Real-Time PCR Assays. *J. Phycol.* **2006**, *42*, 1333–1348. [CrossRef]
69. Nunn, G.B.; Theisen, B.F.; Christensen, B.; Arctander, P. Simplicity-Correlated Size Growth of the Nuclear 28S Ribosomal RNA D3 Expansion Segment in the Crustacean Order Isopoda. *J. Mol. Evol.* **1996**, *42*, 211–223. [CrossRef]
70. Thompson, J.D.; Higgins, D.G.; Gibson, T.J. CLUSTAL W: Improving the Sensitivity of Progressive Multiple Sequence Alignment through Sequence Weighting, Position-Specific Gap Penalties and Weight Matrix Choice. *Nucleic Acids Res.* **1994**, *22*, 4673–4680. [CrossRef]
71. Kumar, S.; Stecher, G.; Li, M.; Knyaz, C.; Tamura, K. MEGA X: Molecular Evolutionary Genetics Analysis across Computing Platforms. *Mol. Biol. Evol.* **2018**, *35*, 1547–1549. [CrossRef]
72. Huelsenbeck, J.P.; Ronquist, F. MRBAYES: Bayesian Inference of Phylogenetic Trees. *Bioinformatics* **2001**, *17*, 754–755. [CrossRef]
73. Ronquist, F.; Huelsenbeck, J.P. MrBayes 3: Bayesian Phylogenetic Inference under Mixed Models. *Bioinformatics* **2003**, *19*, 1572–1574. [CrossRef]
74. Nylander, J.A.A. *MrModeltest v2. Program. Distributed by the Author*; Evolutionary Biology Centre, Uppsala University: Uppsala, Sweden, 2004.
75. Kumar, S.; Stecher, G.; Tamura, K. MEGA7: Molecular Evolutionary Genetics Analysis Version 7.0 for Bigger Datasets. *Mol. Biol. Evol.* **2016**, *33*, 1870–1874. [CrossRef] [PubMed]
76. Keller, M.D.; Selvin, R.C.; Claus, W.; Guillard, R.R.L. Media for the Culture of Oceanic Ultraphytoplankton. *J. Phycol.* **1987**, *23*, 633–638. [CrossRef]
77. McNabb, P.; Selwood, A.I.; Holland, P.T. Collaborators: Multiresidue Method for Determination of Algal Toxins in Shellfish: Single-Laboratory Validation and Interlaboratory Study. *J. AOAC Int.* **2005**, *88*, 761–772. [CrossRef]

Article

Sub-Acute Feeding Study of Saxitoxin to Mice Confirms the Effectiveness of Current Regulatory Limits for Paralytic Shellfish Toxins

Sarah C. Finch [1,*], Nicola G. Webb [1], Michael J. Boundy [2], D. Tim Harwood [2], John S. Munday [3], Jan M. Sprosen [1], Vanessa M. Cave [1], Ric B. Broadhurst [1] and Jeane Nicolas [4]

[1] AgResearch Ltd. Ruakura Research Centre, Private Bag 3123, Hamilton 3240, New Zealand; nikki.webb@agresearch.co.nz (N.G.W.); jan.sprosen@agresearch.co.nz (J.M.S.); vanessa.cave@agresearch.co.nz (V.M.C.); ric.broadhurst@agresearch.co.nz (R.B.B.)
[2] Cawthron Institute, Private Bag 2, Nelson 7042, New Zealand; Michael.Boundy@cawthron.org.nz (M.J.B.); Tim.Harwood@cawthron.org.nz (D.T.H.)
[3] Department of Pathobiology, School of Veterinary Science, Massey University, Private Bag 11 222, Palmerston North 4442, New Zealand; J.Munday@massey.ac.nz
[4] Ministry for Primary Industries–Manatu Ahu Matua, P.O. Box 2526, Wellington 6021, New Zealand; Jeane.Nicolas@mpi.govt.nz
* Correspondence: sarah.finch@agresearch.co.nz

Abstract: Regulatory limits for shellfish toxins are required to protect human health. Often these limits are set using only acute toxicity data, which is significant, as in some communities, shellfish makes up a large proportion of their daily diet and can be contaminated with paralytic shellfish toxins (PSTs) for several months. In the current study, feeding protocols were developed to mimic human feeding behaviour and diets containing three dose rates of saxitoxin dihydrochloride (STX.2HCl) were fed to mice for 21 days. This yielded STX.2HCl dose rates of up to 730 μg/kg bw/day with no effects on food consumption, growth, blood pressure, heart rate, motor coordination, grip strength, blood chemistry, haematology, organ weights or tissue histology. Using the 100-fold safety factor to extrapolate from animals to humans yields a dose rate of 7.3 μg/kg bw/day, which is well above the current acute reference dose (ARfD) of 0.5 μg STX.2HCl eq/kg bw proposed by the European Food Safety Authority. Furthermore, to reach the dose rate of 7.3 μg/kg bw, a 60 or 70 kg human would have to consume 540 or 630 g of shellfish contaminated with PSTs at the current regulatory limit (800 μg/kg shellfish flesh), respectively. The current regulatory limit for PSTs therefore seems appropriate.

Keywords: saxitoxin; feeding study; toxicology; paralytic shellfish toxins

Key Contribution: The feeding of saxitoxin dihydrochloride to mice for 21 days using a protocol to mimic human feeding behaviour showed no adverse effects at a dose rate of up to 730 μg STX.2HCl/kg bw/day. These results suggest that the current regulatory limit for PSPs is protective of human health.

Citation: Finch, S.C.; Webb, N.G.; Boundy, M.J.; Harwood, D.T.; Munday, J.S.; Sprosen, J.M.; Cave, V.M.; Broadhurst, R.B.; Nicolas, J. Sub-Acute Feeding Study of Saxitoxin to Mice Confirms the Effectiveness of Current Regulatory Limits for Paralytic Shellfish Toxins. *Toxins* **2021**, *13*, 627. https://doi.org/10.3390/toxins13090627

Received: 6 August 2021
Accepted: 5 September 2021
Published: 7 September 2021

Publisher's Note: MDPI stays neutral with regard to jurisdictional claims in published maps and institutional affiliations.

1. Introduction

Paralytic shellfish poisoning (PSP) is induced by the ingestion of shellfish contaminated with paralytic shellfish toxins (PSTs). These toxins are produced by the marine dinoflagellates of the genera *Alexandruim*, *Gymnodinium* and *Pyrodinium* [1–3], and are accumulated by filter-feeding shellfish. The major PST is saxitoxin (STX), although over fifty structurally related analogues make up this toxin class [4]. PSP is characterised by tingling and numbness around the lips, incoordination and muscle weakness, as well as neurological symptoms such as headaches [5]. In severe cases, muscular paralysis will be marked and respiratory paralysis can result in death. This intoxication is not location specific and throughout history PSP outbreaks have been regularly reported worldwide [6,7].

The first such report was documented in a ship captain's diary in 1793 when five crew members became ill after eating mussels harvested off the coast of British Columbia which resulted in one death [8]. In Alaska, the first case of PSP was in 1799, and between 1973 and 1994, there were 54 outbreaks resulting in 117 people falling ill. Of those affected, one person died, four required intubation and 29 required emergency treatment [9]. In order to protect human health, a regulatory limit for PSTs has been adopted by many countries. Currently the regulatory limit for the European Union (EU) is set at 800 μg STX.2HCl eq/kg shellfish flesh (regulation (EC) No 853/2004) [10]. The same limit was adopted at the twenty-eighth session of the Codex Committee on Fish and Fishery Products (CCFFP) in 2006 [11], resulting in the development of the Standard for Live and Raw Bivalve Molluscs (CODEXSTAN 292-2008, rev. 2015) [12]. This standard is used in many countries, including New Zealand. Because STX hydrate (the free base) is of low stability, the STX limit is expressed as saxitoxin dihydrochloride (STX.2HCl). It is of vital importance that whenever PST mass concentrations are mentioned, it is specified whether this refers to the salt (STX.2HCl) or to the free base (STX hydrate). This information is needed to interpret results and to allow the comparison of concentrations quoted in different studies [13]. The regulatory limit is also expressed in terms of STX equivalents (STX.2HCl eq) as the PSTs include more than just STX, and other derivatives may simultaneously be present in the shellfish ingested by humans. To accommodate this, the toxicity data of each analogue is compared to that of STX on a molar basis to yield a toxicity equivalence factor (TEF). Using the TEF values, the concentrations of each analogue, as determined by analytical methods, can be converted into STX.2HCl equivalents such that the overall toxicity of a given sample can be evaluated against the regulatory limit.

Estimating the quantity of PSTs that have been responsible for reported cases of human illness is very difficult as a number of different pieces of information are required including an accurate determination of the amount of toxin in the food (STX.2HCl eq), the quantity of contaminated shellfish eaten, and the bodyweight of the person. The quantity of shellfish eaten is often not known and it is rare for the weight of the patient to be reported in the literature. Furthermore, there is often a time lag between the consumption of contaminated shellfish and the diagnosis of PSP, meaning that remnant food is rarely available for analysis. While shellfish can be harvested from the affected area, this is not necessarily representative of the food ingested and it has been demonstrated that the concentrations of PSP toxins can change very quickly in shellfish [14]. In addition, the effects of different cooking processes on PST concentrations are uncertain and whether broth or cooking liquid is consumed or discarded adds another layer of complexity [15,16].

By considering the available data on human PSP cases, the European Food Safety Authority (EFSA) determined that the quantity of STX which could be consumed in a 24 h period without adverse effect, the acute reference dose (ARfD), was 0.5 μg STX.2HCl eq/kg bw [7]. To relate this to the quantities which could be ingested by consumers of shellfish, three pieces of information are required: the STX concentration in the shellfish, the quantity of shellfish consumed (portion size), and the bodyweight of the consumer. In calculations of risk, the concentration of STX in shellfish is assumed to be at the current NZ/Codex/EU regulatory limit (800 μg STX.2HCl eq/kg shellfish flesh). Understandably, portion size will vary greatly throughout the population, making it difficult to select an appropriate value. A large portion size of 400 g has been proposed by EFSA which represents the 95th percentile of shellfish consumption in Germany and the Netherlands [17]. However, this value appears high as the 97.5th percentile for the portion size of shellfish eaten by adults is 133 g in Japan, 181 g in Australia, 225 g in the USA and 263 g in New Zealand, and a portion size of 250 g would cover 97.5% of the consumers of most countries for which data are available [8]. Although the bodyweight of adults is also highly variable, it is imperative that risk assessments are done using conservative estimates, thus ensuring that the majority of the population is covered. EFSA has used an adult bodyweight of 60 kg [7], but a further EFSA committee has argued that a standard bodyweight of 70 kg would be more appropriate [18]. Looking at the range of possible portion sizes of shellfish in particular, but also bodyweights, illustrates that the estimate of STX which could be ingested by a human varies widely. Using the worst-case scenario of shellfish at the regulatory limit (800 µg STX.2HCl eq/kg shellfish flesh), a portion size of 400 g and a 60 kg adult, the dose rate consumed would be 5.3 µg STX.2HCl eq/kg bw. However, consumption of shellfish at the regulatory limit with a portion size of 250 g by a 70 kg adult would yield a dose rate of 2.9 µg STX.2HCl eq/kg bw. Although significantly different, both of these figures are above the ARfD of 0.5 µg STX.2HCl eq/kg bw. This difference was acknowledged by EFSA who concluded "there is a concern for the health for the consumer at the present regulatory limit" [7].

Given the uncertainties, generating regulatory limits using human cases of poisoning is difficult and as an alternative animal models can be used. Oral toxicity is the most relevant route of administration and, when available, these data are used in preference to toxicity determined by intraperitoneal injection (i.p.) [19]. To extrapolate from animal data to human, uncertainty factors are applied to animal toxicity data. Those generally applied are a ten-fold safety factor to allow for the species difference, and a further ten-fold safety factor to allow for possible susceptibility variations within a human population [20]. An acute no observable adverse effect level (NOAEL) for STX of 473 µg STX.2HCl eq/kg bw has been determined in mice using oral administration [21]. When the combined safety factors of 100 are applied to this figure, the concentration which would be expected to induce no adverse effects in humans is 4.7 µg STX.2HCl eq/kg bw. This dose rate is much higher than the ARfD suggested by EFSA (0.5 µg STX.2HCl eq/kg bw). Comparing the Health Based Guidance Value (HBGV) to the current STX.2HCl regulatory limit of 800 µg STX.2HCl eq/kg shellfish flesh is difficult because this is dependent on an estimate of portion size and human bodyweight.

While there is still debate around the validity of the current regulatory limit for PSTs, of further concern is the fact that people who eat shellfish often do so on a regular basis, and for some communities, shellfish constitutes a high proportion of their diet. Since algal blooms can be present for several months, shellfish may be contaminated with PSTs for an extended time [22–24]. Despite this, as noted by the FAO/IOC/WHO (2004) Committee, no repeated oral toxicity studies have been performed to assess whether STX has a cumulative effect or whether more subtle indications of toxicity may arise with regular sub-lethal doses [8]. To fill this significant knowledge gap, a sub-acute study with daily dosing of STX.2HCl to mice was performed. Feeding protocols using meal times were developed in order to more closely represent average human feeding behaviour.

2. Results

2.1. Development of Experimental Protocols

2.1.1. Development of Suitable Diet

Previous mouse feeding studies have been conducted by incorporating the test ingredient, such as ground endophyte-infected ryegrass seed [25] or the endophyte-expressed metabolite chanoclavine [26] with ground mouse food. In these studies, the diet portions weighed approximately 7 g and the compounds were found to be homogenous in the diet. However, using the same protocol, STX was found to be unevenly distributed in the diet with the STX.2HCl concentration 24–40% higher at the edges of the diet portion in comparison to the middle. This difference is likely due to the high water solubility of STX which allows it to be redistributed in the diet during drying, a process which would not be possible for the seed or highly lipophilic compound used in the other studies. Modifications to the protocol where smaller portions of diet (approximately 1 g each) were prepared using a lower drying temperature for a shorter time rectified this issue and STX.2HCl was confirmed to be homogenously distributed in this diet.

The stability of STX.2HCl in laced diet was assessed by storing individual portions (1 g) for a number of days in the fridge (4 °C) or at room temperature (20 °C). This showed that STX was stable in the diet for up to five days at both temperatures (Table 1). Since fresh diet was prepared every 2–4 days during the feeding study, the stability of STX in the mouse diet was judged to be adequate.

Table 1. Saxitoxin concentrations in mouse diet stored at different temperatures and for different time periods.

Sample	Concentration (µg STX.2HCl eq/g) [1]	Recovery [2]
Day 1 Control	3.30 ± 0.24	95%
Day 2 20 °C	3.30 ± 0.48	95%
Day 2 4 °C	3.26 ± 0.36	94%
Day 5 20 °C	3.33 ± 0.24	96%
Day 5 4 °C	3.26 ± 0.45	94%

[1] ± 95% confidence interval, [2] Compared to theoretical maximum of 3.48 µg STX.2HCl eq/g.

2.1.2. Establishment of a Suitable STX Dose Rate

Many studies on shellfish toxins using mice have successfully utilised a method of oral administration whereby the test toxin is mixed with a very small amount of cream cheese [27,28]. After training with unlaced cream cheese for a few days prior, mice will consume this laced cream cheese within 30 s. Using this method, an acute oral dose rate of 400 µg STX.2HCl/kg bw induced mild toxic effects in one out of three mice, so it was anticipated that this could be an appropriate dose rate for the high STX treatment group in the feeding study. However, it was found that mice with unlimited access to STX-laced diet, delivering a daily dose of 476 µg STX.2HCl/kg bw for two weeks, showed no adverse effects. Furthermore, mice with unlimited access to STX-laced diet to deliver a daily dose rate of 620 µg STX.2HCl/kg bw for four days were also unaffected. This demonstrated that the toxicity of STX could be influenced by the feeding protocols used in the study, highlighting the importance of using a feeding protocol that is relevant to humans. In contrast to the human feeding behaviour of meal times, the feeding behaviour of mice is that of continuous grazing. To develop a relevant feeding protocol, pairs of individually caged mice were offered either unrestricted access to control diet (no STX) or access to the same food between only 9–11 a.m. and 3–5 p.m., with food consumption being measured in both groups. This showed that for two days, mice with limited feeding times had a low daily food intake (0.029 and 0.176 g food/g bw on days 1 and 2, respectively) compared to those with unlimited access to food (0.314 and 0.296 g food/g bw on days 1 and 2, respectively). However, by day 3, the two groups had an equivalent total food consumption (0.256 compared to 0.240 g food/g bw, for unlimited and limited, respectively). For mice fed between 9–11 a.m. and 3–5 p.m., it was observed that, at the most, they ate for only half of their 2 h feeding period before going to sleep. Meal times of a 1 h duration were therefore chosen for the feeding study.

Since no toxicity was observed on feeding STX.2HCl in laced diet, the possible effect of the matrix was investigated. A mouse with unrestricted access to control diet dosed with STX.2HCl in cream cheese to deliver a 400 µg STX.2HCl/kg bw dose showed a hunched posture and splayed back legs 2 h post-dosing, symptoms characteristic of STX toxicity. Using the meal time feeding protocol, a further two mice were taken, with one fed a STX.2HCl-laced diet and the other fed a control diet, but half way through each feeding period STX.2HCl laced cream cheese was offered, and consumed by the mouse within 30 s. Each of these latter two mice ingested 400 µg STX.2HCl/kg bw at each feeding period (800 µg/kg bw daily total) and neither showed any adverse effects. This demonstrates that the lack of STX.2HCl toxicity is the same whether fed via mouse food diet laced with STX or in laced cream cheese, eliminating the possibility of a matrix effect.

2.2. Results of the 21-Day Feeding Study

2.2.1. Diet Analysis

Preliminary work and palatability studies allowed the STX.2HCl concentrations for the feeding study to be chosen. Laced diet from each treatment group was analysed to ensure that the STX.2HCl concentrations were as expected. A comparison between the theoretical and actual concentrations of STX.2HCl in each of the three treatment groups showed this to be the case (Table 2).

Table 2. Saxitoxin concentrations in mouse diet for the three different treatments.

Treatment	Measured Concentration (μg STX.2HCl eq/g) [1]	Theoretical Concentration (μg STX.2HCl eq/g)	Recovery [2]
Low STX	1.12 ± 0.20	1.15	97%
Mid STX	2.34 ± 0.41	2.27	103%
High STX	3.19 ± 0.56	3.41	94%

[1] ± 95% confidence interval, [2] Compared to theoretical.

2.2.2. Dose Rates

By combining the daily food intake and bodyweight data for each individual animal, the dose rate of STX.2HCl consumed each day could be calculated, thereby allowing the mean daily dose rate of STX.2HCl to be determined for each treatment group. The dose rates showed good consistency throughout the 21-day feeding period with means (± standard error) of 253 ± 5 and 248 ± 5 μg/kg bw/day for female and male mice fed with low-dose STX-laced diet, 494 ± 9 and 486 ± 8 μg/kg bw/day for female and male mice fed mid-dose STX-laced diet, and 730 ± 13 and 699 ± 11 μg/kg bw/day for female and male mice fed high-dose STX-laced diet.

2.2.3. Clinical Observations and Appearance

The appearance, movement and behaviour of all mice remained normal throughout the 21-day experimental period.

2.2.4. Bodyweight and Food Consumption

Statistical analysis of the daily food intake data showed that there was no evidence of an interaction between gender and treatment (gender.treatment.day, $p = 0.940$; gender.treatment, $p = 0.198$). A graph could therefore be created for the temporal treatment effect pooled over gender (Figure 1).

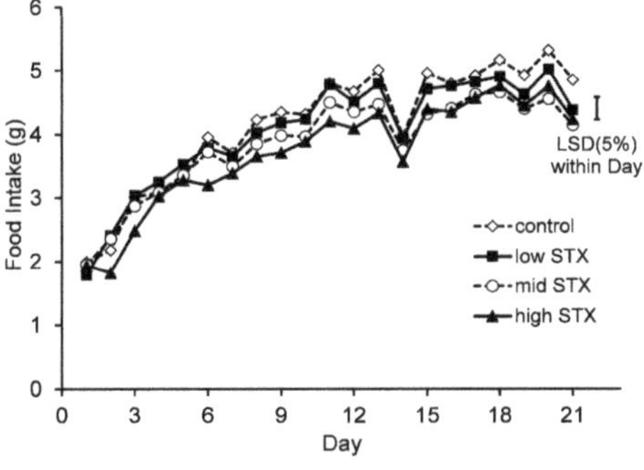

Figure 1. Temporal trend in food intake of mice fed control (··◇··), low STX (-■-), mid STX (··○··) or high STX (-▲-) diets (1.14, 2.28 and 3.48 μg STX.2HCl/g, respectively) over the 21-day feeding study. The error bar denotes the Fisher's unprotected least significant differences at the 5% level (LSD (5%)).

The food consumption of all mice was low at the start of the study as they adapted to the meal time feeding protocol. The dips in food intake on days 14 and 21 were observed in all treatment groups and coincided with the days that motor coordination, blood pressure, heart rate and grip strength were measured. Although still present, this observed dip in food intake was less marked on day 7 when only motor coordination and grip strength were measured. This demonstrated that the measurement of blood pressure and heart rate was the major driver of this disruption of mouse feeding. The statistical analysis of the pooled data showed that although feeding of mice in all treatment groups was good, the presence of STX did cause a dose-dependent statistical effect on food intake. Mice fed a high STX diet ate a statistically lower amount of food than mice fed a control diet on 16 of the 21 days. Similarly, mice fed with a mid and low STX diet ate a statistically lower amount of food than mice fed a control diet on 9 and 1 of the 21 days, respectively.

Statistical analysis of the bodyweights of mice showed no evidence of an interaction between gender and treatment (gender.treatment.day, $p = 0.918$; gender.treatment, $p = 0.393$). A graph could therefore be created for the temporal treatment effect pooled over gender (Figure 2).

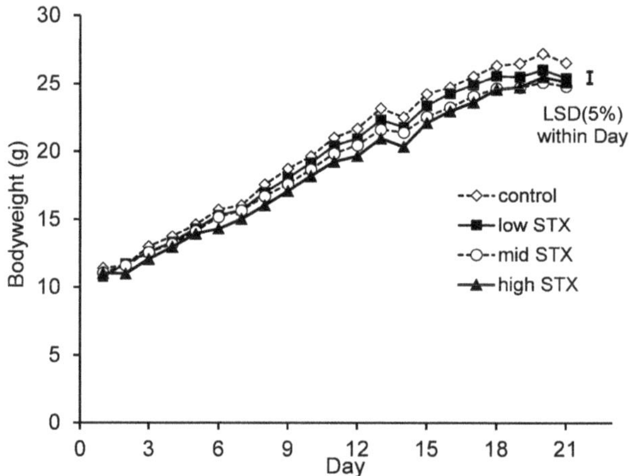

Figure 2. Temporal trend in bodyweights of mice fed control (··◇··), low STX (-■-), mid STX (··○··) or high STX (-▲-) diets (1.14, 2.28 and 3.48 µg STX.2HCl/g, respectively) over the 21-day feeding study. The error bar denotes the Fisher's unprotected least significant differences at the 5% level (LSD (5%)).

This bodyweight data showed that all mice gained weight over the experimental period but mice fed a high, mid and low STX diet gained a lower amount of weight on average than mice fed a control diet on 17, 13 and 3 of the 21 days of the study, respectively.

2.2.5. Motor Coordination

Motor coordination of all mice was analysed on days 7, 14 and 21 using an accelerating rotarod. Statistical analysis of the motor coordination data showed no evidence of an interaction between gender and treatment (gender.treatment.day, $p = 0.113$; gender.treatment, $p = 0.761$). A graph could therefore be created for the temporal treatment effect pooled over gender (Figure 3). There were no statistically significant treatment effects on the motor coordination of mice.

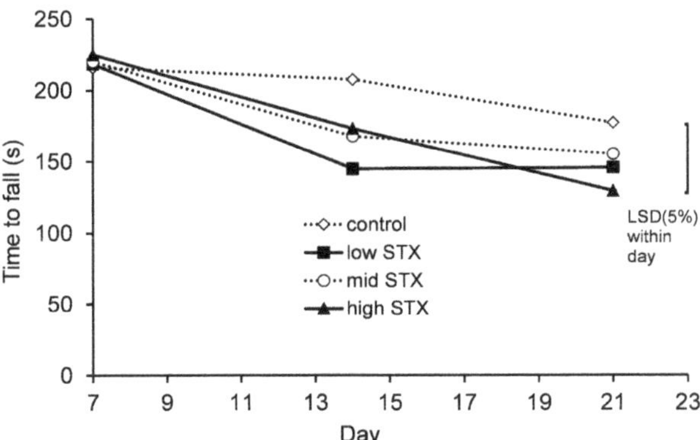

Figure 3. Temporal trend in motor coordination (time to fall) of mice fed control ("◇"), low STX (-■-), mid STX ("○") or high STX (-▲-) diets (1.14, 2.28 and 3.48 µg STX.2HCl/g, respectively) on days 7 and 14 of the feeding study. The error bar denotes the Fisher's unprotected least significant differences at the 5% level (LSD (5%)).

2.2.6. Grip Strength

The grip strength of all mice was analysed on days 7, 14 and 21. Statistical analysis of the grip strength data showed no evidence of an interaction between gender and treatment (gender.treatment.day, $p = 0.574$; gender.treatment, $p = 0.334$). A graph could therefore be created for the temporal treatment effect pooled over gender (Figure 4). The data showed no statistically significant treatment effects on the grip strength of mice.

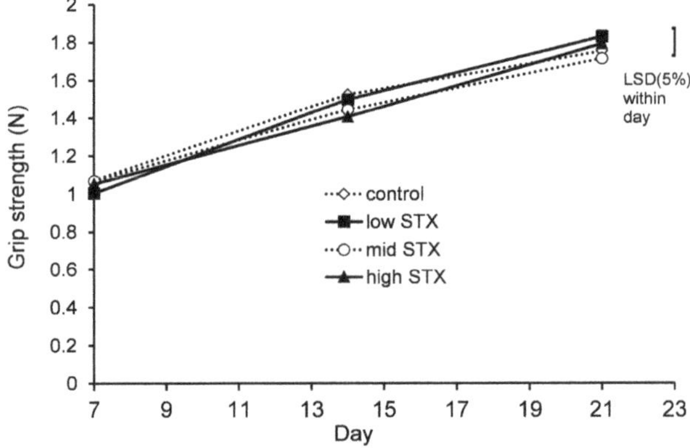

Figure 4. Temporal trend in grip strength of mice fed control ("◇"), low STX (-■-), mid STX ("○") or high STX (-▲-) diets (1.14, 2.28 and 3.48 µg STX.2HCl/g, respectively) on days 7, 14 and 21 of the feeding study. The error bar denotes the Fisher's unprotected least significant differences at the 5% level (LSD (5%)).

2.2.7. Blood Pressure and Heart Rate

The blood pressure and heart rate were measured in all mice on days 14 and 21 (Table 3). Measurements were not possible prior to day 14 as the mice were too small to fit into the blood pressure analysis system. There were no statistically significant differences observed in systolic blood pressure, diastolic blood pressure or heart rate between any of the treatment groups.

Table 3. Heart rate and blood pressure of mice fed control, low STX, mid STX or high STX diets (1.14, 2.28 and 3.48 µg STX.2HCl/g, respectively) on days 14 and 21 of the feeding study.

	Heart Rate [1] (BPM)		Systolic BP [1] (mmHg)		Diastolic BP [1] (mmHg)	
	Day 14	Day 21	Day 14	Day 21	Day 14	Day 21
Females						
control	729 ± 16.7 [a]	710 ± 16.7 [a]	112.6 ± 6.89 [a]	119.0 ± 6.89 [a]	45.8 ± 4.20 [a]	52.0 ± 4.20 [a]
low	732 ± 16.7 [a]	679 ± 16.7 [a]	110.2 ± 6.89 [a]	118.4 ± 6.89 [a]	54.4 ± 4.20 [a]	57.2 ± 4.20 [a]
mid	753 ± 16.7 [a]	690 ± 16.7 [a]	105.4 ± 6.89 [a]	111.0 ± 6.89 [a]	52.6 ± 4.20 [a]	49.4 ± 4.20 [a]
high	754 ± 16.7 [a]	699 ± 16.7 [a]	99.4 ± 6.89 [a]	106.0 ± 6.89 [a]	43.6 ± 4.20 [a]	52.8 ± 4.20 [a]
Males						
control	709. ± 16.7 [a]	704 ± 16.7 [a]	108.4 ± 6.89 [a]	111.8 ± 6.89 [a]	48.2 ± 4.20 [a]	48.0 ± 4.20 [a]
low	701. ± 16.7 [a]	692 ± 16.7 [a]	105.2 ± 6.89 [a]	119.4 ± 6.89 [a]	41.0 ± 4.20 [a]	49.2 ± 4.20 [a]
mid	753 ± 16.7 [a]	695 ± 16.7 [a]	101.4 ± 6.89 [a]	116.2 ± 6.89 [a]	42.8 ± 4.20 [a]	50.8 ± 4.20 [a]
high	729 ± 16.7 [a]	680 ± 16.7 [a]	112.6 ± 6.89 [a]	116.0 ± 6.89 [a]	50.8 ± 4.20 [a]	50.8 ± 4.20 [a]

[1] Values are means ± standard error of the mean (n = 5). Fisher's unprotected least significant differences were used to compare the treatment means within each sex. Two means that have no letter in common are statistically different at the 5% level.

2.2.8. Haematological and Serum Biochemical Data

The haematological data of blood samples collected on day 21 are presented in Table 4. For the haemoglobin level (HB), a small, but dose-dependent, decrease was seen for the male mice with the means of the mid STX and high STX treatment groups being statistically significantly different from that of the control group. However, this treatment group effect was less compelling for the female mice, with no statistically significant differences between the control group and any of the three STX treatment groups, making it unlikely that the differences seen in the male mice were of any significance. The mean white blood cell counts were highest in the control group for both genders, although it was only the values for male mice which showed a statistically significant difference. Furthermore, there was no STX dose-dependent effect for either gender, meaning that these results are unlikely to be toxicologically significant.

Table 4. Haematology data of mice fed control, low STX, mid STX or high STX diets (1.14, 2.28 and 3.48 µg STX.2HCl/g, respectively) for 21 days.

Item	Control [1]	Low STX [1]	Mid STX [1]	High STX [1]
Females				
HCT (L/L)	0.45 ± 0.01 [a]	0.47 ± 0.01 [a]	0.46 ± 0.01 [a]	0.44 ± 0.01 [a]
HB (g/L)	136 ± 2.9 [ab]	141 ± 2.9 [b]	138 ± 2.9 [ab]	132 ± 2.9 [a]
RBC ($\times10^{12}$/L)	8.29 ± 0.21 [a]	8.75 ± 0.21 [a]	8.61 ± 0.21 [a]	8.31 ± 0.21 [a]
MCV (fL)	54.5 ± 0.70 [a]	53.9 ± 0.70 [a]	53.9 ± 0.70 [a]	53.3 ± 0.70 [a]
MCH (pg)	16.4 ± 0.18 [a]	16.2 ± 0.18 [a]	16.0 ± 0.18 [a]	16.0 ± 0.18 [a]
MCHC (g/L)	303 ± 3.1 [a]	299 ± 3.1 [a]	297 ± 3.1 [a]	299 ± 3.1 [a]
WBC ($\times10^9$/L)	6.51 ± 0.82 [a]	5.51 ± 0.82 [a]	4.55 ± 0.82 [a]	5.25 ± 0.82 [a]
Neutrophil (%)	9.56 ± 2.73 [a]	11.24 ± 2.73 [a]	6.76 ± 2.73 [a]	10.56 ± 2.73 [a]
Lymphocyte (%)	87.3 ± 3.84 [a]	85.3 ± 3.84 [a]	91.9 ± 3.84 [a]	85.7 ± 3.84 [a]
Monocyte (%)	1.29 ± 0.95 [a]	2.71 ± 0.95 [a]	0.89 ± 0.95 [a]	1.89 ± 0.95 [a]
Eosinophil (%)	0.89 ± 0.586 [a]	0.91 ± 0.586 [a]	0.29 ± 0.586 [a]	1.29 ± 0.586 [a]
Males				
HCT (L/L)	0.46 ± 0.01 [a]	0.47 ± 0.01 [a]	0.46 ± 0.01 [a]	0.45 ± 0.01 [a]
HB (g/L)	137 ± 2.9 [b]	136 ± 2.9 [b]	135 ± 2.9 [ab]	128 ± 2.9 [a]
RBC ($\times10^{12}$/L)	8.58 ± 0.21 [a]	8.57 ± 0.21 [a]	8.57 ± 0.21 [a]	8.12 ± 0.21 [a]
MCV (fL)	54.0 ± 0.70 [a]	54.7 ± 0.70 [a]	53.5 ± 0.70 [a]	54.8 ± 0.70 [a]
MCH (pg)	16.0 ± 0.18 [a]	16.0 ± 0.18 [a]	16.0 ± 0.18 [a]	16.0 ± 0.18 [a]
MCHC (g/L)	295 ± 3.1 [a]	291 ± 3.1 [a]	294 ± 3.1 [a]	288 ± 3.1 [a]
WBC ($\times10^9$/L)	7.36 ± 0.82 [b]	4.51 ± 0.82 [a]	5.25 ± 0.82 [a]	5.62 ± 0.82 [ab]

Table 4. *Cont.*

Item	Control [1]	Low STX [1]	Mid STX [1]	High STX [1]
Neutrophil (%)	13.85 ± 2.73 [b]	13.44 ± 2.73 [b]	11.64 ± 2.73 [ab]	5.95 ± 2.73 [a]
Lymphocyte (%)	82.4 ± 3.84 [a]	84.5 ± 3.84 [a]	85.3 ± 3.84 [a]	90.9 ± 3.84 [a]
Monocyte (%)	2.40 ± 0.95 [a]	1.31 ± 0.95 [a]	1.71 ± 0.95 [a]	2.41 ± 0.95 [a]
Eosinophil (%)	1.00 ± 0.586 [a]	0.91 ± 0.586 [a]	1.51 ± 0.586 [a]	0.78 ± 0.586 [a]

[1] Values are means ± standard error of the mean (n = 5). Fisher's unprotected least significant differences were used to compare the treatment means within each sex. Two means that have no letter in common are statistically different at the 5% level. HCT, haematocrit value; HB, haemoglobin level; RBC, red blood cells; MCV, mean corpuscular volume; MCH, mean corpuscular haemoglobin; MCHC, mean corpuscular haemoglobin concentration; WBC, white blood cells.

The serum biochemical data of blood samples collected on day 21 are presented in Table 5. The aspartate aminotransferase (AST) and alanine aminotransferase (ALT) data were log transformed to stabilise the variance. For female mice, the log ALT is statistically significantly higher for the control than for any of the STX fed mice, but this effect did not show a dose-dependent trend. Furthermore, male mice fed a control diet had the lowest log ALT of any of the treatment groups meaning that this observation is highly unlikely to be of toxicological significance. Creatinine was also statistically significantly higher in the control female group in comparison to the treatment groups fed low STX and mid STX diets. However, no difference was seen between the female control and STX high dose mice, and no effect was seen in the male mice, pointing to random chance rather than any effect of treatment. For the male mice, the log AST was statistically higher in the STX high dose group in comparison to the control, but since the trend in the female mice was the opposite, this is unlikely to be of any importance.

Table 5. Serum biochemical data of mice fed control, low STX, mid STX or high STX diets (1.14, 2.28 and 3.48 µg STX.2HCl/g, respectively) for 21 days.

Item	Control [1]	Low STX [1]	Mid STX [1]	High STX [1]
Females				
log AST (log IU/L)	5.81 ± 0.33 [a]	5.27 ± 0.33 [a]	5.45 ± 0.33 [a]	5.46 ± 0.33 [a]
log ALT (log IU/L)	5.25 ± 0.35 [b]	3.65 ± 0.35 [a]	4.01 ± 0.35 [a]	3.85 ± 0.35 [a]
Urea (mmol/L)	8.96 ± 2.12 [a]	9.50 ± 2.12 [a]	8.94 ± 2.12 [a]	7.72 ± 2.12 [a]
TP (g/L)	47.0 ± 1.7 [a]	47.33 ± 1.7 [a]	49.07 ± 1.7 [a]	51.25 ± 1.7 [a]
ALB (g/L)	28.6 ± 1.03 [a]	28.0 ± 1.03 [a]	29.0 ± 1.03 [a]	30.6 ± 1.03 [a]
Globulin (g/L)	19.8 ± 0.87 [a]	19.5 ± 0.87 [a]	20.5 ± 0.87 [a]	21.3 ± 0.87 [a]
CRN (µmol/L)	10.4 ± 1.04 [b]	7.8 ± 1.04 [a]	7.2 ± 1.04 [a]	9.4 ± 1.04 [ab]
A/G ratio	1.39 ± 0.03 [a]	1.41 ± 0.03 [a]	1.42 ± 0.03 [a]	1.44 ± 0.03 [a]
Na (mmol/L)	151 ± 1.3 [a]	152 ± 1.3 [a]	152 ± 1.3 [a]	149 ± 1.3 [a]
K (mmol/L)	8.07 ± 0.71 [a]	7.81 ± 0.71 [a]	7.53 ± 0.71 [a]	8.35 ± 0.71 [a]
Cl (mmol/L)	114.0 ± 1.02 [a]	113.4 ± 1.02 [a]	114.6 ± 1.02 [a]	113.2 ± 1.02 [a]
Males				
log AST (log IU/L)	4.72 ± 0.33 [a]	5.27 ± 0.33 [ab]	5.05 ± 0.33 [ab]	5.75 ± 0.33 [b]
log ALT (log IU/L)	3.54 ± 0.35 [a]	3.58 ± 0.35 [a]	3.68 ± 0.35 [a]	3.79 ± 0.35 [a]
Urea (mmol/L)	9.08 ± 2.12 [a]	10.00 ± 2.12 [a]	9.86 ± 2.12 [a]	9.65 ± 2.12 [a]
TP (g/L)	50.7 ± 1.7 [a]	50.7 ± 1.7 [a]	51.0 ± 1.7 [a]	49.7 ± 1.7 [a]
ALB (g/L)	28.2 ± 1.03 [a]	28.6 ± 1.03 [a]	28.8 ± 1.03 [a]	28.2 ± 1.03 [a]
Globulin (g/L)	22.5 ± 0.87 [a]	22.7 ± 0.87 [a]	22.3 ± 0.87 [a]	21.7 ± 0.87 [a]
CRN (µmol/L)	7.0 ± 1.04 [a]	5.9 ± 1.04 [a]	6.1 ± 1.04 [a]	7.1 ± 1.04 [a]
A/G ratio	1.24 ± 0.03 [a]	1.26 ± 0.03 [a]	1.28 ± 0.03 [a]	1.30 ± 0.03 [a]
Na (mmol/L)	152 ± 1.3 [a]	151 ± 1.3 [a]	151 ± 1.3 [a]	152 ± 1.3 [a]
K (mmol/L)	8.17 ± 0.71 [a]	8.53 ± 0.71 [a]	9.43 ± 0.71 [a]	8.93 ± 0.71 [a]
Cl (mmol/L)	111.4 ± 1.02 [a]	113.0 ± 1.02 [a]	112.0 ± 1.02 [a]	114.0 ± 1.02 [a]

[1] Values are mean ± standard error of the mean (n = 5). Fisher's unprotected least significant differences were used to compare the treatment means within each sex. Two means that have no letter in common are statistically different at the 5% level. AST, aspartate aminotransferase; ALT, alanine aminotransferase; TP, total protein; ALB, albumin; CRN, creatinine.

2.2.9. Organ Weights

The organ weights of all mice, expressed as the percentage of bodyweight, are presented in Table 6. There was no strong statistical evidence of an overall treatment effect on relative organ weights. Neither the interaction between gender and treatment ($p \geq 0.065$) or the main effect of treatment ($p \geq 0.081$) was statistically significant at the 5% level for any of the organs expressed as percentage of bodyweight. When the pairwise comparisons within gender were examined (Table 6), some statistically significant differences between the means were observed. However, in each case, there was either no dose-dependency or the effect was only seen in one gender. It is therefore unlikely that any of these observations are of any toxicological significance.

Table 6. Organ weights, expressed as percentage of bodyweight, for mice fed control, low STX, mid STX or high STX diets (1.14, 2.28 and 3.48 µg STX.2HCl/g, respectively) for 21 days.

Item	Control [1]	Low STX [1]	Mid STX [1]	High STX [1]
Females				
brain	1.86 ± 0.09 [ab]	1.83 ± 0.09 [a]	1.96 ± 0.09 [ab]	2.01 ± 0.09 [b]
heart	0.58 ± 0.03 [a]	0.55 ± 0.03 [a]	0.60 ± 0.03 [a]	0.57 ± 0.03 [a]
kidneys	1.34 ± 0.07 [a]	1.24 ± 0.07 [a]	1.30 ± 0.07 [a]	1.30 ± 0.07 [a]
liver	4.59 ± 0.10 [a]	4.32 ± 0.10 [a]	4.41 ± 0.10 [a]	4.38 ± 0.10 [a]
spleen	0.39 ± 0.03 [a]	0.42 ± 0.03 [a]	0.37 ± 0.03 [a]	0.38 ± 0.03 [a]
Males				
brain	1.73 ± 0.09 [a]	1.72 ± 0.09 [a]	1.74 ± 0.09 [a]	1.83 ± 0.09 [a]
heart	0.57 ± 0.03 [a]	0.63 ± 0.03 [b]	0.57 ± 0.03 [ab]	0.55 ± 0.03 [a]
kidneys	1.72 ± 0.07 [b]	1.74 ± 0.07 [b]	1.66 ± 0.07 [ab]	1.59 ± 0.07 [a]
liver	4.74 ± 0.10 [a]	4.65 ± 0.10 [a]	4.69 ± 0.10 [a]	4.54 ± 0.10 [a]
spleen	0.37 ± 0.03 [b]	0.32 ± 0.03 [ab]	0.34 ± 0.03 [ab]	0.29 ± 0.03 [a]

[1] Values are mean ± standard error of the mean (n = 5). Fisher's unprotected least significant differences were used to compare the treatment means within each sex. Two means that have no letter in common are statistically different at the 5% level.

2.2.10. Histological Examination

No significant changes in tissues were visible on histological examination in mice from any of the treatment groups.

3. Discussion

In this experiment, using a meal time feeding protocol, male and female mice were fed daily dose rates of up to 699 and 730 µg STX.2HCl/kg bw/day, respectively, for 21 days, without developing any signs of toxicity (average of 715 µg STX.2HCl/kg bw/day). A small, but dose-dependent, effect of STX on the food intake of mice was detected. This may have been due to palatability of the diet to mice or it may have been due to post-digestional malaise. This will be explored further in a future experiment. None of the mice showed any behavioural or physiological effects, the serum biochemistry was normal, and in addition, there were no changes detected at histopathological examination of tissue samples. The NOAEL of STX.2HCl, administered using a meal time feeding protocol, is therefore greater than 715 µg STX.2HCl/kg bw. Taking into account the 100-fold safety factor to extrapolate from animal data to humans, this yields a human safe level equivalent of 7.2 µg STX.2HCl/kg. The dose rates used in this study far exceed the ARfD of 0.5 µg STX.2HCl/kg bw suggested by EFSA. To reach the human equivalent of the dose rates used in this study (7.2 µg STX.2HCl/kg bw), a 60 kg human would have to ingest 540 g of shellfish flesh contaminated with PSTs at the current regulatory limit (800 µg STX.2HCl/kg flesh). For a human weighing a more realistic 70 kg, this portion size increases to 630 g. The highest shellfish portion size proposed for use in risk assessment is that of 400 g by EFSA [7]. However, the FAO/IOC/WHO (2004) Committee noted that a portion size of 250 g would cover 97.5% of the consumers of most countries for which data were available [8].

Animal models, using the oral route of exposure, will reflect the potential human health risk of toxins in food. In this experiment, feeding was chosen as the method of oral administration rather than gavage. This was for a number of reasons. Tingling and numbness of the lips and mouth have been reported as amongst the first symptoms of human PSP cases which demonstrates that local absorption of the toxins through the buccal mucous membranes occurs with oral ingestion of these toxins [29]; this process would be bypassed by using gavage administration. Additionally, numerous studies have shown that gavage dosing overestimates the toxicity of shellfish toxins [21,30]. This is likely due to the consistency of mouse stomach contents as, unlike humans, the stomach contents of rodents are semi-solid which could allow part of the liquid gavage dose to flow around them to be rapidly absorbed by the duodenum [19,31]. In contrast, if the test compound is incorporated with a solid matrix, it mixes with the existing stomach contents. Furthermore, a study by Craig and Elliott [32] using radiolabelled protein showed that 38% of mice dosed by gavage (27/71) did not receive their liquid dose correctly. Mice which had received an incorrect gavage dose showed no changes in behaviour meaning that they could not be distinguished from those which had been successfully dosed. This was attributed to occasional spillover/regurgitation of the dosing material into the lungs and led the authors to conclude that "the common method of gavage feeding mice to assess absorption of orally ingested material can lead to artifacts not seen when the same agent is consumed under more natural circumstances" [32].

As an alternative to gavage, we incorporated STX.2HCl into the diet of mice. Although laced cream cheese has been used successfully in acute toxicity testing [27], this was not suitable for a feeding study as, over time, mice tend to start refusing to eat the dose, an effect which would have a major impact on the study. As an alternative, STX was incorporated into the normal diet of mice, a method which worked well. Mice with unlimited access to STX-containing diet ingested a dose rate of up to 620 µg STX.2HCl/kg bw/day for four days without any symptoms of toxicity. Given that the oral LD_{50} for STX.2HCl in mice is 1063 µg/kg bw and the NOAEL is 473 µg/kg bw [21], this lack of toxicity was surprising. However, a one-off bolus dose, such as that used in the acute toxicity work, is a vastly different scenario to the constant exposure to small quantities of STX over a long period of time as would occur with the unrestricted feeding protocol. The feeding behaviour of mice is one of many bouts of feeding (grazing), with 70% of food consumed in the dark and 30% during the light [33,34]. This pattern is totally different to human feeding behaviour which is that of meal times and which would result in the ingestion of STX over a much shorter time period. To replicate this scenario, a feeding protocol was developed whereby mice were fed for two 1 h feeding periods per day. Mice quickly adapted to this change in feeding regime and after a training period of three days their daily food intake was normal. Preliminary testing of STX.2HCl using this protocol also showed a surprising lack of toxicity. It has been suggested by other authors [35] that the ingestion of PSTs mixed with 150 mg of cream cheese may influence toxicity due to the high fat content of cream cheese. To ensure that the lack of observed toxicity with feeding was not simply a matrix effect (cream cheese versus mouse food), one mouse was fed meals of diet laced with STX, while another was fed meals of control diet along with STX-laced cream cheese. Both methods delivered the same high dose rate of STX.2HCl (400 µg/kg bw twice a day) and both showed no symptoms of toxicity. In contrast, a mouse with unrestricted access to control diet developed severe symptoms of toxicity after just one bolus dose of 400 µg STX.2HCl/kg bw in cream cheese. This difference in toxicity must therefore be caused by the different quantity of food in the stomach of mice at the time of STX ingestion. With unlimited access to food, the stomach of a mouse would be relatively empty in the morning when the STX-laced cream cheese was consumed as a bolus dose. In contrast, the mice fed meals of STX-laced diet consumed half of their daily food intake along with the STX, resulting in the stomach containing considerable solid material. The quantity of matrix (food) in the stomach of mice is therefore impacting the absorption of STX and thus influencing toxicity. Relating this to a human scenario, STX in pure form cannot be eaten, meaning that it is not possible to ingest STX on an empty stomach. Even the standard portion size of 100 g determined by EFSA would result in the ingestion of a significant amount of matrix (food) along with the STX. In this experiment, using the meal time feeding protocol, mice were fed daily dose rates of up to 730 µg STX.2HCl/kg bw/day for 21 days without developing any signs of toxicity.

The high dose rate used in our feeding study, which induced no adverse effects, is at odds with a study by Arnich and Thébault [36] who, on the basis of the quantitative modelling of human PSP cases, proposed a lowest observed adverse effect level (LOAEL) of 0.37 µg STX.2HCl eq/kg bw. After applying the 100-fold safety factor for extrapolation from animals to human, the NOAEL determined in our feeding study was almost 20 times lower than this figure. However, as discussed earlier, the STX.2HCl concentrations in leftover food, the quantity of shellfish eaten and the bodyweight of the human are required to effectively interpret PSP cases. The Arnich and Thébault [36] model was based on 13 outbreaks of PSP which were often missing some of these crucial pieces of information. Furthermore, in all but one of the outbreaks, the PSP toxin analysis was conducted by mouse bioassay (MBA) [37]. This MBA was developed in the 1930s, whereby the relationship between the i.p. dose of pure STX.2HCl and the death time of mice was established to yield a dose-death time curve [38]. This curve was then used to convert the death time of mice injected with shellfish samples into STX.2HCl equivalents. However, PSTs are comprised of a large number of analogues and it has been demonstrated that the different analogues can have a dose-death time curve which is not consistent with that of STX [28]. The MBA approach therefore does not accurately determine the concentrations of PSTs. With the advanced analytical tools now available, along with the push for better information to be collected from patients presenting at hospitals with PSP, better data to input into the model should be available in the future. Another study which appears to contradict our results is that of Boente-Juncal et al. [39] who administered combinations of STX and tetrodotoxin (TTX) by gavage for 28 days. The dose rates used were very low, 44 µg TTX/kg bw along with 5.3, 17 or 54 µg STX.2HCl/kg bw. Surprisingly, one out of the four mice dosed at the lowest dose and two out of five mice dosed at the highest dose rate died. Each of these deaths were described as sudden convulsions and rapid death which perhaps could be attributable to the inherent issues with gavage dosing. Given the mouse deaths, symptoms of toxicity would have been expected in the surviving mice, but all survivors showed no symptoms of toxicity and gained weight normally. The mechanism of action for both STX and TTX is via voltage-gated sodium channels and the toxicities of the two toxins are additive when administered orally [21], resulting in the recommendation that TTX should be added to the PSTs. For this reason, it is valid to add the STX and TTX dose rates given in the Boente-Juncal et al. (2020) study [39] to allow a comparison with those administered in our study. The combined STX and TTX dose rates administered by gavage were 56.6, 68.3 and 105.3 µg STX.2HCl eq/kg bw. In comparison, the STX.2HCl dose rate administered to mice using the meal time feeding protocol in our study was 715 µg STX.2HCl eq/kg bw, seven times higher and with no adverse effects observed.

This feeding study showed no evidence of a cumulative effect of STX in mice, and on the basis of the high dose rates of STX.2HCl given using the meal time feeding protocol, designed to replicate human feeding behaviour, the current regulatory limit for PSTs seems adequate to protect human health. For the determination of this regulatory limit, the PSP cases considered by EFSA were all from the ingestion of non-commercial shellfish samples. Regulation of commercial samples has led to rigorous routine monitoring programmes and it was recognized by the working group convened to "assess the advice from the joint FAO/WHO/IOC Ad Hoc expert consultation on biotoxins and bivalve molluscs" that using the current limit of 800 µg STX.2HCl eq/kg shellfish flesh has resulted in no human illnesses from commercially harvested shellfish for 50 years [11]. This observation, as in our experiment, indicates that the current regulatory level for PSTs is fit for purpose. Since no toxicity was observed in the study, the true NOAEL exceeds the highest dose administered. To determine the true NOAEL, further work, using higher dose rates, would be required. This study highlighted that the feeding regime has a big influence on the toxicity of STX.2HCl. Further work is planned to better define this difference as it is possible that the current methods to determine acute oral toxicity for use in the setting of regulatory limits is overestimating the risk of these toxins to human health. It would be interesting to investigate whether the impact of feeding regime on toxicity is confined to just STX or whether it also occurs with other toxin classes.

4. Materials and Methods

4.1. Purity Assessment of Saxitoxin

STX was supplied by the Cawthron Institute (Nelson, NZ) and calibrated against certified reference material from the National Research Council of Canada (NRC) using HPLC-UV at 210 nm and a method adapted from Rourke et al. [40]. Ion pairing chromatography was performed on a Zorbax Bonus RP 4.6 × 150 mm 3.5 µm column under isocratic conditions over 30 min at 1 mL/min with 11 mM heptane sulfonate, 16.5 mM phosphoric acid 11.5% acetonitrile adjusted to pH 7.1 with ammonium hydroxide. Additionally, trace concentrations of other PSTs were quantified using LC-MS/MS [41], showing the material to contain 99.8% STX, 0.16% decarbamoylSTX and 0.05% neoSTX. The STX material was stored at 4 °C as a stock solution in 3 mM HCl at a concentration of 10.91 mg/mL STX.2HCl. For preparation of mouse diet, working solutions were prepared gravimetrically using 3 mM HCl as the diluting solvent.

4.2. Animals

Swiss albino mice were used for all experimental work. Females were used for the preliminary studies; however, for the feeding study, mice of both genders were used. Mice were individually caged in a temperature-controlled room (21 ± 1 °C) with a 12-h light–dark cycle and with unrestricted access to water. During the feeding study, a housing arrangement was used where the boxes, each containing an individual mouse, were randomised for columns and rows such that the eight combinations of treatment group and gender occurred exactly once along each row (forming the experimental replicate) and no more than once down each column.

4.3. Preparation of Mouse Diets

Mouse diets were based on Teklad Global 2016 mouse food pellets (Harlan UK, Bicester, UK) ground to a fine flour using a cyclone sample mill (Udy Corporation, Fort Collins, CO, USA). To check for STX homogeneity in the diet, a test batch was prepared by mixing ground mouse food (50 g) with water (45 mL) containing STX.2HCl (13.5 µg). This mixture was well combined and seven cookie-shaped portions of diet (approximately 2.5 × 2.5 × 1 cm) were prepared and dried in a fan oven (50 °C for 24 h). After the drying process had been completed, three different areas of one portion of the STX-laced diet were taken and ground using a mortar and pestle. The STX concentration was measured by the analysis of duplicate samples by LC-MS/MS. Since analysis showed that homogeneity was a problem in this diet, a further test batch of STX-laced diet was prepared by taking ground mouse food (25 g) and mixing this with water (20 mL) containing STX.2HCl (87.5 µg). This mixture was well-combined and 25 small cookie-shaped portions of diet (1.5 cm diameter and 0.6 cm thickness) were prepared and dried in a fan oven at a lower temperature and for a shorter time (30 °C for 16 h) compared to the initial batch. The resulting STX-laced diet was extracted and analysed (as detailed in Section 4.4), which showed that the homogeneity issue had been resolved. To check the stability of STX, portions of the laced diet were stored at both 4 °C and 20 °C. Two samples of the stored diet were taken on days 1, 2 and 5, and processed as detailed in Section 4.4. Each sample was analysed in triplicate.

During the feeding trial, 50 g batches of STX-laced diet were prepared as above, resulting in 50 small portions of diet. Each batch was weighed at the end of the drying process so that the moisture contents could be calculated. The average moisture content was 14.1% for the control diet, 15.4% for the low dose STX diet, 13.3% for the mid dose

STX diet and 14.2% for the high dose STX diet. To determine the concentrations of the three different STX diets used in the feeding trial, two samples of each treatment diet were ground using a mortar and pestle and then analysed in quadruplicate using the methods described in Section 4.4. Prepared diets were kept at 4 °C until use and were prepared at least every three days during the feeding study.

4.4. LC-MS Analysis

To determine the STX concentration in laced diet, samples were taken (200 ± 5 mg) and extracted with 0.1 M HCl (10 mL) by placing them in a boiling water bath (5 min). The extracts were then cooled in an ice bath (5 min), briefly mixed on a vortex mixer and then centrifuged (17,000× g for 5 min). An aliquot of the resulting extract (10 µL) was added to 80% acetonitrile with 0.25% acetic acid (490 µL) in a polypropylene autosampler vial and analysed by UHPLC-MS/MS based on the method of Turner et al. [42], using a 6500+ QTRAP tandem quadrupole mass spectrometer coupled to an Exion LC liquid chromatography separations system, with high pH compatibility conversion composed of a multiplate autosampler, binary solvent pumps with low pressure gradient proportioning valve solvent selection, and a column oven with 6-port 2-position column selection valve. Curtain gas was 25 psi, ion source gas 1 was 50 psi, ion source gas 2 was 50 psi, ionspray voltage was −4500 in negative mode, and 5500 V in positive mode, the temperature was 500 °C, and declustering potential was −30 V in negative mode and 30 V in positive mode. Scheduled MRM mode was used for data acquisition with STX monitored in positive ion mode with MRM transitions 300.141 > 204.088 (CE 34 V) and 300.141 > 138.066 (CE 36 V).

4.5. Preliminary Work and Palatability Studies

To determine an appropriate STX dose rate, pairs of mice were housed together and given unrestricted access to STX-laced diet. Food consumption and bodyweight were measured daily to allow the calculation of the STX dose rate ingested. Initially, a pair of mice was fed a diet containing 2.43 µg STX.2HCl/g for 14 days, which delivered an average dose rate of 476 µg STX.2HCl/kg bw/day. In a further trial, a pair of mice was fed a diet containing 2.95 µg STX.2HCl/g for four days, which delivered an average dose rate of 620 µg STX.2HCl/kg bw/day.

To develop experimental protocols to replicate human feeding behaviour, one individually caged mouse was given unrestricted access to control diet and another was fed the same diet between only 8–10 a.m. and 3–5 p.m. for eight days. Daily food intake and bodyweight were measured to allow g food consumed/g bw to be calculated.

To ensure that the administration of STX in the laced diet was not influencing toxicity, three weanling mice were taken and individually caged. Mouse 1 had unrestricted access to control diet and was trained to eat small quantities of cream cheese following a method previously described [21]. Mouse 2 was trained in the meal time feeding protocol using control diet as well as being trained to eat small quantities of cream cheese, and mouse 3 was trained in the meal time feeding protocol using control diet. After four days of training, mouse 1 continued to have unrestricted access to control diet but was fed cream cheese (150 mg) laced with STX. Mouse 2 was fed control diet at meal times along with cream cheese (150 mg) laced with STX half-way through each of its two daily feeding periods, and mouse 3 was fed STX-laced diet at meal times. Mouse 1 ingested 400 µg STX.2HCl/kg bw as a one-off bolus dose, and both mouse 2 and 3 ingested 400 µg STX.2HCl/kg bw at each of their twice daily feeding periods. The mice were observed closely for any signs of toxicity throughout the day.

4.6. Sub-Acute 21-Day Feeding Trial

Sixty weanling mice (30 female and 30 male) were individually caged and trained in the meal time feeding protocol. To allow a more gradual adjustment to the change in feeding regime, mice were given access to food between 9 a.m. and 5 p.m. for one day, and then for the following three days were fed between 9–10 a.m. and 3.30–4.30 p.m.

This period of training was completed with control diet. On each day of the training period, the amount of food consumed by each mouse at each meal time was measured. On each of the two days before the start of the study, mice were trained on an accelerating rotarod (Rotamex 4/8, Columbus Instruments International, Columbus, OH, USA) by placing them on the accelerating rod (13–79 rpm over 12 min) and recording the time to fall. Each mouse was given two attempts per day. Mice which did not adapt to the meal time feeding protocol or those which performed poorly during training on the accelerating rotarod were excluded from the study. From those mice remaining, 20 females and 20 males were selected and randomly assigned to treatment groups. Four treatment groups, each containing five female and five male mice (individually caged) were fed between 9–10 a.m. and 3.30–4.30 p.m. for 21 days with the following diets. Group 1—control diet, Group 2—low-dose STX-laced diet (1.14 µg STX.2HCl/g), Group 3—mid-dose STX-laced diet (2.28 µg STX.2HCl/g), and Group 4—high-dose STX-laced diet (3.39 µg STX.2HCl/g). All animals had unrestricted access to water and food consumption was measured after each twice daily meal time. In addition, after each daily afternoon feed, the bodyweight of each mouse was recorded and posture/appearance noted. On days 14 and 21, heart rate and blood pressure was determined for each mouse using a Visitech Systems BP-2000 blood pressure analysis system (Visitech Systems, Apex, NC, USA). On days 0, 7, 14 and 21, motor coordination was evaluated using the accelerating rotarod as described above and grip strength was measured using a grip strength meter (MK3805S, Muromachi, Tokyo, Japan). On each measurement day, the mice were given two trials on the rotarod and grip strength meter, with the results being averaged. At the end of the 21-day feeding period, mice were euthanised by CO_2 inhalation and blood samples collected by heart puncture with heparin as the anticoagulant. Haematocrit values (HCT), haemoglobin levels (HB), mean corpuscular volumes (MCV), mean corpuscular haemoglobin (MCH), mean corpuscular haemoglobin concentrations (MCHC), and red and white blood cell counts were measured in whole blood. In addition, plasma was analysed for activities of aspartate aminotransferase (AST), alanine aminotransferase (ALT), and for levels of urea, total protein (TP), albumin (ALB), globulin, sodium, potassium, chloride and creatinine (CRN) (IDEXX laboratories, Hamilton, NZ). At necropsy, any macroscopic changes observed were recorded and the weights of brain, heart, kidneys, liver and spleen measured and expressed as percentages of bodyweight. These tissues, together with adrenals, lungs, pancreas, gastrocnemius, jejunum (3 mm section), ovary/uterus or testes, spinal cord (3×2 mm sections), stomach (washed), thymus and urinary bladder were fixed in 4% buffered formaldehyde and routinely processed for histological examination. All samples were assessed by the same pathologist who was blinded to the treatment group of the samples.

4.7. Statistical Analysis

The bodyweight, food consumption, motor coordination, grip strength, blood pressure and heart rate data were analysed using repeated measures formulated as linear mixed effects models and fitted using residual maximum likelihood (REML). The random model comprised of random effects for replicate (i.e., the row in the housing arrangement) and mouse. The basic fixed model comprised of effects for treatment group, day, gender and all two- and three-way interactions. In the analyses of bodyweight and food consumption date, pre-treatment (i.e., day 0) bodyweight was also included as a fixed effect covariate, and in the analyses of motor coordination and grip strength, the pre-treatment measurement on day 0 was included as a fixed effect covariate. Repeated measures on the same mouse over time were assumed to be correlated with a first-order autoregressive structure.

The haematological, serum biochemical and organ weight data were analysed using linear mixed effects models fitted using REML with random effects for replicate and fixed effects for kill day, treatment group, gender and the treatment group by gender interaction. Serum AST and ALT were log transformed to stabilise the variance.

In all analyses, the variance components were constrained to be positive, residual diagnostic plots were inspected for evidence of departures from the residual assumptions of normality and constant variance, a 5% significance level was used when assessing the fixed effects, and Fisher's unprotected least significant differences at the 5% level (LSD (5%)) were used to compare means. Statistical analyses were performed using Genstat 19th Edition (VSN International, Hemel Hempstead, UK).

Author Contributions: Conceptualization, S.C.F., D.T.H., M.J.B. and J.N.; formal analysis, V.M.C.; data curation, S.C.F., N.G.W., J.M.S., R.B.B., J.S.M. and M.J.B.; writing—original draft preparation, S.C.F., M.J.B. and V.M.C.; writing—review and editing, All authors have read and agreed to the published version of the manuscript.

Funding: This work was funded by the New Zealand Ministry for Business, Innovation and Employment—Seafood Safety research programme (Contract CAWX1801).

Institutional Review Board Statement: All animal manipulations were approved by the Ruakura Animal Ethics Committee established under the Animal Protection (code of ethical conduct) Regulations Act, 1987 (New Zealand). The project number for this study was 14673 and the approval date was 10 February 2019.

Data Availability Statement: The data presented in this study are available on request from the corresponding author.

Acknowledgments: We thank Bobby Smith and Genevieve Sheriff, AgResearch Ltd., Ruakura Science Centre, for the care of animals.

Conflicts of Interest: The authors declare no conflict of interest.

References

1. Anderson, D.M.; Alpermann, T.J.; Cembella, A.D.; Collos, Y.; Masseret, E.; Montresor, M. The globally distributed genus *Alexandrium*: Multifaceted roles in marine ecosystems and impacts on human health. *Harmful Algae* **2012**, *14*, 10–35. [CrossRef]
2. Oshima, Y.; Blackburn, S.I.; Hallegraeff, G.M. Comparative study on paralytic shellfish toxin profiles of the dinoflagellate *Gymnodinium catenatum* from three different countries. *Mar. Biol.* **1993**, *116*, 471–476. [CrossRef]
3. Usup, G.; Kulis, D.M.; Anderson, D.M. Growth and toxin production of the toxic dinoflagellate *Pyrodinium bahamense* var. *compressum* in laboratory cultures. *Nat. Toxins* **1994**, *2*, 254–262. [CrossRef]
4. Wiese, M.; D'Agostino, P.M.; Mihali, T.K.; Moffitt, M.C.; Neilan, B.A. Neurotoxic alkaloids: Saxitoxin and its analogs. *Mar. Drugs* **2010**, *8*, 2185–2211. [CrossRef]
5. García, C.; Lagos, M.; Truan, D.; Lattes, K.; Véjar, O.; Chamorro, B.; Iglesias, V.; Andrinolo, D.; Lagos, N. Human intoxication with paralytic shellfish toxins: Clinical parameters and toxin analysis in plasma and urine. *Biol. Res.* **2005**, *38*, 197–205. [CrossRef]
6. Mons, M.N.; Van Egmond, H.P.; Speijers, G.J.A. *Paralytic Shellfish Poisoning: A Review*; RIVM Report 388802 005; National Institute of Public Health and the Environment: Bilthoven, The Netherlands, 1988.
7. EFSA. Scientific opinion of the panel on contaminants in the food chain on a request from the european commission on marine biotoxins in shellfish—Saxitoxin group. *EFSA J.* **2009**, *1019*, 1–76.
8. FAO/IOC/WHO. *Report of the Joint FAO/IOC/WHO Ad Hoc Expert Consultation on Biotoxins in Bivalve Molluscs*; Food and Agriculture Organization of the United Nations: Oslo, Norway, 2004; p. 31.
9. Gessner, B.D.; Middaugh, J.P. Paralytic shellfish poisoning in Alaska: A 20-year retrospective analysis. *Am. J. Epidemiol.* **1995**, *141*, 766–770. [CrossRef]
10. European Union. Regulation (EC) No 853/2004 of the European Parliament and of the Council of 29 April 2004, Laying down specific hygiene rules for food of animal origin. *Off. J. Eur. Union* **2004**, *L 139/55*, 151.
11. Yasumoto, T.; Murata, M.; Oshima, Y.; Sano, M.; Matsumoto, G.K.; Clardy, J. Diarrhetic shellfish toxins. *Tetrahedron* **1985**, *41*, 1019–1025. [CrossRef]
12. Bricelj, V.M.; Shumway, S.E. Paralytic shellfish toxins in bivalve molluscs: Occurrence, transfer kinetics, and biotransformation. *Rev. Fish. Sci.* **1998**, *6*, 315–383. [CrossRef]
13. Turnbull, A.R.; Harwood, D.T.; Boundy, M.J.; Holland, P.T.; Hallegraeff, G.; Malhi, N.; Quilliam, M.A. Paralytic shellfish toxins —Call for uniform reporting units. *Toxicon* **2020**, *178*, 59–60. [CrossRef] [PubMed]
14. Tennant, A.D.; Naubert, J.; Corbeil, H.E. An outbreak of paralytic shellfish poisoning. *Can. Med. Assoc. J.* **1955**, *72*, 436–439. [PubMed]
15. Wong, C.-K.; Hung, P.; Lee, K.L.H.; Mok, T.; Kam, K.-M. Effect of steam cooking on distribution of paralytic shellfish toxins in different tissue compartments of scallops Patinopecten yessoensis. *Food Chem.* **2009**, *114*, 72–80. [CrossRef]
16. Medcof, J.C.; Leim, A.H.; Needler, A.B.; Needler, A.W.H.; Gibbard, J.; Naubert, J. *Paralytic Shellfish Poisoning on the Canadian*; Atlantic coast, Bulletin No. 75; Fisheries Research Board of Canada: Ottawa, ON, USA, 1947; p. 31.
17. EFSA Marine biotoxins in shellfish—Summary on regulated marine biotoxins. *EFSA J.* **2009**, *7*, 1306.

18. EFSA Scientific Committee. Guidance on selected default values to be used by the EFSA Scientific Committee, Scientific Panels and Units in the absence of actual measured data. *EFSA J.* **2012**, *10*, 2579.

19. FAO/WHO. *Technical Paper on Toxicity Equivalency Factors for Marine Biotoxins Associated with Bivalve Molluscs*; FAO: Rome, Italy, 2016; 108p.

20. Interdepartmental Group on Health Risks from Chemicals. *Uncertainty Factors: Their Use in Human Health Risk Assessment by UK Government*; MRC Institute for Environment and Health: Leicester, UK, 2003; p. 69.

21. Finch, S.C.; Boundy, M.J.; Harwood, D.T. The acute toxicity of tetrodotoxin and tetrodotoxin–saxitoxin mixtures to mice by various routes of administration. *Toxins* **2018**, *10*, 423. [CrossRef]

22. Harwood, D.T.; Boundy, M.; Selwood, A.I.; van Ginkel, R.; MacKenzie, L.; McNabb, P.S. Refinement and implementation of the Lawrence method (AOAC 2005.06) in a commercial laboratory: Assay performance during an *Alexandrium catenella* bloom event. *Harmful Algae* **2013**, *24*, 20–31. [CrossRef]

23. MacKenzie, A.L. The risk to New Zealand shellfish aquaculture from paralytic shellfish poisoning (PSP) toxins. *N. Z. J. Mar. Freshw. Res.* **2014**, *48*, 430–465. [CrossRef]

24. Turner, A.D.; Stubbs, B.; Coates, L.; Dhanji-Rapkova, M.; Hatfield, R.G.; Lewis, A.M.; Rowland-Pilgrim, S.; O'Neil, A.; Stubbs, P.; Ross, S.; et al. Variability of paralytic shellfish toxin occurrence and profiles in bivalve molluscs from Great Britain from official control monitoring as determined by pre-column oxidation liquid chromatography and implications for applying immunochemical tests. *Harmful Algae* **2014**, *31*, 87–99. [CrossRef] [PubMed]

25. Finch, S.C.; Munday, J.S.; Munday, R.; Kerby, J.W.F. Short-term toxicity studies of loline alkaloids in mice. *Food Chem. Toxicol.* **2016**, *94*, 243–249. [CrossRef]

26. Finch, S.C.; Munday, J.S.; Sprosen, J.M.; Bhattarai, S. Toxicity studies of chanoclavine in mice. *Toxins* **2019**, *11*, 249. [CrossRef]

27. Munday, R. Toxicology of Seafood Toxins: A Critical Review. In *Seafood and Freshwater Toxins: Pharmacology, Physiology, and Detection*, 3rd ed.; Botana, L.M., Ed.; CRC Press: Boca Raton, FL, USA, 2014; pp. 197–290.

28. Munday, R.; Thomas, K.; Gibbs, R.; Murphy, C.; Quilliam, M.A. Acute toxicities of saxitoxin, neosaxitoxin, decarbamoyl saxitoxin and gonyautoxins 1&4 and 2&3 to mice by various routes of administration. *Toxicon* **2013**, *76*, 77–83. [PubMed]

29. Kao, C.Y. Paralytic shellfish poisoning. In *Algal Toxins in Seafood and Drinking Water*; Falconer, I.R., Ed.; Academic Press: London, UK; San Diego, CA, USA, 1993; pp. 75–86.

30. Selwood, A.I.; Waugh, C.; Harwood, D.T.; Rhodes, L.L.; Reeve, J.; Sim, J.; Munday, R. Acute toxicities of the saxitoxin congeners gonyautoxin 5, gonyautoxin 6, decarbamoyl donyautoxin 2&3, decarbamoyl neosaxitoxin, C-1&2 and C-3&4 to mice by various routes of administration. *Toxins* **2017**, *9*, 73. [CrossRef]

31. Munday, R.; Reeve, J. Risk assessment of shellfish toxins. *Toxins* **2013**, *5*, 2109–2137. [CrossRef]

32. Craig, M.A.; Elliott, J.F. Mice fed radiolabeled protein by gavage show sporadic passage of large quantities of intact material into the blood, an artifact not associated with voluntary feeding. *J. Am. Assoc. Lab. Anim. Sci.* **1999**, *38*, 18–23.

33. Ellacott, K.L.; Morton, G.J.; Woods, S.C.; Tso, P.; Schwartz, M.W. Assessment of feeding behavior in laboratory mice. *Cell Metab.* **2010**, *12*, 10–17. [CrossRef]

34. Minematsu, S.; Hiruta, M.; Taki, M.; Fujii, Y.; Aburada, M. Automatic monitoring system for the measurement of body weight, food and water consumption and spontaneous activity of a mouse. *J. Toxicol. Sci.* **1991**, *16*, 61–73. [CrossRef]

35. Boente-Juncal, A.; Vale, C.; Cifuentes, M.; Otero, P.; Camiña, M.; Rodriguez-Vieytes, M.; Botana, L.M. Chronic in vivo effects of repeated exposure to low oral doses of tetrodotoxin: Preliminary evidence of nephrotoxicity and cardiotoxicity. *Toxins* **2019**, *11*, 96. [CrossRef]

36. Arnich, N.; Thébault, A. Dose-Response modelling of paralytic shellfish poisoning (PSP) in humans. *Toxins* **2018**, *10*, 141. [CrossRef] [PubMed]

37. Anon. AOAC Official Method 959.08. Paralytic shell-fish poisoning. Biological method. In *Official Methods of Analysis of AOAC International*, 18th ed.; Horwitz, W., Latimer, G.W., Eds.; AOAC International: Gaithersburg, MD, USA, 2005; pp. 79–82.

38. Sommer, H.; Meyer, K.F. Paralytic shell-fish poisoning. *Arch. Pathol.* **1937**, *24*, 560–598.

39. Boente-Juncal, A.; Otero, P.; Rodríguez, I.; Camiña, M.; Rodriguez-Vieytes, M.; Vale, C.; Botana, L.M. Oral chronic toxicity of the safe tetrodotoxin dose proposed by the European Food Safety Authority and its additive effect with saxitoxin. *Toxins* **2020**, *12*, 312. [CrossRef] [PubMed]

40. Rourke, W.A.; Murphy, C.J.; Pitcher, G.; van de Riet, J.M.; Burns, B.G.; Thomas, K.M.; Quilliam, M.A. Rapid postcolumn methodology for determination of paralytic shellfish toxins in shellfish tissue. *J. AOAC Int.* **2019**, *91*, 589–597. [CrossRef]

41. Boundy, M.J.; Selwood, A.I.; Harwood, D.T.; McNabb, P.S.; Turner, A.D. Development of a sensitive and selective liquid chromatography–mass spectrometry method for high throughput analysis of paralytic shellfish toxins using graphitised carbon solid phase extraction. *J. Chromatogr. A* **2015**, *1387*, 1–12. [CrossRef]

42. Turner, A.D.; Dhanji-Rapkova, M.; Fong, S.Y.T.; Hungerford, J.; McNabb, P.S.; Boundy, M.J.; Harwood, D.T. Ultrahigh-performance hydrophilic interaction liquid chromatography with tandem mass spectrometry method for the determination of paralytic shellfish toxins and tetrodotoxin in mussels, oysters, clams, cockles, and scallops: Collaborative study. *J. AOAC Int.* **2020**, *103*, 533–562. [CrossRef]

Article

A Comparative Analysis of Methods (LC-MS/MS, LC-MS and Rapid Test Kits) for the Determination of Diarrhetic Shellfish Toxins in Oysters, Mussels and Pipis

Penelope A. Ajani [1,*], Chowdhury Sarowar [2], Alison Turnbull [3], Hazel Farrell [4], Anthony Zammit [4], Stuart Helleren [5], Gustaaf Hallegraeff [3] and Shauna A. Murray [1]

1. School of Life Sciences, University of Technology Sydney, P.O. Box 123, Broadway, NSW 2007, Australia; Shauna.Murray@uts.edu.au
2. Sydney Institute of Marine Science, 19 Chowder Bay Road, Mosman, NSW 2088, Australia; Chowdhury.Sarowar@sims.org.au
3. Institute for Marine and Antarctic Science, University of Tasmania, 15-21 Nubeena Crescent, Taroona, TAS 7053, Australia; alison.turnbull@utas.edu.au (A.T.); gustaaf.hallegraeff@utas.edu.au (G.H.)
4. NSW Food Authority, NSW Department of Primary Industries, P.O. Box 232, Taree, NSW 2430, Australia; Hazel.Farrell@dpi.nsw.gov.au (H.F.); Anthony.Zammit@dpi.nsw.gov.au (A.Z.)
5. Dalcon Environmental, Building 38, 3 Baron-Hay Ct, South Perth, WA 6151, Australia; stuart.helleren@dalconenvironmental.com.au
* Correspondence: Penelope.Ajani@uts.edu.au

Citation: Ajani, P.A.; Sarowar, C.; Turnbull, A.; Farrell, H.; Zammit, A.; Helleren, S.; Hallegraeff, G.; Murray, S.A. A Comparative Analysis of Methods (LC-MS/MS, LC-MS and Rapid Test Kits) for the Determination of Diarrhetic Shellfish Toxins in Oysters, Mussels and Pipis. *Toxins* **2021**, *13*, 563. https://doi.org/10.3390/toxins13080563

Received: 8 July 2021
Accepted: 9 August 2021
Published: 11 August 2021

Publisher's Note: MDPI stays neutral with regard to jurisdictional claims in published maps and institutional affiliations.

Abstract: Rapid methods for the detection of biotoxins in shellfish can assist the seafood industry and safeguard public health. Diarrhetic Shellfish Toxins (DSTs) are produced by species of the dinoflagellate genus *Dinophysis*, yet the comparative efficacy of their detection methods has not been systematically determined. Here, we examined DSTs in spiked and naturally contaminated shellfish–Sydney Rock Oysters (*Saccostrea glomerata*), Pacific Oysters (*Magallana gigas*/*Crassostrea gigas*), Blue Mussels (*Mytilus galloprovincialis*) and Pipis (*Plebidonax deltoides*/*Donax deltoides*), using LC-MS/MS and LC-MS in 4 laboratories, and 5 rapid test kits (quantitative Enzyme-Linked Immunosorbent Assay (ELISA) and Protein Phosphatase Inhibition Assay (PP2A), and qualitative Lateral Flow Assay (LFA)). We found all toxins in all species could be recovered by all laboratories using LC-MS/MS (Liquid Chromatography—tandem Mass Spectrometry) and LC-MS (Liquid Chromatography—Mass Spectrometry); however, DST recovery at low and mid-level concentrations (<0.1 mg/kg) was variable (0–150%), while recovery at high-level concentrations (>0.86 mg/kg) was higher (60–262%). While no clear differences were observed between shellfish, all kits delivered an unacceptably high level (25–100%) of falsely compliant results for spiked samples. The LFA and the PP2A kits performed satisfactorily for naturally contaminated pipis (0%, 5% falsely compliant, respectively). There were correlations between spiked DSTs and quantitative methods was highest for LC-MS (r^2 = 0.86) and the PP2A kit (r^2 = 0.72). Overall, our results do not support the use of any DST rapid test kit as a stand-alone quality assurance measure at this time.

Keywords: LC-MS; rapid test kit; biotoxins; shellfish; diarrhetic shellfish toxins; *Dinophysis*

Key Contribution: LC-MS continues to be a reliable DST detection method across labs and shellfish species; with low to mid-level toxin concentration recovery more variable than high-level. All rapid test kits delivered unacceptably high, falsely compliant results for spiked samples. The Neogen and the PP2A kits performed satisfactorily for naturally contaminated pipis. Overall, our results do not support the use of any DST rapid test kit as a stand-alone quality assurance measure at this time.

1. Introduction

Marine biotoxins are toxic chemical compounds produced by certain microalgae, which can bioaccumulate in shellfish and other marine organisms, and cause poisoning to seafood consumers. As well as seafood related illnesses, marine biotoxin contamination can lead to damaged public perceptions of seafood, direct economic losses and a restriction in the growth of the shellfish industry.

Diarrhetic Shellfish Toxins (DSTs) are produced by dinoflagellates of the planktonic genus *Dinophysis* and *Phalacroma*, and more rarely benthic *Prorocentrum*, and can bioaccumulate in shellfish and cause Diarrhetic Shellfish Poisoning (DSP). With approximately 11,000 human poisonings reported globally over the period 1985–2018 [1], DSP is a gastrointestinal disorder caused by the human consumption of seafood contaminated with DSTs. While symptoms are dose dependent and include diarrhea, nausea, vomiting and abdominal pain, DSTs are potent inhibitors of certain protein phosphatases and may promote tumor/cancer formation [2], although the impact of chronic exposure to DSTs is still not well known.

DSTs are a group of heat stable, polyether toxins consisting of okadaic acid (OA) and its isomer 19-epi-okadaic acid; the OA congeners dinophysistoxin-1 (DTX-1) and dinophysistoxin-2 (DTX-2); and the 7-acyl derivatives of OA, DTX-1 and DTX-2 that are collectively known as DTX-3. Together, they are referred to as the OA group toxins or the 'okadaates' (OAs). While OA, DTX-1 and DTX-2 only differ slightly in their molecular structure, the DTX-3 (group) includes a wide range of derivatives esterified with saturated and unsaturated fatty acids, products of metabolic transformations that occur in the shellfish [3]. Chemical compounds of this group are therefore generally described as either 'free' (unesterified) or 'esterified' [4].

DSP was first described after a large toxin event occurred in Japan in 1976 [5,6], whereby many people became sick after eating scallops (*Patinopecten yessoensis*). This contamination was linked to toxins produced by *Dinophysis fortii*. Following this event, further toxic episodes occurred in Japan, Spain and France, with several thousands of cases of human poisonings occurring over the 1970s and 1980s, and leading to the development of many regional monitoring programs. This monitoring has seen a gradual increase in reported DSP episodes in countries including Chile, Argentina, Mexico, the east coast of North America, Scandinavia, Ireland, Great Britain, Spain, Portugal, Italy, Greece, India, Thailand, Australia and New Zealand [5,7–9].

Dinophysis is common in Australian waters, with 36 species reported [10–12]. Toxic species include *D. acuminata* Claparede and Lachmann, *D. acuta* Ehrenberg, *D. caudata* Saville-Kent, *D. fortii* Pavillard, *D. norvegica* Claparede and Lachmann and *D. tripos* Gourret. There have been three serious DSP events in Australia. The first episode was caused by contamination of Pipis (*Plebidonax deltoides*) in New South Wales in 1997 (NSW) by *D. acuminata* [13]. One hundred and two people were affected and 56 cases of gastroenteritis were reported. A second episode occurred again in NSW in March 1998, this time with 20 cases of DSP poisoning reported [14]. The final event occurred in Queensland in March 2000, which was again linked to the consumption of Pipis [15]. While no human fatalities from DSP are known globally, DSTs continue to be a major food safety challenge for the shellfish industry.

Detection methods for DSTs using liquid chromatography with tandem mass spectrometry (LC-MS/MS and LC-MS) [4,16] and implemented as part of seafood safety programs, are considered the "gold standard" across the globe. These methods replaced the mouse bioassay (MBA), which was previously the most commonly used laboratory analysis tool (e.g., [17]). However, the development of more rapid, cost effective (on farm) testing methods for the presence of DSTs would potentially make harvest management simpler and faster and result in fewer closures. Three types of rapid test kits for the detection of DSTs are currently commercially available. These include an antibody-based enzyme-linked immunosorbent assay (ELISA) test; a functional protein phosphatase inhibition activity (PPIA) assay; and a lateral flow analysis (LFA) rapid test. ELISA assays involve an antigen immobilized on a (micro) plate, which are then complexed with an antibody that is linked to a reporter enzyme. These assays were first developed in the 1960s and 1970s for primarily medical diagnosis purposes [18]. Detection of OA, DTX-1 and DTX-2 (varying analogue cross reactivity depending on kit) is accomplished by assessing the conjugated enzyme activity via incubation with a substrate to produce a quantifiable product. Functional PPIA assays quantify okadaic acid (OA) and DST analogues including DTX-1, DTX-2 and DTX3 by colorimetric phosphatase inhibition, based on the reversible inhibition of protein phosphatase type 2A (PP2A) by the toxin, and the resulting absorbance derived from enzymatic hydrolysis of the substrate. A lateral flow test involves the shellfish extract transported across a reagent zone in which OA, DTX-1, DTX-2 and DTX-3 specific antibodies are combined with colored particles. If a toxin is present, it is captured by the particle-antibody complex, and as its concentration increases, the intensity of the test "line" decreases [19].

In a comprehensive review by McLeod et al. [20] of the currently available field methods for detection of marine biotoxins in shellfish, it was concluded that the ELISAs and LFAs had poor reactivity to the DSP congener DTX-2 and can give false negative results when high levels of DTX-3 are present (and the hydrolysis step is not undertaken to release ester forms). LFAs were also found to give some false positive results when DSP was below the ML (Max Limit), but this was dependent on the toxin profile, geographic region and shellfish species involved. Pectenotoxins (PTXs) are not currently included in Codex Standard for Live and Raw Bivalve Molluscs [21], and therefore are not included in this study. Several other jurisdictions such as Canada, Chile and the European Union do regulate for PTX (but not PTX-2sa), but the European Food Safety Authority has issued an opinion to deregulate PTX [22]. Furthermore, DSP regulation in Australia is governed by Food Standards Australia New Zealand with a maximum regulatory limit of 0.2 mg OA eq/kg [23], while most international standards including the Codex Standard, state a ML of 0.16 mg OA eq/kg [21].

To date, these rapid detection kits have not been tested on various shellfish matrices in a systematic manner, nor a comparison made across multiple analytical laboratories to assess LC-MS/MS or LC-MS detection of DSTs in shellfish. With this in mind, the present study aimed to undertake a comparative study to detect DSTs in differing shellfish matrices using commonly implemented protocols for LC-MS/MS or LC-MS in several different laboratories, as well as compare five commercially available rapid test kits for the detection of DSTs in these same shellfish tissues. The rapid test kits included three quantitative ELISA kits by Beacon™, Eurofins/Abraxis™ and EuroProxima™; a quantitative PP2A kit by Eurofins/Abraxis™, and a qualitative LFA kit by Neogen™.

2. Results

2.1. LC-MS/MS and LC-MS

No toxins were detected in any of the four shellfish species matrices (Sydney Rock Oysters (*Saccostrea glomerata*) (SRO), Pacific Oysters (*Magallana gigas/Crassostrea gigas*) (PO), Blue Mussels (*Mytilus galloprovincialis*) (MUS) and Pipis (*Plebidonax deltoides/Donax deltoides*) (PIPI)) screened before spiking began (see Methods). Of the triplicate SROs spiked with OA at 0.02 mg/kg, Laboratory 1 detected OA in all three samples (x = 0.01, SD ± 0.00, min <0.01, max 0.02 mg/kg), Laboratory 2 and 4 reported concentrations below the detection limit for all samples (<0.01 mg/kg and <0.025 mg/kg respectively), and Laboratory 3 detected OA in all three samples (x = 0.013, SD ± 0.006, min 0.01, max 0.02 mg/kg). In summary, two out of the four laboratories detected OA at this low level, with recoveries between ~50–100% (Table 1).

Of the four shellfish species spiked with OA at 0.02 mg/kg, Laboratory 1 detected this toxin in all four matrices (x = 0.013, SD ± 0.005, min 0.01, max 0.02 mg/kg), Laboratory 2 did not detect OA in SRO or PO; however, it was detected in both MUS and PIPI (x = 0.015, SD ± 0.007; min <0.01, max 0.02 mg/kg), and Laboratory 3 did not detect OA in PO or MUS, but detected it in SRO and PIPI (x = 0.015, SD ± 0.007; min <0.01, max 0.02 mg/kg). Laboratory 4 did not detect OA at this concentration (less than detection limit <0.025 mg/kg). Laboratory 4, however, did detect OA in one PIPI sample at 0.03 mg/kg (>spike concentration). In summary, OA was detected in all matrices at this concentration, although not all laboratories detected toxins in all four matrices. Recovery across all laboratories ranged from ~50–150% (Table 2).

Table 1. Results of LC-MS/MS (Liquid Chromatography—tandem Mass Spectrometry) and LC-MS (Liquid Chroma-tography—Mass Spectrometry) for Sydney Rock Oysters (SRO) spiked with 0.02 mg/kg okadaic acid (no DTX-1 or DTX-2 added).

Replicate	Species	Analyte	Spike	Lab 1	Lab 2	Lab 3	Lab 4
	Code	Code	mg/kg	mg/kg	mg/kg	mg/kg	mg/kg
1	SRO	OA Free	0.02	0.01	<0.01	0.01	<0.025
	SRO	OA Total	0.02	0.01	<0.01	0.01	<0.025
2	SRO	OA Free	0.02	0.02	<0.01	0.01	<0.025
	SRO	OA Total	0.02	<0.01	<0.01	0.01	<0.025
3	SRO	OA Free	0.02	0.01	<0.01	0.02	<0.025
	SRO	OA Total	0.02	0.01	<0.01	0.02	<0.025

<LOR = below limit of reporting; Note: Spike below limit of reporting for Laboratory 4.

Table 2. Results of LC-MS/MS and LC-MS for Australian shellfish—Sydney Rock Oysters (SRO), Pacific Oysters (PO), Blue Mussels (MUS) and Pipis (PIPI) spiked with 0.02 mg/kg okadaic acid (no DTX-1 or DTX-2 added).

Sample	Species	Analyte	Spike	Lab 1	Lab 2	Lab 3	Lab 4
	Code	Code	mg/kg	mg/kg	mg/kg	mg/kg	mg/kg
1	SRO	OA Free	0.02	0.01	<0.01	0.02	<0.025
	SRO	OA Total	0.02	0.01	<0.01	0.02	<0.025
2	PO	OA Free	0.02	0.02	<0.01	<0.01	<0.025
	PO	OA Total	0.02	0.02	<0.01	<0.01	<0.025
3	MUS	OA Free	0.02	0.02	0.01	<0.01	<0.025
	MUS	OA Total	0.02	0.01	0.01	<0.01	<0.025
4	PIPI	OA Free	0.02	0.01	<0.01	0.01	<0.025
	PIPI	OA Total	0.02	0.01	0.02	0.01	0.03

<LOR = below limit of reporting; Note: Spike below limit of reporting for Laboratory 4.

For the shellfish spiked with DTX-1 at 0.04 mg/kg, Laboratory 1 recovered this analogue in all matrices (x = 0.035, SD ± 0.006; min 0.03, max 0.05 mg/kg), with one PIPI sample returning a concentration of 0.01 OA mg/kg. Laboratory 2 detected DTX-1 in all matrices (x = 0.025, SD ± 0.006; min 0.02, max 0.03 mg/kg), also with a detection of OA in PIPI at 0.02 mg/kg. Laboratory 3 detected DTX-1 in all matrices (x = 0.025, SD ± 0.006; min 0.02, max 0.03 mg/kg), while Laboratory 4 did not detected this toxin in MUS (other matrices x = 0.026, min <0.025, max 0.04 mg/kg) (Table 3). In summary, DTX-1 was detected in all shellfish matrices at this concentration; however, one laboratory did not detect DTX-1 in MUS. The overall recovery of this analogue was ~50–100% across laboratories with two detections of OA in PIPIs.

For all shellfish spiked with DTX-2 at 0.01 mg/kg, Laboratory 1 did not recover this analogue in SRO or PIPI, and was only detected it in PO and MUS (both at 0.01 mg/kg). No toxin at this concentration was recovered from either Laboratory 2 nor Laboratory 3, while Laboratory 4 was unable to detect this toxin (below the limit of reporting <0.025 mg/kg) (Table 4). In summary DTX-2 was only detected in PO and MUS at this low concentration, and only at one laboratory. Overall recovery was ~50–100%.

Table 3. Results of LC-MS/MS and LC-MS for Australian shellfish—Sydney Rock Oysters (SRO), Pacific Oysters (PO), Blue Mussels (MUS) and Pipis (PIPI) spiked with 0.04 mg/kg DTX-1 (no OA or DTX-2 added).

Sample	Species	Analyte	Spike	Lab 1	Lab 2	Lab 3	Lab 4
	Code	Code	mg/kg	mg/kg	mg/kg	mg/kg	mg/kg
1	SRO	DTX-1 Free	0.04	0.05	0.02	0.03	0.04
	SRO	DTX-1 Total	0.04	0.04	0.02	0.03	0.026
2	PO	DTX-1 Free	0.04	0.04	0.02	0.02	0.03
	PO	DTX-1 Total	0.04	0.04	0.02	0.02	<0.025
3	MUS	DTX-1 Free	0.04	0.04	0.03	0.02	<0.025
	MUS	DTX-1 Total	0.04	0.03	0.03	0.02	<0.025
4	PIPI	DTX-1 Free	0.04	0.05	0.02	0.03	0.031
	PIPI	DTX-1 Total	0.04	0.03	0.03	0.03	<0.025
	PIPI	OA Total	-	0.01	0.02	<0.01	<0.025

<LOR = below limit of reporting.

Table 4. Results of LC-MS/MS and LC-MS for Australian shellfish—Sydney Rock Oysters (SRO), Pacific Oysters (PO), Blue Mussels (MUS) and Pipis (PIPI) spiked with 0.01 mg/kg DTX-2 (no OA or DTX-1 added).

Sample	Species	Analyte	Spike	Lab 1	Lab 2	Lab 3	Lab 4
	Code	Code	mg/kg	mg/kg	mg/kg	mg/kg	mg/kg
1	SRO	DTX-2 Free	0.01	<0.01	<0.01	<0.01	<0.015
	SRO	DTX-2 Total	0.01	<0.01	<0.01	<0.01	<0.015
2	PO	DTX-2 Free	0.01	0.01	<0.01	<0.01	<0.015
	PO	DTX-2 Total	0.01	<0.01	<0.01	<0.01	<0.015
3	MUS	DTX-2 Free	0.01	0.01	<0.01	<0.01	<0.015
	MUS	DTX-2 Total	0.01	<0.01	<0.01	<0.01	<0.015
4	PIPI	DTX-2 Free	0.01	<0.01	<0.01	<0.01	<0.015
	PIPI	DTX-2 Total	0.01	<0.01	<0.01	<0.01	<0.015

<LOR = below limit of reporting; Note: Spike below limit of reporting for Laboratory 4.

When shellfish were spiked with all toxins (in varying concentrations between 2–10 × LOR depending on toxin analogue; see Methods), laboratory recovery of total toxin per sample for each laboratory was as follows: Laboratory 1: 53–75%; Laboratory 2: 35–88%; Laboratory 3: 13–41%; and Laboratory 4: 0–88% (Table 5). More specifically, all toxins were recovered in all matrices for Laboratory 1, with an individual toxin/sample recovery ranging from 40–200%, with the lowest matrix average recovery in SRO at 57% and the highest in PIPI at 103%. For Laboratory 2, DTX-2 was not detected in SRO or PO, while individual toxin/sample recovery ranged from 40–400%, with the lowest matrix average recovery in SRO at 43%, and the highest in PIPI at 170%. For Laboratory 3, OA was not detected in MUS or PIPI, and DTX-2 was not detected in PIPI. The individual toxin/sample recovery ranged from 20–50%, with the lowest matrix average in PIPI at 40% and the highest in MUS at 47%. Finally, for Laboratory 4, DTX-2 was not detected across all matrices and OA was not detected in MUS. Individual toxin/sample recovery ranged from 50–340% with the lowest matrix average in MUS at 50% and the highest in PIPI at 154%. Overall, most toxins were detected by all laboratories at these concentrations, individual recovery across all labs/matrices ranged from 0–88%, while the recovery across shellfish matrices varied.

In our final analysis to determine the recovery of CRM (OA/DTX-1/DTX-2), all laboratories detected all toxin analogues. Individual toxin recoveries ranged from 88 to 131% for Laboratory 1, 79–81% for Laboratory 2, 83–95% for Laboratory 3 and 101–262% for Laboratory 4 (Table 6). However, considering that these recoveries are the result of one sample per lab, they should be treated as indicative only.

Table 5. Results of LC-MS/MS and LC-MS for Australian shellfish—Sydney Rock Oysters (SRO), Pacific Oysters (PO), Blue Mussels (MUS) and Pipis (PIPI) spiked with a combination of DST analogues-OA 0.1 mg/kg; DTX-1 0.05 mg/kg; and DTX-2 0.02 mg/kg.

Sample	Species	Analyte	Spike	Lab 1	Lab 2	Lab 3	Lab 4
	Code	Code	mg/kg	mg/kg	mg/kg	mg/kg	mg/kg
1	SRO	DTX-1 Free	0.05	0.05	0.01	0.02	0.038
	SRO	DTX-1 Total	0.05	0.04	0.02	0.02	0.03
	SRO	DTX-2 Free	0.02	0.02	<0.01	0.01	<0.015
	SRO	DTX-2 Total	0.02	0.01	<0.01	0.01	<0.015
	SRO	OA Free	0.1	0.05	0.04	0.04	0.089
	SRO	OA Total	0.1	0.04	0.04	0.04	0.062
2	PO	DTX-1 Free	0.05	0.03	0.02	0.02	0.036
	PO	DTX-1 Total	0.05	0.03	0.03	0.02	0.029
	PO	DTX-2 Free	0.02	0.02	<0.01	0.01	<0.015
	PO	DTX-2 Total	0.02	0.02	<0.01	0.01	<0.015
	PO	OA Free	0.1	0.06	0.04	0.04	0.08
	PO	OA Total	0.1	0.04	0.05	0.04	0.067
3	MUS	DTX-1 Free	0.05	0.02	0.03	0.02	0.03
	MUS	DTX-1 Total	0.05	0.03	0.03	0.02	<0.025
	MUS	DTX-2 Free	0.02	0.01	0.01	0.01	<0.015
	MUS	DTX-2 Total	0.02	0.02	0.01	0.01	<0.015
	MUS	OA Free	0.01	0.01	0.01	<0.01	<0.025
	MUS	OA Total	0.01	0.01	<0.01	<0.01	<0.025
4	PIPI	DTX-1 Free	0.05	0.03	0.03	0.01	0.033
	PIPI	DTX-1 Total	0.05	0.03	0.03	0.01	0.036
	PIPI	DTX-2 Free	0.02	0.02	0.02	<0.01	<0.015
	PIPI	DTX-2 Total	0.02	0.01	<0.01	<0.01	<0.015
	PIPI	OA Free	0.01	0.02	0.01	<0.01	<0.025
	PIPI	OA Total	0.01	0.02	0.04	<0.01	0.034

<LOR = below limit of reporting; Note: Spike of OA for MUS and PIPI below limit of reporting for Laboratory 4.

Table 6. Results of LC-MS/MS and LC-MS for Certified Reference Material CRM DSP-Mus-c.

Sample	Species	Analyte	Concentration	Lab 1	Lab 2	Lab 3	Lab 4
	Code	Code	mg/kg	mg/kg	mg/kg	mg/kg	mg/kg
1	+CONT	DTX-1 Free	1.07	1.4	0.87	0.91	1.1
	+CONT	DTX-1 Total	1.1 *	1.4	1.04	2.31	1.3
	+CONT	DTX-2 Free	0.86	0.76	0.68	0.82	0.87
	+CONT	DTX-2 Total	2.2 *	2.0	1.97	1.32	2.6
	+CONT	OA Free	1.07	1.1	0.85	0.89	2.8
	+CONT	OA Total	2.4 *	2.2	2.29	1.79	5.0

* CRM are certified for free toxin; they report higher total toxin concentration post hydrolysis but these are not certified.

2.2. Rapid Test Kits

2.2.1. Wild Harvest Pipis

Prior to rapid test kit screening, OA, DTX-1 and DTX-2 analysis by LC-MS for wild harvest Pipis resulted in a OA toxin range of 0.1 to 0.3 mg/kg (Sample 4A—0.1 mg/kg, 4B—0.1 mg/kg, 4C—0.2 mg/kg, and 4D—0.3 mg/kg). After hydrolysis, no DTX-1 or DTX-2 was detected in any samples. Three batches comprising 10 replicates of each OA toxin concentration of 0.1, 0.2 and 0.3 mg/kg were subsequently screened using each rapid test kit.

2.2.2. LC-MS

Using LC-MS (Laboratory 3), all control shellfish samples (no toxin added) returned a 'not detected' result (Table 7). For OA spiked samples, 43/46 (~93%) returned concentrations at, or slightly above, the spiked toxin concentrations 0.1 and 0.2 mg/kg (Tables 7 and 8). The three samples (7%) that returned concentrations lower that the spiked concentration were all spiked Pipi samples: sample 22 reported 0.09 mg/kg when it was spiked with OA at 0.1 mg/kg; sample 23 reported 0.15 mg/kg when it was spiked with OA at 0.2 mg/kg; and finally, sample 24 reported 0.09 mg/kg when it was spiked with OA at 0.2 mg/kg (Tables 7 and 8). The latter two of these samples were falsely compliant at the regulatory limit (7%, 2/28). A Pearson's correlation analysis between LC-MS results and the concentration of spiked toxin revealed a very strong relationship (r^2 = 0.86) (Figure 1). Subsequently, this method returned a mean recovery of 106.5%, meeting the criteria set out in the AOAC Guidelines for Single Laboratory Validation of Chemical Methods for Dietary Supplements and Botanicals (AOAC 2002).

Table 7. Results of LC-MS and rapid test kits for Okadaic Acid spiked into Australian shellfish (Sydney Rock Oysters [SRO], Pacific Oyster [PO], Blue Mussel [MUS] and Pipis [PIPI]). Note: Neogen qualitative test ($\pm$) with Limit of Quantification = 0.08 mg/kg; Abraxis PP2A Working Range = 0.06 to 0.35 mg/kg; Beacon ELISA Limit of Quantification = 0.1 mg/kg; Abraxis ELISA Working Range = 0.1–5.0 mg/kg; Europroxima ELISA Limit of Quantification = 0.04 mg/kg.

Sample No. and Shellfish Matrix	OA mg/kg	LC-MS	Neogen	Abraxis PP2A	Beacon ELISA	Abraxis ELISA	Europroxima ELISA
Sample 1 (SRO)	-	ND	-	0.02	0.05	0.00	0.03
Sample 2 (SRO)	-	ND	-	0.07	0.06	0.03	0.01
Sample 3 (SRO)	0.1	0.12	-	0.05	0.12	0.00	0.04
Sample 4 (SRO)	0.1	0.13	-	0.02	0.11	0.01	0.19
Sample 5 (SRO)	0.2	0.23	+	0.17	0.18	0.01	0.08
Sample 6 (SRO)	0.2	0.23	-	0.05	0.24	0.01	0.09
Sample 7 (PO)	-	ND	-	0.07	0.07	0.00	0.08
Sample 8 (PO)	-	ND	-	0.03	0.05	0.00	0.02
Sample 9 (PO)	0.1	0.12	-	0.05	0.12	0.01	0.04
Sample 10 (PO)	0.1	0.17	-	0.11	0.18	0.03	0.04
Sample 11 (PO)	0.2	0.23	-	0.12	0.20	0.03	0.04
Sample 12 (PO)	0.2	0.23	-	0.21	0.20	0.03	0.07
Sample 13 (MUS)	-	ND	-	0.02	0.05	0.03	0.01
Sample 14 (MUS)	-	ND	-	0.03	0.06	0.02	0.02
Sample 15 (MUS)	0.1	0.19	-	0.16	0.12	0.01	0.09
Sample 16 (MUS)	0.1	0.17	-	0.06	0.10	0.01	0.02
Sample 17 (MUS)	0.2	0.23	+	0.16	0.21	0.01	0.11
Sample 18 (MUS)	0.2	0.23	-	0.08	0.19	0.02	0.04
Sample 19 (PIPI)	-	ND	-	0.04	0.09	0.01	0.02
Sample 20 (PIPI)	-	ND	-	0.02	0.09	0.01	0.01
Sample 21 (PIPI)	0.1	0.1	-	0.13	0.17	0.02	0.04
Sample 22 (PIPI)	0.1	0.09	-	0.05	0.17	0.04	0.02
Sample 23 (PIPI)	0.2	0.15	+	0.18	0.43	0.01	0.09
Sample 24 (PIPI)	0.2	0.09	-	0.13	0.43	0.01	0.06

ND = not detected (0.01 mg/kg detection limit).

2.2.3. Rapid Test Kits

Qualitative Test

Neogen

The Neogen kit returned negative readings for the eight negative control samples across all species-specific shellfish matrices. However, 23 out of 46 samples (50%) of spiked samples (across all shellfish matrices) returned a negative result when they contained okadaic acid (Tables 7 and 8). Within this group, 18% (5/28 samples again across all matrices) returned a false compliant result when they were spiked at, or above, the regulatory limit (=/> 0.2 mg OA eq/kg), while no naturally contaminated Pipis returned falsely compliant results with this kit.

Quantitative Tests

Abraxis PP2A

The Abraxis PP2A returned 25% (2/8) false positive results, that is, they returned concentrations of toxin within the kit's working (range 0.06 to 0.35 mg/kg), when the samples contained no okadaic acid. Of those shellfish that were spiked, 29% (13/45) of samples returned values that were outside the working range (8 samples below 0.06 mg/kg and 5 samples above 0.35 mg/kg), with 27% (12/45) samples being underestimated and 44% (20/45) returning a concentration which was equal to, or greater than, the spiked toxin concentration (Tables 7 and 8). When samples were spiked at, or above, the regulatory limit, the Abraxis PP2A returned 29% (8/28) falsely compliant results (Table 9). These results were for both spiked and naturally contaminated samples. A Pearson's correlation analysis between the Abraxis PP2A results and spiked toxin concentrations was significant at $r^2 = 0.72$ (Figure 1). This kit returned a mean recovery of 92.2%, again meeting the criteria set out in the AOAC Guidelines [24] (Table 9).

Table 8. Results of LC-MS and rapid test kits for Okadaic Acid in naturally contaminated Pipis [PIPI] Note: Neogen qualitative test ($\pm$) with Limit of Quantification = 0.08 mg/kg; Abraxis PP2A Working Range = 0.06–0.35 mg/kg; Beacon ELISA Limit of Quantification = 0.1 mg/kg; Abraxis ELISA Working Range = 0.1–5.0 mg/kg; Europroxima ELISA Limit of Quantification = 0.04 mg/kg.

Sample No. and Shellfish Matrix	OA mg/kg	LC-MS	Neogen	Abraxis PP2A	Beacon ELISA	Abraxis ELISA	Europroxima ELISA
Sample 25 (PIPI)	0.1	0.1	-	0.04	0.07	0.02	0.03
Sample 26 (PIPI)	0.1	0.1	-	0.10	0.05	0.01	0.02
Sample 27 (PIPI)	0.1	0.1	-	0.04	0.06	0.07	0.02
Sample 28 (PIPI)	0.1	0.1	-	0.07	0.10	0.04	0.03
Sample 29 (PIPI)	0.1	0.1	-	0.07	0.06	0.03	0.03
Sample 30 (PIPI)	0.1	0.1	-	0.05	0.08	0.02	0.02
Sample 31 (PIPI)	0.1	0.1	-	0.15	0.06	0.02	0.02
Sample 32 (PIPI)	0.1	0.1	-	0.10	0.06	0.06	0.02
Sample 43 (PIPI)	0.1	0.1	-	0.06	0.10	0.18	0.03
Sample 44 (PIPI)	0.1	0.1	-	NS	0.08	0.17	0.02
Sample 33 (PIPI)	0.2	0.2	+	0.23	0.08	0.25	0.03
Sample 34 (PIPI)	0.2	0.2	+	0.22	0.13	0.14	0.02
Sample 35 (PIPI)	0.2	0.2	+	0.24	0.10	0.24	0.03
Sample 36 (PIPI)	0.2	0.2	+	0.16	0.11	0.17	0.02
Sample 37 (PIPI)	0.2	0.2	+	0.25	0.13	0.22	0.02
Sample 38 (PIPI)	0.2	0.2	+	0.25	0.04	0.14	0.04
Sample 39 (PIPI)	0.2	0.2	+	0.20	0.06	0.06	0.02
Sample 40 (PIPI)	0.2	0.2	+	0.27	0.05	0.16	0.01
Sample 41 (PIPI)	0.2	0.2	+	0.22	0.10	0.05	0.02
Sample 42 (PIPI)	0.2	0.2	+	0.23	0.11	0.02	0.02
Sample 45 (PIPI)	0.3	0.3	+	0.38	0.05	0.21	0.03
Sample 46 (PIPI)	0.3	0.3	+	0.39	0.06	0.19	0.02
Sample 47 (PIPI)	0.3	0.3	+	0.39	0.05	0.33	0.02
Sample 48 (PIPI)	0.3	0.3	+	0.36	0.09	2.05	0.03
Sample 49 (PIPI)	0.3	0.3	+	0.33	0.07	0.88	0.02
Sample 50 (PIPI)	0.3	0.3	+	0.36	0.10	0.11	0.03
Sample 51 (PIPI)	0.3	0.3	+	0.34	0.17	0.23	0.03
Sample 52 (PIPI)	0.3	0.3	+	0.34	0.06	0.24	0.03
Sample 53 (PIPI)	0.3	0.3	+	0.32	0.08	0.19	0.02
Sample 54 (PIPI)	0.3	0.3	+	0.25	0.05	0.17	0.06

NS = no sample.

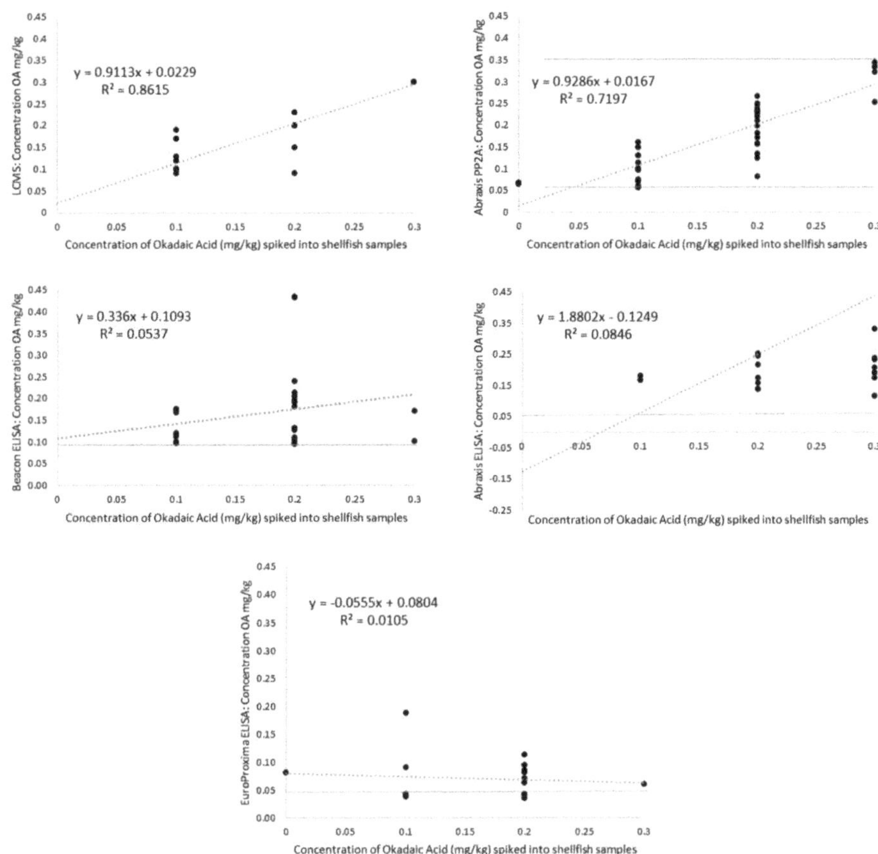

Figure 1. Linear regression plots showing the relationship between spiked toxin concentration with both LC-MS and quantitative rapid test kits results in Australian shellfish samples calculated data within each method's working range. Blue lines represent lower working range and red line upper working range of method. Note: Abraxis PP2A Working Range (WR) = 0.06 to 0.35 mg/kg; Beacon ELISA Limit of Quantification (LOQ) = 0.1 mg/kg; Abraxis ELISA Working Range = 0.1–5.0 mg/kg; EuroProxima ELISA Limit of Quantification = 0.04 mg/kg.

Table 9. List of DST rapid test kits available, their method details and requirements (NR = not reported; ND = not detected). Note: LC-MS Cost ~$300 per sample and ~2 h for analysis. * AU$1 has been added to the cost of each sample for consumables.

Kit No./Name	1. Neogen	2. Abraxis PP2A	3. Beacon ELISA	4. Abraxis ELISA	5. EuroProxima ELISA
Method	Lateral Flow Assay (LFA)—single sample	Protein Phosphatase Inhibition (PPI)—96 well plate	ELISA 96 well plate	ELISA 96 well plate	ELISA 96 well plate
Qualitative or Quantitative	Qualitative	Quantitative	Quantitative	Quantitative	Quantitative
Analogues and Cross reactivity	OA (100%), DTX-1 (89%), DTX-2 (47%) & DTX-3	OA (1.2 nM), DTX-1 (1. 6nM), DTX-2 (1.2 nM), DTX3	OA (100%), DTX-1 (120%), DTX-2 (20%)	OA (100%), DTX-1 (50%), DTX-2 (50%)	OA (100%), DTX-1 (78%), DTX-2 (2.6%)

Table 9. *Cont.*

Kit No./Name	1. Neogen	2. Abraxis PP2A	3. Beacon ELISA	4. Abraxis ELISA	5. EuroProxima ELISA
Limit of Quantification or Working Range	0.08 mg/kg [25]	0.06–0.35 mg/kg	0.1 mg/kg	0.1–5.0 mg/kg	0.04 mg/kg
Standards included	no	0.4, 0.6, 1.0, 1.5 and 2.3 µg/L	0, 0.2, 0.5,1,2, 5 µg/L	0, 0.1, 0.2, 0.5, 1, 2, 5 µg/L	0, 0.2, 0.5, 1.0, 2, 5, 10 µg/L
Hydrolysis step	yes	yes	no	yes	no
Amount of tissue required	2 g	5 g	1 g	1 g	1 g
Samples per kit	24	~35–40 samples	~35–40 samples	~40 samples	~35–40 samples
Cost per kit (AU$)	$974.50	$1277	$849	$848	$999
Cost per sample * (AU$)	$42	$33	$22	$22	$26
Scanner (AU$)	$4000				
Reported False Positives	No false positives compared to ND by LC-MS [25]	14% positive compared to ND by LC-MS [25]	NR	Some false positives [26]	NR
Time for Analysis	~ 1.5 h	~ 3 h	~ 3 h	~ 4 h	~ 3 h

Beacon ELISA

With a limit of quantification reported as 0.1 mg/kg, the Beacon ELISA kit returned 0% (0/8) false positives and 43% (20/46) of spiked samples below the limit of quantification. Of the samples that were spiked (and results above the quantification limit), 22% (10/46) were underestimated, while 35% (16/46) were equal to, or greater than, the spiked toxin concentration (Tables 7 and 8). When samples were spiked at/above the regulatory limit, or were naturally contaminated at/above the regulatory limit, the Beacon ELISA returned 79% (22/28) falsely compliant results (Table 9). A Pearson's correlation analysis between the Beacon ELISA kit test results and the spiked toxin concentrations was extremely weak at $r^2 = 0.05$ (Figure 1). This kit returned a mean recovery of 77%, outside the criteria in the AOAC Guidelines [24] (Table 10).

Table 10. Summary of results comparing LC-MS (Laboratory 3) and five commercially available test kits to spiked Australian shellfish (results are across all species-specific shellfish matrices). Note: Abraxis PP2A Working Range (WR) = 0.06 to 0.35 mg/kg; Beacon ELISA Limit of Quantification (LOQ) = 0.1 mg/kg; Abraxis ELISA Working Range = 0.1–5.0 mg/kg; Europroxima ELISA Limit of Quantification = 0.04 mg/kg; ML = Maximum limit (=Regulatory Limit 0.2 eq OA mg/kg); Repeatability is defined as the standard deviation of the mean (see Methods).

	LC-MS	Neogen	Abraxis PP2A	Beacon ELISA	Abraxis ELISA	Europroxima ELISA
% False Positive (blank matrix)	0 (0/8)	0 (8/8)	25 (2/8)	0 (0/8)	0 (0/8)	13 (1/8)
% False Negative (spiked matrix)	0 (0/54)	50 (23/46)	-	-	-	-
% Results outside WR or LOQ	-	-	29 (13/45)	43 (20/46)	59 (27/46)	65 (30/46)
% Samples Underestimated	7 (3/46)	-	27 (12/45)	22 (10/46)	24 (11/46)	33 (15/46)
% Samples Equal or Overestimated	93 (43/46)	-	44 (20/45)	35 (16/46)	17 (8/46)	2 (1/46)
% Falsely Compliant with ML (overall)	7 (2/28)	18 (5/28)	29 (8/28)	79 (22/28)	71 (20/28)	100 (28/28)
% Falsely Compliant with ML (spiked)	25 (2/8)	63 (5/8)	88 (7/8)	25 (2/8)	100 (8/8)	100 (8/8)
% Falsely Compliant with ML (naturally contaminated)	0 (0/20)	0 (0/20)	5 (1/20)	100 (20/20)	55 (11/20)	100 (20/20)

Table 10. *Cont.*

	LC-MS	Neogen	Abraxis PP2A	Beacon ELISA	Abraxis ELISA	Europroxima ELISA
% Falsely Non-compliant with ML	0 (54/54)	0 (54/54)	0 (53/53)	0 (54/54)	0 (54/54)	0 (54/54)
Mean (SD) Recovery %	106.5 (22.2)	-	92.2 (34.2)	77.7 (51.2)	66.2 (107.9)	26.7 (29.1)
Repeatability (0.1-0.3 eq OA mg/kg PIPI)	0.00	-	0.01	0.00–0.01	0.02–0.18	0.00
Coefficient of Determination (r^2)	0.86	-	0.72	0.05	0.08	0.01

Abraxis ELISA

Similar to the Abraxis PP2A, the Abraxis ELISA reports a working range of 0.01 to 0.5 mg/kg. This kit returned 0% (0/8) false positives and 59% (27/46) of spiked samples below the working range. Of the samples that were spiked (and results within the working range), 24% (11/46) were underestimated and 17% (8/46) were equal to, or greater than, the spiked toxin concentration (Tables 7 and 8). Again, when spiked or naturally contaminated at/above the regulatory limit, the Abraxis ELISA returned 71% (20/28) falsely compliant results (Table 10). A Pearson's correlation analysis between the Abraxis ELISA kit test results and the spiked toxin concentrations was weak at $r^2 = 0.08$ (Figure 1). Subsequently, this kit returned a mean recovery of 66%, well outside the criteria in the AOAC Guidelines [24] (Table 10).

EuroProxima ELISA

With a limit of quantification reported as 0.04 mg/kg, the EuroProxima ELISA kit returned 13% (1/8) false positives and 65% (30/46) of spiked samples returning results outside the limit of quantification (<0.04 mg/kg). Of the samples that were spiked (and results reported were above the limit of quantification), 33% (15/46) were underestimated, while only 2% (1/46) were equal to, or greater than, the spiked toxin concentration (Tables 7 and 8). When either spiked or naturally contaminated at, or above, the regulatory limit, the EuroProxima returned 100% (28/28) falsely compliant results (Table 10). A Pearson's correlation analysis between this rapid kit test and the spiked toxin concentrations was extremely weak at $r^2 = 0.01$ (Figure 1). This kit returned a very low mean recovery of 26.7%, well outside the criteria set in the AOAC Guidelines [24] (Table 10).

Repeatability of Kits

The repeatability/reliability of all kits was high (standard deviations of the mean ranged from 0.00 to 0.01, with the lower the variation, the higher the reliability of the results). The only exception to this was the Abraxis ELISA kit. From the naturally contaminated Pipi batch with the highest toxin concentration (0.3 OA mg/kg), the repeatability of this kit was low at 0.02 (based on a relatively low number of samples however) (Table 9).

3. Discussion

3.1. DSTs in Australia

Toxic *Dinophysis* blooms and their impacts remain one of the most problematic HABs worldwide, especially in Mediterranean and European waters [1]. Positive DST detections periodically occur in Australian shellfish, although these events remain largely unstudied [1,27]. Using the official analytical method of LC-MS/MS and LC-MS, shellfish data spanning 2012 to 2017 from four Australian states (Tasmania, Victoria, South Australia and Western Australia) showed that 53 (0.65%) shellfish samples out of the 8156 analyzed exceeded the domestic regulatory limit (0.2 mg OA eq/kg). Exceedances, across all samples combined, for cockles/pipis, clams, mussels, oysters and scallops were 4.9, 1.1, 1.1, 0.03 and 0%, respectively. Of those that exceeded this threshold, OA was the most commonly detected toxin analogue, with only one sample containing DTX-1, and no samples containing DTX-2 (unpublished data).

3.2. LC-MS/MS and LC-MS Laboratory Comparison

In the present study, we spiked four different shellfish matrices (SRO, PO, MUS, PIPI) with fixed volumes of relevant CRM to determine the ability of laboratories to quantify DSTs in shellfish using LC-MS/MS and LC-MS. We found that all spiked analogues, OA, DTX-1, and DTX-2, were recovered in all shellfish species across all laboratories, but the results were not consistent across all samples. For example, low and mid-concentration toxin recovery was variable both within and between laboratories (0–150%), while high concentration toxin recovery, which included CRM, was higher, between 60–262%. Two false positives were reported in Pipi samples in which OA was detected at 0.01 and 0.02 mg/kg (Laboratory 1 and Laboratory 2, respectively), and one anomalously high concentration of 2.8 mg/kg was reported from CRM that was submitted at a concentration of 1.07 mg/kg (Table 6). These results need to be interpreted in light of each laboratory's measurement uncertainty (MU), which was reported as ~10–26%, dependent on the analogue detected (Appendix A). Another issue that must be considered is the homogeneity of toxin within the shellfish, and how that may contribute to the variability in results, particularly at the low- to mid-level spiked concentration.

Finally, we cannot completely discount that there may have been some very low toxin concentrations in these samples which were not detected by the original LC-MS screening. Lab 3, in fact, had the highest level of detection (0.006–0.007 mg/kg for analogues OA, DTX-1 and ±DTX-2) across all the labs used in this study.

In a single laboratory validation study to detect and quantify six lipophilic toxins (azaspiracid, domoic acid, gymnodimine, okadaic acid, pectenotoxin and yessotoxin) in Greenshell mussel, Pacific Oyster, cockle and scallop roe, McNabb et al. [4] reported mean OA recoveries between 92% (from a toxin concentration of 0.5–1.0 mg/kg) and 99% (from a toxin concentration of 0.05–0.10 mg/kg). All six toxins recoveries ranged from 71–99%. As discussed above, this variability was also apparent in our results, albeit in a converse way, whereby shellfish with a higher spiked toxin concentration generally reported a better recovery than those at lower concentrations. McNabb's study concluded that with some slight methodological adjustments (methanol-water $\cong$ 9 + 1; 18 mL for 2 g of shellfish tissue), the LC-MS/MS method provides good precision/accuracy and high specificity, and is therefore suitable for the quantification of biotoxins in shellfish for regulatory purposes.

In another study to compare the mouse bioassay (MBA) to electrospray ionization (ESI) LC-MS/MS for the quantification of lipophilic toxins in ~200 samples of shellfish, Suzuki and Quilliam [28] similarly concluded that LC-MS/MS was a powerful tool for both the identification and structure elucidation of many toxins including OA/DTX analogues, but also for the discovery of unknown toxin analogues. Furthermore, studies have shown that LC-MS/MS demonstrates linearity, specificity, repeatability and reproducibility in shellfish samples collected from the environment [29], and is able to resolve the toxin profiles of OA analogues in various *Dinophysis* species isolated from bloom samples [30].

There are, however, disadvantages to using LC-MS/MS and LC-MS for the detection of toxins in shellfish. LC-MS/MS (and LC-MS) is expensive, particularly for farmers in low-risk areas who have a regulatory requirement to undertake marine biotoxin testing using LC-MS/MS at regular intervals (e.g., weekly). The cost is also high for farmers in remote areas, where transport of samples to specialized laboratories is expensive. The LC-MS/MS and LC-MS method is also complex, requiring expert analyst training in dedicated laboratories for sophisticated instrument maintenance and performance. Time delays are another concern, and it can take between 2–7 days to obtain results from a contract laboratory, potentially causing a loss in harvest time and profits to shellfish farmers, and risk to consumers. Finally, high quality and expensive reference material is required to calibrate the method. Despite these disadvantages, and in the absence of a more reliable, sensitive and rapid test, there remains an international acceptance that LC-MS/MS and

LC-MS continue to be the standard operating procedure (along with the MBA in many Latin American and Asian countries), for the determination of lipophilic marine biotoxins in mollusks [31].

3.3. Rapid Test Kits Comparison

In the search for an inexpensive and reliable alternative method to LC-MS/MS or LC-MS that could be used for screening purposes to serve as an early warning for the shellfish industry, we compared five Rapid test kits against the LC-MS/MS and LC-MS methods. Fifty-five shellfish samples (24 spiked and 30 naturally contaminated pipis) were screened with four quantitative (Beacon, Abraxis and EuroProxima ELISA kits and the Abraxis PP2A kit) and one qualitative (Neogen LFA) rapid test kit to detect OA in Sydney Rock Oysters, Pacific Oysters, Blue Mussels and Pipis. Okadaic acid was the only DST analogue to be tested with these kits for multiple reasons: (i) It has been the dominant analogue detected in Australian shellfish to date; (ii) The cost of purchasing sufficient CRM for spiking all other analogues to detection levels is high; and (iii) Rapid test kit results are reported as µg OA eq/kg, and a spike of varying DST analogues will not reveal individual analogue concentrations (noting the Neogen rapid test kit is qualitative only). Furthermore, each kit reports a level of cross reactivity to the various analogues, and while in most cases this is 100% for OA, it varies for DTX-1 and DTX-2 between kits. For example, if three samples were individually spiked with the same concentration of okadaic acid, DTX-1 and DTX-2, the concentration of okadaic acid from the Abraxis ELISA kit would read as double the concentrations of the other two compounds. This is because DTX-1 (50%) and DTX-2 (50%) only give half of the response that okadaic acid does with this technology.

With this in mind, all quantitative kits should theoretically provide a comparable concentration of OA to that obtained using the LC-MS method. Regression analyses showed the correlations between the ELISA Rapid test kits and LC-MS in our study were all very low (0.002–0.19), while the correlation between the PP2A Abraxis kit and LC-MS was moderate to high (0.72) (Figure 1). The observed variations between these methods could not be attributed to matrix effects however, as no clear differences were observed between spiked samples across methods. Certain kits nonetheless performed better on naturally contaminated samples (Pipis only) compared to spiked samples (Neogen and Abraxis PP2A). The reasons for this remain unclear, but support the assertion by Turner et al. [32] that validation studies need to include both relevant shellfish species and naturally contaminated shellfish samples, so that any rapid test kit performance is measured using local toxin profiles.

After the development of the first ELISA method by Dubois et al. [33], a comparison across assay techniques was undertaken whereby cell counts, LC-MS/MS, the newly developed Abraxis ELISA and PP2A Okatests were compared. Naturally contaminated samples of edible Blue Mussels (*Mytilus edulis*) were examined for total DST toxin content including esters and DTX-3. The ELISA showed matrix effects on hydrolyzed samples, which had both high and low levels of toxins, while the PP2A adequately detected both low and high DST concentrations in mussel samples. While the Okatest was recommended in preference to the ELISA, it was concluded to be a specific assay (could not detect other regulated DSTs), and therefore could not replace LC-MS/MS or LC-MS. Subsequent to these findings, three further studies—a single laboratory validation and an interlaboratory study on the PP2A Okatest [34,35], and a comparison across three RTKs (the lateral flow (Jellett/Scotia), ELISA (Abraxis) and PPIA (Okatest) kits) [26] were undertaken. Considering issues such as an unacceptable number of false negatives (Jellett), and low cross-reactivity with DTX-1 (the dominant toxin profile in the shellfish tested) by the ELISA, Eberhart et al. concluded that the PP2A was the most promising kit on the market. It is these differences in toxin profiles, the inclusion (or not) of a hydrolysis step, and whether the shellfish tested is spiked or naturally contaminated, that prevents a direct comparison between these studies and the present study, although it highlights the issues that must be standardized in any future validation study.

In 2015, Jawaid et al. reported on the development and validation of a new rapid test kit, the Neogen LFA, this time a qualitative test strip/reader for the OA group toxins in shellfish [19]. This validation method tested both spiked (OA, DTX-1, DTX-2 and DTX-3 with hydrolysis procedure) and naturally contaminated shellfish (mussels, scallops, oysters, and clams) and compared the results to LC-MS/MS. While our study showed only minor differences in shellfish matrices (low number of samples tested however) and zero falsely compliant results in naturally contaminated samples, Jawaid et al. showed no matrix effects, false compliant results or false noncompliant results at <50% MPL (maximum permitted level). Both Jawaid and the present study suggest that this method, with some further work, may be an effective early warning tool for the shellfish industry. The results reported in this study, however, do not support the use of any DST rapid test kit as a stand-alone quality assurance measure at this time, and further research and development work is needed.

Since the development of the LFA technology, two additional studies generated rapid test kit comparisons [25,36]. The first study compared DSTs in shellfish from Argentina using two qualitative lateral flow kits (Scotia and Neogen), the quantitative PPIA kit (OkaTest), and the ELISA kit (Max Signal—no longer commercially available) and compared the results to LC-MS/MS. The specificity was reported as good for all kits, with no false compliant results against the ML of <0.16 mg OA eq/kg). The second study screened four RTKs, again on naturally contaminated shellfish, but this time from Great Britain. The quantitative PP2A (OkaTest) was the only test to show the complete absence of falsely compliant results (i.e., mussel samples containing OA-group toxins above the MPL of 0.16 mg OA eq/kg which returned negative results) and showed a fair correlation to LC-MS/MS but with an overall overestimation of sample toxicity with some indication of matrix effect, particularly in oysters [36]. The quantitative ELISA (MaxSignal) gave a reasonable correlation with LC-MS/MS, no evidence of overestimation, accuracy at low concentrations and only one falsely compliant result (as above, a mussel samples containing OA-group toxins above the MPL of 0.16 mg OA eq/kg which returned a negative result). The two lateral flow assays (Neogen and Scotia) were observed to show high agreement with LC-MS/MS and no indications of false positives), although both returned one false negative [36].

In the present study, all four quantitative kits showed varying levels of over/underestimation (many at the regulatory limit). Many results were outside the working range or limit of these kits. This ranged from 29% of samples using the Abraxis PP2A to 65% with the Euro-Proxima ELISA (Table 10). Two kits also showed false positives from blank matrices (i.e., samples that did not contain toxins), these being the Abraxis PP2A and EuroProxima ELISA at 25% and 13% respectively. We cannot, however, discount the fact that there may have been some very low toxin concentrations in these samples which were not detected by LC-MS. All methods (quantitative and qualitative) delivered high levels (25% to 100%) of falsely compliant results for spiked samples. The Neogen and Abraxis PP2A performed satisfactorily (0%, 5% falsely compliant at the regulatory limit or above, respectively) for naturally contaminated pipis. The mean percent recovery ranged from 27% (EuroProxima ELISA) to 107% (LC-MS), while only the LC-MS method and the Abraxis PP2A kit (92%) fell within the "acceptable recovery" range of 80–100% as set by the AOAC Guidelines [24].

4. Conclusions

Overall, considering the highly varied, and sometimes erroneous results, along with other factors such as method cost, preparation time, test complexity, and extra equipment required, the disadvantages of using the currently available rapid test kits are considerable (Table 9. Quantitatively, the Abraxis PP2A kit outperformed all other rapid test kits (notably in naturally contaminated pipis)) and may be suitable for screening purposes. In using this kit, however, one sample took ~3 h to complete. This kit also requires more rigorous testing to determine the statistics around its false compliant results. Continued collaboration with the manufacturer to refine this test procedure should be undertaken to improve its potential. Qualitatively, the Neogen test kit performed well for naturally contaminated Pipis (0%

falsely compliant results at the regulatory level) but appeared much less reliable (63% false negative results at regulatory level) for spiked pipis, oysters, and mussels. These results suggest possible differences in kit performance dependent on the shellfish matrix analyzed, or whether the shellfish is naturally contaminated or artificially spiked. The reason(s) for differing results between naturally contaminated shellfish and spiked samples, however, remains unclear, particularly when toxin determination using LC-MS did not result in any significant difference between these two routes in the present study. The Neogen kit is, however, relatively simple to use, returns a faster result than other kits, and, as discussed above, shows promising results for naturally contaminated shellfish. A single laboratory validation study, such as the one carried out by for paralytic shellfish toxins in mussels and oysters [37], followed by an international validation study, is recommended prior to approval of any rapid test kit for regulatory purposes.

5. Materials and Methods

5.1. Interlaboratory Comparison for LC-MS/MS and LC-MS

5.1.1. Shellfish Preparation

Sample preparation was based on the standard operating procedure for the determination of lipophilic marine biotoxins in molluscs by LC-MS/MS and LC-MS. Specifically, raw samples of Sydney Rock Oysters (*Saccostrea glomerata*), Pacific Oysters (*Magallana gigas/Crassostrea gigas*), Blue Mussels (*Mytilus galloprovincialis*) and Pipis (*Plebidonax deltoides/Donax deltoides*) were sourced from the Sydney Fish Markets on 6/6/2019. From here on, these matrices are referred to as SRO, PO, MUS and PIPI, respectively. These were stored at 4–8 °C and transported immediately to the laboratory for processing. All shellfish were washed thoroughly with fresh water, shucked (if necessary) and tissue was removed. Stock material of each species was made by pooling the tissue of 3–6 individuals (for each spike treatment) of that species, homogenizing and spiking with fixed volumes of relevant standards (see below) and homogenizing again. Subsamples of this species-specific tissue homogenate were then accurately weighed (~3 g) and aliquoted into 5 mL polypropylene Bacto sample jars (Model No. SCP5014UU) and frozen at −20 °C until they were dispatched to contract laboratories for toxin determination by LC-MS/MS and LC-MS.

5.1.2. Standard Reference Materials

Certified reference materials (CRMs) were purchased from the National Research Council Canada (NRC) for shellfish spiking and quality control testing. These included: (i) CRM DSP-Mus-c which is a thermally sterilized homogenate (4.0 ± 0.75 g) of mussel tissue (*Mytilus edulis*) and the dinoflagellate *Prorocentrum lima*, with toxin levels of okadaic acid (OA), dinophysistoxin-1 (DTX-1) and dinophysistoxin-2 (DTX-2) at 1.07 ± 0.08 μg/g, 1.07 ± 0.11 μg/g and 0.86 ± 0.08 μg/g, respectively (positive control); (ii) CRM-OA-d which contained ~0.5 mL of a solution of OA in methanol at a concentration of 8.4 ± 0.4 μg/mL; (iii) CRM-DTX-1-b which contained ~0.5 mL of a solution of dinophysistoxin 1 (DTX-1) in methanol at a concentration of 7.8 ± 0.5 μg/mL; and (iv) CRM-DTX-2-b which contained ~0.5 mL of a solution of dinophysistoxin-2 (DTX-2) in methanol at a concentration of 3.8 ± 0.2 μg/mL.

5.1.3. Spiking of Shellfish Matrices

A subsample (3 g) of each pooled, species-specific matrix (SRO, PO, MUS and PIPI) was first analyzed by LC-MS at Laboratory 3 (see below) to ensure each matrix contained no DSTs before the experiment began (limit of detection (LOD) = 0.006–0.007 mg/kg for analogues OA, DTX-1 and ± DTX-2) (Appendix A).

Spiking of each species-specific homogenate with a range of DST concentrations then followed for both LC-MS/MS and LC-MS. These concentrations were chosen based on the capability of most laboratories to achieve a limit of reporting (LOR) of ~0.01 mg/kg (Table 11, Appendix A). In brief, one batch of each matrix was spiked with OA (≅7.2 μL/3g, which is equivalent to 2 × LOR (0.02 mg/kg); the second one with DTX-1(≅14.0 μL/3g, which is 4 × LOR (0.04 mg/kg), and the third with DTX-2 (≅8 μL/3g, which is equivalent to the LOR (0.01 mg/kg). While increasing the spiking concentration of this latter analogue would provide a more rigorous comparison of the laboratories capabilities, our decision to spike DTX-2 at the LOR was based on cost and the infrequency of this analogue identified in Australian shellfish to date. A ~3 g aliquot of each of these species-specific homogenates was then sent to each laboratory to test their LOR and any matrix effect (Table 11).

Table 11. List of Australian shellfish samples, toxin volume of CRM added per 3 g of homogenised shellfish tissue, and OA equivalent concentrations (shaded) dispatched to each laboratory for DST determination using LC-MS.

Matrix	DST Spiking Volumes				Total
	OA Only	DTX-1 Only	DTX-2 Only	OA/DTX-1/DTX-2	
Sydney Rock Oysters	7.2 μL/3 g (3) *	14 μL/3 g	8 μL/3 g	35, 17.6, 16 μL/3 g	6
Pacific Oyster	7.2 μL/3 g	14 μL/3 g	8 μL/3 g	35, 17.6, 16 μL/3 g	4
Mussel	7.2 μL/3 g	14 μL/3 g	8 μL/3 g	7.2, 17.6, 16 μL/3 g	4
Pipi	7.2 μL/3 g	14 μL/3 g	8 μL/3 g	7.2, 17.6, 16 μL/3 g	4
Concentration mg/kg	0.02 mg	0.04 mg	0.01	0.02 or 0.1 [#], 0.05, 0.02	
Positive Control (CRM DSP-Mus-c)	-	-	-	-	1
Total Samples					N = 19

* n = 3 for reproducibility/repeatability; [#] 0.02 mg/kg for mussel and pipi; 0.1 mg/kg for Sydney Rock Oysters and Pacific Oyster.

Next, a second species-specific homogenate was spiked with a combination of all three toxins: 35 μL/3 g OA for SRO and PO which is 10 × LOR (0.1 mg/kg) or 7.2 μL/3g OA for MUS and PIPI which is equivalent to 2 × LOR (0.02 mg/kg); 17.6 μL/3 g DTX-1 which is 5 × LOR (0.05 mg/kg) into all shellfish species; and 16 μg/3 g DTX-2 which is 2 × LOR (0.02 mg/kg) again into all shellfish species. These combination-spiked samples were then aliquoted (~3 g) and sent to each laboratory to test toxin profile detection capability and also any matrix effect (Table 11).

Furthermore, to test the reproducibility/repeatability of each laboratory, a third batch of the SRO homogenate was spiked with OA (≅7.2 μL/3 g which is equivalent to 2 × LOR (0.02 mg/kg) and three replicate aliquots of this stock material (3 g) were dispatched to each laboratory. Finally, one sample (~3 g) of the CRM DSP-Mus-c was sent to each laboratory as a positive control. In total, 19 samples (randomly numbered 1–19) were dispatched frozen to each of four laboratories (Table 11).

5.1.4. LC-MS/MS and LC-MS Toxin Determination

Four commercial and/or government analytical laboratories with experience in conducting LC-MS/MS and LC-MS of marine biotoxins in shellfish were engaged to determine DSTs in spiked shellfish, identified only as Laboratories 1–4. The aim of this part of the study was to determine an inter-laboratory comparison of standardized samples, in order to obtain a baseline result using currently mandated seafood safety procedures in Australia [38]. The LC-MS/MS and LC-MS methods were engaged by each of the laboratories, and their limits of detection and limits of reporting/quantification are shown in Appendix A. No recovery corrections were applied to the final results reported from any of the labs.

5.2. Rapid Test Kit Comparison

5.2.1. Shellfish Preparation

Raw samples of SRO, PO, MUS and PIPI (same species as above), were sourced from the Sydney Fish Markets on 29/4/2020. These were stored at 4–8 °C and transported immediately to the University of Technology Sydney laboratory for processing. Again, all shellfish were washed thoroughly with fresh water, shucked and tissue was removed. Bulk material of each species was then made by pooling the tissue of individuals of that species up to 90 g, homogenizing and separating into 3 batches for downstream processing. The first batch served as unspiked controls and were first examined by LC-MS at Laboratory 3 (see above) to ensure each matrix was clear of toxins before the experiment began. The second batch was spiked with CRM-OA-d at ~12 µL/g (0.1 OA eq. mg/kg), which is half the regulatory limit, and the third batch was spiked at ~24 µL/g, which is equal to the regulatory limit. Once prepared all batches were returned to the freezer ($-20\ ^\circ$C) until further processing.

Additionally, during Oct/Nov 2019, DSTs were detected in wild harvest Pipis from Sydney Fish Markets (~400 mg/kg), and a recall was immediately actioned. A batch of these naturally contaminated Pipis were obtained and prepared as positive controls: Sample 4A—14/11/19 Stockton 4–6 km; 4B—7/11/19 Stockton 4 km; 4C—31/10/19 Stockton 2–4 km; and 4D—Sydney Fish Market Stockton recall Nov 2019. Once the OA toxin concentration was determined using LC-MS for these environmentally contaminated samples, samples with toxin level closest to the regulatory level (0.2 mg OA eq/kg) were chosen, and 10 replicates of these positive controls were run on each kit to test the reliability/repeatability of each kit.

A subsample (3 g) of each pooled, species-specific matrix was first examined by LC-MS (Laboratory 3) to ensure each matrix was clear of toxins before the experiment began (unspiked controls). All remaining batches (spiked and positive controls) were then subsampled and prepared according to the rapid test kit protocols for each kit or for LC-MS analysis. Duplicate samples of each treatment/shellfish were tested using both LC-MS and the five test kits.

5.2.2. Rapid Test Kits

A list of DST rapid test kits were screened, their method details including their limit of quantification or working range, amount of tissue required, cost, time for analysis etc., are summarized in Table 9.

Qualitative Test
Neogen

Neogen Reveal 2.1 DSP Test strips (Lot: 9561-49, Neogen Corporation, Scotland, UK) and DSP hydrolysis packs (Lot: 9555-09) were stored at room temperature until experiments began. Each shellfish sample (2 g) was defrosted to room temperature (20–25 $^\circ$C), then transferred to the extraction bag provided before being homogenized with 8 mL analytical grade methanol (Sigma-Aldrich, Sydney, Australia). The sample extract was then poured from each extraction bag (from opposite side of mesh divider) into a 15 mL falcon tube, prior to filtration using a 0.45 µm sterile Minisart® syringe filter into another clean 15 mL tube. Eighty µL of filtered extract was then transferred to a clean glass vial, followed by 100 µL of 2.5 M NaOH, before being capped tightly and mixed using a vortex on full speed for 30 s. The sample vial was then transferred to a heater block set at 76 $^\circ$C for 40 min, after which time the sample was cooled on ice. At room temperature, 100 µL of 2.5 M HCl was added to the sample extract, mixed by hand for 30 s, before 100 µL transferred into a DSP buffer A vial (provided). The sample was again vigorously mixed, before 100 µL was transferred to a microwell plate. A DSP strip was then placed into the microwell plate for 15 min before being immediately placed into the AccuScan® PRO 2.0 scanner for result interpretation.

Quantitative Tests
Abraxis PP2A

The Eurofins/Abraxis Okadaic Acid (PP2A) Microtiter Plate kit Product No. 520025, Lot No. 19/1259, Eurofins Abraxis, Warminster, PA, USA) was stored at 4 $^\circ$C prior to use. Upon opening, the solutions were prepared as per the manufacturer's protocols and allowed to reach room temperature before analysis began. Each shellfish subsample (5 g) was defrosted and 25 mL methanol (Sigma-Aldrich, Sydney, Australia) added before homogenization in a tube shaker for 2 min. The sample was then centrifuged at 2000× *g* for 10 min at 4 °C and 640 µL of the methanolic extract removed and transferred to a clean 15 mL falcon tube. The extract was then mixed with 100 µL of 2.5 N NaOH, sealed, and placed in a water bath at 76 ± 2 °C for 40 min. After removal from the water bath, 80 µL of 2.5 N HCl was added to each sample, followed by 20 mL buffer solution.

For the test protocol, a volume of 50 µL of each OA standard (provided at 0.5, 0.8, 1.2, 1.8, and 2.8 nM) and each shellfish sample was added to the 96 well-plate provided. To each of these wells, 70 µL phosphatase solution was added. The plate was then tapped gently to ensure mixing, before being covered with parafilm and incubated for 20 min at 30 ± 2 °C. Immediately after this incubation period, 90 µL of chromogenic substrate was added to each well, and again, the plate was tapped gently to ensure mixing. The plate was then incubated (covered) for a further 30 min at 30 °C ± 2 °C, after which 70 µL of stop solution was added to each well. Absorbance was immediately read at 405 nm using a Tecan Infinite M1000 PRO plate reader.

For data analysis, a standard curve was obtained by plotting the absorbance values in a linear *y*-axis and the concentration of okadaic acid in a logarithmic *x*-axis. The OA concentration contained in the sample (Cs) was then calculated using the following equation:

$$x = \text{EXP}\,((y - b)/a), \tag{1}$$

where x was the OA concentration in the sample (Cs) and y the absorbance of the sample. The concentration of DSTs in tissue (Ct) was then determined as:

$$\text{Ct (mg/kg)} = ((\text{Cs (nM)} \times \text{FD} \times \text{MW (g/mol)} \times \text{Ve (L)})/\text{Mt (g)})/1000 \tag{2}$$

where Ct: DST concentration in tissue expressed as equivalents of OA; Cs: toxins concentration in sample; FD: Methanolic extract dilution factor (i.e., 640 µL/20 mL → × 31.25); MW: Okadaic acid molecular weight = 805; Ve: Methanolic extract volume (0.025 L); Mt: Tissue weight (5 g).

Beacon ELISA

The Beacon Okadaic Acid (ELISA) Plate kit (Cat. No. 20-0184, Lot No. 6289J, Beacon Analytical Systems Inc., Sako, ME, USA) was stored at 4 °C and all reagents brought to room temperature before use. Each shellfish sample (1 g) was defrosted and 2 mL 80% methanol (Sigma-Aldrich, Sydney, Australia)/water was added before homogenization and transfer to a clean 15 mL falcon tube. A further 8 mL of 80% methanol/water was then added, before vortexing for 5 min followed by centrifugation at 3000 rpm for 5 min. The supernatant was then filtered into a clean 15 mL tube through a 0.45 µm sterile Minisart® syringe filter and the extract diluted 1:50 into 10% methanol/10 mM PBS (Sigma-Aldrich, Sydney, Australia) (i.e., 40 µL of filtered extract into 1.96 mL of 10% methanol/10 mM PBS).

For the test procedure, 50 µL of enzyme conjugate was added into each test well, followed by 100 µL of each OA calibrator (provided at 0, 0.2, 0.5,1.2 and 5 µg/L) or shellfish sample, and 50 µL of antibody. Wells were then mixed for 30 s using gentle shaking, followed by incubation at room temperature for 30 min. The content of the well plates were then decanted, and well plates were washed four times using Milli-Q water, and inverting the plate onto absorbent paper between each wash. After the final wash, 100 µL of substrate was added to each well, before incubation for 30 min at room temperature. Finally, 100 µL of stop solution was added to each well and absorbance read at 450 nm using the Tecan Infinite M1000 PRO plate reader.

For quantitative interpretation of the absorbance readings, a standard curve was then constructed by plotting the absorbance of the calibrators (standards) on the *y*-axis versus the concentration of okadaic acid in a logarithmic *x*-axis. The OA concentration (ppb) contained in the sample (Cs) was then calculated using Equation (1) above. Finally, to obtain the final DST (mg/kg) in each sample, a factor of ×500 to account for the dilution during the shellfish extraction step was applied.

Abraxis ELISA

The Eurofins/Abraxis Okadaic Acid (DSP) ELISA, Microtiter Plate (Product No. 520021, Lot No. 19/1178, Eurofins Abraxis, Warminster, PA, USA) was stored at 4 °C and brought to room temperature before use. All solutions were prepared as per the manufacturer's protocols. Each shellfish subsample (1 g) was defrosted and 6 mL methanol (Sigma-Aldrich, Sydney, Australia)/Milli-Q water (80/20) added before homogenization for 2 min. Each sample was then centrifuged for 10 min at 3000× g and the supernatant was transferred to a clean 15 mL falcon tube. A further 2 mL methanol/Milli-Q was added to the shellfish residue, the sample centrifuged again for 10 min at 3000× g, and the supernatant added to the first portion. The final volume was brought up to 10 mL with methanol/Milli-Q, before filtration into a clean 15 mL tube through a 0.45 μm sterile Minisart® syringe filter. For the hydrolysis step, 500 μL of each sample extract was added to a 2 mL glass vial, and 100 μL of 1.25 N NaOH added. The sample was then vortexed for 15–20 s before incubation on a heat block at 80 °C for 40 min. Each sample was then cooled and 100 μL of 1.25 N HCl added and vortexed for 15–20 s. Finally, 10 μL of the hydrolyzed extract was mixed with 990 μL of 1× sample diluent (1:100 dilution) in a 2 mL glass vial with a cap and vortexed again.

For the assay procedure, a volume of 100 μL of each OA standard (provided at 0, 0.1, 0.2, 0.5, 1, 2, 5 ppb) and shellfish sample was added to each strip well and placed into the well plate provided. To each of these, 50 μL of enzyme conjugate and 50 μL of antibody solution was added. The plate was then covered with parafilm, rotated carefully to mix and left to incubate for 60 min at room temperature, after which the covering was removed and the contents were decanted by inverting the plate onto a paper towel. Each well was then thoroughly washed three times using the diluted wash buffer (~25 μL for each wash/each well), blotting after each step. Following the final washing step, 150 μL of substrate solution was added to each well, before covering with parafilm, rotating gently to mix, and incubating at room temperature for 30 min. Finally, 100 μL of stop solution was added to each well plate prior to immediate absorbance reading at 450 nm using the Tecan Infinite M1000 PRO plate reader.

Kit performance was evaluated by calculating %B/Bo for each standard by dividing the absorbance value for each standard by the Zero standard mean absorbance. A standard curve was then constructed by plotting the %B/Bo for each standard on the y-axis versus the concentration of okadaic acid in a logarithmic x-axis. The OA concentration (ppb) contained in the sample (Cs) was then calculated using Equation (1) above. Finally, to account for hydrolysis sample extraction, hydrolysis and dilutions during the hydrolysis step, all results were multiplied × 1400 to obtain the DSP concentration (ppb) before conversion to mg/kg.

EuroProxima ELISA

The EuroProxima Okadaic Acid ELISA (Catalogue No. 5191OKA, Lot No. UN6635, Arnhem, The Netherlands) was stored at 4 °C before use and subsequently brought to room temperature before use. Reagents were prepared as specified in the manufacturer's protocol. To begin, 1 mL of water was added to each 1 g of shellfish, the sample vortexed for 1 min, and a further 2 mL of 100% methanol (Sigma-Aldrich, Sydney, Australia) was added. The sample was again vortexed for 1 min followed by centrifugation at 2000× g for 10 min. The clear supernatant was then filtered using a 0.45 μm sterile Minisart® syringe filter into a clean 15 mL falcon tube and the sample subsequently diluted 1:50 with the sample dilution buffer provided.

For the assay procedure, 100 μL of the zero standard (0 ng/mL) was pipetted into the first well, and 50 μL thereafter of each OA standard (provided at 0, 0.2, 0.5, 1.0, 2.0, 5.0 10.0 ng/mL) and shellfish samples into the 96 well-plate provided. Following on, 25 μL of enzyme conjugate and 25 μL of antibody was added to each well, except A1. The plate was then sealed with parafilm and gently shaken for 1 min before incubation at room temperature for 30 min. Parafilm was subsequently removed, the well contents discarded onto absorbent paper, and all wells were washed three times with a rinsing buffer. After the final rinse, 100 µL of substrate solution was added to each well, mixed thoroughly and left to incubate for 15 min in the dark prior to 100 mL of stop solution being added. Absorbance was read at 450 nm using the Tecan Infinite M1000 PRO plate reader.

For data interpretation, the mean optical density (OD) value of the wells A1 and A2 were subtracted from the individual OD reading from each of the standards and samples. The OD values of the six standards and samples are then divided by the OD value of the zero standard (well no. B1) and multiplied by 100. The zero standard is then equal to 100% (maximum OD) and the other OD values are % of the maximal OD. A calibration curve was then constructed with the values (% maximal OD) plotted on the y-axis versus the concentration of okadaic acid (ng/mL) in a logarithmic x-axis. The OA concentration (ng/mL) contained in the sample (Cs) was then calculated using Equation (1) above, but this time where x was the OA concentration in the sample (Cs) and y the % max OD of the sample. Finally, to obtain OA equivalents in the final shellfish, a factor of $\times$ 200 (and /1000) was applied.

5.3. Data Assessment

Toxin recovery from samples analyzed using LC-MS/MS and LC-MS were assessed in four ways. 1. Where sample replication was available, mean ($\pm$SD) toxin recoveries were calculated and compared to the spiked concentration and LOR, and finally compared across laboratories. 2. To determine each analogue recovery using LC-MS/MS and LC-MS, toxin results from each shellfish species were compared to the spiked toxin concentration, and then compared across laboratories. 3. For shellfish that were spiked with a combination of OA analogues, the results were compared to both spiked concentration and the ML (0.2 mg/kg OA), as well as across laboratories. 4. Finally, the recovery of toxins in certified reference material CRM (DSP-Mus-c) were compared across laboratories.

To examine the performance of the rapid test kits, firstly, we assessed the performance of the qualitative Neogen kit by comparison to the spiked toxin concentration in each sample (% false positives/% false negatives). Secondly, the performance and recovery of all quantitative methods (including LC-MS) were compared (% overestimated; % underestimated; % recovery; Pearson's correlation using Excel 2016) to the spiked concentration of each sample. For those samples spiked at, or above, the ML adopted by the Food Standards Australia New Zealand (0.2 OA mg/kg), we also determined whether they were "falsely compliant" or "falsely non-complaint" with the ML. These terms refer to the comparison of the results obtained to the maximum regulatory limit. For example, if a sample was spiked above the regulatory limit but resulted in a concentration below the regulatory limit, it was referred to as "falsely compliant". Conversely, if a sample was spiked below the regulatory limit but returned a concentration above the regulatory limit, it was referred to as "falsely non-compliant". Thirdly, a comparison across species-specific matrices was undertaken to assess the suitability of rapid test kits across a range of shellfish species. Finally, the reliability or repeatability of each kit was assessed (defined as the standard deviation of the mean, Excel 2016) from the replicate positive controls (naturally contaminated Pipi samples) across all quantitative kits.

Author Contributions: Conceptualization: P.A.A., A.T., H.F., A.Z., S.H., G.H. and S.A.M. Methodology: all authors. Formal analysis and data curation: P.A.A., C.S. and S.A.M. Writing: P.A.A. Manuscript review and editing: all authors. Project administration: P.A.A. and S.A.M. All authors have read and agreed to the published version of the manuscript.

Funding: This research was funded by Fisheries Research and Development Corporation (FRDC) Project No. 2017-203 Risk from Diarrhetic Shellfish Toxins and Dinophysis to the Australian Shellfish Industry.

Institutional Review Board Statement: Not applicable.

Informed Consent Statement: Not applicable.

Data Availability Statement: Data is contained within the article.

Acknowledgments: The authors would like to acknowledge Sydney Fish Markets, University of Technology, and Sydney Institute of Marine Science for laboratory and research support.

Conflicts of Interest: The authors declare no conflict of interest.

Appendix A

Table A1. Methods, detection limits, limit of quantification/reporting and measurement uncertainty (standard uncertainty at the LOR) as reported by each laboratory for LC-MS/MS and LM-MS determination of DSTs in shellfish.

	Method	Limit of Detection	Limit of Quantification (LOQ)/Limit of Reporting (LOR)	Measurement Uncertainty
Lab 1 LC-MS/MS	LC-MS/MS Method similar to McNabb (2005) and Villar-Gonzalez et al. (2011) and the EU-Harmonised method from the EU Reference Lab. That is, an 80% MeOH extraction, with two portions of the extract analysed after (1) hexane-cleanup, (2) alkaline hydrolysis (to convert esters to acids).	0.004 mg/kg OA, DTX-1, DTX-2	0.01 mg/kg OA, DTX-1, DTX-2	25% OA 26% DTX-1 24% DTX-2 (at a confidence level of 95%)
Lab 2 LC-MS/MS	Multitoxin LC-MS/MS method for lipophilic toxins based on McNabb 2005 with IANZ (ISO 17025) accreditation.	0.001–0.002 mg/kg OA, DTX-1, DTX-2	0.01 mg/kg OA, DTX-1, DTX-2	21% at 0.01 mg/kg
Lab 3 LC-MS	Sample extraction was performed using the method as described by McNabb et al. (2005). OA analysis was conducted using a Thermo Scientific™ Q EXACTIVE™ high resolution mass-spectrometer equipped with an electrospray ionization. Chromatographic separation was performed on a Thermo Scientific™ ACCELA™ UPLC system.	0.006 mg/kg OA 0.007 mg/kg DTX-1 0.007 mg/kg DTX-2	0.021 mg/kg OA 0.023 mg/kg DTX-1 0.024 mg/kg DTX-2	19% OA 21% DTX-1 12 % DTX-2
Lab 4 LC-MS/MS	LC-MS/MS using the instrument AB ScieX Triple Quad 6500.	~5–10 × lower than the LOQ/LOR	0.025 mg/kg OA, DTX-1 0.015 mg/kg DTX-2	20% Total OA 20% Total DTX-1 20% Total DTX-2 15% Free OA 15% Free DTX-1 10% Free DTX-2

References

1. Hallegraeff, G.; Enevoldsen, H.; Zingone, A. Global harmful algal bloom status reporting. *Harmful Algae* **2021**, *102*, 101992. [CrossRef] [PubMed]

2.   Lee, T.C.-H.; Fong, F.L.-Y.; Ho, K.-C.; Lee, F.W.-F. The Mechanism of Diarrhetic Shellfish Poisoning Toxin Production in *Prorocentrum* spp.: Physiological and Molecular Perspectives. *Toxins* **2016**, *8*, 272. [CrossRef]
3.   Reguera, B.; Riobó, P.; Rodríguez, F.; Díaz, P.A.; Pizarro, G.; Paz, B.; Franco, J.M.; Blanco, J. Dinophysis Toxins: Causative Organisms, Distribution and Fate in Shellfish. *Mar. Drugs* **2014**, *12*, 394–461. [CrossRef]
4.   McNabb, P.; Selwood, A.I.; Holland, P.T.; Aasen, J.; Aune, T.; Eaglesham, G.; Hess, P.; Igarishi, M.; Quilliam, M.; Slattery, D.; et al. Multiresidue Method for Determination of Algal Toxins in Shellfish: Single-Laboratory Validation and Interlaboratory Study. *J. AOAC Int.* **2005**, *88*, 761–772. [CrossRef]
5.   Yasumoto, T.; Oshima, Y.; Sugawara, W.; Fukuyo, Y.; Oguri, H.; Igarashi, T.; Fujita, N. Identification of Dinophysis fortii as the causative organism of diarrhetic shellfish poisoning. *Bull. Jpn. Soc. Sci. Fish.* **1980**, *46*, 1405–1411. [CrossRef]
6.   Yasumoto, T.; Oshima, Y.; Yamaguchi, M. Occurrence of a new type of shellfish poisoning in the Tohoku district. *Bull. Jpn. Soc. Sci. Fish.* **1978**, *44*, 1249–1255. [CrossRef]
7.   Lembeye, G. DSP Outbreak in Chilean Fjords. In *Toxic Phytoplankton Blooms in the Sea*; Smayda, T.J., Shimizu, Y., Eds.; Elsevier: Amsterdam, The Netherlands, 1993; pp. 525–529.
8.   Taylor, M.; McIntyre, L.; Ritson, M.; Stone, J.; Bronson, R.; Bitzikos, O.; Rourke, W.; Galanis, E.; Team, O.I. Outbreak of Diarrhetic Shellfish Poisoning Associated with Mussels, British Columbia, Canada. *Mar. Drugs* **2013**, *11*, 1669–1676. [CrossRef]
9.   Whyte, C.; Swan, S.; Davidson, K. Changing wind patterns linked to unusually high Dinophysis blooms around the Shetland Islands, Scotland. *Harmful Algae* **2014**, *39*, 365–373. [CrossRef]
10.   Ajani, P.; Ingleton, T.; Pritchard, T.; Armand, L. Microalgal blooms in the coastal waters of New South Wales, Australia. *Proc. Linn. Soc. N. S. W.* **2011**, *133*.
11.   Hallegraeff, G.M.; Lucas, I.A.N. The marine dinoflagellate genus Dinophysis (Dinophyceae)—Photosynthetic, neritic and non-photosynthetic, oceanic species. *Phycologia* **1988**, *27*, 25–42. [CrossRef]
12.   McCarthy, P.M. Census of Australian Marine Dinoflagellates. 2013. Available online: http://www.anbg.gov.au/abrs/Dinoflagellates/index_Dino.html (accessed on 1 May 2021).
13.   Quaine, J.; Kraa, E.; Holloway, J.; White, K.; McCarthy, R.; Delpech, V.; Trent, M.; McAnulty, J. Outbreak of gastroenteritis linked to eating pipis. *N. S. W. Public Health Bull.* **1997**, *8*, 103–104.
14.   Madigan, T.L.; Lee, K.G.; Padula, D.J.; McNabb, P.; Pointon, A.M. Diarrhetic shellfish poisoning (DSP) toxins in South Australian shellfish. *Harmful Algae* **2006**, *5*, 119–123. [CrossRef]
15.   Burgess, V.; Shaw, G. Pectenotoxins—An issue for public health: A review of their comparative toxicology and metabolism. *Environ. Int.* **2001**, *27*, 275–283. [CrossRef]
16.   Quilliam, M.A.; Xie, M.; Hardstaff, W.R. Rapid Extraction and Cleanup for Liquid Chromatographic Determination of Domoic Acid in Unsalted Seafood. *J. AOAC Int.* **1995**, *78*, 543–554. [CrossRef]
17.   Christian, B.; Luckas, B. Determination of marine biotoxins relevant for regulations: From the mouse bioassay to coupled LC-MS methods. *Anal. Bioanal. Chem.* **2007**, *391*, 117–134. [CrossRef]
18.   Lequin, R.M. Historical background of the invention of EIA and ELISA—Response. *Clin. Chem.* **2006**, *52*, 1431. [CrossRef]
19.   Jawaid, W.; Meneely, J.P.; Campbell, K.; Melville, K.; Holmes, S.J.; Rice, J.; Elliott, C.T. Development and Validation of a Lateral Flow Immunoassay for the Rapid Screening of Okadaic Acid and All Dinophysis Toxins from Shellfish Extracts. *J. Agric. Food Chem.* **2015**, *63*, 8574–8583. [CrossRef]
20.   Macleod, C.; Burrell, S.; Holland, P. *Review of the Currently Available Field Methods for Detection of Marine Biotoxins in Shellfish Flesh*; Seafood Safety Assessment Ltd.: Wiltshire, UK, 2015.
21.   Codex Alimentarius. In *Standard for Live and Raw Bivalve Molluscs*; World Health Organisation: Rome, Italy, 2015.
22.   EFSA. Scientific Opinion of the Panel on Contaminants in the Food Chain on a Request from the European Commission on Marine Biotoxins in Shellfish—Pectenotoxin Group. *ESFA J.* **2009**, *1109*, 1–47.
23.   FSANZ, Maximum Levels of Non-Metal Contaminants. F2017C00333 S19-5. In Australia New Zealand Food Standards Code—Schedule 19. Available online: https://www.legislation.gov.au/Details/F2016C00197 (accessed on 1 May 2021).
24.   AOAC. AOAC International Guidelines for Validation of Qualitative Binary Chemistry Methods. *J. AOAC Int.* **2019**, *97*, 1492–1495.
25.   Turner, A.D.; Goya, A.B. Comparison of four rapid test kits for the detection of okadaic acid-group toxins in bivalve shellfish from Argentina. *Food Control.* **2016**, *59*, 829–840. [CrossRef]
26.   Eberhart, B.-T.L.; Moore, L.K.; Harrington, N.; Adams, N.G.; Borchert, J.; Trainer, V.L. Screening Tests for the Rapid Detection of Diarrhetic Shellfish Toxins in Washington State. *Mar. Drugs* **2013**, *11*, 3718–3734. [CrossRef] [PubMed]
27.   Farrell, H.; Ajani, P.; Murray, S.; Baker, P.; Webster, G.; Brett, S.; Zammit, A. Diarrhetic Shellfish Toxin Monitoring in Commercial Wild Harvest Bivalve Shellfish in New South Wales, Australia. *Toxins* **2018**, *10*, 446. [CrossRef] [PubMed]
28.   Suzuki, T.; Quilliam, M. LC-MS/MS Analysis of Diarrhetic Shellfish Poisoning (DSP) Toxins, Okadaic Acid and Dinophysistoxin Analogues, and Other Lipophilic Toxins. *Anal. Sci.* **2011**, *27*, 571. [CrossRef] [PubMed]
29.   Schirone, M.; Berti, M.; Visciano, P.; Chiumiento, F.; Migliorati, G.; Tofalo, R.; Suzzi, G.; Di Giacinto, F.; Ferri, N. Determination of Lipophilic Marine Biotoxins in Mussels Harvested from the Adriatic Sea by LC-MS/MS. *Front. Microbiol.* **2018**, *9*, 152. [CrossRef] [PubMed]
30.   Uchida, H.; Watanabe, R.; Matsushima, R.; Oikawa, H.; Nagai, S.; Kamiyama, T.; Baba, K.; Miyazono, A.; Kosaka, Y.; Kaga, S.; et al. Toxin Profiles of Okadaic Acid Analogues and Other Lipophilic Toxins in Dinophysis from Japanese Coastal Waters. *Toxins* **2018**, *10*, 457. [CrossRef]

31. European Union. *Harmonised Standard Operating Procedure for determination of Lipophilic Marine Biotoxins in Molluscs by LC-MS/MS*; Working Group LC-MS for Lipophilic Toxins of the European Network of National Reference Laboratories (NRL) for Marine Biotoxins: Vigo, Spain, 2015; pp. 1–33.
32. Turner, A.D.; Tarnovius, S.; Hatfield, R.G.; Teixeira-Alves, M.; Broadwater, M.; Van Dolah, F.; Garcia-Mendoza, E.; Medina, D.; Salhi, M.; Goya, A.B.; et al. Application of Six Detection Methods for Analysis of Paralytic Shellfish Toxins in Shellfish from Four Regions within Latin America. *Mar. Drugs* **2020**, *18*, 616. [CrossRef]
33. Dubois, M.; Demoulin, L.; Charlier, C.; Singh, G.; Godefroy, S.B.; Campbell, K.; Elliott, C.T.; Delahaut, P. Development of ELISAs for detecting domoic acid, okadaic acid, and saxitoxin and their applicability for the detection of marine toxins in samples collected in Belgium. *Food Addit. Contam. Part A Chem. Anal. Control. Expo. Risk Assess.* **2010**, *27*, 859–868. [CrossRef]
34. Smienk, H.; Domínguez, E.; Rodríguez-Velasco, M.L.; Clarke, D.; Kapp, K.; Katikou, P.; Cabado, A.G.; Otero, A.; Vieites, J.M.; Razquin, P.; et al. Quantitative Determination of the Okadaic Acid Toxins Group by a Colorimetric Phosphatase Inhibition Assay: Interlaboratory Study. *J. AOAC Int.* **2013**, *96*, 77–85. [CrossRef]
35. Smienk, H.G.F.; Calvo, D.; Razquin, P.; Domínguez, E.; Mata, L. Single Laboratory Validation of A Ready-to-Use Phosphatase Inhibition Assay for Detection of Okadaic Acid Toxins. *Toxins* **2012**, *4*, 339–352. [CrossRef]
36. Johnson, S.; Harrison, K.; Turner, A.D. Application of rapid test kits for the determination of Diarrhetic Shellfish Poisoning (DSP) toxins in bivalve molluscs from Great Britain. *Toxicon* **2016**, *111*, 121–129. [CrossRef]
37. Turnbull, A.R.; Tan, J.Y.; Ugalde, S.C.; Hallegraeff, G.M.; Campbell, K.; Harwood, D.T.; Dorantes-Aranda, J.J. Single-Laboratory Validation of the Neogen Qualitative Lateral Flow Immunoassay for the Detection of Paralytic Shellfish Toxins in Mussels and Oysters. *J. Aoac Int.* **2018**, *101*, 480–489. [CrossRef] [PubMed]
38. ASQAAC. *The Australian Shellfish Quality Assurance Program Operations Manual*; A.S.Q.A.A. Committee: Canberra, Australia, 2016.

Review

Critical Review and Conceptual and Quantitative Models for the Transfer and Depuration of Ciguatoxins in Fishes

Michael J. Holmes [1], Bill Venables [2] and Richard J. Lewis [3,*]

[1] Queensland Department of Environment and Science, Brisbane 4102, Australia; michael.holmes@des.qld.gov.au
[2] CSIRO Data61, Brisbane 4102, Australia; bill.venables@gmail.com
[3] Institute for Molecular Bioscience, The University of Queensland, Brisbane 4072, Australia
* Correspondence: r.lewis@imb.uq.edu.au

Abstract: We review and develop conceptual models for the bio-transfer of ciguatoxins in food chains for Platypus Bay and the Great Barrier Reef on the east coast of Australia. Platypus Bay is unique in repeatedly producing ciguateric fishes in Australia, with ciguatoxins produced by benthic dinoflagellates (*Gambierdiscus* spp.) growing epiphytically on free-living, benthic macroalgae. The *Gambierdiscus* are consumed by invertebrates living within the macroalgae, which are preyed upon by small carnivorous fishes, which are then preyed upon by Spanish mackerel (*Scomberomorus commerson*). We hypothesise that *Gambierdiscus* and/or *Fukuyoa* species growing on turf algae are the main source of ciguatoxins entering marine food chains to cause ciguatera on the Great Barrier Reef. The abundance of surgeonfish that feed on turf algae may act as a feedback mechanism controlling the flow of ciguatoxins through this marine food chain. If this hypothesis is broadly applicable, then a reduction in herbivory from overharvesting of herbivores could lead to increases in ciguatera by concentrating ciguatoxins through the remaining, smaller population of herbivores. Modelling the dilution of ciguatoxins by somatic growth in Spanish mackerel and coral trout (*Plectropomus leopardus*) revealed that growth could not significantly reduce the toxicity of fish flesh, except in young fast-growing fishes or legal-sized fishes contaminated with low levels of ciguatoxins. If Spanish mackerel along the east coast of Australia can depurate ciguatoxins, it is most likely with a half-life of $\leq$1-year. Our review and conceptual models can aid management and research of ciguatera in Australia, and globally.

Keywords: ciguatera; ciguatoxin; maitotoxin; 44-methylgambierone; toxin depuration; *Gambierdiscus*; *Fukuyoa*; Platypus Bay; Great Barrier Reef; *Scomberomorus commerson*; Spanish mackerel; *Plectropomus*; coral trout; turf algae; surgeonfish; *Ctenochaetus*; *Acanthurus*

Key Contribution: Conceptual models were developed for the transfer of ciguatoxins across marine trophic levels, and for the dilution and depuration of ciguatoxins from fishes. For commercial-sized fishes, growth is unlikely to dilute ciguatoxin concentrations to levels safe for human consumption.

Citation: Holmes, M.J.; Venables, B.; Lewis, R.J. Critical Review and Conceptual and Quantitative Models for the Transfer and Depuration of Ciguatoxins in Fishes. *Toxins* **2021**, *13*, 515. https://doi.org/10.3390/toxins13080515

Received: 3 June 2021
Accepted: 16 July 2021
Published: 23 July 2021

1. Introduction

Ciguatera is a disease in humans caused by eating normally edible warm water fishes contaminated with a class of potent, lipid-soluble toxins called ciguatoxins (CTX). It is estimated to poison >25,000 people annually [1], with ~500,000 people poisoned across the Pacific basin between 1973 and 2008 [2]. The largest single outbreak of ciguatera possibly occurred in 1748 in the invasion fleet of British Admiral Boscawen at Rodrigues Island in the Indian Ocean, when >1500 men reportedly died before his failed assault on the then French island of Mauritius [3]. Until recently, ciguatera was mostly confined to tropical and subtropical coastal communities, but with tropical fish now part of the global food supply chain, cases can occur anywhere. Victims typically experience a range of gastrointestinal and neurological symptoms with diagnosis based upon clinical presentation after recently eating a suspect fish species [4–7]. Paraesthesia and cold allodynia are diagnostic neurological symptoms for ciguatera across much of the Pacific basin [1,5,8].

Ciguatoxins activate voltage sensitive sodium channels at pM to nM concentrations and are the most potent, orally active, mammalian sodium channel toxins known [9,10]. They cause a hyperpolarizing shift in the voltage-dependence of channel activation resulting in channels opening at resting membrane potentials. The pathophysiological effects of the ciguatoxins are thought to be defined by their ability to cause the persistent activation of these channels and to also inhibit neuronal potassium channels, leading to increased neuronal excitability and neurotransmitter release, impaired synaptic vesicle recycling and modified Na^+-dependent mechanisms in numerous cell types [11]. They share a common binding site (site 5) on sodium channels with the structurally similar polyether brevetoxins [12] and effect the cathepsin S-protease activated receptor-2 pathway that contributes to the sensory effects of the toxins [13].

Ciguatera is an uncommon but underreported disease in Australia with most cases caused by Spanish mackerel (*Scomberomorus commerson*) caught from the east coast of Australia or demersal reef fish from the Great Barrier Reef [5,6,14,15]. It is a notifiable disease in Queensland, Australia, but many cases go unreported, either because health care professionals are not familiar with the disease, or patients do not associate their illness with eating fish given the delayed onset of symptoms, which are often mild. There is no confirmatory test available to assist doctors with their diagnosis, and treatment is symptomatic and supportive only, although intravenous D-mannitol has shown promise as an early treatment for severe cases [16–18]. Between 2014 and 2020, there were 182 ciguatera cases reported to the health department for the State of Queensland, Australia (Queensland Health, notifiable conditions annual reporting: https://www.health.qld.gov.au/clinical-practice/guidelines-procedures/diseases-infection/surveillance/reports/notifiable/annual (accessed on 10 July 2019, 13 July 2020, 20 July 2021)). While there are sensitive and reliable laboratory methods for detecting and quantifying ciguatoxins from the flesh of fishes (reviewed by [19]), no rapid, reliable, cost-effective screening method is available to test commercial quantities of fishes prior to consumption.

Ciguatoxin was the name given by Scheuer et al. [20] to the toxin in a partially purified extract from Pacific Ocean moray eels. Ciguatoxins are now considered to be a family of large, ladder-like, cyclic polyether toxins (Figure 1) with the structure of the first ciguatoxin isolated from fishes from the Pacific Ocean, Pacific-ciguatoxin-1 (P-CTX-1, also known as CTX-1B), determined by Murata et al. [21] using ciguatoxin that was isolated and purified in French Polynesia. The current nomenclature for the various toxin congeners is confusing and the subject of debate [22], with standard naming conventions often ignored. More than 20 analogs have been determined from fishes and the causative dinoflagellates from the Pacific with P-CTX-4A (Figure 1, previously designated as GTX-4A) being the presumed precursor of P-CTX-1 [22–25]. These analogs are generally referred to as belonging to either the P-CTX-4A or P-CTX3C family of toxins (Figure 1, reviewed by [22]).

P-CTX-1, or its 54-dexoy analog, P-CTX-2 (52-*epi*-54-deoxy P-CTX-1) (Figure 1), are the major toxins so far found in ciguateric fishes from the east coast of Australia and the Northern Territory [26–29]. A stereoisomer of P-CTX-2, named P-CTX-3 (52-*epi*-P-CTX-2), has also been extracted from ciguateric fishes [26,30–32] with P-CTX-3 having the lower energy configuration for the terminal furan M-ring attached to C-52 [33]. Lewis and Holmes [33] proposed a pathway where P-CTX-4A produced by dinoflagellates is biotransformed through marine food chains to P-CTX-2/-3, and then to P-CTX-1. Evidence supporting this pathway was found from the Republic of Kiribati, where the flesh of carnivorous fishes generally had higher P-CTX-1 concentrations relative to P-CTX-2/-3, but P-CTX-2/-3 mostly dominated in the flesh of herbivorous and omnivorous fishes (relative to P-CTX-1) [32]. However, P-CTX-2 can sometimes be the dominant ciguatoxin in the flesh of carnivorous fishes in Australia and Kiribati [28,32], and Yogi et al. [31] found that P-CTX-3 was dominant in the flesh of a carnivorous blue-spot coral trout, *Plectropomus laevis* from Okinawa. Due to the different toxicities of the various ciguatoxin analogs, the toxicity of samples containing unknown or mixtures of toxins is often expressed in toxicity equivalents, especially as P-CTX-1 or P-CTX3C toxicity equivalents [22].

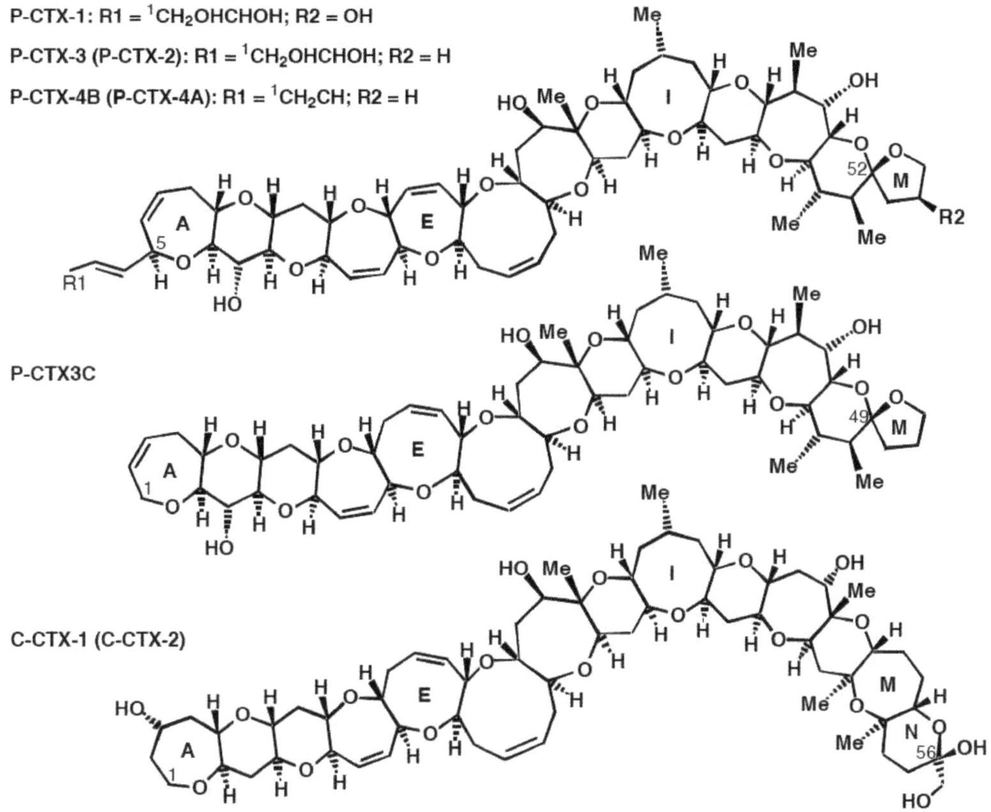

Figure 1. Pacific-ciguatoxin-1 (P-CTX-1), P-CTX-2 and P-CTX-3 (52-*epi*-P-CTX-2) are the major ciguatoxins found in ciguateric fishes from the east coast of Australia, with P-CTX-4A their presumed precursor produced by *Gambierdiscus* and *Fukuyoa* spp. P-CTX3C has been found in dinoflagellates and fishes from the Pacific Ocean, but not yet reported from Australia. Caribbean-ciguatoxin-1 (C-CTX-1) and C-CTX-2 are two of the major ciguatoxins found in fishes from the Caribbean Sea. P-CTX-2 and -3 differ by the stereochemistry of the H on C54, as does P-CTX-4A and -4B. C-CTX-1 and -2 differ by the stereochemistry of the CH$_2$-OH sidechain on C56. P-CTX-1 is also known as CTX1B, P-CTX-2 as 52-epi-54-deoxyCTX1B, and P-CTX-3 as 54-deoxyCTX1B.

In addition to Pacific ciguatoxins, two other structural families of ciguatoxins are known based upon geographical location, the Caribbean ciguatoxins (C-CTX, Figure 1) and Indian Ocean ciguatoxins (I-CTX). Structures of four C-CTX analogs have been determined [34–36], along with additional toxic variants characterised by mass [34] but those from the Indian Ocean have to-date only been characterized by liquid chromatography-mass spectroscopy, with structure elucidation remaining elusive due to poor recovery during purification [37–39].

The origin of the ciguatoxins are microscopic, benthic dinoflagellate species belonging to the genera *Gambierdiscus* and *Fukuyoa*, normally found as epiphytes on macroalgae and a range of other benthic substrates [22,40,41]. They attach to substrates using mucous threads or "webs" [42] but can swim for short periods to be dispersed by local currents (tycoplanktonic). *Gambierdiscus toxicus* was the first microorganism linked to production of ciguatoxins by a French-Japanese collaboration working in the ciguatera-endemic Gambier Islands in French Polynesia [40,42,43]. To-date, 18 anterior-posteriorly flattened (discus-shaped) *Gambierdiscus* species, and 4 globular-shaped *Fukuyoa* species have been identified (Table S1). Most of the *Gambierdiscus* species and *F. ruetzleri* appear to be capable of producing ciguatoxins [22,44], although the cellular concentrations can vary greatly. The highest cellular concentrations of ciguatoxins appear to be produced by *G. polynesiensis* and *G. excentricus* [45–47]. It is now thought that the original description of *G. toxicus* [40] was based upon a sample containing more than one species, so Litaker et al. [48] redesignated the type species for *G. toxicus* based upon the GTT-91 isolate by Chinain et al. [49]. However, it is difficult to know how much of the literature published after 2009 referring to ciguatoxins from *G. toxicus*, is consistent with the revised description of Litaker et al. [48].

Most *Gambierdiscus* and possibly *Fukuyoa* species also produce more polar, water-soluble toxins with the first of these identified as maitotoxin (MTX). The name maitotoxin comes from the Tahitian word "Maito" for the lined-bristletooth surgeonfish (*Ctenochaetus striatus*) from which the putative toxin was first extracted [50]. The structure of the MTX extracted from a French Polynesian strain of *Gambierdiscus* was elucidated by Murata et al. [51] and renamed MTX-1 by Holmes et al. [52] to distinguish it from other maitotoxins. MTX-1 has a polyether structure unrelated to the ciguatoxins, and maitotoxins have not yet been shown to have a role in human poisoning. MTX-1 is one of the largest non-polymeric chemical structures ever determined and is the most potent marine toxin known when injected intraperitoneally into mice [51], but ciguatoxins are significantly more toxic when dosed orally to match the route of exposure in humans [53]. Two other maitotoxins, named MTX-2 and -3 were characterised from Queensland isolates of *Gambierdiscus* [52,54,55] and MTX-4 from *G. excentricus* [56].

The structure of MTX-3 was recently suggested to be 44-methylgambierone [57,58], and has been detected in many *Gambierdiscus* and *Fukuyoa* species [59–64]. However, Murray et al. [62] reported the toxicity (LD_{50}) of 44-methylgambierone injected intraperitoneally (i.p.) into mice at between 20–38 mg/kg. The low toxicity and signs displayed by mice injected with 44-methylambierone [62] are inconsistent with the potency and signs induced by MTX-3 (Figures S1 and S2, Table S2). Indeed, while Holmes and Lewis [54] did not isolate sufficient MTX-3 to obtain a weight for pure material, a chromatographically enriched fraction containing MTX-3 had an $LD_{50} < 50$ μg/kg (Table S2), indicating ~1000-fold greater toxicity than 44-methylgambierone. Critical to its characterization, purified MTX-3 produced similar signs in mice (i.p.) to MTX-1 ($LD_{50} \approx 50$ ng/kg, [51]) and MTX-2 ($LD_{50} \approx 80$ ng/kg, [54]), all three toxins lost toxicity upon desulphation [54,65], could cause death in mice in <2 h when injected i.p. [52,54] Figure S1, and produced similar pharmacological responses in cardiac and smooth muscle tissues [54]. This characterisation shows that MTX-3 is not 44-methylgambierone.

Much of the early literature on the toxicity of *Gambierdiscus* from cultured isolates failed to use methods that could differentiate ciguatoxins from maitotoxins and therefore should be interpreted cautiously [22,23,66,67]. Early research on benthic dinoflagellates in Australia found only 2 of 13 laboratory clones of *Gambierdiscus* isolated from Queensland produced detectable concentrations of ciguatoxins by mouse bioassay [68].

Speculation on the dietary accumulation of toxins by fish to cause human poisoning has appeared in the literature from at least the 16th century. However, the marine food chain transfer of toxins causing ciguatera appears to have been first suggested by Mills [69] and later expanded by Randall [70] into a comprehensive hypothesis for the accumulation and bio-concentration of toxins from a benthic source into herbivorous fishes and then carnivorous fishes that prey on these herbivores. Lewis and Holmes [33] expanded on this model to incorporate ciguatoxin bio-transformations and dilution in fish through growth and toxin depuration. However, recent evidence suggests that ciguatoxins accumulate, but do not always bio-concentrate across trophic levels, at least in juvenile marine fishes [71,72] and freshwater goldfish [73]. The aim of this review is to develop conceptual models for the trophic transfers, accumulation, and loss of ciguatoxins in fishes for two different ecosystems that produce ciguateric fishes based upon our knowledge of ciguatera in Platypus Bay and the Great Barrier Reef on the east coast of Australia. We do this by modelling the transfer of ciguatoxins from their production in benthic dinoflagellates, to invertebrates/herbivores and then carnivorous fish for each ecosystem. Fisheries stock assessment and catch data are then used to model the dilution of ciguatoxins by growth in carnivorous fish species responsible for causing ciguatera from each ecosystem, Spanish mackerel (*Scomberomorus commerson*) from Platypus Bay, and common coral trout (*Plectropomus leopardus*) from the Great Barrier Reef. We also model the dilution of ciguatoxin from Spanish mackerel by a combination of somatic growth and depuration, and compare hypothetical depuration rates with the incidence of ciguatera from Spanish mackerel from the east coast of Australia. The relative risk of ciguatera from Spanish mackerel and coral trout is then compared. We use our review to identify knowledge gaps and make recommendations for future research directions.

2. A Conceptual Model for a Ciguateric Food-Chain in Platypus Bay

Platypus Bay is a small (<400 km^2), sheltered, sandy bay mostly <20 m in depth, which is bounded by a line between Rooney Point and Coongul Point on the north-west coast of Fraser Island (known as K'gari by the indigenous Butchulla people), Queensland (Figure 2). It forms the eastern boundary of the much larger Hervey Bay, and lies within the Great Sandy Marine Park. In recent times, it has become a destination for observing humpback whales as they transit on their annual southward migration from the Coral Sea to Antarctica. Two of the authors (RJL, MJH) began investigating Platypus Bay in the 1980's because of the frequency of ciguatera from Spanish mackerel (*S. commerson*) and barracuda (*Sphyraena jello*) caught in its vicinity [5,74,75]. Platypus Bay is the only known location on the east coast of Australia where ciguateric fishes have been repeatedly caught [76,77], with many cases of ciguatera from Spanish mackerel reported across the Hervey Bay region in the late 1970's and 1980's [78,79]. However, because of Platypus Bay's small area relative to that of the Great Barrier Reef (>340,000 km^2), absence of a coral reef environment more typically associated with ciguatera, the fact that the fishes causing ciguatera are highly mobile pelagic species and the paradigm that fish retain toxicity [70], it was initially suspected that Spanish mackerel and barracuda might accumulate ciguatoxins elsewhere along the coast of Queensland [78,80]. It was thought that Platypus Bay was where toxic pelagic fish and mackerel fishers occasionally intersected, with fishers taking advantage of the shelter that Fraser Island provides from the predominate easterly winds. This was supported by the observation that extracts of livers from rabbitfish (*Siganus spinus*), a common herbivore that consumes the dominant macroalgae in the bay, were non-toxic and dredged macroalgal samples collected at 5 m depth contained no detectable *Gambierdiscus* [80].

However, later evidence pointed to Platypus Bay as the source of ciguatoxins contaminating Spanish mackerel. Firstly, in 1987 the Queensland Government banned the taking of Spanish mackerel and barracuda from Platypus Bay, after which, ciguatera cases in Queensland were no longer dominated by Spanish mackerel, but instead were mostly caused by demersal reef fish caught from the Great Barrier Reef [77,81]. Secondly, there was no obvious increase in frequency of ciguatera caused by demersal fishes from the Great Barrier Reef, but ciguatera cases from Spanish mackerel decreased. It is difficult to believe that prohibiting the taking of Spanish mackerel from such a small bay could make such a difference for a species capable of making annual migrations of more than a 1000 km [82]. It also raised questions about the accepted logic that fish retained toxicity for life, since toxic fish were not captured elsewhere given the high and previously unsustainable catches of this fish along the east coast of Queensland [83]. Thirdly, a family of recreational fishers were poisoned after eating blotched-javelin fish (*Pomadasys maculatus*) they caught in Platypus Bay [84]. The blotched-javelin is a small carnivorous grunter bream that is abundant in tropical estuaries and can grow to a length of 25 cm, but is not commercially important in Australia [85]. What was surprising about this ciguatera outbreak, was that all five family members were poisoned despite eating separate fish. In addition, when batches of the remaining 54 uneaten fish were assayed, every batch of fish was toxic [26]. While up to 100% toxicity of high-risk fish species is not unusual in ciguateric hot spots such as the islands of Tarawa and Marakei in the Republic of Kiribati [32,86,87], this remains a unique occurrence in Australia.

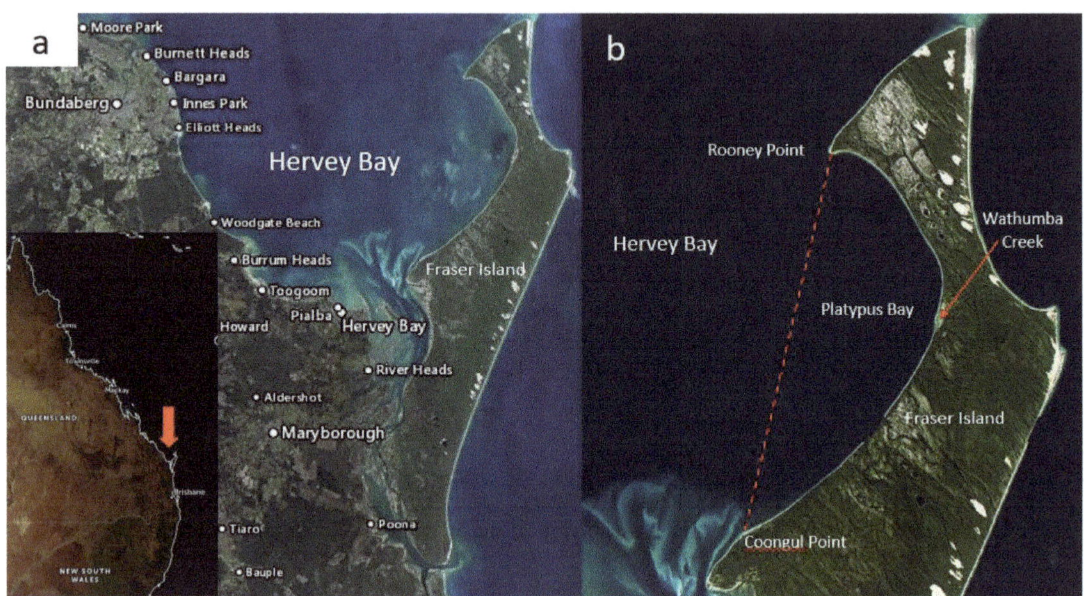

Figure 2. (**a**) Hervey Bay and Fraser Island, Queensland, Australia. Image adapted from QGlobe (https://qldglobe. information.qld.gov.au/ (accessed on 20 July 2021); Creative Commons Attribution 4.0 International (CC BY 4.0) licence). Inset: Outline of Queensland and New South Wales with arrow showing the location of Hervey Bay. Image adapted from Zoom Earth (https://zoom.earth/ (accessed on 20 July 2021)). (**b**) Platypus Bay is a smaller bay within Hervey Bay, and lies between Rooney Point and Coongul Point on Fraser Island. Image adapted from Zoom Earth (https://zoom.earth/ (accessed on 20 July 2021)).

2.1. Food Chain Links Leading to Ciguatera in Platypus Bay, Trophic Level 1

In contrast to earlier studies using a dredge [80], hand collections of the benthic, free-living green filamentous macroalgae (*Cladophora* sp.) that covered large areas of the sandy bottom of Platypus Bay identified *Gambierdiscus* epiphytes at up to 556 cells per gram of wet weight *Cladophora* [77]. Observations of the benthos between 5–20 m depth using scuba diving found that the bottom of this ~50–100 mm thick macroalgal layer typically transitioned into blackened sand consistent with an anoxic layer at the interface between the sand and the overlaying blanket of macroalgae. Analysis of cultured clones of *Gambierdiscus* isolated from the *Cladophora* from Platypus Bay revealed that one of four produced detectable levels of ciguatoxins [68,88]. At the time, *Gambierdiscus* was a monotypic genus, so Holmes et al. [68] and Holmes and Lewis [88] attributed the species to *G. toxicus* and referred to the ciguatoxins they identified as gambiertoxins (GTX), following the naming convention originally proposed by Murata et al. [21] to distinguish these toxins from the ciguatoxins found in fish. Gambiertoxins were subsequently renamed as ciguatoxins by Satake et al. [23]. Based on ciguatoxin production, this suggests the presence of at least two species of *Gambierdiscus* in Platypus Bay, but we cannot exclude the possibility of multiple strains of one species with different toxin-producing capabilities. Holmes et al. [77] also assayed bio-detrital fractions sieved from bulk samples of *Cladophora* collected by scuba divers at ~15 m depth over a 22-month period. Extracts of the size fraction containing *Gambierdiscus* (45–250 µm), from one of these six bulk collections produced bioassay signs in mice consistent with ciguatoxins. While this sample had the highest *Gambierdiscus* populations [77], it presumably contained ciguatoxin-producing and non-, or low-ciguatoxin-producing species/strains, indicating that simple monitoring of *Gambierdiscus* populations alone may not accurately identify ciguatera risk even in ciguateric areas [22,89]. This conclusion is supported by studies at Flinders Reef in southern Queensland which had some of the highest population densities of *Gambierdiscus* so far found in Australia [90], but wild cells and cultured clones isolated from Flinders Reef did not produce detectable ciguatoxins [91], and demersal fishes, including herbivores, from this reef were also not toxic [92]. *Gambierdiscus* populations on Flinders Reef [90], in Platypus Bay [77], as well as in New South Wales [93] appeared to bloom seasonally, as observed elsewhere [94–97].

The cellular concentration of ciguatoxin in wild *Gambierdiscus* collected from Platypus Bay (as determined by mouse bioassay) was similar to those found in wild cells collected from Marakei Island in the Republic of Kiribati [68,77]. However, these concentrations were ~100-fold greater than those extracted from a cultured clonal isolate of *Gambierdiscus* collected from Platypus Bay [68], suggesting that ciguatoxin production in *Gambierdiscus* may be enhanced under natural conditions, or that higher ciguatoxin-producing species/strains (super-producers) remain to be isolated from Platypus Bay [77]. Elsewhere, there are few direct comparisons of cellular ciguatoxin production from wild and cultured populations of benthic dinoflagellates isolated from the same location. Attributing cellular toxin production to a wild sample is complicated by a potential mix of species/strains in the sample which may include cells having a range of toxin concentrations. Recently, Liefer et al. [97] found that *Gambierdiscus* populations in the U.S. Virgin Islands peaked in summer, but cellular ciguatoxin loads peaked in cooler months during lower population densities. Repeated sampling and toxin analysis of *Gambierdiscus* populations in Platypus Bay did not find evidence for such asynchronous cell toxin loads [77] but these studies need be repeated using the more sensitive toxin assays now available. Future studies in Platypus Bay could be facilitated by assays that allow the detection and species-identification of single *Gambierdiscus* and *Fukuyoa* cells [98]. It could also be interesting to grow cultured cells back in the wild to determine if ciguatoxin production of isolates match those from wild cells, possibly using mesocosm experiments or using dialysis chambers to grow cultured cells at a hot spot such as Platypus Bay. Alternatively, comparative intra- and inter-specific omics studies on ciguatoxin and non-ciguatoxin producing strains may provide new understandings of these differences [99,100]. Ciguatoxin concentrations in *Gambierdiscus* may also be altered by quorum-sensing bacteria [101].

The evidence indicates that some *Gambierdiscus* in Platypus Bay are a source of ciguatoxins [68,77] but that for much of the time, only low levels of ciguatoxins are produced that limit the contamination of fish in Platypus Bay with these toxins. However, a major gap in our knowledge is we do not know the species of *Gambierdiscus* living in Platypus Bay and the chemical profile of the toxins they produce. We only know that a cultured clone and possibly wild cells from Platypus Bay produced at least two ciguatoxins of different polarity [68,77,88]. Recent studies on the accumulation of ciguatoxins in fishes feeding on cultured *Gambierdiscus* used the TB-92 strain of *G. polynesiensis* isolated from French Polynesia [71,102]. The ciguatoxin profile of this strain is dominated by P-CTX3C and its isomers, with lesser amounts of P-CTX-4A [71,103], the precursor of the major ciguatoxin (P-CTX-1) found in ciguateric Platypus Bay fishes [26,33]. Ledreux et al. [71] suggested that the two toxin classes have different tissue depositions and retention in fishes, with P-CTX3C analogs possibly being poorly retained.

We also do not know the spatial extent of the *Cladophora* layer and if it extends beyond Platypus Bay into Hervey Bay proper, or how much of this substrate that supports benthic dinoflagellates changes significantly through time. Archived aerial photography (QImagery: https://qimagery.information.qld.gov.au/ (accessed on 20 July 2021)) from

1994 and 2000 show that at times the *Cladophora* can be an almost continuous benthic layer in the eastern part of the bay (Figure 3), as it appeared to be during the studies by Holmes et al. [77] and Lewis et al. [84], whereas more recent satellite imagery (Queensland Globe: https://qldglobe.information.qld.gov.au/ (accessed on 20 July 2021)) suggests that the *Cladophora* layer near the beach can sometimes thin and become patchier (data not shown). The earliest available aerial photography images of Platypus Bay in QImagery are from 1958 and show dark streaks in the water consistent with *Cladophora*. Tidal currents form a large gyre in Platypus Bay [104] which may be a mechanism that helps maintain the unattached macroalgae within the bay. However, strong westerly winds can push considerable amounts of the macroalgae (presumably from shallower waters) up onto the western beach of Fraser Island, creating problems for vehicles driving on the beach that sink through a light covering of sand hiding a thick layer of rotting *Cladophora*. This wind-wave-driven transport of *Cladophora* onto the beach of Platypus Bay likely creates space for new *Cladophora* growth that would provide new substrate for epiphytic benthic dinoflagellates, especially in the shallower near-shore areas. It may also cause major, localized reductions in the benthic dinoflagellate populations in these shallow near-shore areas, reducing the risk of production of ciguatoxins within Platypus Bay. At other times, especially in deeper waters, a near continuous macroalgal substrate that can support *Gambierdiscus* populations likely covers many km^2.

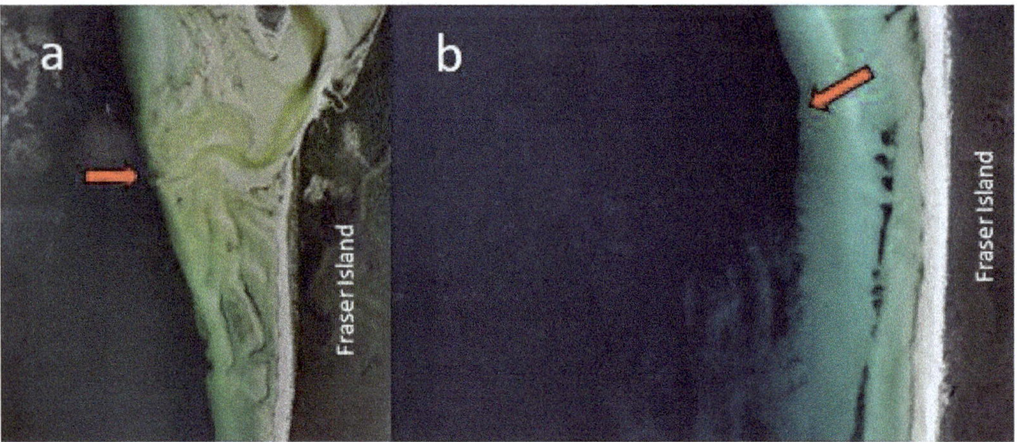

Figure 3. Benthic unattached macroalgae (*Cladophora*) lying on sand in Platypus Bay near the mouth of Wathumba Creek. Aerial photography from QImagery (Creative Commons Attribution 4.0 International (CC BY 4.0) licence): https://qimagery.information.qld.gov.au/ (accessed on 20 July 2021). (**a**) Arrow shows the eastern limit of a near-continuous, dark, benthic layer of *Cladophora* on 28 May 1994 at the entrance to Wathumba Creek. (**b**) Arrow shows limit of *Cladophora* near beach just south of Wathumba Creek on 24 June 2000.

Much of Platypus Bay has no reef structure, so in the past, fishers would dump objects such as old washing machines at sites within the bay to try and create secret artificial reefs. One such location is about 1.5 nautical miles west of Wathumba Creek (Figure 2) where benthic dinoflagellates and fish were collected [77]. While the site consistently held a range of fishes such as trevallies (*Caranx* spp.) and yellowtail kingfish (*Seriola lalandi*), assays of liver extracts from fish caught at this site were nontoxic (RJL unpublished results), so we have no evidence that these sites had any role in exacerbating the production of ciguateric fishes in Platypus Bay.

Hervey Bay supports the largest seagrass meadows (>2000 km^2) on the east coast of Australia [105] and subtidal mapping by the Queensland Government suggest that some deep- and shallow-water seagrass beds cover the south-western part of Platypus Bay

(Wetland*Maps*: https://wetlandinfo.des.qld.gov.au/wetlandmaps/ (accessed on 20 July 2021)). There have been some studies of the epiphytes on Hervey Bay seagrasses but none that report any benthic dinoflagellate populations. Hervey Bay seagrasses support one of the largest populations of dugong (*Dugong dugon*) on the east coast of Australia [106], but as yet there are no reports of illness or death of these marine mammals linked to possible poisoning by feeding on seagrasses with epiphytic benthic dinoflagellates. *Gambierdiscus* has been found on seagrass (*Zostera*) on the east coast of Australia [93,107], and the deaths of manatees [108] and dolphins [109] have been linked in the U.S.A. to poisoning by brevetoxins produced by the dinoflagellate *Karenia brevis*. Brevetoxins have a similar mechanism of action as ciguatoxins and compete for the same receptor binding site on Na^+-channels [12], suggesting that dugongs may also be sensitive to ciguatoxins. This hypothesis is supported by ciguatoxins being detected from the tissues of living and dead Hawaiian Monk seals [110]. However, there is no evidence currently for the poisoning of dugong from feeding on these seagrasses, so we think they are unlikely to be a major source of ciguatoxins into local marine food chains.

An examination of the origin of nutrients that support growth of *Cladophora* and the associated benthic dinoflagellates in Platypus Bay suggest that there are likely three major sources. Firstly, recycling of nutrients in situ from dead and decaying biomass of *Cladophora* and the invertebrates living within this macroalgal layer (including gastropods, crustaceans, polychaetes and nematodes; [84]). Secondly, episodic upwelling where summer north-west winds can drive wind- or Ekman-driven coastal and shelf-edge upwelling to replace the surface waters transported offshore by the wind, with intrusions of upwelling waters flowing into northeast Hervey Bay [111,112]. Thirdly, catchment derived nutrients transported offshore in flood plumes that can extend well out into Hervey Bay. For example, in 1992 sediment transport from two major floods and a cyclone in a 3-week period caused the loss of more than 1000 km^2 of seagrass in Hervey Bay [113]. However, outside of flood events, there is little freshwater flow into the bay [104]. Even though Hervey Bay is not within the Great Barrier Reef Marine Park, the Burnett-Mary region is adjacent to Hervey Bay and is the southernmost Natural Resource Management (NRM) region for the Great Barrier Reef. Many Great Barrier Reef ecosystems continue to be in poor condition due largely to the collective impact of land run-off associated with past and ongoing catchment development, coastal development activities, extreme weather events and climate change impacts [114]. To reduce these impacts, targets have been developed from a 2013 baseline for the reduction by 2025 of end-of-catchment loads of nutrients and sediments, and concentrations of pesticides (mostly herbicides) that flow from the major river catchments into the Great Barrier Reef lagoon ([115], Great Barrier Reef 2050 Water Quality Improvement Plan, https://www.reefplan.qld.gov.au/ (accessed on 20 July 2021)). This includes targets for the largest three rivers that discharge into Hervey Bay, the Mary, Burrum and Burnett rivers (Figure S3). The largest load reduction targets are for the Mary River, the flood plume of which is the most likely to reach Platypus Bay [116]. If the Burnett-Mary region load reduction targets are met, it could reduce catchment-derived nutrients reaching Platypus Bay and its benthic layer of *Cladophora* and epiphytic dinoflagellates. If this reduction in nutrients led to reduced populations of ciguatoxic *Gambierdiscus*, it would be an unintended, but direct benefit from the 2050 Great Barrier Reef Water Quality Improvement Plan. However, a study at reef sites in the US Virgin Islands found that enriched nutrient levels do not always produce significantly greater populations of *Gambierdiscus* [117].

Fraser Island is the largest of the five large barrier sand islands along the southern coast of Queensland and may also be a local source of nutrients to Platypus Bay. Wathumba Creek is one the largest of the western flowing creeks on Fraser Island and it flows into the middle of Platypus Bay (Figure 2). However, these barrier sand islands tend to be low in nutrients with oligotrophic lakes and streams fed by underground sand aquifers. The majority of Fraser Island is also World Heritage listed National Park and the only permanent human structure along the coast of Platypus Bay from which anthropogenic nutrients could be exported is a camping toilet block near to Wathumba Creek. It is also possible that nutrients are occasionally input from aeolian sources such as smoke and ash from infrequent but large bushfires that can persist in much of the inaccessible wilderness of Fraser Island, such as the recent fire that burned for 6 weeks on Fraser Island in late 2020.

A pesticide reduction target has been set under the Great Barrier Reef 2050 Water Quality Improvement Plan based upon keeping concentrations below that required to protect at least 99% of aquatic species from all pesticides at river mouths ([115], Great Barrier Reef 2050 Water Quality Improvement Plan, https://www.reefplan.qld.gov.au/ (accessed on 20 July 2021)). Pesticide reduction is not a priority for catchments flowing into Hervey Bay; however, at moderate flows of the Mary River, nearshore seagrass areas are at risk of being exposed to concentrations of herbicides that are known to inhibit photosynthesis [118]. For example, during a low flow period in the Mary River, the photosystem-II inhibitor diuron was detected near the edge of Platypus Bay at 5 ng/L [118]. However, herbicide concentrations tend to be highest during "first-flush" events during high river flows, not during periods of low river flows [119]. Community composition changes in benthic microalgae communities can occur at low μg/L concentrations of diuron [120,121]. Currently there is insufficient information to know if catchment derived herbicide concentrations could be having an impact on populations of *Gambierdiscus* in Platypus Bay.

Several species of benthic dinoflagellate other than *Gambierdiscus* also occur as epiphytes on the *Cladophora* in Platypus Bay, including *Prorocentrum* sp. and *Coolia* sp. [77], but these have not yet received much study in Platypus Bay. Many benthic dinoflagellate species produce toxins but apart from *Gambierdiscus* and *Fukuyoa*, there is limited evidence for their involvement in the ciguatera food chain [33,67,77]. Holmes et al. [122] did isolate a new toxin from a clonal culture of *Coolia* collected from the *Cladophora* that was distinct from ciguatoxins and named it cooliatoxin. At the time, *Coolia* was a monotypic genus, so Holmes et al. [122] attributed the species to *C. monotis*. A subsequent review based mostly upon old scanning electron micrographs suggested that the species may have been *C. tropicalis* [123]. 44-methylgambierone has recently been detected from *C. tropicalis* and *C. malayensis* [61–63,124] along with an additional isomer from *C. tropicalis* [124]; however, the greater mass and toxicity of cooliatoxin show that it is distinct from 44-methylgambierone [63,122]. *Coolia tropicalis*, *C. malayensis*, *C. canariensis* and *C. palmyrensis* are now known to occur in Queensland waters [125,126].

2.2. Food Chain Links Leading to Ciguatera in Platypus Bay, Trophic Level 2

The most obvious herbivorous fish in Platypus Bay are rabbitfish (*Siganus spinus*) [80,84] that rarely exceed 24 cm in length (Fishes of Australia: http://fishesofaustralia.net.au/home/species/1886 (accessed on 20 July 2021)). Rabbitfishes are known to accumulate ciguatoxins in the Pacific Ocean [32,87], with the highest concentration recorded being 0.13 μg/kg P-CTX-1 equivalents from the Republic of Kiribati [87]. Rabbitfishes are grazers that feed on filamentous algae or browse on larger macroalgae and are abundant around tropical reefs [127]. They are also abundant in Platypus Bay with stomachs often tightly packed full of *Cladophora* [80,84]. Even though these rabbitfishes presumably consume significant numbers of *Gambierdiscus*, to our knowledge they rarely cause ciguatera [84] and extraction and analysis of the dissected livers from 50 rabbitfish caught from Platypus Bay failed to detect any ciguatoxins [80]. While rabbitfishes are generally not targeted by recreational fishers in Queensland, small quantities are netted from Platypus Bay and sold by commercial fishers (Table 1). The consumption of rabbitfishes from Platypus Bay, apparently without harm, is not consistent with the classical food chain hypothesis for the bioaccumulation and bio-concentration of ciguatoxins [70], unless the toxins are not efficiently absorbed by rabbitfishes, and/or any toxins bio-accumulated are rapidly depurated. Assuming an average weight of ~0.5 kg per rabbitfish, the total commercial catch of more than 45 tonnes for the 25 years between 1994 and 2018 (Table 1) equates to the sale and presumed human consumption of >90,000 fish over this time frame, apparently without harm. Similarly, in the ciguatera endemic Cook Islands, *S. spinus* is a targeted food fish that rarely causes ciguatera [128].

Table 1. Catch (tonnes) of rabbitfishes (*Siganus* spp.) caught by Queensland commercial fishers from Platypus Bay between 1994 and 2018. Data derived from commercial fishers' compulsory logbooks and retrieved from the Queensland Government's QFish database (http://qfish.fisheries.qld.gov.au/ (accessed on 20 July 2021)). See Figure S4 for details.

Years	Rabbitfish Catch (Tonnes)
1994–1998	22.12
1999–2003	6.35
2004–2008	13.59
2009–2013	3.19
2014–2018	0.16
Total for all years	45.41

Platypus Bay appears to be a case where the classical transfer of ciguatoxins from benthic dinoflagellates to herbivorous fish may not apply. Kelly et al. [129] proposed a model for an expanded marine food chain for ciguatera that included invertebrates in the bio-transfer of ciguatoxins. Lewis et al. [84] found evidence for this transfer of ciguatoxins through invertebrates in Platypus Bay (especially Alpheid shrimps), to small carnivorous fish (blotched-javelin fish) that feed on invertebrates in the *Cladophora*. Based on this finding, they suggested that Spanish mackerel might accumulate ciguatoxins by preying on these blotched-javelin fish. Even though ciguatera is generally considered a disease caused by fishes, ciguatoxins have also been detected in molluscs, crustaceans, and echinoderms, reinforcing the evidence for the potential transfer of these toxins through invertebrates [32,130–134].

Cells of cultured *Gambierdiscus* were lethal when fed to laboratory reared brine shrimp [129,135]. The brine shrimp may have been intoxicated by ciguatoxins and/or water-soluble toxins, as the latter are produced by most *Gambierdiscus* species/strains, unlike the ciguatoxins which are produced in detectable quantities by fewer species/strains [46,53,68]. Behavioural changes in invertebrates caused by intoxication from feeding on benthic dinoflagellates could increase their probability of being preyed upon [33,135], and invertebrates feeding upon ciguatoxin producing dinoflagellates could be a mechanism that concentrates ciguatoxins into predatory fish [84]. It appears that this concentration step in the food chain is required for ciguatoxins to accumulate in higher trophic level to cause ciguatera from Platypus Bay fishes.

2.3. Food Chain Links Leading to Ciguatera in Platypus Bay, Trophic Levels 3 and 4

Spanish mackerel, barracuda and blotched-javelin are three carnivorous fish species caught from Platypus Bay known to cause ciguatera [74,75,78,84]. The ciguatoxic blotched-javelin and barracuda caught from Platypus Bay were relatively small fish, <0.5 and ~2.0 kg, respectively [78,84], whereas toxic Spanish mackerel tend to be much larger (typically >10 kg). Lewis et al. [84] suggested that barracuda and Spanish mackerel acquire ciguatoxins from preying on blotched-javelin fish, with Spanish mackerel also possibly preying on barracuda. Blotched-javelin can be considerably more toxic than Spanish mackerel, and both fish species carry the same suite of major ciguatoxins [26]. These results are consistent with the prey fish species (blotched-javelin) frequently feeding within an area producing ciguatoxins, and sometimes being more toxic relative to its predator because the larger Spanish mackerel are only transiently feeding within the toxic area on a mix of toxic and non-toxic prey fishes. This reduction of toxin concentration up the food chain is also consistent with ciguatera in French Polynesia where the lower trophic level fishes (herbivores) tend to be more toxic than the higher trophic level piscivorous fish that feed upon them [136,137]. However, this is contrary to the conceptual model for the bio-concentration of toxins along the ciguatera food chain from herbivorous to carnivorous fishes, and from small to large carnivorous fishes [70]. The highest known ciguatoxin concentrations from the Pacific Ocean are found in predatory fishes, and in ciguatera endemic regions such as the Republic of Kiribati, the pattern of toxicity in fish appears to be consistent with toxin concentration along the food chain [32,87].

Some fish are susceptible to intoxication by ciguatoxins [71,73,138–140]. If blotched-javelin are affected by ciguatoxins, then they may be more vulnerable to predation making this a mechanism by which blotched-javelin contaminated with ciguatoxins are selectively preyed upon by larger predators such as Spanish mackerel [84]. This increases the potential for transfer of ciguatoxins along the food chain into larger fish that are more likely to enter the human food chain in Australia. However, blotched-javelin fish do not always carry detectable levels of ciguatoxins in Platypus Bay, as Lewis et al. [84] caught non-toxic fish <1 year after finding toxic fish.

Overall, these results are consistent with significant levels of ciguatoxins only intermittently entering the marine food chain in Platypus Bay because:

- The spatial extent of the *Cladophora* layer varies over time, limiting the available substrate for benthic dinoflagellates;
- Most of the *Gambierdiscus* species/strains present are not those that produce significant levels of ciguatoxins [68,77];
- And/or, the population of ciguatoxin-producers are sometimes too low for significant amounts of toxin to enter food chains in the bay [77];
- And/or, the profile of the ciguatoxins being produced varies over time, with possibly P-CTX3C congeners less likely to be retained by fishes [71];
- And/or, ciguatoxins are produced, but are not concentrated/transferred into the invertebrate populations being eaten by small predatory fish; and
- And/or, ciguatoxins are not accumulated by small predatory fish (e.g., blotched-javelin) at high enough concentrations to increase their risk of predation.

Any of the above factors would act as a rate-limiting step to the input and transfer of ciguatoxins along food chains in Platypus Bay and explain the observation by Lewis [78] that ciguatera cases from Spanish mackerel cycle through periods of high and low incidences.

After the prohibition on the taking of Spanish mackerel and barracuda from Platypus Bay was scheduled into law in 1987, most ciguatera cases in Queensland were caused by demersal reef fish caught from the Great Barrier Reef [76,77,81]. However, even after this ban, recreational and commercial fishers continued to take Spanish mackerel from non-prohibited waters beyond Platypus Bay, with commercial fishers' logbooks showing Spanish mackerel continue to be taken from within the two 30 × 30 nautical mile logbook grids/squares (W32, W33) that include Platypus Bay (Figure S4), presumably from waters to the north and east of Fraser Island. Between 1990 and 2019, commercial fishers caught an average of 11.9 ± 4.8 tonnes annually (mean ± standard deviation) of Spanish mackerel from within the logbook grids that include Platypus Bay (W32 and W33; range 3.5 to 24.3 tonnes; QFish [141]). In addition, an average of 2.5 ± 1.4 tonnes of Spanish mackerel (range < 1–5.6 tonnes) were caught annually between 1990 and 2019 from Hervey Bay in the two logbook grids to the west of Platypus Bay (V32, V33, Figure S4). The combined catch from these four grids (V32, V33, W32, W33) that encompass Platypus Bay and its adjacent waters, showed considerable annual variation ranging from a low of 4 tonnes in 1994 to 26 tonnes in 1999 (Figure S5). Assuming an average weight of 7.7 kg for commercially caught Spanish mackerel [83], this equates to between ~500 and 3300 fish caught annually between 1990 and 2019 from waters surrounding Platypus Bay, apparently without causing the regional levels of ciguatera reported previously by Lewis [78].

Historically, most cases of ciguatera along the east coast of Australia have been caused by demersal reef fishes caught from along the Great Barrier Reef north of about Mackay (~21° S latitude), and by Spanish mackerel caught south of this latitude [6,74,76], including in recent years from New South Wales [14]. This north-south divide is not absolute, with Spanish mackerel from north Queensland occasionally causing ciguatera [15,79] and demersal reef species more usually associated with the Great Barrier Reef sometimes causing ciguatera in southern waters, e.g., a mild case of ciguatera was caused in 2016 by a blue-spot coral trout (*Plectropomus laevis*), caught off Moreton Island (~27° S) (MJH personal communication). The ciguatera suffered by people eating toxic Spanish mackerel has ranged from very mild cases through to critical poisonings requiring hospitalization, occasionally leading to death [5,78,142].

Spanish mackerel from the east coast of Australia form a single genetic stock in coastal shelf waters between Cape York Peninsula and northern New South Wales, but some travel long distances within this range with the longest recorded from tagging studies being up to 1000 nautical miles from northern Queensland to New South Wales [79]. Their movement patterns depend on spawning and feeding behaviours, water temperatures and currents [79]. Fishers often interpret Spanish mackerel movement as a southward's coastal migration of fish from Queensland waters to northern and central New South Wales waters in summer, and then a northern return to Queensland waters in late autumn. Movement is likely more complex than this but may in part relate to fish attaining favourable environmental or feeding conditions within a 24 °C isotherm [79]. There is a correlation between the distance moved southwards and fish size, with the larger sized fish usually being females [79].

Before the prohibition on taking Spanish mackerel from Platypus Bay, ciguatoxic Spanish mackerel were caught in southern Queensland during most months with possibly a gradual increase through winter and early spring to a maximum in October ([78], Figure S6). This was for fish caught between Maryborough and Gladstone (i.e., the waters of Hervey Bay and coastal waters to the north of Hervey Bay) between 1976 and 1983. Spanish mackerel from the east coast of Australia show predictable, seasonal migratory behaviour with annual spawning concentrated over a two lunar month period between September and November near a spatially discrete group of inner reefs northeast of Townsville (Figure S7) on the central Great Barrier Reef [143]. It is one of the most predictable spawning aggregations of fish on the Great Barrier Reef [83], but this does not mean there is a salmonid-like return to the same spawning reefs [79]. Most of Queensland's annual commercial catch of Spanish mackerel is taken each year in spring between September and November from these central Great Barrier reefs [144]. In contrast, most of the Spanish mackerel caught by commercial fishers from the waters around Platypus Bay are taken between April and August (Figure S6). The predominance of ciguatoxic Spanish mackerel caught between April and October [78] likely reflects when most fish are being caught in this region, either of resident fishes before they move north for spawning, or as part of the general northwards' movement of fish along the east coast of New South Wales and Queensland from late summer and autumn. However, spawning may also occur on isolated reefs as far south as Hervey Bay and fishers in this region believe that some fish do not always leave the area [79].

If all Spanish mackerel, including those feeding in Platypus Bay, showed high fidelity to the spawning aggregations northeast of Townsville, then we might have expected a spike in ciguatera cases from Spanish mackerel caught from the spring spawning aggregations, especially given the very high exploitation rates in this fishery [79], but this has not been noticeable. However, a recent analysis did find a slightly increased frequency of ciguatera associated with mackerel species (*Scomberomrus* spp.) between November-March, corresponding to the Austral wet season from late spring to early autumn [15]. East coast Queensland stocks of Spanish mackerel were previously considered either fully-fished or overfished relative to maximum sustainable levels, initially reaching annual landings of around 1000 tonnes during the 1970's [143]. The east coast stock is now thought to be sustainably fished, but maximally exploited [83]. However, the recent increase in ciguatera cases from Spanish mackerel caught in New South Wales [14], might be consistent with fish feeding in Platypus Bay and becoming contaminated with ciguatoxins before moving south along the east coast of Australia into New South Wales. The distribution of Spanish mackerel along the east coast of Australia has been found to be especially sensitive to the environmental effects of climate change with southward range shifts into southern New South Wales exceeding 200 km per decade [145].

A range of large carnivorous fishes other than Spanish mackerel are often captured from Platypus Bay and apparently eaten without harm by recreational fishers. We know this from personal observations (MJH, RJL) and regional reports published in recreational fishing publications from charter boat operators and fishing guides. The carnivorous fish species caught by recreational and charter fishers from Platypus Bay include at least five species of trevally (*Caranx* spp.), yellowtail kingfish (*Seriola lalandi*), spotted mackerel (*Scomberomorus munroi*), school mackerel (*S. queenslandicus*), mackerel-tuna (*Euthynnus affinis*), longtail tuna (*Thunnus tonggol*) and juvenile black marlin (*Makaira indica*). Spotted mackerel, school mackerel, trevally and yellowtail kingfish have caused ciguatera in Australia, but not many cases are known [1,5] and we are not aware of any cases of ciguatera in Australia from longtail tuna, marlin or mackerel-tuna (mackerel-tuna are not often caught for food in Australia and black marlin are legally protected from commercial fishing in the Australian Fishing Zone). Possibly, these large predatory species do not often feed within the ciguatera-food chain in Platypus Bay, with some instead preying upon small pelagic, plankton-feeding fishes such as scads (Carangids) and herrings (Clupeids). Indian scad (*Decapterus russelli*) are common along the east-coast of Queensland, including Platypus Bay and may be one such small prey species. Lewis [80] assayed Indian scad captured from Platypus Bay at the same time as toxic barracuda were captured but could not detect any ciguatoxin. It is therefore likely that there are a range of non-toxic food chains in Platypus Bay, into which many predatory fishes feed. Stable isotope analysis of fishes from the Great Barrier Reef suggests that there is little dietary niche overlap of Spanish mackerel with predatory demersal reef fishes [146]. If dietary overlap is the exception, as suggested by Espinoza et al. [146], and if it also operates outside of Great Barrier Reef waters, then separate, non-toxic food chains may be a mechanism that prevents ciguatoxins from benthic dinoflagellates contaminating most predatory fishes in Platypus Bay. The availability of toxic and non-toxic food chains to a predatory fish species in highly ciguatoxic areas may exert selection pressure against ciguateric fish populations if the behaviour or health of these fish are affected by the ciguatoxins.

There are also several fish species caught from Platypus Bay that appear to be inconsistent with our hypothesis for the food chain production of ciguateric fishes in Platypus Bay. For example, snapper (*Chrysophrys auratus*) is a large, long-lived (>30 years) predatory Sparid species often caught by recreational fishers from Platypus Bay (reports from recreational fishing magazines and weekly fishing reports for Hervey Bay). This prized sporting and table fish typically does not move large distances [147] with the species so heavily exploited in southern Queensland that stocks are classified as overfished past maximum sustainable yield [147]. Inshore sheltered habitats such as Hervey Bay provide important nursery grounds for juvenile snapper, which feed mainly on worms, crustaceans, and other invertebrates, while adults have a wider diet including small fishes and hard-shelled invertebrates that they crush with their molar-like teeth [147]. Surprisingly, we are not aware of any cases of ciguatera caused by snapper, despite their feeding preference overlapping those of blotched-javelin fish. Given the targeting of this species by fishers, and their high level of exploitation, it is likely that snapper are rarely ciguateric.

A range of small, carnivorous fish species normally associated with shallow water estuaries and open beaches are also caught by recreational fishers from along the shoreline of Platypus Bay, including whiting (*Sillago* spp.), flathead (*Platycephalus* spp.) and yellowfin bream (*Acanthopagrus australis*). Despite these species eating a range of benthic invertebrates and small fishes [148], to our knowledge they have never caused ciguatera. In the past, these species were also netted from the beach of Platypus Bay and sold by commercial fishers (RJL personal communication). Possibly these species do not feed within the *Cladophora*, otherwise it is difficult to understand why they would not be exposed to ciguatoxins in their diet in Platypus Bay.

2.4. Summarizing the Production and Food Chain Transfer of Ciguatoxins in Platypus Bay

Platypus Bay appears to be a unique ecosystem for production, transfer, and accumulation of ciguatoxins into higher trophic level fishes. It is the only location known that repeatedly produces ciguateric fishes along the east coast of Australia. Ciguatoxins are produced by *Gambierdiscus* spp. growing epiphytically on unattached macroalgae (*Cladophora* sp.) lying over an unconsolidated sandy substrate. We suggest these benthic dinoflagellates are consumed by invertebrates living within the macroalgae, principally Alpheid shrimps, which are in turn preyed upon by blotched-javelin fish (*Pomadasys maculatus*). The blotched-javelin are eaten by transient pelagic carnivores, especially Spanish mackerel (*Scomberomorus commerson*) (Figure 4). If these Spanish mackerel are caught and eaten before they can depurate ciguatoxins, ciguatera poisoning occurs. However, further research is required to determine the chemical profile of the ciguatoxins produced by Platypus Bay *Gambierdiscus* and the bio-transformations that occur through at least three trophic transfers.

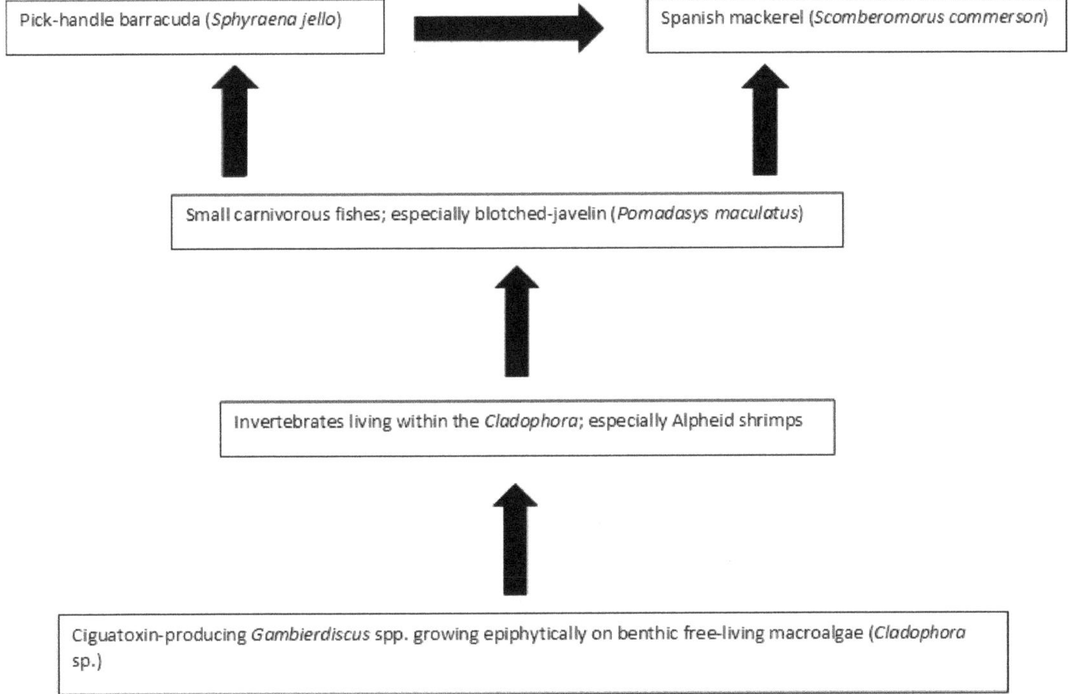

Figure 4. Conceptual model for the food chain transfer (arrows) of ciguatoxins through at least three trophic transfers in Platypus Bay, Fraser Island.

Platypus Bay is a simpler ecosystem compared to the diversity and complexity generally found on coral reefs, and this relative simplicity may facilitate future studies on ciguatera food chains, which should include:

- Identifying the resident *Gambierdiscus*/*Fukuyoa* species in Platypus Bay;
- Determining the profile and quantities of ciguatoxins produced by these species;
- Identifying the profile and quantities of ciguatoxins that bio-transfer across Platypus Bay trophic levels; and

- Determining the relationship between the spatial extent of the *Cladophora* substrate and the risk of ciguatera to allow the development of remote sensing imagery as a monitoring tool (e.g., [149]).

3. Model for the Dilution of Ciguatoxins in the Flesh of Spanish Mackerel (*S. commerson*) through Growth

Lewis and Holmes [33] developed a model for the factors that influence the accumulation of ciguatoxins through marine food chains, hypothesising that somatic growth could dilute the concentrations of ciguatoxins contaminating the flesh of fishes over time. Similar suggestions were also made by Yang et al. [150]. However, the suggestion that fish could lose toxicity was made as early as Halstead and Bunker [151] who suggested the toxin could metabolise in the liver of fishes and that higher concentrations of toxin could be expected in the liver and intestine, and lower concentrations in the muscle if the fish had been captured soon after feeding on the source material. Randall [70] did not agree with this hypothesis. However, Banner et al. [152] also suggested that fish could lose toxicity, in part to account for some detailed observations by Cooper [153] of localized changes in fish toxicity on reefs in the Gilbert Islands (Republic of Kiribati). Clausing et al. [102] recently provided experimental support for the concept of somatic growth diluting ciguatoxin concentrations in juvenile herbivorous unicornfish (*Naso brevirostris*) fed on a gel diet containing a cultured clone of *G. polynesiensis* that produces a ciguatoxin profile dominated by P-CTX3C.

We therefore constructed growth-based models to explore the potential for dilution of ciguatoxin concentrations in the flesh of commercial (legal) sized Spanish mackerel to below that which could cause human poisoning, i.e., below the precautionary action concentration of 0.01 µg/kg of P-CTX-1 equivalents suggested by the U.S. Food and Drug Administration (FDA) for the safe consumption of seafood [154]. We only consider ciguatoxin concentration in the muscle (flesh) of fishes, because in Australia the potentially more toxic viscera (including gonads) of Spanish mackerel and reef fish are generally discarded and not sold commercially. Our first model is based only upon dilution of toxicity through somatic growth and assumes that Spanish mackerel do not depurate or metabolize ciguatoxins from their tissues. This is the simplest model and was a mechanism proposed by Lewis et al. [86] for reducing fish toxicity over time. The model assumes that if there is no loss or gain of toxin over time, then the ciguatoxin concentration in muscle will decrease in proportion to the relative increase in mass from somatic growth. That is, if $[CTX]_i$ is the concentration of ciguatoxin in year i in µg/kg, then the concentration in year $i + 1$ can be described by:

$$[CTX]_{i+1} = [CTX]_i \times \mathrm{mass}_i / \mathrm{mass}_{i+1} \tag{1}$$

Spanish mackerel are the largest of the mackerel species found in Australian waters and are targeted by both commercial and recreational fishers because of their large size, good eating qualities and because they are fun to catch. Spanish mackerel can reportedly grow to 2.4 m in length and weigh up to 70 kg [82], although fish of this size are rarely caught now. One of the largest fish captured in recent years weighed 54 kg and was caught off Fraser Island in 2015. Fisheries scientists estimated its age at 26 years (Queensland Department of Agriculture and Fisheries, personal communication). Spanish mackerel are a fast-growing fish, especially in the first year, with females reaching sexual maturity at ~89 cm total length between 2- and 4-years of age [83,144]. Males and females have different growth rates with females growing faster and larger [82]. Both Queensland and New South Wales have the same minimum legal size of 75 cm (total length) for the taking of Spanish mackerel by commercial and recreational fishers. This means fish become vulnerable to harvesting along the east coast of Australia mostly between 1- to 2-years of age with the estimated age at which fish become 50% vulnerable to fishing being ~1.5 years, and 95% vulnerable by ~2.1 years [83].

We constructed our model from von Bertalanffy growth curves for male and female Spanish mackerel (Figure 5) based upon fish fork lengths at age [83]. The von Bertalanffy growth curve is one of the most widely used models for describing fish growth in fishery science, although it often performs poorly for young ages for which there are usually little data. Often, the observed data shows a broad distribution around the modelled growth curve because of considerable variation in the lengths that individual fish can reach for the same age [155]. O'Neill et al. [83] reported most Spanish mackerel caught by commercial and recreational fishers belong to annual cohorts up to 7-years of age. We therefore modelled changes in ciguatoxin concentrations in fish between 0.5- to 10-years of age to ensure we cover most of the fish sizes caught and eaten from the east coast of Australia. While fast-growing fish may reach the minimum-legal-size for harvesting before 1-year of age, most only reach this total length between 1–2 years of age [83]. The minimum-legal-size of 75 cm total length corresponds to about 67 cm fork length (fisheries scientists more commonly use fork length than total length as the measure for fish length, but legal limits for harvesting fish are based upon total length). Total length for Spanish mackerel is related to fork length by the equation of Begg et al. [156]:

$$\text{Total length (cm)} = 4.275 + (1.06\ \text{Fork length (cm)}) \tag{2}$$

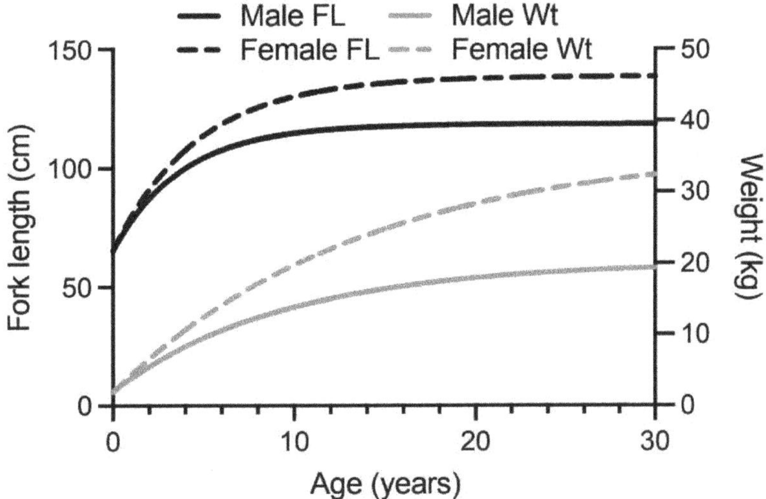

Figure 5. Growth curves for Spanish mackerel from the east coast of Australia (data supplied by Michael O'Neill from O'Neill et al., [83]). The black lines show von Bertalanffy growth curves for fork length-at-age for male and female Spanish mackerel. The grey lines show weight-at-age for growth of male and female Spanish mackerel.

Our models rely upon conversion of Spanish mackerel fork length-at-age values, to weight-at-age values (Figure 5), using the equation of Campbell et al. [157]:

$$\text{Weight (kg)} = 2.35 \times 10^{-6}\ (\text{Fork Length (cm)}^{3.2766}) \tag{3}$$

The consistent increase in weight-for-age that this mathematical function produces' does not reflect the reality of seasonal changes in weight that animals in the wild undergo, especially in relation to gonad development and spawning. Prior to spawning, the gonads can contribute a significant amount to the total weight of the fish with ovaries of female Spanish mackerel ranging from <0.1% to 13% of fish body weight, and male testes ranging between <0.1% to 7% of fish body weight [156].

Our growth dilution model assumes the concentration of ciguatoxins are homogeneous throughout the flesh of fish, and this is supported by recent findings that P-CTX-1,

-2 and -3 were similarly distributed throughout the fillets of yellow-edge coronation trout (*Variola louti*) caught from Okinawa [158]. However, it is possible that the concentrations of ciguatoxins across the tissues of Spanish mackerel fillets could vary, as Li et al. [72] have recently shown that much higher concentrations of ciguatoxins can accumulate in the skin compared to the muscle in goldspotted rockcod (*Epinephelus coioides*). In Queensland, Spanish mackerel fillets or "steaks" are normally sold with the skin still attached, whereas demersal reef fish fillets are normally sold skinned (Queensland Department of Agriculture and Fisheries, personal communication). Previous studies in the Caribbean found that toxicity was evenly distributed throughout the flesh of a ciguatoxic trevally (*Caranx bartholomaei*) [159]. In the Pacific, Helfrich et al. [160] reported that there was no difference in toxicity of the flesh of red bass (*Lutjanus bohar*) tested from different parts of the body, and Banner [161] found no statistical difference in the toxicity between the flesh of the anterior and posterior halves of moray eels (*Gymnothorax javanicus*) and red bass. However, there are no data for the distribution of ciguatoxins in the flesh of Spanish mackerel or any reef fish from Australia.

Our model also assumes that the ratio for flesh weight to total weight for fish greater than 0.5-year of age is constant. This appears a reasonable assumption as there is a linear relationship ($r^2 = 0.98$) between fillet (flesh) weight and whole weight for Spanish mackerel from Western Australia [162]. However, if the flesh weight to total weight ratio increases over time, then the "dilution" effect of growth will be greater, and our results will underestimate the reduction in fish toxicity over time. The corollary is that if the ratio decreases over time, then the "dilution" effect will be less, and our results will overestimate the reduction in toxicity.

Our model assumes that the accumulation of ciguatoxin does not affect the long-term growth characteristics of Spanish mackerel and therefore the age-weight model for growth is consistent for non-toxic fish as well as fish contaminated with ciguatoxins. However, growth may be affected, at least in the short-term, by accumulation of ciguatoxins as Davin et al. [139] reported behavioural changes in demersal reef fishes (*Epinephelus* and *Lutjanus* spp.) fed extracts from ciguatoxic barracuda, including changes in feeding behaviour. O'Toole et al. [163] suggested that the home-range of ciguatoxic barracuda in the Caribbean may be less than for non-toxic individuals, indicating possible behavioural changes. Ciguatoxins are potent ichthyotoxins [140], that can produce similar electrophysiological effects on fish and mammalian nerves [138], so it would not be surprising if the physiology of Spanish mackerel were affected by feeding on ciguatoxic prey. Lewis [140] suggested that the lethal effects on fish may impose an upper limit on the concentrations of toxins that some fish can accumulate. However, there is conflicting evidence about the effect of ciguatoxins on fishes. Early studies feeding toxic fish flesh to red bass (*Lutjanus bohar*) and an omnivorous surgeonfish (*Acanthurus xanthropterus*) did not report any signs of intoxication [152,164]. More recently, juvenile sea mullet (*Mugil cephalus*) fed gel pellets containing *Gambierdiscus polynesiensis* displayed signs of intoxication with a mix of abnormal hyperactive and hypoactive behaviours [71]. In contrast, juvenile unicornfish (*Naso brevirostris*) fed the same cultured strain of *G. polynesiensis* showed no signs of abnormal behaviour or poisoning [102]. The *G. polynesiensis* strain (TB-92) used for these experiments has a ciguatoxin profile dominated by the P-CTX3C family of toxins but also includes lesser amounts of P-CTX-4A, the precursor to P-CTX-1 and -2 [71,102,103]. Growth was also reduced in juvenile goldspotted rockcod (*Epinephelus coioides*) fed fish-pellets contaminated with P-CTX-1, -2 and -3 [72], and freshwater goldfish (*Carassius auratus*) fed C-CTX-1 became lethargic after 2-weeks of daily ingestion of toxin [73]. Unicornfish (surgeonfish) and goldspotted rockcod (grouper) are fish species that are often found in a coral reef environment typical for ciguatera, whereas sea mullet is an estuarine and coastal species that does not (at least in Australia). Intoxication likely depends upon the dose ingested, the fish species consuming the toxins and possibly the chemical profile of the ciguatoxins being consumed (as well as water-soluble toxins such as maitotoxins ingested by herbivorous and detritivorous fishes but not bio-accumulated across trophic levels). Our models would overestimate the dilution of ciguatoxins from the flesh of Spanish mackerel by somatic growth, if fish growth is reduced because of the physiological or behavioural effects of accumulating ciguatoxins.

Finally, our model assumes that the accumulation of ciguatoxins does not affect the susceptibility of Spanish mackerel to be caught by commercial or recreational fishers (e.g., through enhancing or deterring feeding behaviours as most fish are caught on lines using baits or lures, not by netting). Any increased bias towards catchability would be a mechanism that funnels ciguatera into the human food chain. In contrast, reduced catchability would suggest that the actual incidence rate for ciguatoxic Spanish mackerel in the wild (see [29]), is underestimated.

To simplify the interpretation of our model, we have only modelled fish that have accumulated the indicated ciguatoxin concentration in their flesh from single age points (0.5, 1, 2 and 4 years of age). Any additional toxin uptake after these time points would complicate the modelling, and result in the current models being an underestimate of the time required for dilution of ciguatoxins below the precautionary action concentration of 0.01 µg/kg of P-CTX-1 equivalents suggested by the U.S. Food and Drug Administration (FDA) for the safe consumption of seafood [154].

Our modelling assumes that the severity of ciguatera poisoning suffered by most people is proportional to the concentration of ciguatoxins in the fish flesh eaten [1]. We model four different initial concentrations for contamination of Spanish mackerel flesh (P-CTX-1 equivalents of 5.0, 1.0, 0.1 and 0.03 µg/kg), and assume that people eat a relatively "standard" portion size, and do not intentionally eat small (<50 g) portions as a personal risk minimization measure [1].

We consider a flesh ciguatoxin concentration of 5.0 µg/kg P-CTX-1 equivalents or more as an extremely toxic fish for Australia. This is the maximum ciguatoxin concentration that Lehane and Lewis [1] reported for the range of toxin concentrations found in ciguateric fishes from across the Pacific. However, for our models it is a hypothetical concentration as we do not have data for the maximum concentration of ciguatoxins that can accumulate into the flesh of Spanish mackerel or any Australian reef fish. The highest P-CTX-1 flesh concentrations we know of from Australia are 3.9 µg/kg from the flesh of a Coral Cod (*Cephalopholis miniata*) that caused ciguatera in the Northern Territory [27], 3.5 µg/kg P-CTX-1 from the flesh of a sawtooth barracuda (*Sphyraena putnamiae*) that caused three cases of ciguatera including the death of an elderly male in Queensland [28], and 1.0 µg/kg from a ciguatoxic Spanish mackerel caught from northern New South Wales [14]. The maximum concentration reported from the muscle of fishes from the Pacific Ocean is 81.8 µg/kg of P-CTX-1 equivalents from moray eel (*Gymnothorax* sp.) from the Republic of Kiribati [87]. Such high toxin concentrations are likely life-threatening, with probably more deaths from ciguatera in the Pacific caused by moray eels than any other fish [165]. Possibly moray eels can accumulate such high concentrations through a resistance mechanism [87], not possessed to the same degree by other fish species [140,166,167]. The accumulation of such high ciguatoxin concentrations in moray eels is even more surprising if Lewis et al. [86] are correct with their hypothesis for the depuration of ciguatoxins from fishes. There is no market for moray eels as food fishes in Australia and they are not targeted by commercial or recreational fishers, so there is no basis to assess the toxicity of moray eels from the Great Barrier Reef or east coast of Australia.

We consider highly toxic fish from Australia as having flesh P-CTX-1 concentration of 1.0 µg/kg P-CTX-1 equivalents, i.e., 100-times the U.S. FDA precautionary action concentration of 0.01 µg/kg of P-CTX-1 equivalents. We consider lowly toxic fish as having flesh P-CTX-1 concentration equivalents of 0.1 µg/kg, as Lehane and Lewis [1] estimated that 2 out of 10 people would be poisoned by this concentration. Precautionary action concentrations incorporate a safety margin, so the actual no-adverse-effect concentration likely lies between 0.01 and 0.1 µg/kg of P-CTX-1 equivalents. However, Hossen et al. [168] suggested that Caribbean fishes contaminated with as little as 0.02 µg/kg of P-CTX-1 equivalents of Caribbean ciguatoxins (C-CTX) could cause human poisoning, but this may correspond to a higher toxin load, as C-CTX-1 is less toxic than P-CTX-1 [34]. Toxic fish that cause acute or life-threatening poisonings are rare in Australia [5], although the death of a healthy young woman in Queensland has been attributed to ciguatera from eating Spanish mackerel fillets [142]. Therefore, modelling the dilution of ciguatoxins from the flesh of Australian fishes is probably most relevant for lowly to highly toxic fish. Based upon the symptom profile and time to onset of symptoms, Lewis et al. [76] concluded that mackerel were "on-average" more toxic than non-mackerel species in Queensland.

Ciguatera is uncommon in Australia [5], so repeat poisonings are even rarer. This contrasts with communities with much greater reliance on tropical seafood, such as some Pacific island communities where a fatalistic attitude to ciguatera poisoning can develop [7,136,169]. We have therefore assumed that our modelled concentrations for causing clinical symptoms of ciguatera ($\geq$0.01 µg/kg P-CTX-1 equivalents) is for a naive human population not previously exposed to ciguatera, as it is thought that people who have been poisoned previously can become more sensitive to the ciguatoxins [4].

Our model suggests that for 0.5-year-old Spanish mackerel contaminated with 0.03 µg/kg of P-CTX-1 equivalents, somatic growth could reduce the toxicity of the flesh below the U.S. FDA precautionary action concentration of 0.01 µg/kg equivalents by 4-years of age for female fish, and 5-years for male fish (Figure 6). However, a toxin concentration of 0.03 µg/kg P-CTX-1 equivalents may not always cause human poisoning [1]. We conclude that somatic growth cannot reduce the toxicity of the flesh of fish of $\geq$0.5-years age contaminated with 0.1 µg/kg P-CTX-1 equivalents (i.e., a lowly toxic fish), to less than the U.S. FDA precautionary level within 10-years (Figure 6 and Figure S8). As most Spanish mackerel caught from along the east coast of Australia are less than 8-years old [83] this effectively means that somatic growth on its own is unlikely to reduce the ciguatoxin burden in the flesh of ciguateric Spanish mackerel of legal size to levels considered safe for consumption. However, for fish that accumulate a significant toxin burden at a young age, the reduction in toxin concentration from rapid growth during its first year (Figure 5), could reduce the toxin concentration sufficiently to reduce the severity of the illness.

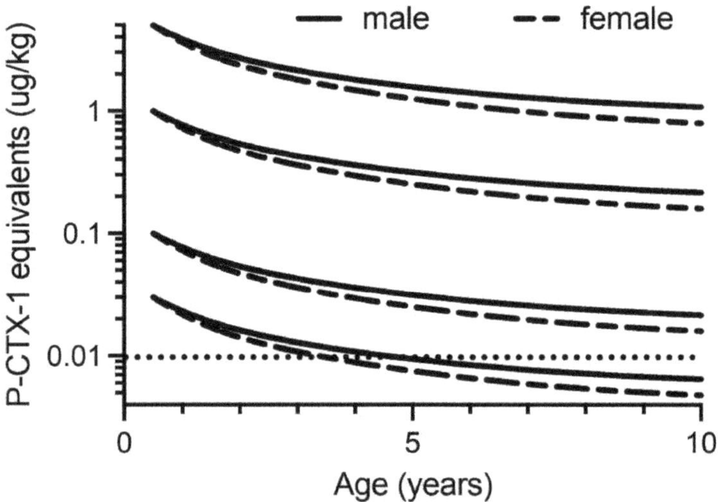

Figure 6. Modelled dilution of 5.0, 1.0, 0.1 and 0.03 µg/kg P-CTX-1 equivalents from Spanish mackerel flesh by somatic growth, for male and female fish contaminated with ciguatoxins at 0.5 years of age. Dotted line = USFDA precautionary action concentration of 0.01 µg/kg of P-CTX-1 equivalents. (See Figure S8 for expanded figure with modelled dilution of fish contaminated with ciguatoxins at 0.5, 1, 2 and 4 years of age).

4. Model for the Dilution of Ciguatoxins in the Flesh of Spanish Mackerel (*S. commerson*) through Growth and Depuration

Our second quantitative model hypothesizes that Spanish mackerel can depurate ciguatoxins. Halstead and Bunker [151] thought that ciguateric fish could detoxify their tissues, but Randall [70] did not. Thirty years later, Tosteson et al. [170] suggested that barracuda (*Sphyraena barracuda*) may be able to detoxify ciguatoxins based upon a seasonal reduction in frequency of toxic fish caught in the Caribbean. Populations of ciguateric surgeonfish were reported to lose toxicity over several months in French Polynesia [94], and Lewis et al. [86] suggested moray eels from Tarawa in the Republic of Kiribati could depurate ciguatoxins with a half-life of <1 year. Li et al. [72] have recently demonstrated rapid depuration of ciguatoxins from the muscle of juvenile goldspotted rockcod (*Epinephelus coioides*), with depuration consistent with a mono-phasic exponential decay, and half-lives for P-CTX-1, -2 and -3 of 28, 26 and 33 days, respectively. Previously, Ledreux et al. [71] were the first to experimentally demonstrate the depuration of ciguatoxins from the muscle of fishes. They found that juvenile sea mullet (*Mugil cephalus*) fed a gel diet containing *Gambierdiscus polynesiensis* depurated ciguatoxins from muscle tissue with a half-life in hours. The depuration of ciguatoxins from muscle was suggested to be through the liver via the bile to the intestine [71]. However, it is likely that the depuration rates for xenobiotics such as ciguatoxins are faster in juvenile fish compared to those of adults. In humans, the half-life for pharmacokinetic clearance of drugs can be faster in children than adults [171]. It is difficult to see how ciguatera would be a global health problem if ciguatoxins could depurate from the muscle of large fish with a half-life of <1 day.

We find the use of sea mullet (*M. cephalus*) as a model for the uptake and depuration of ciguatoxins [71] interesting as sea mullet is one of the most important commercial fisheries in Australia with the annual catch exceeding that of all other fin-fish species in New South Wales and Queensland [172]. Along the east coast of Australia, sea mullet is commercially fished from southern New South Wales to just north of Hervey Bay, including the northwestern beach of Fraser Island to the immediate north of Platypus Bay (Queensland Department of Agriculture and Fisheries, personal communication). In recent years, the most valuable part of the mullet fishery is for the ovaries (roe) which are exported to East Asia [173] for production of high value comestibles such as Karasumi/Bottarga. Sea mullet are a species never associated with ciguatera in Australia and given that the viscera of ciguateric reef fishes can be up to 50-fold more toxic per unit mass than the flesh [43,161], it is likely that we would know of poisoning cases from mullet flesh or roe if they occurred. The depuration of ciguatoxins from sea mullet reported by Ledreux et al. [71] suggests that the enzymatic pathways responsible for depuration are common to many fish species including those not normally associated with ciguatera. Ikehara et al. [25] have shown that liver enzymes from both toxic and non-toxic fish species can bio-oxidize ciguatoxin P-CTX-4A to its more toxic form (P-CTX-1), as well as produce M-*seco*-analogs, which they interpreted as liver detoxification products.

Yogi et al. [31] suggested the absence of P-CTX3C (Figure 1) from a range of predatory reef fish from Okinawa containing P-CTX-1, could indicate that the local *Gambierdiscus* did not produce the P-CTX3C family of toxins. Alternatively, P-CTX3C analogs may depurate rapidly from higher trophic level fish as suggested by Ledreux et al. [71]. However, both P-CTX-1 and P-CTX3C analogs have been detected from red bass (*Lutjanus bohar*) [31], amberjack (*Seriola dumerili*) [174] and moray eel (*Gymnothorax javanicus*) [22], showing that both families of toxins can bio-accumulate into higher trophic level fishes. Oshiro et al. [158] could not detect P-CTX3C from five yellow-edge coronation trout (*Variola louti*) caught from Okinawa and contaminated with P-CTX-1, -2 and -3. As yet, P-CTX3C analogs have not been detected from dinoflagellates or fish from Australia.

Lewis et al. [86] suggested that Pacific moray eels from the Republic of Kiribati could depurate ciguatoxins with a half-life of <1 year. We constructed our second model for dilution of ciguatoxins in the muscle (flesh) of Spanish mackerel using a combination of growth and depuration. We have modelled a range of potential half-life's, from 0.5 to 4 years

(Figure 7 and Figures S9–S12). We recognize that this modelling is speculative without any direct evidence for depuration in Spanish mackerel. However, there is circumstantial evidence for depuration, and we cautiously interpret the modelling results in the context of the biology and fishery for Spanish mackerel and the frequency of ciguatera cases from the east coast of Australia. Spanish mackerel are heavily exploited along the east coast of Australia with unsustainable total annual catches (~1000 tonnes) occurring in the late 1970's and early 1980's, and again in the early 2000's, with commercial catches remaining high at other times before the introduction of a total allowable commercial catch in 2004 [83]. Given this high exploitation rate, we hypothesize that after the ban on taking Spanish mackerel from Platypus Bay came into force in 1987, which appeared to produce a reduction in the frequency of ciguatera along the east coast of Australia, fish that accumulated ciguatoxins in Platypus Bay, left the bay and depurated ciguatoxins relatively quickly before they were caught elsewhere or died of natural causes. Alternatively, it is possible that the ban may have coincided with a long-term decrease in ciguatoxin production from Platypus Bay. Even without direct evidence for depuration of ciguatoxins in Spanish mackerel, our model provides a conceptual pilot for the use of fishery science in developing such risk assessments. Our model for depuration assumes that the depuration rate is constant over the lifetime of adult fish as we have no basis for considering the up- or down-regulation of the enzymes involved in depuration pathways. We have used the same four initial ciguatoxin concentrations that we used for our previous growth dilution model (5.0, 1.0, 0.1 and 0.03 µg/kg P-CTX-1 equivalents).

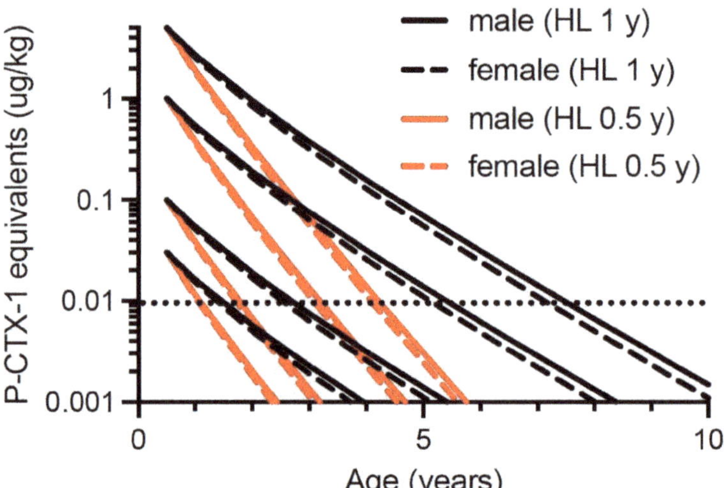

Figure 7. Modelled dilution of 5.0, 1.0, 0.1 and 0.03 µg/kg P-CTX-1 equivalents from the flesh of male and female Spanish mackerel of 0.5 years of age by a combination of somatic growth and depuration. Red lines = depuration half-life (HL) = 0.5 year. Black lines = depuration half-life (HL) = 1 year. Dotted line = USFDA precautionary action concentration of 0.01 µg/kg of P-CTX-1 equivalents. (See Figures S9–S12 for expanded figure with modelled dilution from somatic growth and depuration for fish contaminated with ciguatoxins at 0.5, 1, 2 and 4 years of age, and with hypothetical half-lives for depuration of 0.5, 1, 2 and 4 years).

Our model suggests that somatic growth in combination with a hypothetical half-life for ciguatoxin depuration of six months could reduce the toxicity of the flesh of even an extremely toxic (5.0 µg/kg P-CTX-1 equivalents) six-month-old fish to below the U.S. FDA precautionary action threshold (0.01 µg/kg P-CTX-1 equivalents) by the time the fish reaches 5 years of age (Figure 7). This is for both male and female fish which would weigh ~9.5 and ~12.4 kg at 5-years of age, respectively (Figure 5). In contrast, the model suggests that with a half-life of 1-year, although the 5.0 µg/kg toxin concentration would quickly reduce, the flesh concentration would not drop below the 0.01 µg/kg threshold for male or female fish until they were between 7- and 8-years of age (Figure 7). As most commercially caught Spanish mackerel are less than 8-years of age [83], this would effectively be for most Spanish mackerel caught from Queensland. A half-life of 4-years would effectively mean that a highly toxic 0.5-year-old fish may not become safe to eat over most of its lifetime (Figure S12), and even a 0.5-year-old lowly toxic fish (0.1 µg/kg P-CTX-1 equivalents) would be between 5- and 7-years of age before it reached the U.S. FDA threshold concentration (Figure S12a). All these conclusions are based upon the assumption that the fish does not accumulate any additional ciguatoxin after six months of age as this would significantly delay the time for the fish to reach a flesh concentration that was safe to eat.

We can use the relationships shown in Figure 7 and Figures S9–S12 to estimate the half-life most consistent with ciguatera outbreaks in southern Queensland and northern New South Wales for low and high ciguatoxin concentrations contaminating Spanish mackerel to reach the FDA threshold concentration of 0.01 µg/kg P-CTX-1 equivalents (although the safe-level for consumption is likely between 0.01 and 0.1 µg/kg P-CTX-1 equivalents). Ciguatera cases caused by Spanish mackerel decreased in Queensland after the 1987 ban on their capture from Platypus Bay, possibly indicating that fish had depurated toxicity before being caught elsewhere or dying of natural causes. If we assume commercial fishers were catching 2- and 3-year old Spanish mackerel from Platypus Bay before the ban (Table 2), and we assume an initial P-CTX-1 concentration of 1.0 or 0.1 µg/kg equivalents contaminating Spanish mackerel [14,26], then it would take a further ~10–11 years before 2- or 3-year old highly-toxic fish contaminated with 1.0 µg/kg P-CTX-1 equivalents would reach the U.S. FDA threshold concentration with a depuration half-life of 2-years (Table 2). For lowly toxic fish (0.1 µg/kg P-CTX-1 equivalents), it would take between a further 4 and 6 years (Table 2). As most Spanish mackerel caught in Queensland are less than 8-years of age [83], a half-life of 2-years or longer for depuration seems inconsistent with the observed reduction in ciguatera cases from Spanish mackerel after 1987. In contrast, a half-life of 1-year would take between a further 2 and 6 years for 2- and 3-year-old fish to depurate 1.0 or 0.1 µg/kg P-CTX-1 equivalents to the U.S. FDA threshold concentration (Table 2), and a half-life of 0.5-year would take between a further 1 and 3 years (Table 2). If Spanish mackerel along the east coast of Australia can depurate ciguatoxins, then we believe that it is most likely with a half-life of 1-year or less, because this rate of depuration is consistent with the incidence of ciguatera from Spanish mackerel reducing after the enactment of the ban on their capture from Platypus Bay. A half-life of ≤1-year is also consistent with the depuration rate suggested for moray eels by Lewis et al. [86].

The circumstantial [86,151,170,175] and experimental [71,72] evidence for depuration of ciguatoxins in fishes is strong, but the rate of depuration is likely to vary between tissues, species, life stage, toxin profile and metabolic regulation of detoxifying enzyme levels. Given the co-occurrence of rabbitfish and blotched-javelin feeding within the same ciguatoxic food web in Platypus Bay, but the former being generally non-toxic, it is possible that blotched-javelin depurate ciguatoxins more slowly than rabbitfishes. We have hypothesised that Spanish mackerel can depurate ciguatoxins, but ideally this requires experimental proof. Unfortunately, Spanish mackerel, especially larger specimens, do not often survive capture and release [144], so capturing commercial-sized fish and keeping them alive for direct feeding and depuration experiments may be difficult. In such cases, the use of fisheries science to model depuration as used in this review, helps provide a conceptual framework for future hypothesis testing.

Table 2. Age at which 2- or 3-year-old Spanish mackerel contaminated with 0.1 or 1.0 µg/kg of P-CTX-1 equivalents are estimated to reach the U.S. FDA threshold concentration of 0.01 µg/kg P-CTX-1 equivalents, assuming a depuration half-life of 0.5, 1.0, 2.0 or 4.0 years.

Gender and Approx. Weight (kg) of 2- or 3-Year-old Ciguatoxic Spanish Mackerel	Initial [P-CTX-1] µg/kg Equivalents	Modelled Fish Age (Years) at Which P-CTX-1 Concentrations of 1.0 or 0.1 µg/kg Reach the Threshold of 0.01 µg/kg Equivalents, from Somatic Growth and Depuration (for Depuration Half-Life's of 0.5, 1.0, 2.0 or 4.0 Years)			
		Half-Life = 0.5 Year	Half-Life = 1.0 Year	Half-Life = 2.0 Year	Half-Life = 4.0 Year
Male 2-year-old, 5.5 kg	1.0	4.9	7.5	12.4	21.6
Male 3-year-old, 7.0 kg	1.0	6.0	8.8	13.9	23.9
Female 2-year-old, 6.6 kg	1.0	4.9	7.3	11.8	20.1
Female 3-year-old, 8.7 kg	1.0	6.0	8.6	13.4	22.5
Male 2-year-old, 5.5 kg	0.1	3.4	4.6	6.6	10.0
Male 3-year-old, 7.0 kg	0.1	4.5	5.7	8.0	11.9
Female 2-year-old, 6.6 kg	0.1	3.4	4.5	6.3	9.2
Female 3-year-old, 8.7 kg	0.1	4.5	5.7	7.7	11.2

If ciguatoxic fish can develop mature gonads, then spawning could be another mechanism for dilution of ciguatoxins from fish. The relative toxicity of red bass ovaries was reported to be slightly greater than muscle (flesh), but testes were more than tenfold more toxic [160]. Colman et al. [176] suggested that barracuda can transfer Caribbean ciguatoxin (C-CTX-1) to their eggs at much higher concentrations than found in muscle, and Yan et al. [177] have recently demonstrated transfer of P-CTX-1 to eggs in marine medaka (*Oryzias melastigma*). Prior to spawning, the gonads can contribute a significant amount to the total weight of Spanish mackerel with ovaries being up to 13% of fish body weight, and testes up to 7% of fish body weight [156]. Spawning occurs by the broadcasting of eggs and sperm into water, which may quickly shed a considerable tissue burden of toxin from the fish. It is interesting to speculate if such a fish survives until the next spawning season, would the re-maturing gonads again accumulate ciguatoxins from any residual toxin in other tissues? The embryos produced by broadcast sperm and eggs from spawning ciguateric fishes may not survive as ciguatoxins have deleterious effects on developing fish embryos [176–179].

5. A conceptual Model for Ciguateric Food-Chains on the Great Barrier Reef

The Great Barrier Reef extends offshore of the east coast of Queensland from the northern tip of Cape York to Bundaberg in the south, is about 2300 km long and made up of ~3000 separate reefs, ~600 mainland islands and ~300 coral cays. It sits within the Great Barrier Reef Marine Park, which covers an area of ~344,400 km^2 consisting of 70 broadscale habitats (bioregions) and was enacted in 1975 and World Heritage Listed in 1981. Our conceptual model for ciguatera food-chains on the Great Barrier Reef is simplistic given the complexity of habitats and communities produced through spatial gradients across the marine park, both in a general east-west direction across the continental shelf from in-shore to mid-shelf and outer-shelf reefs, as well as the large north-south latitudinal gradient. For example, predatory fish assemblages differ markedly across the continental shelf with some but weaker differences along the north-south latitudinal gradient [180].

No individual reefs or regions are known that regularly produce ciguateric fishes, and none are known that produce more toxic fish than any other reef or region. Ciguatera cases are mostly caused by demersal predatory fishes caught from apparently anywhere along its length or breadth [5,6,181]. There is also almost no information on the production and transfer of toxins across trophic levels along the Great Barrier Reef. We know more about the food chain transfers that cause ciguatera from Platypus Bay fishes than from the Great Barrier Reef. Without more information, the food chains discussed in this review for the development of ciguatera along the Great Barrier Reef are hypotheses based mostly upon the general models developed for Pacific Island nations and territories (e.g., [70]) and contextualised within the food chains and reef fish fisheries found on the Great Barrier Reef.

In French Polynesia, herbivorous fishes are often more toxic than their higher trophic level predators, and there is often a considerable delay after the herbivorous fishes in a region become toxic before the predatory fishes become toxic [136,137,182]. This delay has been interpreted as a time-dependant transformation of analogs such as P-CTX-4A in the food chain before the appearance of P-CTX-1 in higher trophic level fishes [71]. An alternative hypothesis is that depuration of ciguatoxins could delay the accumulation of significant toxin concentrations in fishes exposed to ongoing low-concentration sources of toxins. A similar situation may exist in the Cook Islands where four of the top five fishes that cause ciguatera are herbivores [128]. The higher toxicity of herbivorous fishes in French Polynesia contrasts with the much greater toxicity of higher trophic level fishes in the Republic of Kiribati [32,87].

Herbivorous reef fishes generally have a "stronger" flavour than demersal carnivorous fishes and in many Pacific cultures this can be a desirable characteristic, whereas in Australia, herbivorous fishes from the Great Barrier Reef are generally not preferred and are rarely targeted by recreational or commercial fishers [183,184], with less than 4 tonnes of rabbitfish harvested by commercial fishers from the entire Great Barrier Reef marine park between 1990 to 2019 [141]. This is unlikely to change in the future with the organization responsible for managing the Great Barrier Reef Marine Park (GBRMPA), now actively discouraging the harvesting of rabbitfishes, parrotfishes and surgeonfishes (including unicornfishes) to reduce macroalgal growth on reefs impacted by coral bleaching and crown of thorns starfish (http://www.gbrmpa.gov.au/__data/assets/pdf_file/0006/2 47848/Coral-Recovery-A4-Flyer_4Print.pdf (accessed on 20 July 2021)). We therefore have no basis for comparison of the relative frequencies of toxic herbivorous and carnivorous fishes or the concentrations of the ciguatoxin congeners that are accumulated by them.

5.1. Food Chain Links Leading to Ciguatera along the Great Barrier Reef, Trophic Level 1

Benthic dinoflagellates, including *Gambierdiscus* species are common epiphytes on many macroalgae and turf algae (the epilithic algal matrix/community) found along the Queensland coast and the Great Barrier Reef [15,90,175,185]. They are mostly found in low population densities and on all major classes of macroalgae [15,185,186]. Six *Gambierdiscus* and one *Fukuyoa* species have been so far confirmed from the waters of the Great Barrier Reef (Table S1), with *G. carpenteri* [93] and *F. paulensis* [187] also isolated from New South Wales. Of these species, low concentrations of ciguatoxins have been detected from cultures of *G. carpenteri* [47], *G. belizeanus* [45,47,188] and *F. paulensis* [189] isolated from regions outside of Australia. However, known ciguatoxins could not be detected from cultures of *G. carpenteri*, *G. lapillus*, *G. lewisii* or *G. holmesii* isolated from the Great Barrier Reef [107,190,191], or *G. carpenteri* from New South Wales [93]. Larsson et al. [107] did report finding ciguatoxin-like activity from *G. lewisii* and *G. holmesii*, and trace amounts from *G. lapillus*, all isolated from Heron Island on the southern Great Barrier Reef (*G. lewisii* and *G. holmesii* were reported as unidentified species by Larsson et al., [107], with the species subsequently described by Kretzschmar et al., [191]). Earlier studies from the Great Barrier Reef detected ciguatoxin by mouse bioassay from a cultured clone of *Gambierdiscus* isolated from Arlington Reef adjacent to the Wet Tropics region in north Queensland [68]. At the time, *Gambierdiscus* was a monospecific genus so Holmes et al. [68] attributed the clone to *G. toxicus*, but with the description of many new species in recent years (Table S1), we are unable to assign this clone to a species. As yet, we don't know which species from the Great Barrier Reef are responsible for producing P-CTX-4A, the precursor to P-CTX-1 and P-CTX-2, the major toxins that cause ciguatera along the east coast of Australia [26,28,29].

Most monitoring programs for benthic dinoflagellates, including those in Australia, have been based upon sampling a diverse array of benthic foliose or calcareous macroalgal species and expressing the epiphytic dinoflagellate cell density shaken from these substrates per gram of wet weight macroalgae [192,193]. Adaptations of this method have been widely adopted because it is easy to use. However, unless the same macroalgal species is sampled from different locations, or repeatedly from the same location, cell density comparisons based upon substrate weight are of limited value. For epiphytic species, cell density per unit surface area of substrate is probably more useful as a basis for comparison of populations, especially given the dramatic differences between surface area and unit weight of macroalgae [194]. The use of artificial substrates (settling plates) may overcome many of these deficiencies and provide a better basis for spatial and temporal comparisons [117,195–198], but have not yet been used for monitoring on the Great Barrier Reef. In locations such as Platypus Bay where a single substrate (*Cladophora*) is known to support the source of ciguatoxins, cell densities based upon wet weight of macroalgae are sufficient for internal comparisons, but if a standardised assessment method can be validated [196,198,199] it would allow for better comparisons between sites. However, Parsons et al. [200,201] suggest caution as they found poor correlation of *Gambierdiscus* densities on artificial substrates and macroalgae in the Florida Keys and U.S. Virgin Islands; although, from a ciguatera perspective, what is important is whether the populations being quantified are those likely to enter food chains leading to ciguatera.

There are 600–700 species of macroalgae on the Great Barrier Reef and they have considerable latitudinal, cross-shelf and within reef variation [202]. In contrast to inshore reefs, offshore reefs usually have low standing biomass of foliose macroalgae, but high cover of crustose calcareous algae and turf assemblages [202]. Where foliose macroalgae do survive on coral reefs, it is often because of their unpalatability to herbivores [203] or where herbivore behaviour is modified through other mechanisms, such as fear of predation [204–206]. Macroalgae may act as "reservoirs" for epiphytic dinoflagellates on coral reefs, but unless they and their epiphytes are consumed in what Cruz-Rivera and Villareal [203] termed their shared doom, they do not become part of the food chain leading to ciguatera. In the Cook Islands, increased ciguatera cases were associated with increased turf algae cover on coral reefs, whereas decreases in ciguatera cases occurred during a period of increased macroalgal cover [207]. Turf assemblages are ubiquitous and diverse on the Great Barrier Reef where 1 cm^2 of space can host more than 20 species [202] and they can be 15 times more productive than foliose macroalgae on reefs [208]. They are one of the most abundant benthic habitats on the Great Barrier Reef [209] and are grazed by herbivorous and detritivorous fish, and invertebrates. We believe that epiphytic *Gambierdiscus* or *Fukuyoa* on turf algae are the most likely to enter food chains leading to ciguatera on many reefs [33,197,198,201,203], including the mid- and outer-shelf reefs of the Great Barrier Reef, but this is yet to be tested. Lewis et al. [175] developed an air-lift suction device to sample 0.8 m^2 of turf algae for benthic dinoflagellates, in part to mimic the feeding process of the detritivorous, lined-bristletooth surgeonfish (*Ctenochaetus striatus*), one of the major vectors of ciguatoxins on Pacific reefs [210]. Parsons et al. [96] used a similar monitoring approach to sampling turf algae in Hawaii. While the method is suitable for sampling turf algae for benthic dinoflagellates, a way to compare population densities between sites still needs to be validated, possibly in conjunction with direct monitoring of dissolved ciguatoxins from the water [211–213].

Runoff of fine sediments and nutrients from land catchments are considered major threats to the health and resilience of the Great Barrier Reef [114]. Nutrients have been hypothesized to promote the growth of macroalgae and crown-of-thorns starfish outbreaks, and fine sediments reduce light and can smother seagrass meadows and inshore coral reefs [114]. To reduce the impacts of these stressors, targets have been developed for reducing the end-of-catchment loads of nutrients and sediments, and concentrations of pesticides (mostly herbicides) that flow from the major river catchments into the Great Barrier Reef lagoon (Great Barrier Reef 2050 Water Quality Improvement Plan, https://www.reefplan.qld.gov.au/ (accessed on 20 July 2021)). While there have been major improvements in land practices in recent years (Great Barrier Reef Report Card for 2017 and 2018, https://www.reefplan.qld.gov.au/tracking-progress/reef-report-card/2017-2018

(accessed on 20 July 2021)), at present they are not happening fast enough to be able to meet the 2050 targets [114].

Most catchment sourced nutrients and sediments flow to the Great Barrier Reef lagoon during high rainfall events during the monsoon (wet) season [214]. Fine sediments (<16 μm) can be carried in suspension by flood plumes and reach inshore and sometimes mid- and outer-shelf reefs and be easily resuspended by wind and wave action [215]. Whether sediments and nutrients reach mid- or outer-shelf reefs depends upon the size of the contributing catchments, the size of the flood plumes generated and the distance from river mouths to mid- and outer-shelf reefs. In addition, south-easterly winds and the Coriolis force will tend to push flood plumes northwards along the coast. However, in the Wet Tropics NRM region of north Queensland (from north of Townsville to south of Cooktown, approximately 15.5° S to 18.5° S, Figure S7), the continental shelf is narrow and nutrients and fine sediments probably reach mid-shelf reefs most years [216]. Riverine inputs dominate nutrient inputs to inshore reefs in the Wet Tropics, but upwelling, mixing events and nitrogen-fixation are more important for offshore reefs in the Wet Tropics [214]. In the relatively oligotrophic waters of the Great Barrier Reef, increased nitrogen concentrations are short-lived due to rapid biological uptake [214]. Benthic dinoflagellates living within turf algae matrices may face considerable competition for nitrogen resources given the rapid assimilation of inorganic nitrogen by their host algae [217,218]. However, some *Gambierdiscus* species may partially offset these limiting inorganic nitrogen concentrations through mixotrophy [219,220].

Sedimentation can drive declines in the productivity of algal turfs as well as suppress feeding on them by herbivores such as surgeonfishes [221–224], with feeding by *C. striatus* especially sensitive to small increases in sedimentation [225]. Sediments can cause the length of turf algae to increase while reducing their productivity and the proportion of detrital particulates within them [226]. The filling of interstitial spaces of turf algae by sediments not only limits feeding on them by surgeonfishes [222,224,225], it may also reduce space for growth of benthic dinoflagellates, with fine sediments (<16 μm) being smaller than the minimum diameter of *Gambierdiscus* and *Fukuyoa* species. Both the reduction in habitat for benthic dinoflagellates and in herbivory caused by deposition of sediments could be mechanisms that limit the production of ciguatoxins into coral reef food chains. Sediment loads on inshore reefs can overwhelm turf algae [224,227], and deter grazing by parrotfish on inner-shelf reefs where they are often the dominant herbivore/grazer [228]. It would be ironic if the major investment by Australian and Queensland Governments to restore the health of the Great Barrier Reef led to reductions in sediment loads that produced more grazing of turf algae supporting populations of *Gambierdiscus*. However, there are many links in marine food chains, and it is not possible to predict what the outcome would be from any such change with respect to the frequency or severity of ciguatera.

Reducing mixtures of pesticides flowing off agricultural catchments into the Great Barrier Reef lagoon during high rainfall events is a priority action for improving the health and resilience of the Great Barrier Reef, especially for certain catchments in the Mackay-Whitsunday, Burdekin and Wet Tropics NRM regions (Great Barrier Reef 2050 Water Quality Improvement Plan, https://www.reefplan.qld.gov.au/ (accessed on 20 July 2021)). While the impacts of pesticides most likely occur on inshore reefs, in the Wet Tropics NRM region, the photosystem II inhibitor diuron may reach more than 40 km offshore from river mouths at concentrations with inhibitory effects [229]. Shaw et al. [230] reported that herbicides could be transported in plumes at concentrations that affected zooxanthellae dinoflagellates on inshore coral reefs. Community composition changes in tropical benthic microalgae communities can occur at low μg/L concentrations of diuron [120]. However, there is not enough information to know if catchment derived herbicide concentration could be having an impact on populations of *Gambierdiscus* and other benthic dinoflagellates along the Great Barrier Reef.

Halimeda banks (bioherms) are deep water reefs (>20 m) with a surface layer of live *Halimeda* adjacent to the inside of outer barrier reefs of the Great Barrier Reef and covering many thousands of km^2 [209,231]. *Halimeda* species are calcareous green macroalgae also common on shallow reefs along the Great Barrier Reef [202], which often host low population densities of epiphytic benthic dinoflagellates, including *Gambierdiscus* [90,185,186]. It is possible these deep-water reefs may be a reservoir for benthic dinoflagellate populations, but we are unaware of any sampling from these banks or other deep-water habitats. *Gambierdiscus* species are photosynthetic but may be able grow under low light conditions by acquiring nutrients through mixotrophy [219,220]. Herbivorous fish are rare below 50 m off the Great Barrier Reef shelf-break, but macroalgae have been found to 194 m depth and *Halimeda* to 150 m [232].

It is possible that there may be more than one source of ciguatoxins produced among the diverse microalgal species growing amongst the turf community matrix, as bio-synthetic pathways can be common to different algal groups. For example, paralytic shellfish poisoning toxins are produced by a diverse mixture of dinoflagellate and cyanobacterial species [233], and 44-methylgambierone has been isolated from species of *Gambierdiscus*, *Fukuyoa* and *Coolia* [60–63,124]. Ciguatoxin-like activity has been reported from the planktonic cyanobacteria *Trichodesmium* [234,235] but there has been little follow-up of these results. The recent suggestion that parrotfishes are microphages that acquire their nutrition by feeding mainly on photosynthetic microorganisms, predominantly cyanobacteria that are epilithic, epiphytic, endolithic or endosymbionts [236,237] requires further study with respect to their accumulation of toxins that cause ciguatera. Laurent et al. [238] previously linked ciguatoxin-like activity extracted from parrotfish and cyanobacterial mats from New Caledonia, in the apparent absence of *Gambierdiscus*.

Gambierdiscus is often only one of several toxin-producing benthic dinoflagellates found as epiphytes on macroalgae and turf algae on coral reefs. Due to this, many authors have suggested that these other species may play a role in ciguatera, but benthic dinoflagellate toxins are only part of the milieu of secondary metabolites produced on coral reefs [239]. For example, we found that between 50–100% of the solvent extracts of different sized fractions sieved from epiphytes and detritus from macroalgae and turf algae collected from along the east coast of Queensland were lethal to mice by intraperitoneal injection [77,91,175]. However, without evidence for their accumulation into the flesh of carnivorous reef fishes, there is no basis for suggesting these toxic fractions have a role in human poisoning along the east coast of Australia.

Ostreopsis, Prorocentrum, Amphidinium and *Coolia* are dinoflagellate genera containing toxic benthic species often found with *Gambierdiscus* as epiphytes on macroalgae on the Great Barrier Reef [15,90,125,126,185,186,240]. *Prorocentrum lima* is often found with *Gambierdiscus* and produces analogs of okadaic acid which are lipid-soluble toxins also produced by several planktonic dinoflagellate species of *Dinophysis* [241,242]. Shellfish accumulate these toxins by filter feeding on toxic *Dinophysis* which causes the disease Diarrhetic Shellfish Poisoning [243]. Okadaic acid has been reported from the flesh of a ciguatoxic Caribbean barracuda [244] but this result has never been repeated from a toxic fish and requires validation. However, okadaic acid has been shown to accumulate in some temperate fish species (reviewed by [245]). Some isolates of *Ostreopsis siamensis*, *O. ovata* and *O. mascarenensis* produce analogs of the extremely potent water-soluble toxin called palytoxin [246–248], as does the cyanobacteria *Trichodesmium* [249]. Tropical planktivorous Clupeid fishes such as herrings and sardines as well as various invertebrates are thought to sometimes accumulate palytoxins in their viscera and/or gut [250,251]. Some communities eat Clupeid fishes' whole, exposing them to the risk of the visceral and gut contents of the fish, and tropical clupeids contaminated with palytoxin analogs have been linked to human poisoning [250,251]. The disease is called clupeotoxin poisoning to distinguish it from ciguatera. It is much rarer than ciguatera but apparently has higher morbidity [250]. We detected a water-soluble toxin from *O.* c.f. *siamensis* isolated from the Great Barrier Reef [240] and palytoxin-like toxins have been detected from *O.* c.f. *siamensis* strains isolated from along the east coast of Australia [252,253]. Relatively high populations of *Ostreopsis* have been found on macroalgae from inshore and mid-shelf reefs of the Great Barrier Reef [15,185].

5.2. The Food Chain Links Leading to Ciguatera along the Great Barrier Reef, Trophic Level 2

There is considerable species richness (178 species in 9 families) of herbivorous and nominally herbivorous (e.g., detritivorous) fishes on the Great Barrier Reef [254]. The surgeonfishes comprise a major group of these herbivores, especially on mid- and outer-shelf reefs [255]. The lined-bristletooth (*Ctenochaetus striatus*) and the brown surgeonfish (*Acanthurus nigrofuscus*) are among the most abundant herbivorous fish on Indo-Pacific reefs, including the Great Barrier Reef [255–257]. *Ctenochaetus striatus* has long been considered a major vector for ciguatoxins in the Pacific [210], as well as also causing ciguatera poisoning where it is eaten [128,258] and responsible for up to 65% of ciguatera cases in Tahiti [259]. It is a detritovore that does not consume macroalgae and uses its bristle-like teeth to comb or brush algal turfs and consumes considerable quantities of detritus and sediment [257,260], including epiphytic benthic dinoflagellates [42,43]. In contrast, the teeth of *A. nigrofuscus* are designed for cropping and consuming turf algae [257], resulting in the presumed consumption of the associated epiphytic dinoflagellates. To date, the direct evidence for consumption of *Gambierdiscus* by surgeonfish in the wild is the initial research by Yasumoto et al. [42,43] in the Gambier Islands of French Polynesia, and *Gambierdiscus* being found in the gut contents of *C. strigosus* from Hawaii [261]. Magnelia et al. [262] found that Atlantic Ocean surgeonfish (*A. bahianus* and *A. chirurgas*) did not avoid eating *Gambierdiscus* when offered a choice of foods, suggesting that at least some surgeonfish do not find *Gambierdiscus* unpalatable.

At Heron Island, two species of *Ctenochaetus*, *C. striatus* and *C. binotatus* are the most abundant grazers on turf algae with *C. striatus* more abundant in shallow waters [263]. Heron Island is part of the Capricorn Bunker group of coral atolls in the southern Great Barrier Reef (Figure S13) and is a healthy reef system that at least up until 2019, has been spared many of the major issues impacting other sections of the Great Barrier Reef, coral bleaching, damage from cyclones and crown-of-thorn-starfish outbreaks [209]. Marshell and Mumby [263] found that surgeonfish feeding on turf algae at Heron Island make up 74% of the herbivorous fish biomass and remove 73% of the daily productivity of the turf algae in the most productive habitat for turf algae. However, the turfs quickly recover because of their fast growth rates. However, given that the standing crop appears to be constant without major differences between habitats, Marshell and Mumby [263] suggest that grazers concentrate their feeding in habitats where turf algae are most productive. Generally, fish grazer biomass on the Great Barrier Reef correlates more strongly with turf algae production than with turf biomass [264].

There are only limited studies of benthic dinoflagellates on turf algae on the Great Barrier Reef [90,175,186] but these found that *Gambierdiscus* were often present. This suggests that benthic dinoflagellate growth rates, possibly along with imports from local population reservoirs such as those on calcareous and foliose macroalgae, are enough to maintain cell populations on turf algae with the daily removal and/or "combing" of much of the daily productivity of these by surgeonfishes. *Gambierdiscus* species are slow growing, with growth rates under laboratory conditions up to ~0.65 divisions/day [265], although most reported growth rates are considerably slower than this [45,46,266–268]. Growth rates for *G. carpenteri* isolated from the Great Barrier Reef were up to 0.17 divisions/day [269] while Holmes et al. [52] reported up to 0.25 divisions/day for a *Gambierdiscus* isolate from the Great Barrier Reef. Growth is dependent on many interacting environmental factors including temperature, light, and nutrients, all of which can vary with depth and aspect on the reef. Mustapa et al. [270] reported growth rates for Malaysian *Gambierdiscus* isolates growing in culture on turf algae that ranged between ~0.1 to <0.3 divisions/day. Growth rates of 0.1–0.25 divisions/day (i.e., population doubling every 4–10 days) could possibly maintain a *Gambierdiscus* population on turf algae if, as suggested by Marshell and Mumby [263] and references therein, that turf algae biomass can be turned over by grazers every 4 to 25 days (especially as reported growth rates for *Gambierdiscus* are often from experiments trying to produce optimal laboratory growth conditions). High grazing rates may be a feedback mechanism that keep low benthic populations from dramatically increasing. Even though grazing is unlikely to be 100% efficient at removing epiphytic benthic dinoflagellates from turf algae, high grazing pressure would result in only a trickle of ciguatoxin entering into the second trophic level, assuming the *Gambierdiscus* species on the turf algae produce ciguatoxins. Any possible depuration of ciguatoxins from surgeonfish [175] could then further limit toxin transfer to higher trophic levels. The dramatic increase in abundance of *Gambierdiscus* found in the US Virgin Islands when turf algae was caged to exclude grazers provides experimental support for this hypothesis [117].

If populations of turf algal grazers were rapidly reduced, for example through fishing, then the grazing rate on turf algae might be significantly reduced. This could allow time for populations of benthic dinoflagellates to increase on the turf algae, as found in the US Virgin Islands when grazers were excluded by caging turf algae growing on sandstone tiles [117]. If these populations consisted of species of *Gambierdiscus* and/or *Fukuyoa* that produce ciguatoxins, then this could lead to increased loads of ciguatoxins being available to flow into the remaining, smaller populations of herbivores/detritovores. High grazing pressure on turf algae in natural (healthy) reef ecosystems could limit the opportunity for production of significant ciguatoxin loads, while low grazing pressure, possibly from overharvesting of herbivores, might allow for the development of smaller numbers of highly toxic herbivores (Figure 8). This feedback hypothesis could be partially explored through field cage experiments on turf algae to exclude herbivores similar to those used by Loeffler et al. [117]. Such experiments are often used for studying turf algae productivity [221,223,226,263,264], but have not yet been used on the Great Barrier Reef to study changes in epiphytic benthic dinoflagellate populations on turf algae in the presence or absence of macro-grazers. Using artificial surfaces to develop flat turf layers for cage experiments could also make it simpler to standardize benthic dinoflagellate counts using an underwater vacuum or similar device [84]. If our hypothesis is correct, it would provide more support for the campaign by the Great Barrier Reef Marine Park Authority to discourage the harvesting of herbivorous fishes from the Great Barrier Reef (www.gbrmpa.gov.au/__data/assets/pdf_file/0006/247848/Coral-Recovery-A4-Flyer_4Print.pdf (accessed on 20 July 2021)).

Surgeonfishes, including *Ctenochaetus striatus* are harvested for food in the Cook Islands although they are considered high-risk for ciguatera [128]. Between 2006 and 2011, after major disturbances to coral reefs from crown-of-thorns outbreaks and several severe cyclones, macroalgae cover increased, turf algae cover and surgeonfish density decreased (mainly *Acanthurus nigrofuscus* but also *C. striatus*) and ciguatera cases declined, the latter from a peak in 2004 [207]. Before the peak in ciguatera cases, there were increases in turf algae cover but unfortunately there was no information on the abundance of surgeonfish between 1999 and 2006, that is, the years prior to the sudden increase in ciguatera cases in 2004. So, although Rongo and van Woesik [207] concluded that high densities of *C. striatus* were a good predictor of ciguatera cases, their data showing concurrent decreases in ciguatera cases with turf algae cover may be consistent with our feedback hypothesis for production of greater ciguatoxin loads in herbivores under reduced grazing pressure.

To date, long-term monitoring of herbivorous parrotfish and surgeonfish indicate that abundances across the Great Barrier Reef have generally remained stable over time, with some cross-shelf variability [209]. In part, this is likely because herbivorous fishes are not generally harvested by recreational or commercial fishers on the Great Barrier Reef [183,184]. The zoning of the Great Barrier Reef into areas that are permanently open or closed to fishing has therefore created an on-going experiment on the effect of top-down control of herbivory by predatory fish, especially the mostly meso-predators targeted by fishers such as Serranid, Lethrinid and Lutjanid fishes. As expected, the meso-predatory fish targeted by fishers have much greater abundance on reefs closed to fishing compared to those open to fishing [271,272], but this has not led to increased herbivore abundance on fished reefs or an increase in turf algae [272–275]. On the Great Barrier Reef, herbivore abundance appears to be weakly controlled by predators [273,274,276], at least at the current level of fish harvesting. Similar results were found by Mellin et al. [277] in an analysis of the effects of disturbance from storms, disease, crown-of-thorns starfish and coral bleaching on fished and un-fished reefs of the Great Barrier Reef, with the exception that all four types of disturbance led to increases in turf algae. This resilience of herbivore fish populations to predatory control may keep herbivore populations high and help prevent the production and transfer of ciguatoxins to predatory fish on the Great Barrier Reef, if the feedback conceptual model (Figure 8) for production of toxic herbivores is correct. In the only study to date of ciguatoxins from surgeonfish from the Great Barrier Reef, Lewis et al. [175] found only low levels of ciguatoxins from *C. striatus* collected from John Brewer and Davies Reefs off Townsville.

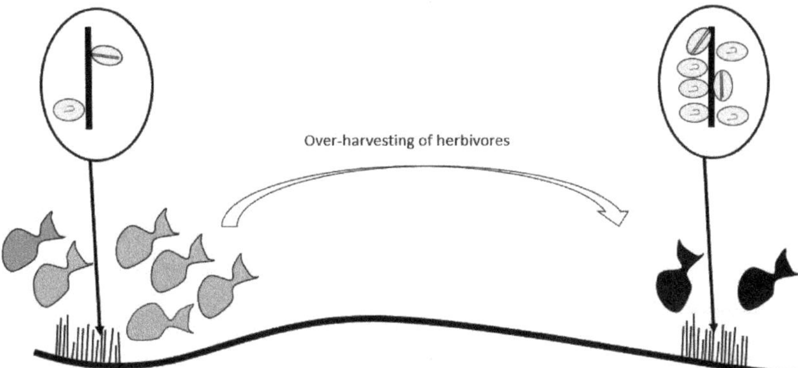

Figure 8. Conceptual model for hypothesised transfer of ciguatoxins from ciguatoxin-producing *Gambierdiscus* or *Fukuyoa* growing as epiphytes on turf algae before and after overharvesting of herbivorous fishes (especially surgeonfishes). On the left, grazing from surgeonfishes prevent benthic dinoflagellate populations from blooming (as indicated by the low number of dinoflagellates shown in the magnified view of turf algae in the oval frame). This limits the amount of ciguatoxin being transferred to the many herbivorous fishes grazing on the turf community and individual fish are generally less toxic (indicated by lighter shading of fish). Overharvesting of surgeonfishes produces the outcome on the right, where surgeonfish abundance is reduced, allowing space and time for benthic dinoflagellate populations to bloom. A higher ciguatoxin load can then be accumulated by the smaller surgeonfish population producing more toxic individual fish (indicated by the darker shading of fish).

The coral reef fisheries of many island nations are thought to be overfished or being unsustainably fished [278]. If the feedback conceptual model for development of ciguatoxic fish (Figure 8) is correct, then a sudden loss of herbivores could be a process that concentrates the trophic transfers of ciguatoxins into the remaining smaller fish populations (herbivores and then carnivores), possibly leading to local increases in ciguatera if there are simultaneous increases in ciguatoxin-producing *Gambierdiscus*/*Fukuyoa* species. The loss of herbivore populations may produce phase shifts in ecosystems from coral to macroalgal dominated states [279,280], although the top-down control of macroalgae by parrotfish has been questioned [281]. Ciguatera research has tended to interpret a shift to macroalgae dominated reefs as an increase in the amount of substrate that supports growth of benthic dinoflagellate populations. The underlying assumption to this interpretation is that lower amounts of macroalgal foliage is limiting benthic dinoflagellate populations and therefore the production and transfer of ciguatoxins into fishes consumed by people. However, it may be the change in ecosystem functioning brought about by the reduction of herbivory that leads to increases in ciguatera, rather than the change to a macroalgal dominated ecosystem per se.

We have hypothesized that *Gambierdiscus* or *Fukuyoa* growing on turf algae are the major contributor to the flow of ciguatoxins into food chains on the Great Barrier Reef because:

- *Gambierdiscus* are common epiphytes of turf algae on the Great Barrier Reef;
- Turf algae are ubiquitous on reefs; and
- Turf algae have high rates of productivity which support high grazing pressure by surgeonfishes on the Great Barrier Reef.

However, most sampling of benthic dinoflagellates around the world to-date, including ours (MJH, RJL), has focused on macroalgal substrates. Often this is because it is physically easier to sample macroalgae than turf algae, and because of the difficulty in interpreting and comparing dinoflagellate populations from turf algae. However, the assumption underlying most sampling of *Gambierdiscus* and *Fukuyoa* is that the population numbers will relate to the risk of ciguatera (although this is not always explicitly stated). Researchers often focus on macroalgae that produce the highest density of benthic dinoflagellates, without considering the likelihood of those macroalgal species being consumed by herbivores and becoming part of a ciguateric food chain.

Some macroalgae species on the Great Barrier Reef are consumed by herbivorous fishes but often consumption is locally dominated by a single fish species which varies between locations [282–284]. At Lizard Island on the Great Barrier Reef, the dominant herbivore of macroalgae (the unicornfish *Naso unicornis*), was the eighth most abundant of the local herbivore species by numbers and second most abundant by biomass [284]. This indicates that the fish controlling macroalgae abundance may not be obvious based upon simple observations of the local fish community when sampling macroalgae for benthic dinoflagellates. Studies that identify functional processes within local reef ecosystems, e.g., [284,285], may help ciguatera studies target the appropriate fish species to understand the transfer of ciguatoxins along localized food chains. This is especially so for the Great Barrier Reef, where fishers mostly target higher trophic level fishes [184,286,287] so only carnivorous species from this ecosystem tend to cause ciguatera.

Many herbivorous fishes, including surgeonfishes void their faeces over the reef with much of the algal material, including turf algae, remaining structurally intact and capable of growth [288]. *Ctenochaetus* and other surgeonfish feeding over turf algae likely consume a considerable amount of this faecal material [257], which if it contains undigested or damaged *Gambierdiscus*, could be a re-circulation mechanism for concentrating ciguatoxins into surgeonfish. Defecation by surgeonfishes moves sediments across reefs [289] and may also be a mechanism for dispersing benthic dinoflagellates, but this has not been studied.

Small benthic carnivorous fishes, and parrotfish, consume large numbers of crustaceans and other invertebrates living within turf algae [290]. Nothing is known about the potential involvement of these invertebrates in ciguateric food chains on the Great Barrier Reef. However, given our hypothesis for the transfer of ciguatoxins through invertebrates to blotched-javelin fishes in Platypus Bay, it is worthy of study. This could include examining the role for the transfer of ciguatoxins from invertebrates into some 17 families of cryptobenthic reef fishes [291]. It was recently suggested that cryptobenthic fishes account for almost 60% of the fish biomass consumed by reef predators on the Great Barrier Reef [292]. To date, ciguatera research has focussed on the trophic transfer of ciguatoxin to carnivorous fishes from larger herbivorous fishes such as surgeonfishes, while most of the marine food chain on coral reefs may be driven through relatively small fish that grow fast but suffer extreme mortality, such as gobies, blennies, and cardinal fish [292].

5.3. The Food Chain Links Leading to Ciguatera along the Great Barrier Reef, Trophic Levels 3 and 4

Ciguatera from the Great Barrier Reef is mostly caused by demersal, meso-predatory fish species such as coral trouts (*Plectropomus* and *Variola* spp.), emperors/tropical snappers (*Lethrinus* and *Lutjanus* spp.) and cods (*Epinephelus* spp.) [5,293], all of which are targeted by commercial and recreational line fishers [184,286,287]. The adults of many of these demersal predatory reef fishes are thought to have relatively high reef fidelity with little range movement on the Great Barrier Reef [294,295], in contrast with Spanish mackerel which are pelagic and often travel large distances. Stable isotope analysis suggests that there is little dietary niche overlap of Spanish mackerel with demersal reef fishes such as coral trout on the Great Barrier Reef [146,296]. If dietary overlap is the exception, then possibly different food chains are responsible for the transfer of ciguatoxins from the causative benthic dinoflagellates *Gambierdiscus* and *Fukuyoa* spp., to demersal and pelagic fish predators. Alternatively, it is possible that the intoxicating effects of ciguatoxins and/or water-soluble toxins such as maitotoxins on some herbivorous fishes might produce opportunistic feeding for both demersal and pelagic reef predators on the Great Barrier Reef, creating space for dietary niche overlap and toxin transfer.

Coral trout is the common name for a species complex of demersal Serranid fishes (groupers) on the Great Barrier Reef, made up mostly of the common coral trout (*Plectropomus leopardus*), bar-cheek coral trout (*P. maculatus*), blue-spot coral trout (*P. laevis*), passionfruit coral trout (*P. areolatus*), vermicular cod (*P. oligacanthus*), yellow-edge coronation trout (*Variola louti*) and white-edge coronation trout (*V. albimarginata*) [297]. Of these, the common coral trout (*P. leopardus*) is the main target species of the commercial coral reef, fin-fish fishery [286]. Coral trout are considered premier and high-priced food fishes in Australia, in contrast to Spanish mackerel which is only a medium-priced fish (Sydney Fish Market website). On the east coast of Australia, a far greater weight of coral trout is caught and sold commercially than of Spanish mackerel with most fish caught commercially after about 1996 exported live to overseas markets in Asia [298]; although this was interrupted in 2020 through export restrictions on live species caused in-part by the human coronavirus COVID-19 disease pandemic. On the Great Barrier Reef, most of the commercial catch of coral trout (*P. leopardus*) comes from mid- and outer-shelf reefs whereas recreational catches of *P. maculatus* dominate in-shore reefs [286]. In contrast with Spanish mackerel, where catches are almost equal between commercial and recreational sectors [83], the commercial catch dominates the coral trout harvest [286,299]. *Plectropomus leopardus* form a single stock on the east coast of Queensland and are protogynous hermaphrodites, beginning life as a female, with many later changing to male [298]. Stocks are considered sustainably fished on the Great Barrier Reef [286] with the population level in 2020 about 59% of the unfished spawning biomass [299].

We have presumed that predatory demersal fish such as coral trout accumulate ciguatoxins by feeding on ciguatoxic herbivores such as surgeonfishes [70], but there may be alternate food chains for accumulation of ciguatoxins, such as through cryptobenthic fish species. Coral trout are high trophic level predators that eat mostly fish from a wide range of prey families plus a small proportion of invertebrates [300,301]. Gut analysis of speared *P. leopardus* from the Great Barrier Reef indicated that herbivores (parrotfishes, surgeonfishes, and rabbitfishes) make up <10% of prey numbers, with surgeonfishes <3% [301]. Recent studies using a combination of gut analysis, DNA metabarcoding and stable isotope analysis suggest that many prey species of coral trout may not be detected by gut analysis alone. However, this study also confirmed that herbivores make up only a small component of their diet, and relative to their abundance, surgeonfishes appear to be not selected as prey [302]. Instead, prey is dominated by planktivorous fishes (fusiliers, Caesionidae and damselfishes, Pomacentridae) and other carnivores [302]. Surgeonfishes such as *Ctenochaetus* and *Acanthurus* appear to be only a small part of the diet of both *P. leopardus* and *P. maculatus*, with *P. maculatus* feeding on more benthic prey than *P. leopardus* [296,302]. How do coral trout become poisonous based upon a predominant diet of planktivorous fishes? Their diet is not thought to change seasonally [303]. Coral trout are opportunistic generalist carnivores, so if surgeonfishes were intoxicated by ciguatoxins and/or maitotoxins when feeding on high populations of *Gambierdiscus* or *Fukuyoa*, then this could provide a mechanism for temporary diet switching to surgeonfishes by coral trout. Two Caribbean surgeonfishes (*Acanthurus bahianus* and *A. chirurgus*) fed a gel diet containing a *Gambierdiscus* species became disorientated and lost equilibrium [262], behaviour that in the wild would greatly increase their chance of predation. However, Magnelia et al. [262] found that surgeonfish could acclimate to feeding on a fixed dose of *Gambierdiscus* suggesting that the intoxicating effects, and risk of opportunistic predation from coral trout and many other carnivorous fish species, may be greatest after they feed on a sudden population increase (bloom) of *Gambierdiscus/Fukuyoa*. This may be a mechanism by which occasional blooms of ciguatoxin-producing benthic dinoflagellates lead to the production of ciguatoxic predatory reef fish in ecosystems such as the Great Barrier Reef where herbivorous fish populations are largely unaffected by fishing pressure (Figure 9). Marshell and Mumby [263] reported a considerable range in the rate of turnover of turf algae by surgeonfish at Heron Island (4–25 days). Slower turnover rates may allow time for benthic dinoflagellates to occasionally bloom in patches across the Great Barrier Reef, producing the occasional (stochastic) opportunity for ciguatoxins to flow through herbivores/detritovores into meso-predatory fishes such as coral trout. Alternatively, the patchy nature of surgeonfish grazing [304], may allow "islands" of un-grazed turf for benthic dinoflagellates to proliferate for a time.

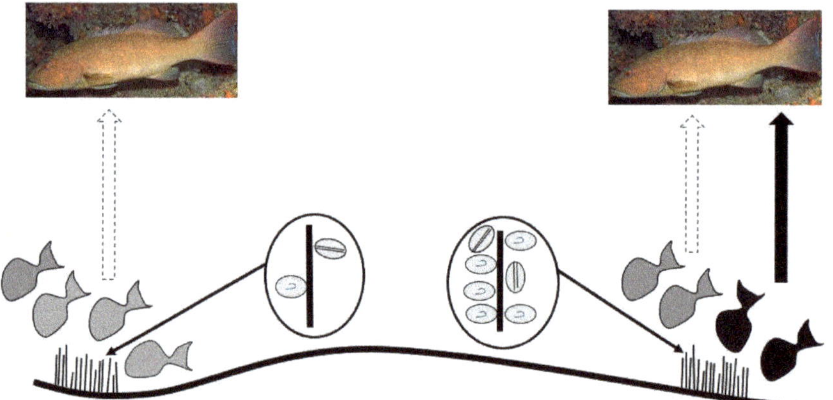

Figure 9. Conceptual model for hypothesised transfer of ciguatoxins from ciguatoxin-producing *Gambierdiscus* or *Fukuyoa* to herbivorous fish (such as surgeonfish) and then to predatory reef fish such as coral trout (*Plectropomus* spp.), in an ecosystem where herbivorous fish populations are mostly unimpacted (such as the Great Barrier Reef). On the left, surgeonfish graze on turf algae with low populations of benthic dinoflagellate populations (as indicated by low numbers of dinoflagellates in oval frame). This limits the amount of ciguatoxin being transferred to the many herbivorous fishes grazing on the turf community, and individual fish are lowly or non-toxic (indicated by lighter shading of surgeonfish). There is little transfer of ciguatoxin to coral trout because these lowly-toxic surgeonfish make up a small proportion of their diet (broken arrow). On the right, the occasional bloom of benthic dinoflagellates leads to a higher ciguatoxin load being accumulated by some surgeonfish (indicated by the darker shading). The intoxicating effects of the dinoflagellate toxins on these surgeonfish renders them more likely to be preyed upon by opportunistic predators such as coral trout leading to toxin transfer. (Coral trout *P. leopardus* image: Graham Edgar, www.reeflifesurvey.com (accessed on 20 July 2021), Creative Commons by Attribution licence for non-commercial use).

If benthic dinoflagellates growing on macroalgae are a source of ciguatoxins on the Great Barrier Reef that poison people, they need to be consumed by herbivores that are preyed upon by carnivorous fish eaten by people, such as Serranid, Lethrinid or Lutjanid meso-predatory fishes. *Naso* spp. (unicornfishes) are surgeonfishes that feed on macroalgae, so may be one such intermediate, with *N. unicornis* being the dominant grazer of macroalgae at Lizard Island on the Great Barrier Reef [284]. However, similar to other surgeonfishes, unicornfishes appear not be selected as prey by coral trout species [296,300,302].

Opportunistic predation on intoxicated herbivores could facilitate toxin transfer; however, Clausing et al. [102] reported that the unicornfish, *N. brevirostris*, did not display any abnormal behaviour when fed a gel diet containing *G. polynesiensis*. Possibly there are species-specific differences in the response of herbivores to the various suite of toxins produced by *Gambierdiscus/Fukuyoa*. Otherwise, even if grazers of macroalgae accumulate ciguatoxins, if they are not preyed upon by carnivorous fishes, their toxin load is unlikely to become part of the food chain leading to ciguatera from the Great Barrier Reef.

Matley et al. [302] found that damselfishes (Pomacentridae) make up a large proportion of the prey of coral trout which includes species such as *Stegastes nigricans* that "farm" algal turfs. The turf algae within the Great Barrier Reef territories of these farming damselfishes have longer lengths and higher detritus and lower sediment loads than those outside [305], so may be favourable habitats for growth of benthic dinoflagellates, although we are not aware of any studies. Damselfishes can consume the detritus fraction directly as well as invertebrates within the algal matrix [306], so this could be a mechanism for toxin transfer from epiphytic dinoflagellates. Tebbett et al. [305] suggest that the farmed turfs of damselfish may also incentivise the raiding of them by schools of other herbivorous/detritivorous fishes, providing multiple mechanisms for the transfer of ciguatoxins if damselfish territories offer scope for growth of *Gambierdiscus/Fukuyoa*. In regions where herbivores are fished, biomass and abundance of territorial algal farming damselfishes increases [307]. Coral trout preying upon other carnivorous fishes that have accumulated ciguatoxins such as Lutjanids and Lethrinids [302], might be a secondary route for them to become ciguatoxic.

5.4. Summarizing the Production and Food Chain Transfer of Ciguatoxins from the Great Barrier Reef

The Great Barrier Reef marine park covers an area >340,000 km^2 with cross-shelf and latitudinal differences in ecosystem structure. There are no local-scale studies of trophic transfers of ciguatoxins to extrapolate generalizations of the food chain links for communities across these large-scale differences. At present, we can only base our conceptual model (Figure 10) on the knowledge that *Gambierdiscus* is a common component of turf and macroalgae communities across the Great Barrier Reef, together with recent research on the biology and fisheries of Great Barrier Reef fishes. We believe that *Gambierdiscus* or *Fukuyoa* species growing on turf algae are likely the main source of ciguatoxins entering marine food chains on the Great Barrier Reef to cause ciguatera, but this remains to be tested. There are at least three possible food chains for the transfer of ciguatoxins from *Gambierdiscus/Fukuyoa* through invertebrates, cryptobenthic fish or larger herbivorous fish species to demersal predators such as coral trout (Figure 10). We suggest that the abundance of surgeonfish that feed on turf algae on the Great Barrier Reef is a feedback mechanism controlling the flow of ciguatoxins through the marine food chain and suggest that this hypothesis could be explored through cage experiments. The intoxicating effects of ciguatoxins and/or water-soluble toxins on herbivores, may be greatest after feeding on a sudden population increase such as a *Gambierdiscus/Fukuyoa* bloom, which likely increases the risk of opportunistic predation on them. This may be a mechanism by which occasional blooms of ciguatoxin-producing *Gambierdiscus/Fukuyoa* lead to the production of ciguatoxic predatory reef fish in ecosystems such as the Great Barrier Reef, where herbivorous fish populations are largely unaffected by fishing pressure. This likely concentrates ciguatoxins into the human food chain.

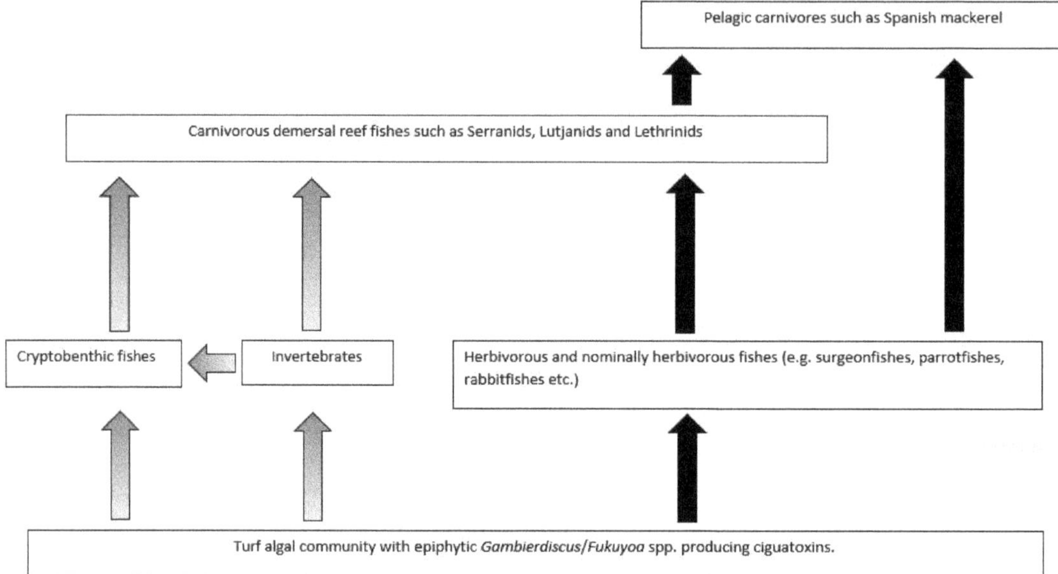

Figure 10. Conceptual model for food chain transfer (arrows) of ciguatoxins on the Great Barrier Reef. Solid arrows indicate transfer of ciguatoxins consistent with the general paradigm for the accumulation of ciguatoxins (Randall 1958). Shaded arrows indicate possible additional linkages.

However, as with Platypus Bay, some of the same fundamental questions remain:

- What are the *Gambierdiscus/Fukuyoa* species that produce ciguatoxins on the Great Barrier Reef?
- What is the profile of ciguatoxins produced by these species?
- What ciguatoxins are transferred/transformed along the food chain that leads to the ciguatoxin profile found in demersal reef fish such as coral trout?
- Which of the potential food chains (Figure 10) operate to produce demersal ciguatoxic fish such as coral trout?

6. Model for Dilution of Ciguatoxins in the Flesh of the Common Coral Trout (*P. leopardus*) through Growth

We constructed growth models to explore the potential for dilution of ciguatoxin concentrations (5.0, 1.0, 0.1 and 0.03 μg/kg P-CTX-1 equivalents) in the flesh of the common coral trout to below the U.S. FDA threshold concentration of 0.01 μg/kg P-CTX-1 equivalents. Our modelling is based upon the same assumptions and limitations we outlined for Spanish mackerel with regard to; the homogeneity of ciguatoxins in the flesh, the ratio of flesh (fillet) weight to whole fish weight, any effect of ciguatoxins on growth, and any negative or positive bias for catchability as coral trout are mostly caught by line fishers using baits or lures.

The minimum legal size for taking all coral trout species in Queensland is 38 cm total length, except for the blue-spot trout (*P. laevis*) which has minimum size of 50 cm and a maximum size of 80 cm. Common and bar-cheek coral trout are long-lived species with common coral trout reaching 38 cm total length between 2–3 years of age, and most commercially harvested fish being 3–5 years old and averaging 1.58 kg [286]. However, the large variation in growth between individual fishes [155], adds considerable uncertainty to any estimation of the dilution of toxicity from growth of *P. leopardus*. It has been suggested that only about 5% of the population eventually reach 10-years or older [308].

The annual weight-at-age data to construct the growth curve for common coral trout (Figure 11) was supplied by the Queensland Department of Agriculture and Fisheries from data published in Campbell et al. [298] (their Figure 3). We used this data to estimate the annual rate of dilution of ciguatoxin concentrations by somatic growth from 1-, 2- and 4-year-old fish (Figure 12), as we did for Spanish mackerel. The model assumes that the ciguatoxin concentration in muscle decreases in proportion to the relative increase in mass from somatic growth. We did not include 0.5-year-old coral trout as young ages are poorly represented by the weight-at-age data (Figure 11).

Somatic growth can reduce the ciguatoxin concentrations contaminating young ($\leq$1-year of age) coral trout by ~10-fold (Figure 12), before the fish reaches the legal size for harvesting at 2–3 years of age [286]. As with Spanish mackerel, this would likely reduce the severity of the disease if the fish did not accumulate additional toxin before it was harvested and eaten. However, somatic growth cannot reduce the ciguatoxin concentration of a 1-year-old fish contaminated with 1.0 µg/kg P-CTX-1 equivalents to below the U.S. FDA threshold of 0.01 µg/kg over a 10-year modelled lifetime (Figure 12). For fish >2years-old, somatic growth is unlikely to reduce the frequency of toxic fish except for those contaminated with very low (e.g., 0.03 µg/kg P-CTX-1 equivalents) concentrations of ciguatoxins (Figure 12), especially as most fish are harvested between 3–5 years of age [286]. Therefore, as with Spanish mackerel, somatic growth on its own is unlikely to significantly reduce the toxicity of ciguateric coral trout, especially if the fish accumulates the toxin after 1–2 years of age. We have not explored the possible depuration of ciguatoxins from coral trout as we have no evidence to support this hypothesis, although we think it likely.

Other reef fish species such as red bass (*Lutjanus bohar*) and the lined-bristletooth (*Ctenochaetus striatus*) are frequently implicated in ciguatera throughout the Pacific, and the former is a no-take fish in Queensland because of the risk of ciguatera. Red bass is often found on the mid- and outer-shelf of the Great Barrier Reef [180], as is *C. striatus* [255]. Both are long-lived species and very-slow growing after an initial period of rapid growth, with red bass living up to 43-years [309] and *C. striatus* between 30–45 years [310]. Based upon our models for the dilution of ciguatoxins by somatic growth in Spanish mackerel and common coral trout, growth on its own is unlikely to significantly reduce toxicity in red bass or lined- bristletooth surgeonfish, except if they accumulate toxins at a young age.

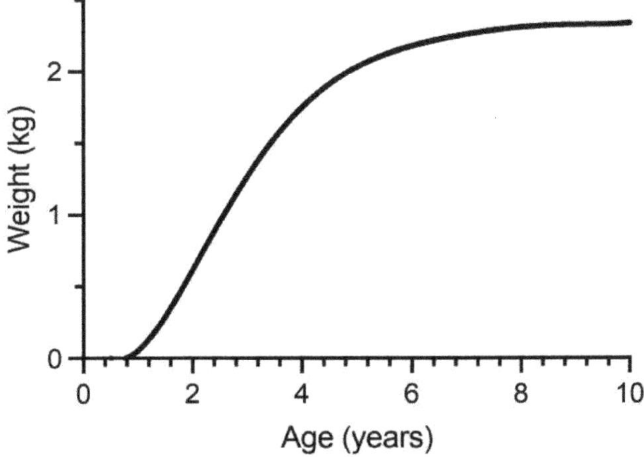

Figure 11. Growth curve (weight-at-age) for common coral trout from the east coast of Australia (data provided by the Queensland Department of Agriculture and Fisheries from Campbell et al., [298]).

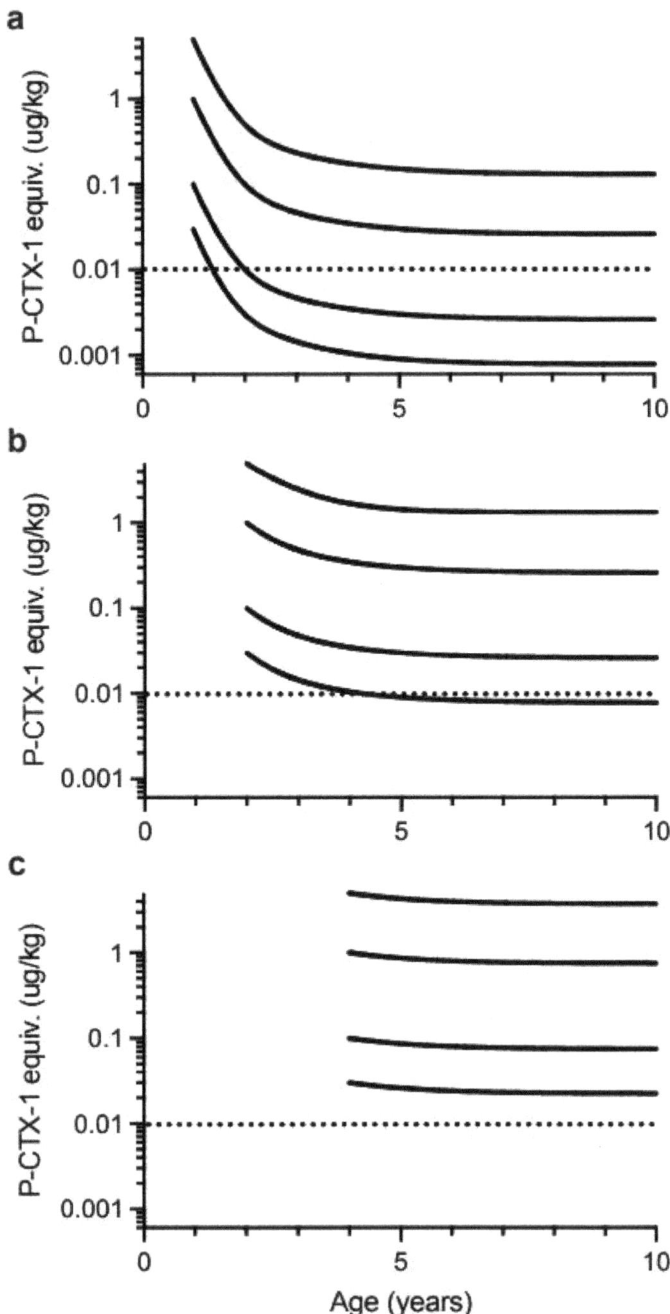

Figure 12. Modelled dilution of 5.0, 1.0, 0.1 and 0.03 µg/kg P-CTX-1 equivalents from the flesh of common coral trout (*P. leopardus*) by somatic growth, for fish contaminated with ciguatoxins at; (**a**) 1 year of age, (**b**) 2 years of age and (**c**) 4 years of age. Dotted line = USFDA precautionary action concentration of 0.01 µg/kg of P-CTX-1 equivalents.

7. Comparative Risk of Ciguatera Estimated from Catches of Spanish Mackerel and Coral Trout (*Plectropomus* spp.) along the East Coast of Australia

Based upon the detection of a single toxic sample from the testing of the flesh of 71 Spanish mackerel caught in northern New South Wales waters in 2015, Kohli et al. [29] estimated the frequency of ciguatoxin contamination at 1.4%. This was the first study to estimate incidence rates for any fish from Australia. However, 1.4% seems high if this translated into fishes with the potential to cause ciguatera, given how heavily the stock is targeted by both commercial and recreational fishers. The commercial harvest of Spanish mackerel from the east coast of Queensland in 2015 was 299 tonnes, with about 46 tonnes taken from southern Queensland [141]. The commercial catch from northern New South Wales waters ranges between 3 and 40 tonnes with an average of 15 tonnes since 2000 [83]. The combined east coast Queensland and New South Wales annual recreational catch since 2004 is estimated to be between 250–300 tonnes with the New South Wales recreational catch being ~13% of the recreational harvest from Queensland waters [83]. If we use the lower range of this catch (250 tonnes) to estimate the New South Wales recreational catch, then 13% of 250 tonnes = 32.5 tonnes. Spanish mackerel caught in New South Wales tend to be larger fish [79], so it would not be appropriate to use an average fish size of 7.7 kg [83] to estimate population numbers from catch weight. The 71 New South Wales fish assayed by Kohli et al. [29] had a mean weight (± 1 standard deviation) of 13.4 ± 3.9 kg (range 4.5–23.6 kg). Assuming a combined commercial (15 tonnes) and recreational catch (32 tonnes) of Spanish mackerel in New South Wales waters of 47 tonnes, with an average fish size of 13.4 kg, would suggest the capture of ~3500 fish. However, the Sydney Fish Market refuses to accept fish above 10 kg because of the risk of ciguatera so commercial fishers may discard many of these larger fish. If only a recreational harvest of about 32 tonnes was considered then this would equate to about 2350 fish, with recreational fishers possibly less likely to discard large fish. This is a conservative figure because the estimate for the annual recreational catch of Spanish mackerel in New South Wales waters during financial years 2010–2011 and 2014–2015 by O'Neill et al. [83] are more than double the above at ~5000 fish each year.

An incidence rate of 1.4% for ciguatoxic Spanish mackerel [29] would equate to 33 toxic fish per year from 2350 recreationally caught fish, or 70 toxic fish caught from 5000 fish. Even small, legal-sized Spanish mackerel can be processed into many meals, so the number of cases resulting from 33 toxic fish in one year would be expected to be a multiple of this number depending upon how many people ate the first meal from the toxic fish, i.e., before the fish was recognized as being poisonous. Most likely this would be at least two people within the family and friends of recreational fishers, but the multiplier could likely be much higher for a commercially sold fish. For example, more than 30 poisoning cases were attributed to a single 21 kg commercial consignment of Spanish mackerel [311]. In either case, 33 toxic Spanish mackerel could be expected to create >50 ciguatera poisonings per year in New South Wales alone. Ciguatera is an under-reported disease in Australia [293,312], but we think such a high incidence rate from Spanish mackerel unlikely given the media interest generated by the 9 outbreaks that poisoned 37 individuals between 2014 and 2017 in New South Wales [14]. Five of these outbreaks were from Spanish mackerel caught in northern New South Wales waters in 2014 (2 fish), 2015 (1 fish) and 2016 (2 fish). In total, they poisoned 24 individuals [14].

The commercial catch of 46 tonnes of Spanish mackerel from southern Queensland in 2015 equates to ~6200 fish based upon an average weight of 7.4 kg (mean weight of Spanish mackerel caught by commercial fishers in southern Queensland during the 2014–2015 financial year; MJH personal communication from Queensland Department of Agriculture and Fisheries). Applying an incidence rate of 1.4% [29] would suggest ~86 toxic Spanish mackerel entering the human food chain during 2015 from commercial sources alone. Ciguatera is a notifiable disease in Queensland but only 11 ciguatera poisonings were recorded for 2015 from the entire State (Queensland Health, Notifiable condition reports: https://www.health.qld.gov.au/clinical-practice/guidelines-procedures/diseases-infection/surveillance/reports/notifiable/weekly (accessed on 26 August 2019 and 20 July 2021)). The fish species causing these 11 poisonings are not available on the public database but are unlikely to have all been due to Spanish mackerel, and the number of fish causing these poisonings is likely to be less than the number of people poisoned. It is not possible to know the true incidence rate of ciguatoxic fishes without a better estimate for the actual number of ciguatera cases that occur, given that many are thought to go un-reported to authorities. However, based upon the analysis above for 2015, we suggest that the true incidence of ciguatoxic Spanish mackerel in southern Queensland and northern New South Wales is likely much less than the 1.4% suggested by Kohli et al. [29].

Campbell et al. [298] assessed the current annual Queensland harvest of common coral trout (*P. leopardus*) by commercial fishers at ~829 tonnes and ~171 tonnes by recreational fishers. As the average weight of legally caught common coral trout is 1.58 kg [286], this equates to >634,000 fish harvested annually from the Great Barrier Reef. This compares with recent estimates for the combined annual commercial and recreational harvest of Spanish mackerel from the east coast of Queensland and New South Wales at ~600–700 tonnes [83]. Using a conservative average fish size of 7.7 kg for the combined catch [83], this equates to between 77,000 and 91,000 Spanish mackerel. However, since 1996, most commercially caught common coral trout are exported live to Asia and we have no information of any ciguatera cases from these fish; although, between 2004 and 2013, *P. leopardus* and red bass (*Lutjanus bohar*) caused most ciguatera cases in Hong Kong [313].

If just the recreational component (171 tonnes; [298]) of the Queensland catch of common coral trout (*P. leopardus*) is compared with the total catch of Spanish mackerel, it suggests that >108,000 common coral trout are eaten annually from the east coast of Australia compared with a conservative estimate of 77,000–91,000 Spanish mackerel. However, most of the recreational catch of coral trout from the Great Barrier Reef is bar-cheek coral trout (*P. maculatus*) [286], so more than twice the number of coral trout of all species is likely consumed on the east coast of Australia compared to Spanish mackerel. If the risk of ciguatera from Spanish mackerel and demersal reef fish were the same, we would expect more than twice the number of outbreaks from the east coast of Australia caused by coral trout than Spanish mackerel.

Reconstructed historical catches of Spanish mackerel and common coral trout from the east coast of Australia for 1986, the year before the ban on taking Spanish mackerel from Platypus Bay and before the development of the live export market for coral trout, indicate total catches of about 680 tonnes (>88,000 fish) and 1200 tonnes (>759,000 fish), respectively [83,298], suggesting that at times there was at least an order of magnitude more coral trout (of all *Plectropomus* and *Variola* species) consumed on the east coast of Australia than Spanish mackerel. This does not consider the many other demersal species caught by commercial and recreational fishers from the Great Barrier Reef that can cause ciguatera. For example, between 1990 and 2018, an average of 524 ± 243 tonnes of red-throat emperor (*Lethrinus miniatus*) were harvested annually by commercial fishers from the Great Barrier Reef [141]. At an average weight of 1.17 kg [314] this suggests that >448,000 of these reef fish are eaten annually in Australia from the commercial food chain; not including the considerable annual recreational catch of this species which was estimated at 65,000 fish in a 2010 survey [297].

We do not know the true incidence of ciguatera for any fish species from Australia, but Gillespie et al. [5] reported 30 outbreaks of ciguatera caused by Spanish mackerel and another 51 by unknown mackerel species between 1965 and 1984 (before the ban on harvesting Spanish mackerel from Platypus Bay came in force in 1987). If just half of the unknown outbreaks were Spanish mackerel, this would equate to ~55 outbreaks from Spanish mackerel whereas over the same period there were only 18 outbreaks from coral trout or 54 from all non-pelagic fish species [5]. This suggests that the probability of ciguatoxins being accumulated is much greater for a Spanish mackerel on the east coast Queensland than a demersal reef fish, but the considerable underreporting of ciguatera makes it difficult to be certain. Unfortunately, there are no recent studies linking fish species with ciguatera cases in Queensland to make better comparisons using fisheries data. If this were available, it would be possible to use fishery models to estimate the ciguatera risk from the catch of populations of individual fish species from the Great Barrier Reef and the east coast of Queensland. For some species, this could extend to estimating the risk from the entire population of legal-sized fish on the Great Barrier Reef as Leigh et al. [286] were able to estimate the number of legal-sized common coral trout in fishable zones of the Great Barrier Reef in the 1980's at >5.3 million fish, with another >2.1 million fish in no-fishing zones.

A general paradigm of ciguatera research is that larger fish carry a higher risk of ciguatera because the toxins bio-accumulate through the marine food web [70], leading to general advice to be cautious about eating large reef fish [5]. Early research suggested that red bass (*Lutjanus bohar*) could remain toxic for up to 30 months when fed on a non-toxic diet [152]. It is therefore understandable that the Sydney fish market's response to previous cases of ciguatera from large Spanish mackerel is to not accept any fish over 10 kg, which corresponds to a 3- to 4-year-old female fish or a 5- to 7-year-old male fish (Figure 5). However, the evidence for a relationship between fish size and toxicity is contradictory. Early research by Hessel et al. [315] and Banner et al. [316] in the Line Islands in the Pacific found an increasing frequency of toxic red bass with fish size, as well as an increasing toxicity with size. Vernoux [317] found toxicity correlated with size for one species of trevally (*Caranx*) in the Caribbean, but not another. Chan et al. [87] and Mak et al. [32] found a positive correlation between body weight and ciguatoxin concentration for moray eels from the Republic of Kiribati, whereas previously, Lewis et al. [86] had found no such relationship. Bravo et al. [318] suggested that weight was a risk factor for amberjack (*Seriola* spp.) from the Canary Islands in the Atlantic Ocean, whereas Gaboriau et al. [137] found little relationship between toxicity and reef fish size in French Polynesia (with the possible exception of red bass). All the above suggests that any relationship between fish size and ciguatera risk varies between species and their environment. In an environment with continuous or periodic input of ciguatoxins into the food chain, larger fish of the same species are likely older and could have accumulated a higher ciguatoxin burden assuming that the rate of toxin input is greater than losses/dilution through growth and depuration. For some species, the risk could also change with age if this was associated with ontogenetic changes in feeding behaviour or diet. Ciguatera in Australia is rare, and Kohli et al. [29] could not detect any relationship between size and frequency of toxic Spanish mackerel from the east coast of Queensland and New South Wales. It is possible that for many species, fish size is not a higher risk of ciguatera in Australia, but fish size carries a higher community risk because of the larger number of people that can be poisoned by a large toxic fish compared to a small one, as happened in 1994 in New South Wales, when more than 30 people were poisoned by 21 kg of Spanish mackerel caught from Queensland [311].

Large, minimally toxic fish caught by recreational fishers have an additional risk for the fisher and their family and friends, as they are more likely to eat repeat meals from the same fish, than people who eat single fish portions sold commercially through restaurants and other food outlets. While restaurant clients eating a single portion of a lowly toxic fish may avoid being poisoned, people eating repeat meals from a lowly toxic fish may become eventually succumb to poisoning [4], as was the case with a recreational spearfisher that captured a lowly toxic blue-spot coral trout off Moreton Island and only became sick after eating repeat meals (MJH personal communication).

Our modelling shows that dilution of ciguatoxin concentration through somatic growth of fishes is a slow process (Figures 6 and 12), and that fish can retain toxicity for considerable time even if they depurate toxins (Figure 7 and Figures S9–S12). If Spanish mackerel can depurate ciguatoxins with a half-life of six months, it would still take more than a year for the flesh of a lowly toxic fish (0.1 µg/kg P-CTX-1 equivalents) to reach the U.S. FDA recommended safe concentration (Figure 7, Table 2). For Spanish mackerel undertaking annual migrations along the southern Queensland and northern New South

Wales coasts, a year could allow for fish to re-visit the same toxic areas and accumulate an additional toxin burden.

Reducing the absolute risk of ciguatera for the fishing industry and the consumers of wild-caught tropical and temperate fishes may require the development of a rapid, sensitive, reliable, and low-cost assay, preferably one that could be used in the field or in a non-laboratory environment. Cost will be an important factor for deployment of such an assay in both poor and wealthy communities, as poorer consumers may not be able to afford the additional cost of testing individual fish or fish fillets, and richer consumers generally have choice between a range of alternate sources of animal protein. The recent development and commercialization of an ELISA that can detect a range of ciguatoxins including P-CTX-1 at 0.01 ppb [319] offers hope for the future development of portable kits for the detection of ciguateric fishes. An antibody-based approach has also been developed using an electrochemical immunosensor [320]. Alternatively, the relative long-term risk of ciguatera from different fish species along the east coast of Australia could be estimated with the use of currently available fisheries data, if better epidemiological data on ciguatera cases and the causative fishes was available.

8. Disturbance and the New Surface Hypothesis for Ciguatera

The Great Barrier Reef has been impacted by an increasing range of major disturbances from cyclones, crown-of-thorns starfish, catchment runoff from past and on-going development, and coral bleaching events due to increasing water temperatures from climate change [114]. Disturbance to reef environments has long been suggested to be linked with increased ciguatera in the Pacific [70,153,182,207,258,321]. In what became known as the "new surface hypothesis", Randall [70] suggested that disturbance created new surfaces for the causative organism to colonise and proliferate (implying that space for growth is otherwise limited). Fleshy macroalgae and/or turf algae can colonise and dominate coral reefs after major disturbances [277,279,285,322–324], although reefs and their fish communities vary in their resilience to these ecological phase shifts [285,324,325]. After the discovery of *Gambierdiscus* [40], it was suggested that this epiphytic benthic dinoflagellate attached to, and proliferated on, algae colonizing new surfaces [193]. However, there are few direct studies that have tested this hypothesis, and over the past decade, ecologists have found that the response of reefs to any one pressure is often non-linear [326]. Kaly and Jones [327] could not find supporting evidence for the new surface hypothesis from studies on reef disturbance sites in Tuvalu. An unpublished study (Holmes and Lewis) at a small marina development site on a fringing reef at Hayman Island on the Great Barrier Reef in the 1980's failed to detect any increase in *Gambierdiscus* populations near the development site (data archived in [186]). In the only study of the toxicity of surgeonfishes (*Ctenochaetus striatus*) from the Great Barrier Reef, Lewis et al. [175] found only low concentrations of ciguatoxins from *C. striatus* speared from John Brewer and Davies reefs. At the time, John Brewer reef had been damaged by crown-of-thorns starfish whereas Davies Reef had only been lightly impacted [175].

The impact of disturbance to coral reefs on fish populations is not clear and may vary with local conditions. Morais et al. [285] studied changes in fish populations across a mid-shelf coral reef at Lizard Island on the Great Barrier Reef that had suffered major coral degradation over 14–15 years from cyclones and coral bleaching events. These stressors led to increases in turf algae but not macroalgae, with coral loss associated with increases in total herbivorous fish biomass and productivity but with decreased turnover. There was a shift in the size structure of parrotfish, surgeonfish, and rabbitfish populations to larger bodied individuals, with an overall decline in abundance of surgeonfishes. Grazing pressure for a given biomass tends to be greatest with smaller fish sizes [328] so a population shift to larger sizes may reduce overall grazing pressure. McClure et al. [329] also found an increase in herbivore biomass for mid- and outer-shelf reefs of the Great Barrier Reef after disturbance from cyclones and coral bleaching events. However, in contrast to Morais et al. [285], they found an increased abundance of surgeonfishes, with an increase in *Acanthurus nigrofuscus* on mid-shelf reefs and increases in *Ctenochaetus striatus* and *A. lineatus* on outer-shelf reefs. Morais et al. [285] suggested that severe loss of coral can produce a more productive assemblage of reef fish herbivores but with reduced energy flow across trophic levels.

There are suggestions that increasing water temperatures through climate change could be linked with increasing ciguatera [330–334]. Much of this concern is about possible range expansion of benthic dinoflagellate species [335], for example southwards along the east coast of Australia into nominally temperate waters [93,253]; although, it is also possible that warmer seawater temperatures may reduce the incidence of ciguatera [97,332]. However, at the species level, some free-living dinoflagellates can have broad temperature tolerances. For example, the Paralytic Shellfish Poisoning dinoflagellate *Gymnodinium catenatum* can bloom in Tasmanian waters at >40° S latitude [336], but also grows in Singapore waters at ~1° N [337]. Higher temperatures may lead to higher latent growth rates [335] and references therein; however, this does not necessarily translate into higher cell numbers in the wild. The maximum cell densities of *Gambierdiscus* species recorded from the wild along the east coast of Australia occur quasi-seasonally but at seawater temperatures below the local annual maximum for each location [77,90,93]. However, climate driven range shifts could also increase the possibility of ciguatoxic fish species being caught by fishers from waters previously outside their range, especially highly mobile pelagic species. For example, the distribution of Spanish mackerel along the east coast of Australia has been found to be especially sensitive to the environmental effects of climate change with southward range shifts exceeding 200 km per decade [145].

Ciguatera can increase in South Pacific communities during seawater warming driven by the El Niño Southern Oscillation [330], although an earlier analysis failed to find a simple correlation [338]. Recently, Zheng et al. [339] modelled relationships between seawater surface temperature anomalies and monthly prevalence of ciguatera in French Polynesia and the Cook Islands. In both regions, there were time-lagged delays before the increase in ciguatera, although these time delays were very different between the two regions [339]. This suggests the possibility of using sea surface temperature anomalies in a risk assessment tool for ciguatera. Warming temperature (degree heating week) is routinely used to predict coral bleaching on the Great Barrier Reef, so it would be interesting to determine if there is sufficient data on ciguatera cases from the east coast of Australia to conduct modelling similar to that used by Zheng et al. [339]. However, there was an average of only ~28 ciguatera cases per year reported between 2014–2019 from Queensland (Queensland Health, notifiable conditions annual reporting: https://www.health.qld.gov.au/clinical-practice/guidelines-procedures/diseases-infection/surveillance/reports/notifiable/annual (accessed on 26 August 2019 and 20 July 2021)), compared to ~174 cases per year from the Cook Islands and ~428 per year from French Polynesia [339]. If the statistical relationship between sea surface temperature anomalies and monthly changes in ciguatera is predictive [339], it would suggest that changes in fish toxicity can occur rapidly, with increases and reductions in ciguatera cases occurring over monthly time periods. It is possible that such rapid reductions in cases may be driven through rapid toxin depuration from fish.

To date, the increasing impacts to the Great Barrier Reef have not produced any major increase in ciguatera in Australia. However, as outlined in this review, there are many links in the food chain that need to connect to produce ciguateric fishes, and the breaking of anyone could limit the bioaccumulation of ciguatoxins into the higher trophic level fishes to cause ciguatera along the east coast of Australia. Each individual link in the food chain may be described deterministically, but once linked, the chain may be chaotic.

9. Mitigation of Ciguatera

Lewis and Holmes [33] speculated on the potential for development of mitigation programs that can reduce the incidence of ciguatera. The apparent discrete nature of the source of ciguatera in Platypus Bay may offer the opportunity to identify key controllable factors that would underpin such strategies. Trawling or dredge removal of *Cladophora*, the substrate hosting the *Gambierdiscus* source of ciguatoxins in Platypus Bay, could be trialled for reducing ciguatera from Platypus Bay, should there be enough pressure in the future to justify such an intervention. Large scale, expensive, ecosystem interventions are being currently trialled on the Great Barrier Reef by the Great Barrier Reef Marine Park Authority and the Great Barrier Reef Foundation through manual and robotic culling of crown-of-thorns starfish. Research projects are also being funded to study the potential of even larger-scale interventions to mitigate some of the effects of climate change on the Great Barrier Reef (Great Barrier Reef Foundation; https://www.barrierreef.org/ (accessed on 20 July 2021)). The removal of *Cladophora* would be consistent with the current (2006) General Use zoning of the Great Sandy Marine Park, except for a small area along the beach which is zoned for Habitat Protection (Queensland Department of Environment and Science, https://parks.des.qld.gov.au/parks/great-sandy-marine/ (accessed on 20 July 2021)).

Supplementary Materials: The following are available online at https://www.mdpi.com/article/10.3390/toxins13080515/s1, Table S1: Known species of *Gambierdiscus* and *Fukuyoa* from the Great Barrier Reef (GBR) and species found on the east coast of Australia outside the GBR, Table S2: Toxicity (~LD_{50}) of chromatographic fractions from experiments conducted during the purification and characterisation of MTX-3. Figure S1: Death-time vs. dose for MTX-3, Figure S2: Elution profiles for MTX-1, -2 and -3 from a PRP-1 reverse-phase HPLC column, Figure S3: Map of Hervey Bay and Fraser Island, Figure S4: Logbook map grids for the QFish database, Figure S5: Annual combined catches of Spanish mackerel extracted from QFish 30 × 30 nm logbook grids W32, W32, V32, V33 between 1990 and 2019, Figure S6: Summed monthly catches of Spanish mackerel extracted from QFish logbook grids W32 and W33 between 1990 and 2018, and monthly numbers of ciguatoxic Spanish mackerel caught in the Hervey Bay region between 1976 and 1983, Figure S7: Map showing the central and north Queensland coasts between Proserpine and Cooktown, Figure S8: Modelled dilution of 5.0, 1.0, 0.1 and 0.03 µg/kg P-CTX-1 equivalents from Spanish mackerel flesh by somatic growth, for fish contaminated with ciguatoxins at 0.5, 1, 2, and 4 years of age, Figure S9: Modelled dilution of 5.0, 1.0, 0.1, and 0.03 µg/kg P-CTX-1 equivalents from the flesh of female and male Spanish mackerel by a combination of somatic growth and depuration half-life of 0.5 year, Figure S10: Modelled dilution of 5.0, 1.0, 0.1, and 0.03 µg/kg P-CTX-1 equivalents from the flesh of female and male Spanish mackerel by a combination of somatic growth and depuration half-life of 1.0 year, Figure S11: Modelled dilution of 5.0, 1.0, 0.1, and 0.03 µg/kg P-CTX-1 equivalents from the flesh of female and male Spanish mackerel by a combination of somatic growth and depuration half-life of 2.0 years, Figure S12: Modelled dilution of 5.0, 1.0, 0.1, and 0.03 µg/kg P-CTX-1 equivalents from the flesh of female and male Spanish mackerel by a combination of somatic growth and depuration half-life of 4.0 years, Figure S13: Map showing the position of some of the Capricorn-Bunker group of reefs and coral cays off the coast of Gladstone, Queensland.

Author Contributions: Conceptualization, original draft preparation and editing, M.J.H.; writing, review and editing, R.J.L.; mathematical modelling, B.V. All authors have read and agreed to the published version of the manuscript.

Funding: This research received no external funding.

Institutional Review Board Statement: Not applicable. Animal (mouse) studies described in Supplementray material were from published studies conducted before 1994 under then Australian National Health and Medical Research Council guidelines.

Informed Consent Statement: Not applicable.

Data Availability Statement: Data sources are outlined in the article.

Acknowledgments: We thank Michael O'Neill, Alex Campbell, Prasadini Salgado, Jo Langstreth, Jason McGilvray and Jonathan Staunton-Smith from the Queensland Department of Agriculture and Fisheries for data on Queensland fisheries, including the Spanish mackerel fishery, weight-at-age data for common coral trout, and catch data for Siganids from Platypus Bay. We thank Michaela Larsson for her review and comments on an earlier draft.

Conflicts of Interest: The authors declare no conflict of interest.

References

1. Lehane, L.; Lewis, R.J. Ciguatera: Recent advances but the risk remains. *Int. J. Food Microbiol.* **2000**, *61*, 91–125. [CrossRef]
2. Skinner, M.P.; Brewer, T.D.; Johnstone, R.; Fleming, L.E.; Lewis, R.J. Ciguatera fish poisoning in the Pacific Islands (1998 to 2008). *PLoS Negl. Trop. Dis.* **2011**, *5*, e1416. [CrossRef] [PubMed]
3. Grant, C. *The History of Mauritius, or the Isle of France, and the Neighbouring Islands, from Their First Discovery to the Present Time*; Bulmer and Co.: London, UK, 1801; p. 571. Available online: https://books.google.com.au/books?id=0kpBAAAAcAAJ&pg=PR9&source=gbs_selected_pages&cad=2#v=onepage&q&f=false (accessed on 1 July 2021).
4. Bagnis, R.; Kuberski, T.; Langier, S. Clinical observations of 3009 cases of ciguatera (fish poisoning) in the South Pacific. *Am. J. Trop. Med. Hyg.* **1979**, *28*, 1067–1073. [CrossRef] [PubMed]
5. Gillespie, N.C.; Lewis, R.J.; Pearn, J.H.; Bourke, A.T.C.; Holmes, M.J.; Bourke, J.B.; Shields, W.J. Ciguatera in Australia: Occurrence, clinical features, pathophysiology and management. *Med. J. Aust.* **1986**, *145*, 584–590. [CrossRef]
6. Lewis, R.J. Ciguatera: Australian perspectives on a global problem. *Toxicon* **2006**, *48*, 799–809. [CrossRef]
7. Chinain, M.; Gatti, C.M.; Roué, M.; Darius, H.T. Ciguatera poisoning in French Polynesia: Insights into the novel trends of an ancient disease. *New Microbes New Infect.* **2019**, *31*, 565. [CrossRef]
8. Vetter, I.; Touska, F.; Hess, A.; Hinsbey, R.; Sattler, S.; Lampert, A.; Sergejeva, M.; Sharov, A.; Collins, L.S.; Eberhardt, M.; et al. Ciguatoxins activate specific cold pain pathways to elicit burning pain from cooling. *EMBO J.* **2012**, *31*, 3795–3808. [CrossRef]
9. Lewis, R.J. The changing face of ciguatera. *Toxicon* **2001**, *39*, 97–106. [CrossRef]
10. Inserra, M.C.; Israel, M.R.; Caldwell, A.; Castro, J.; Deuis, J.R.; Harrington, A.M.; Keramidas, A.; Garcia-Caraballo, S.; Maddern, J.; Erickson, A.; et al. Multiple sodium channel isoforms mediate the pathological effects of Pacific ciguatoxin-1. *Sci. Rep.* **2017**, *7*, 42810. [CrossRef] [PubMed]
11. Vetter, I.; Zimmermann, K.; Lewis, R.J. Ciguatera Toxins: Pharmacology, Toxicology, and Detection. In *Seafood and Freshwater Toxins. Pharmacology, Physiology and Detection*; Botana, L., Ed.; CRC Press: Boca Raton, FL, USA, 2014; pp. 925–950.
12. Lombet, A.; Bidard, J.N.; Lazdunski, M. Ciguatoxin and brevetoxins share a common receptor site on the neuronal voltage-dependant Na$^+$ channel. *FEBS Lett.* **1987**, *219*, 355–359. [CrossRef]
13. L'Herondelle, K.; Pierre, O.; Fouyet, S.; Leschiera, R.; Le Gall-Ianotto, C.; Phillippe, R.; Buscaglia, P.; Mignen, O.; Talgas, M.; Lewis, R.J.; et al. PAR2, kearatinocytes, and cathepsin S mediate the sensory effects of ciguatoxins responsible for ciguatera poisoning. *J. Investig. Dermatol.* **2020**, *140*, 648–658. [CrossRef]
14. Farrell, H.; Murray, S.A.; Zammit, A.; Edwards, A.W. Management of ciguatera risk in eastern Australia. *Toxins* **2017**, *9*, 367. [CrossRef]
15. Sparrow, L. Key Factors Influencing the Occurrence and Frequency of Ciguatera. Ph.D. Thesis, James Cook University, Townsville, Australia, 2017.
16. Palafox, N.A.; Jain, L.G.; Pinano, A.Z.; Gulick, T.M.; Williams, R.K.; Schatz, I.J. Successful treatment of ciguatera fish poisoning with intravenous mannitol. *JAMA* **1988**, *259*, 2740–2742. [CrossRef]
17. Pearn, J.H.; Lewis, R.J.; Ruff, T.; Tait, M.; Quinn, J.; Murtha, W.; King, G.; Mallett, A.; Gillespie, N.C. Ciguatera and mannitol: Experience with a new treatment regimen. *Med. J. Aust.* **1989**, *151*, 77–80. [CrossRef] [PubMed]
18. Mullins, M.E.; Hoffman, R.S. Is mannitol the treatment of choice for patients with ciguatera poisoning? *Clin. Toxicol.* **2017**, *55*, 947–955. [CrossRef] [PubMed]
19. Pasinszki, T.; Lako, J.; Dennis, T.E. Advances in detecting ciguatoxins in fishes. *Toxins* **2020**, *12*, 494. [CrossRef]
20. Scheuer, P.J.; Takahashi, W.; Tsutsumi, J.; Yoshida, T. Ciguatoxin: Isolation and chemical nature. *Science* **1967**, *155*, 1267–1268. [CrossRef] [PubMed]
21. Murata, M.; Legrand, A.M.; Ishibashi, Y.; Fukui, M.; Yasumoto, T. Structure and configurations of ciguatoxin from the moray eel *Gymnothorax javanicus* and its likely precursor from the dinoflagellate *Gambierdiscus toxicus*. *J. Am. Chem. Soc.* **1990**, *112*, 4380–4386. [CrossRef]
22. FAO; WHO. Report of the Expert Meeting on Ciguatera Poisoning: Rome, Italy, 19–23 November 2018. *Food Saf. Qual.* **2020**. [CrossRef]
23. Satake, M.; Ishibashi, Y.; Legrand, A.-M.; Yasumoto, T. Isolation and structure of ciguatoxin-4A, a new ciguatoxin precursor from cultures of dinoflagellate *Gambierdiscus toxicus* and parrotfish *Scarus gibbus*. *Biosci. Biotechnol. Biochem.* **1997**, *60*, 2103–2105. [CrossRef] [PubMed]
24. Yasumoto, T.; Igarashi, T.; Legrand, A.-M.; Cruchet, P.; Chinain, M.; Fujita, T.; Naoki, H. Structural elucidation of ciguatoxin congeners by fast-atom bombardment tandem mass spectrometry. *J. Am. Chem. Soc.* **2000**, *122*, 4988–4989. [CrossRef]
25. Ikehara, T.; Kuniyoshi, K.; Oshiro, N.; Yasumoto, T. Biooxidation of ciguatoxins leads to species-specific toxin profiles. *Toxins* **2017**, *9*, 205. [CrossRef]
26. Lewis, R.J.; Sellin, M. Multiple ciguatoxins in the flesh of fishes. *Toxicon* **1992**, *30*, 915–919. [CrossRef]
27. Lucas, R.E.; Lewis, R.J.; Taylor, J.M. Pacific ciguatoxin-1 associated with a large common-source outbreak of ciguatera in east Arnhem Land, Australia. *Nat. Toxins* **1997**, *5*, 136–140. [CrossRef] [PubMed]
28. Hamilton, B.; Whittle, N.; Shaw, G.; Eaglesham, G.; Moore, M.R.; Lewis, R.J. Human fatality associated with Pacific ciguatoxin contaminated fish. *Toxicon* **2010**, *56*, 668–673. [CrossRef]

29. Kohli, G.S.; Haslauer, K.; Sarowar, C.; Kretzschmar, A.L.; Boulter, M.; Harwood, D.T.; Laczka, O.; Murray, S.M. Qualitative and quantitative assessment of the presence of ciguatoxin, P-CTX-1b, in Spanish mackerel (*Scomberomorus commerson*) from waters in New South Wales (Australia). *Toxicol. Rep.* **2017**, *4*, 328–334. [CrossRef] [PubMed]

30. Lewis, R.J.; Sellin, M.; Poli, M.A.; Norton, R.S.; MacLeod, J.K.; Sheil, M.M. Purification and characterization of ciguatoxins from moray eel (*Lycodontis javanicus*, Muraenidae). *Toxicon* **1991**, *29*, 1115–1127. [CrossRef]

31. Yogi, K.; Oshiro, N.; Inafuku, Y.; Hirama, M.; Yasumoto, T. Detailed LC-MS/MS analysis of ciguatoxins revealing distinct regional and species characteristics in fish and causative alga from the Pacific. *Anal. Chem.* **2011**, *83*, 8886–8891. [CrossRef] [PubMed]

32. Mak, Y.L.; Wai, T.-C.; Murphy, M.B.; Chan, W.H.; Wu, J.J.; Lam, J.C.W.; Chan, L.L.; Lam, P.K.S. Pacific ciguatoxins in food web components of coral reef systems in the Republic of Kiribati. *Environ. Sci. Technol.* **2013**, *47*, 14070–14079. [CrossRef]

33. Lewis, R.J.; Holmes, M.J. Origin and transfer of toxins involved in ciguatera. *Comp. Biochem. Physiol.* **1993**, *106*, 615–628. [CrossRef]

34. Vernoux, J.P.; Lewis, R.J. Isolation and characterisation of Caribbean ciguatoxins from the horse-eye jack (*Caranx latus*). *Toxicon* **1997**, *35*, 889–900. [CrossRef]

35. Lewis, R.J.; Vernoux, J.-P.; Brereton, I.M. Structure of Caribbean ciguatoxin isolated from *Caranx latus*. *J. Am. Chem. Soc.* **1998**, *120*, 5914–5920. [CrossRef]

36. Kryuchkov, F.; Robertson, A.; Miles, C.O.; Mudge, E.M.; Uhlig, S. LC-HRMS and chemical derivatization strategies for the structure elucidation of Caribbean ciguatoxins: Identification of C-CTX-3 and -4. *Mar. Drugs* **2020**, *18*, 182. [CrossRef] [PubMed]

37. Hamilton, B.; Hurbungs, M.; Vernoux, J.P.; Jones, A.; Lewis, R.J. Isolation and characterisation of Indian Ocean ciguatoxin. *Toxicon* **2002**, *40*, 685–693. [CrossRef]

38. Hamilton, B.; Hurbungs, M.; Jones, A.; Lewis, R.J. Multiple ciguatoxins present in Indian Ocean reef fish. *Toxicon* **2002**, *40*, 1347–1353. [CrossRef]

39. Diogène, J.; Reverté, L.; Rambla-Alegre, M.; del Río, V.; de la Iglesia, P.; Campàs, M.; Palacios, O.; Flores, C.; Caixach, J.; Ralijaona, C.; et al. Identification of ciguatoxins in a shark involved in a fatal food poisoning in the Indian Ocean. *Sci. Rep.* **2017**, *7*, 8240. [CrossRef]

40. Adachi, R.; Fukuyo, Y. The thecal structure of a marine toxic dinoflagellate Gambierdiscus toxicus gen. et sp. nov. collected in a ciguatera-endemic area. *Bull. Jpn. Soc. Sci. Fish.* **1979**, *45*, 67–71. [CrossRef]

41. Holmes, M.J. *Gambierdiscus yasumotoi* sp. nov. (Dinophyceae), a toxic benthic dinoflagellate from southeastern Asia. *J. Phycol.* **1998**, *34*, 661–668. [CrossRef]

42. Yasumoto, T.; Nakajima, I.; Bagnis, R.; Adachi, R. Finding of a dinoflagellate as a likely culprit for ciguatera. *Bull. Jpn. Soc. Sci. Fish.* **1977**, *43*, 1021–1026. [CrossRef]

43. Yasumoto, T.; Bagnis, R.; Thevenin, S.; Garcon, M. A survey of comparative toxicity in the food chain of ciguatera. *Bull. Jpn. Soc. Sci. Fish.* **1977**, *43*, 1015–1019. [CrossRef]

44. Chinain, M.; Gatti, C.M.; Roué, M.; Darius, H.T. Ciguatera-Causing Dinoflagellates in the Genera *Gambierdiscus* and *Fukuyoa*: Distribution, Ecophysiology and Toxicology. In *Dinoflagellates*; Subba Rao, D.V., Ed.; Nova Science: New York, NY, USA, 2020; pp. 405–457.

45. Chinain, M.; Darius, H.T.; Ung, A.; Cruchet, P.; Wang, Z.; Ponton, D.; Laurent, D.; Pauillac, S. Growth and toxin production in the ciguatera-causing dinoflagellate *Gambierdiscus polynesiensis* (Dinophyceae) in culture. *Toxicon* **2010**, *56*, 739–750. [CrossRef]

46. Pisapia, F.; Holland, W.C.; Hardison, D.R.; Litaker, R.W.; Fraga, S.; Nishimura, T.; Adachi, M.; Nguyen-Ngoc, L.; Séchet, V.; Amzil, Z.; et al. Toxicity screening of 13 *Gambierdiscus* strains using neuro-2a and erythrocyte lysis bioassays. *Harmful Algae* **2017**, *63*, 173–183. [CrossRef]

47. Litaker, R.W.; Holland, W.C.; Hardison, D.H.; Pisapia, F.; Hess, P.; Kibler, S.R.; Tester, P.A. Ciguatoxicity of *Gambierdiscus* and *Fukuyoa* species from the Caribbean and Gulf of Mexico. *PLoS ONE* **2017**, *12*, e0185776. [CrossRef] [PubMed]

48. Litaker, R.W.; Vandersea, M.W.; Faust, M.A.; Kibler, S.R.; Chinain, M.; Holmes, M.J.; Holland, W.C.; Tester, P.A. Taxonomy of *Gambierdiscus* including four new species, *Gambierdiscus caribaeus*, *Gambierdiscus carpenteri* and *Gambierdiscus ruetzleri* (Gonyaucales, Dinophyceae). *Phycologia* **2009**, *48*, 344–390. [CrossRef]

49. Chinain, M.; Faust, M.A.; Pauillac, S. Morphology and molecular analyses of three toxic species of *Gambierdiscus* (Dinophyceae): *G. pacificus*, sp. nov., *G. australes*, sp. nov., and *G. polynesiensis*, sp. nov. *J. Phycol.* **1999**, *35*, 1282–1296. [CrossRef]

50. Yasumoto, T.; Bagnis, R.; Vernoux, J.-P. Toxicity of the surgeonfishes –II Properties of the principal water-soluble toxin. *Bull. Jpn. Soc. Sci. Fish.* **1976**, *42*, 359–365. [CrossRef]

51. Murata, M.; Naoki, H.; Iwashita, T.; Matsunaga, S.; Sasaki, M.; Yokoyama, A.; Yasumoto, T. Structure of maitotoxin. *J. Am. Chem. Soc.* **1993**, *115*, 2060–2062. [CrossRef]

52. Holmes, M.J.; Lewis, R.J.; Gillespie, N.C. Toxicity of Australian and French Polynesian strains of *Gambierdiscus toxicus* (Dinophyceae) grown in culture: Characterization of a new type of maitotoxin. *Toxicon* **1990**, *28*, 1159–1172. [CrossRef]

53. Munday, R.; Murray, S.; Rhodes, L.L.; Larsson, M.E.; Harwood, D.T. Ciguatoxins and maitotoxins in extracts of sixteen *Gambierdiscus* isolates and one *Fukuyoa* isolate from the South Pacific and their toxicity to mice by intraperitoneal and oral administration. *Mar. Drugs* **2017**, *15*, 208. [CrossRef] [PubMed]

54. Holmes, M.J.; Lewis, R.J. Purification and characterisation of large and small maitotoxins from cultured *Gambierdiscus toxicus*. *Natural Toxins* **1994**, *2*, 64–72. [CrossRef]

55. Lewis, R.J.; Holmes, M.J.; Alewood, P.F.; Jones, A. Ionspray mass spectrometry of ciguatoxin-1, maitotoxin-2 and -3, and related marine polyether toxins. *Nat. Toxins* **1994**, *2*, 56–63. [CrossRef] [PubMed]

56. Pisapia, F.; Sibat, M.; Herrenknecht, C.; Lhaute, K.; Gaini, G.; Ferron, P.-J.; Fessard, V.; Fraga, S.; Nascimento, S.M.; Litaker, R.W.; et al. Maitotoxin-4, a novel MTX analog produced by *Gambierdiscus excentricus*. *Mar. Drugs* **2017**, *15*, 220. [CrossRef]

57. Boente-Juncal, A.; Álvarez, M.; Antelo, Á.; Rodríguez, I.; Calabro, K.; Vale, C.; Thomas, O.P.; Botana, L.M. Structure elucidation and biological evaluation of maitotoxin-3, a homologue of gambierone, from Gambieridiscus belizeanus. *Toxins* **2019**, *11*, 79. [CrossRef] [PubMed]

58. Murray, J.S.; Selwood, A.I.; Harwood, D.T.; van Ginkel, R.; Puddick, J.; Rhodes, L.L.; Rise, F. Wilkins Al 44-Methylgambierone, a new gambierone analogue isolated from *Gambierdiscus australes*. *Tetrahedron Lett.* **2019**, *60*, 621–625. [CrossRef]

59. Longo, S.; Sibat, M.; Viallon, J.; Darius, H.T.; Hess, P.; Chinain, M. Intraspecific variability in the toxin production and toxin profiles of in vitro cultures of *Gambierdiscus polynesiensis* (Dinophyceae) from French Polynesia. *Toxins* **2019**, *11*, 735. [CrossRef]

60. Rhodes, L.L.; Smith, K.F.; Murray, J.S.; Nishimura, T.; Finch, S.C. Ciguatera fish poisoning: The risk from an Aotearoa/New Zealand perspective. *Toxins* **2020**, *12*, 50. [CrossRef] [PubMed]

61. Yan, M.; Leung, P.T.Y.; Gu, J.; Lam, V.T.T.; Murray, J.S.; Harwood, D.T.; Wai, T.-C.; Lam, P.K.S. Hemolysis associated toxicities of benthic dinoflagellates from Hong Kong waters. *Mar. Pol. Bull.* **2020**, *155*, 111114. [CrossRef]

62. Murray, J.S.; Nishimura, T.; Finch, S.C.; Rhodes, L.L.; Puddick, J.; Harwood, D.T.; Larsson, M.E.; Doblin, M.A.; Leung, P.; Yan, M.; et al. The role of 44-methylgambierone in ciguatera fish poisoning: Acute toxicity, production by marine microalgae and its potential as a biomarker for *Gambierdiscus* spp. *Harmful Algae* **2020**, *97*, 101853. [CrossRef] [PubMed]

63. Murray, J.S.; Finch, S.C.; Puddick, J.; Rhodes, L.L.; Harwood, D.T.; van Ginckel, R.; Prinsep, M. Acute Toxicity of gambierone and quantitative analysis of gambierones produced by cohabitating benthic dinoflagellates. *Toxins* **2021**, *13*, 333. [CrossRef]

64. Estevez, P.; Sibat, M.; Leão-Martins, J.M.; Tudó, A.; Rambla-Alegre, M.; Aligizaki, K.; Diogène, J.; Gago-Martinez, A.; Hess, P. Use of mass spectrometry to determine the diversity of toxins produced by *Gambierdiscus* and *Fukuyoa* species from Balearic Islands and Crete (Mediterranean Sea) and the Canary Islands (Northeast Atlantic). *Toxins* **2020**, *12*, 305. [CrossRef]

65. Yokoyama, A.; Murata, M.; Oshima, Y.; Iwashita, T.; Yasumoto, T. Some chemical properties of maitotoxin, a putative calcium channel agonist from a marine dinoflagellate. *J. Biochem.* **1988**, *104*, 184–187. [CrossRef]

66. Holmes, M.J.; Lewis, R. Toxin-Producing Dinoflagellates. In *Perspectives in Molecular Toxinology*; Ménez, A., Ed.; John Wiley & Sons Ltd.: Chichester, UK, 2002; pp. 39–65.

67. Holmes, M.J.; Brust, A.; Lewis, R.J. Dinoflagellate Toxins: An overview. In *Seafood and Freshwater Toxins. Pharmacology, Physiology and Detection*; Botana, L., Ed.; CRC Press: Boca Raton, FL, USA, 2014; pp. 3–38.

68. Holmes, M.J.; Lewis, R.J.; Poli, M.A.; Gillespie, N.C. Strain dependent production of ciguatoxin precursors (gambiertoxins) by *Gambierdiscus toxicus* (Dinophyceae) in culture. *Toxicon* **1991**, *29*, 761–775. [CrossRef]

69. Mills, A.R. Poisonous fishes in the South Pacific. *J. Trop. Med. Hyg.* **1956**, *59*, 99–103.

70. Randall, J.E. A review of ciguatera, tropical fish poisoning, with a tentative explanation of its cause. *Bull. Mar. Sci.* **1958**, *8*, 236–267.

71. Ledreux, A.; Brand, H.; Chinain, M.; Bottein, M.-Y.D. Dynamics of ciguatoxins from *Gambierdiscus polynesiensis* in the benthic herbivore *Mugil cephalus*: Trophic transfer implications. *Harmful Algae* **2014**, *39*, 165–174. [CrossRef]

72. Li, J.; Mak, Y.L.; Chang, Y.-H.; Xiao, C.; Chen, Y.-M.; Shen, J.; Wang, Q.; Ruan, Y.; Lam, P.K.S. Uptake and depuration kinetics of Pacific ciguatoxins in orange-spotted grouper (*Epinephelus coioides*). *Environ. Sci. Technol.* **2020**, *54*, 4475–4483. [CrossRef] [PubMed]

73. Sanchez-Henao, A.; García-Àlvarez, N.; Padilla, D.; Ramos-Sosa, M.; Sergent, F.S.; Fernández, A.; Estévez, P.; Gago-Martínez, A.; Diogène, J.; Real, F. Accumulation of C-CTX1 in muscle tissue of goldfish (*Carassius auratus*) by dietary exposure. *Animals* **2021**, *11*, 242. [CrossRef] [PubMed]

74. Lewis, R.J.; Endean, R. Purification of ciguatoxin-like material from *Scomberomorus commersoni* and its effect on the rat phrenic nerve-diaphragm. *Toxicon Suppl.* **1983**, *3*, 249–252. [CrossRef]

75. Lewis, R.J.; Endean, R. Ciguatoxin from the flesh and viscera of the barracuda, *Sphyraena jello. Toxicon* **1984**, *22*, 805–810. [CrossRef]

76. Lewis, R.J.; Chaloupka, M.Y.; Gillespie, N.C.; Holmes, M.J. An Analysis of the Human Response to Ciguatera in Australia. In Proceedings of the 6th International Coral Reef Symposium Executive Committee, Townsville, Australia, 8–12 August 1988; Choat, J.H., Barnes, D., Borowitzka, M.A., Coll, J.C., Davies, P.J., Flood, P., Hatcher, B.G., Hopley, D., Hutchings, P.A., Kinsey, D., et al., Eds.; Townsville, Australia, 1988; Volume 3, pp. 67–72.

77. Holmes, M.; Lewis, R.J.; Sellin, M.; Street, R. The origin of ciguatera in Platypus Bay, Australia. *Mem. Qld Mus.* **1994**, *34*, 505–512.

78. Lewis, R.J. Ciguatera in South-Eastern Queensland. In *Toxic Plants and Animals: A Guide for Australia*; Covacevich, J., Davie, P., Pearn, J., Eds.; Queensland Museum: Brisbane, Australia, 1987; pp. 181–187.

79. Buckworth, R.C.; Newman, S.J.; Ovenden, J.R.; Lester, R.J.G.; McPherson, G.R. *The Stock Structure of Northern and Western Australian Spanish Mackerel*; Final Report, Fisheries Research & Development Corporation Project 1988/159; Fishery Report 88; Department of Primary Industry, Fisheries and Mines: Northern Territory Government, Australia, 2007; Volume i–vi, p. 225.

80. Lewis, R.J. Ciguatera and Ciguatoxin-Like Substances in Fishes, Especially *Scomberomorus commersoni* from Southern Queensland. Ph.D. Thesis, University of Queensland, Brisbane, Australia, 1985.

81. Chaloupka, M.Y.; Lewis, R.J.; Sellin, M. The changing face of ciguatera prevalence. *Mem. Qld Mus.* **1994**, *34*, 554.

82. McPherson, G.R. Age and growth of the narrow-barred Spanish mackerel (*Scomberomorus commerson* Lacépède, 1800) in northeastern Queensland waters. *Aust. J. Mar. Freshw. Res.* **1992**, *43*, 1269–1282. [CrossRef]
83. O'Neill, M.F.; Langstreth, J.; Buckley, S.M.; Stewart, J. *Stock Assessment of Australian East Coast Spanish Mackerel: Predictions of Stock Status and Reference Points*; Queensland Government Report: Brisbane, Australia, 2018; p. 103.
84. Lewis, R.J.; Holmes, M.J.; Sellin, M. Invertebrates implicated in the transfer of gambiertoxins to the benthic carnivore *Pomadasys maculatus*. *Mem. Qld Mus.* **1994**, *34*, 561–564.
85. McKay, R.J. Classification of the grunters and javelin-fishes of Australia. *Aust. Fish.* **1984**, *43*, 37–40.
86. Lewis, R.J.; Sellin, M.; Street, R.; Holmes, M.J.; Gillespie, N.C. Excretion of ciguatoxin from Moray eels (Muraenidae) of the central Pacific. In *Proceedings of the Third International Conference on Ciguatera Fish Poisoning, La Parguera, Puerto Rico*; Tosteson, T.R., Ed.; Polyscience Publications: Quebec City, QC, Canada, 1992; pp. 131–143.
87. Chan, W.H.; Mak, Y.L.; Wu, J.J.; Jin, L.; Sit, W.H.; Lam, J.C.W.; de Mitcheson, Y.S.; Chan, L.L.; Lam, P.K.S.; Murphy, M.B. Spatial distribution of ciguateric fish in the Republic of Kiribati. *Chemosphere* **2011**, *84*, 117–123. [CrossRef]
88. Holmes, M.J.; Lewis, R.J. Multiple Gambiertoxins (ciguatoxin precursors) from an Australian Strain of *Gambierdiscus toxicus* in Culture. In *Recent Advances in Toxinology Research*; Gopalakrishnakone, P., Tan, C.K., Eds.; National University of Singapore: Singapore, 1992; Volume 2, pp. 520–529.
89. Holmes, M.J.; Lewis, R.J. The origin of ciguatera. *Mem. Qld Mus.* **1994**, *34*, 497–504.
90. Gillespie, N.C.; Holmes, M.J.; Burke, J.B.; Doley, J. Distribution and Periodicity of *Gambierdiscus toxicus* in Queensland, Australia. In *Toxic Dinoflagellates*; Anderson, D.M., White, A.W., Baden, D.G., Eds.; Elsevier: Oxford, UK, 1985; pp. 183–188.
91. Gillespie, N.C.; Lewis, R.J.; Burke, J.; Holmes, M. The Significance of the Absence of Ciguatoxin in a Wild Population of *G. toxicus*. In Proceedings of the Fifth International Coral Reef Congress, Tahiti, France, 27 May–1 June 1985; Gabrie, C., Salvat, B., Eds.; Antenne Museum-Ephe: Moorea, France, 1985; pp. 437–441.
92. Lewis, R.J.; Gillespie, N.C.; Holmes, M.J.; Burke, J.B.; Keys, A.B.; Fifoot, A.T.; Street, R. Toxicity of Lipid-Soluble Extracts from Demersal Fishes at Flinders Reef, Southern Queensland. In Proceedings of the 6th International Coral Reef Symposium Executive Committee, Townsville, Australia, 8–12 August 1988; Volume 3, pp. 61–65.
93. Kohli, G.S.; Murray, S.A.; Neilan, B.A.; Rhodes, L.L.; Harwood, T.; Smith, K.F.; Meyer, L.; Capper, A.; Brett, S.; Hallegraeff, G.M. High abundance of the potentially maitotoxic dinoflagellate *Gambierdiscus carpenteri* in temperate waters of New South Wales, Australia. *Harmful Algae* **2014**, *39*, 134–145. [CrossRef]
94. Bagnis, R.; Bennett, J.; Prieur, C.; Legrand, A.M. The Dynamics of Three Benthic Dinoflagellates and the Toxicity of Ciguateric Surgeonfish in French Polynesia. In *Toxic Dinoflagellates*; Anderson, D.M., White, A.W., Baden, D.G., Eds.; Elsevier: Oxford, UK, 1985; pp. 177–182.
95. Chinain, M.; Germain, M.; Deparis, X.; Pauillac, S.; Legrand, A.-M. Seasonal abundance and toxicity of the dinoflagellate *Gambierdiscus* spp. (Dinophyceae), the causative agent of ciguatera in Tahiti, French Polynesia. *Mar. Biol.* **1999**, *135*, 259–267. [CrossRef]
96. Parsons, M.L.; Settlemier, C.J.; Bienfang, P.K. A simple model capable of simulating the population dynamics of *Gambierdiscus*, the benthic dinoflagellate responsible for ciguatera fish poisoning. *Harmful Algae* **2010**, *10*, 71–80. [CrossRef]
97. Liefer, J.D.; Richlen, M.L.; Smith, T.B.; DeBose, J.L.; Xu, Y.; Anderson, D.M.; Robertson, A. Asynchrony of *Gambierdiscus* spp. abundance and toxicity in the U.S. Virgin Islands: Implications for monitoring and management of ciguatera. *Toxins* **2021**, *13*, 413. [CrossRef]
98. Gaiani, G.; Toldrà, A.; Andree, K.B.; Rey, M.; Diogène, J.; Alcaraz, C.; O'Sullivan, C.K.; Campàs, M. Detection of *Gambierdiscus* and *Fukuyoa* single cells using recombinase polymerase amplification combined with a sandwich hybridization assay. *J. App. Phycol.* **2021**. [CrossRef]
99. Wu, Z.; Luo, H.; Yu, L.; Lee, W.H.; Li, L.; Mak, Y.L.; Lin, S.; Lam, P.K.S. Characterizing ciguatoxin (CTX)- and non-CTX-producing strains of *Gambierdiscus balechii* using comparative transcriptomics. *Sci. Total Environ.* **2020**, *717*, 137184. [CrossRef] [PubMed]
100. Van Dolah, F.M.; Morey, J.S.; Milne, S.; Ung, A.; Anderson, P.E.; Chinain, M. Transcriptomic analysis of polyketide synthases in a highly ciguatoxic dinoflagellate, *Gambierdiscus polynesiensis* and low toxicity *Gambierdiscus pacificus*, from French Polynesia. *PLoS ONE* **2020**, *15*, e0231400. [CrossRef]
101. Wang, B.; Yao, M.; Zhou, J.; Tan, S.; Jin, H.; Zhang, F.; Mal, Y.L.; Wu, J.; Chan, L.L.; Cai, Z. Growth and toxin production of *Gambierdiscus* spp. can be regulated by quorum-sensing bacteria. *Toxins* **2018**, *10*, 257. [CrossRef]
102. Clausing, R.J.; Losen, B.; Oberhaensli, F.R.; Darius, H.T.; Sibat, M.; Hess, P.; Swarzenski, P.W.; Chinain, M.; Bottein, M.-Y.D. Experimental evidence of dietary ciguatoxin accumulation in an herbivorous coral reef fish. *Aquat. Toxicol.* **2018**, *200*, 257–265. [CrossRef]
103. Yon, T.; Sibat, M.; Réveillon, D.; Bertrand, S.; Chinain, M.; Hess, P. Deeper insight into *Gambierdiscus polynesiensis* toxin production relies on specific optimization of high-performance liquid chromatography-high resolution mass spectrometry. *Talanta* **2021**, *232*, 122400. [CrossRef]
104. Gräwe, U.; Wolff, J.-O.; Ribbe, J. Mixing, hypersalinity and gradients in Hervey Bay. *Ocean Dyn.* **2009**, *59*, 643–658. [CrossRef]
105. McKenzie, M.; Roder, C.A.; Roelofs, A.J.; Lee Long, W.J. *Post-Flood Monitoring of Seagrasses in Hervey Bay and the Great Sandy Strait, 1999: Implications for Dugong, Turtle & Fisheries Management*; DPI Information Series QI00059: Cairns, Australia, 2000; p. 46.
106. Meager, J.J.; Limpus, C.J.; Sumpton, W. *A Review of the Population Dynamics of Dugongs in Southern Queensland 1830–2012*; Department of Environment and Heritage Protection, Queensland Government: Brisbane, Australia, 2013; p. 29.

107. Larsson, M.E.; Laczka, O.F.; Harwood, D.T.; Lewis, R.J.; Himaya, S.W.A.; Murray, S.A.; Doblin, M.A. Toxicology of *Gambierdiscus* spp. (Dinophyceae) from tropical and temperate Australian waters. *Mar. Drugs* **2018**, *16*, 7. [CrossRef] [PubMed]

108. Landsberg, J.H.; Steidinger, K.A. A Historical Review of *Gymnodinium brevis* Red Tide Implicated in Mass Mortalities of the Manatee (*Trichechus manatus latirostris*) in Florida, USA. In *Harmful Algae, Proceedings of the VIII International Conference on Harmful Algae, Vigo, Spain, 25–29 June 1997*; Reguera, B., Blanco, J., Fernández, M.L., Wyatt, T., Eds.; Xunta de Galicia and International Oceanographic Commission of UNESCO: Santiago de Compostela, Spain, 1998; pp. 97–100.

109. Fire, S.E.; Flewelling, L.J.; Naar, J.; Twiner, M.J.; Henry, M.S.; Pierce, R.H.; Gannon, D.P.; Wang, Z.; Davidson, L.; Wells, R.S. Prevalence of brevetoxins in prey fish of bottlenose dolphins in Sarasota Bay, Florida. *Mar. Ecol. Prog. Ser.* **2008**, *368*, 283–294. [CrossRef]

110. Bottein, M.-Y.D.; Kashinsky, L.; Wang, Z.; Littnan, C.; Ramsdell, J.S. Identification of ciguatoxins in Hawaiian Monk seals *Monachus schauinslandi* from the northwestern and main Hawaiian islands. *Env. Sci. Technol.* **2011**, *45*, 5403–5409. [CrossRef] [PubMed]

111. Brieva, D.; Ribbe, J.; Lemckert, C. Is the East Australian Current causing a marine ecological hot-spot and an important fisheries near Fraser Island, Australia? *Est. Coast. Shelf. Sci.* **2015**, *153*, 121–134. [CrossRef]

112. Steinberg, C.; Lawrey, E. Circulation and Upwelling, Introduction to Circulation and Upwelling and Why It Is Important. 2018. Available online: https://eatlas.org.au/ne-aus-seascape-connectivity/circulation-upwelling (accessed on 20 July 2021).

113. Preen, A.R.; Lee Long, W.J.; Coles, R.G. Flood and cyclone related loss, and partial recovery, of more than 1000 km^2 of seagrass in Hervey Bay, Queensland, Australia. *Aquat. Bot.* **1995**, *52*, 3–17. [CrossRef]

114. Waterhouse, J.; Schaffelke, B.; Bartley, R.; Eberhard, R.; Brodie, J.; Star, M.; Thorburn, P.; Rolfe, J.; Ronan, M.; Taylor, B.; et al. *Scientific Consensus Statement: Land Use Impacts on Great Barrier Reef Water Quality and Ecosystem Condition*; Queensland Government: Brisbane, Australia, 2017. Available online: https://www.reefplan.qld.gov.au/__data/assets/pdf_file/0029/45992/2017-scientific-consensus-statement-summary.pdf (accessed on 20 July 2021).

115. Brodie, J.E.; Baird, M.; Mongin, M.; Skerrat, J.; Waterhouse, J.; Robillot, C.; Smith, R.; Mann, R.; Warne, M. *Development of Basin Specific Ecologically Relevant Targets*; Report Submitted to the Reef Plan Independent Science Panel; TropWater, James Cook University: Townsville, QLD, Australia, 2017; p. 53.

116. Baird, M.E.; Mongin, M.; Skerratt, J.; Margvelashvili, N.; Tickell, S.; Steven, A.D.L.; Robillot, C.; Ellis, R.; Waters, D.; Kaniewska, P.; et al. Impact of catchment-derived nutrients and sediments on marine water quality on the Great Barrier Reef: An application of the eReefs marine modelling system. *Mar. Pol. Bull.* **2021**, *167*, 112297. [CrossRef]

117. Loeffler, C.R.; Richlen, M.L.; Brandt, M.E.; Smith, T.B. Effects of grazing, nutrients, and depth on the ciguatera-causing dinoflagellate *Gambierdiscus* in the US Virgin Islands. *Mar. Ecol. Prog. Ser.* **2015**, *531*, 91–104. [CrossRef]

118. McMahon, K.; Nash, S.B.; Eaglesham, G.; Müller, J.F.; Duke, N.C.; Winderlich, S. Herbicide contamination and the potential impact to seagrass meadows in Hervey Bay, Queensland, Australia. *Mar. Poll. Bull.* **2005**, *51*, 325–334. [CrossRef]

119. Davis, A.M.; Lewis, S.E.; Bainbridge, Z.T.; Glendenning, L.; Turner, R.D.R.; Brodie, J.E. Dynamics of herbicide transport and portioning under event flow conditions in the lower Burdekin region, Australia. *Mar. Poll. Bull.* **2012**, *65*, 182–193. [CrossRef]

120. Magnusson, M.; Heimann, K.; Ridd, M.; Negri, A.P. Chronic herbicide exposures affect the sensitivity and community structure of tropical benthic microalgae. *Mar. Poll. Bull.* **2012**, *65*, 363–372. [CrossRef]

121. Wood, R.J.; Mitrovic, S.M.; Lim, R.P.; Warne, M.S.t.J.; Dunlop, J.; Kefford, B.J. Benthic diatoms as indicators of herbicide toxicity in rivers–A new Species at Risk (SPEARherbicides) index. *Ecol. Indic.* **2019**, *99*, 203–213. [CrossRef]

122. Holmes, M.J.; Lewis, R.J.; Jones, A.; Hoy, A.W. Cooliatoxin, the first toxin from *Coolia monotis* (Dinophyceae). *Nat. Toxins* **1995**, *3*, 355–362. [CrossRef] [PubMed]

123. Mohammad-Noor, N.; Moestrup, Ø.; Lundholm, N.; Fraga, S.; Adam, A.; Holmes, M.J.; Saleh, E. Autecology and phylogeny of *Coolia tropicalis* and *Coolia malayensis* (Dinophyceae), with emphasis on taxonomy of *C. tropicalis* based on light microscopy, scanning electron microscopy and LSU rDNA(1). *J. Phycol.* **2013**, *49*, 536–545. [CrossRef]

124. de Azevedo Tibiriçá, C.E.J.; Sibat, M.; Fernandes, L.F.; Bilien, G.; Chomerat, N.; Hess, P. Diversity and toxicity of the genus *Coolia* Meunier in Brazil, and detection of 44-methyl gambierone in *Coolia tropicalis*. *Toxins* **2020**, *12*, 327. [CrossRef]

125. Momigliano, P.; Sparrow, L.; Blair, D.; Heimann, K. The diversity of *Coolia* spp. (Dinophyceae Ostreopsidaceae) in the Central Great Barrier Reef Region. *PLoS ONE* **2013**, *8*, e79278. [CrossRef] [PubMed]

126. Larsson, M.E.; Smith, K.F.; Dobin, M.A. First description of the environmental niche of the epibenthic dinoflagellate species *Coolia palmyrensis*, *C. malayensis*, and *C. tropicalis* (Dinophyceae) from Eastern Australia. *J. Phycol.* **2019**, *55*, 565–577. [CrossRef]

127. Horn, M. Biology of marine herbivorous fishes. *Oceanogr. Mar. Biol. Annu. Rev.* **1989**, *27*, 167–172.

128. Rongo, T.; van Woesik, R. Ciguatera poisoning in Rarotonga, southern Cook Islands. *Harmful Algae* **2011**, *10*, 345–355. [CrossRef]

129. Kelly, A.M.; Kohler, C.C.; Tindall, D.R. Are crustaceans linked to the ciguatera food chain? *Environ. Biol. Fishes* **1992**, *33*, 275–286. [CrossRef]

130. Yasumoto, T.; Kanno, K. Occurrence of toxins resembling ciguatoxin, scaritoxin and maitotoxin in a turban shell. *Bull. Jpn. Soc. Sci. Fish.* **1976**, *42*, 1399–1404. [CrossRef]

131. Roué, M.; Darius, H.T.; Picot, S.; Ung, A.; Viallon, J.; Gaertner-Mazouni, N.; Sibat, M.; Amzil, Z.; Chinain, M. Evidence of the bioaccumulation of ciguatoxins in giant clams (*Tridacna maxima*) exposed to *Gambierdiscus* spp. Cells. *Harmful Algae* **2016**, *57*, 78–87. [CrossRef]

132. Silva, M.; Rodriguez, I.; Barreiro, A.; Kaufmann, M.; Neto, A.I.; Hassouani, M.; Sabour, B.; Alfonso, A.; Botana, L.M.; Vasconcelos, V. First report of ciguatoxins in two starfish species: *Ophidiaster ophidianus* and *Marthasterias glacialis*. *Toxins* **2015**, *7*, 3740–3757. [CrossRef]

133. Darius, H.T.; Roue, M.; Sibat, M.; Viallon, J.; Gatti, C.M.; Vandersea, M.W.; Tester, P.A.; Litaker, R.W.; Amzil, Z.; Hess, P.; et al. *Tectus niloticus* (Teguilidae, Gastropod) as a novel vector of ciguatera poisoning: Detection of Pacific ciguatoxins in toxic samples from Nuku Hiva island (French Polynesia). *Toxins* **2018**, *10*, 2. [CrossRef]

134. Díaz-Ascenio, L.; Clausing, R.J.; Vandersea, M.; Chamero-Lago, D.; Gómez-Batista, M.; Hernández-Albernas, J.I.; Chomérat, N.; Rojas-Abrahantes, G.; Litaker, R.W.; Tester, P.; et al. Ciguatoxin occurrence in food-web components of a Cuban coral reef ecosystem: Risk-assessment implications. *Toxins* **2019**, *11*, 722. [CrossRef]

135. Neves, R.A.F.; Fernandes, T.; dos Santos, L.N.; Nascimento, M. Toxicity of benthic dinoflagellates on grazing, behavior and survival of the brine shrimp *Artemia salina*. *PLoS ONE* **2017**, *12*, e0175168. [CrossRef]

136. Chinain, M.; Darius, H.T.; Ung, A.; Fouc, M.T.; Revel, T.; Cruchet, P.; Paullac, S.; Laurent, D. Ciguatera risk management in French Polynesia: The case study of Raivavae Island (Australes Archipelago). *Toxicon* **2010**, *56*, 674–690. [CrossRef]

137. Gaboriau, M.; Ponton, D.; Darius, H.T.; Chinain, M. Ciguatera fish toxicity in French Polynesia: Size does not always matter. *Toxicon* **2014**, *84*, 41–50. [CrossRef]

138. Capra, M.F.; Cameron, J.; Flowers, A.E.; Coombe, I.F.; Blanton, C.G.; Hahn, S.T. The Effects of Ciguatoxin on Teleosts. In Proceedings of the 6th International Coral Reef Symposium, Townsville, Australia, 8–12 August 1988; Volume 3, pp. 37–41.

139. Davin, W.T.; Kohler, C.C.; Tindall, D.R. Ciguatera toxins adversely affect piscivorous fishes. *Trans. Am. Fish. Soc.* **1988**, *117*, 374–384. [CrossRef]

140. Lewis, R.J. Ciguatoxins are potent ichthyotoxins. *Toxicon* **1992**, *30*, 207–211. [CrossRef]

141. QFish. Available online: http://qfish.fisheries.qld.gov.au/ (accessed on 20 July 2021).

142. Tonge, J.I.; Battey, Y.; Forbes, J.J.; Grant, E.M. Ciguatera poisoning: A report of two outbreaks and a probable fatal case in Queensland. *Med. J. Aust.* **1967**, *2*, 1088–1090. [CrossRef]

143. Tobin, A.; Heupel, M.; Simpfendorfer, C.; Buckley, S.; Thustan, R.; Pandolfi, J. *Utilizing Innovative Technology to Better Understand Spanish Mackerel Spawning Aggregations and the Protection Offered by Marine Protected Areas*; Centre for Sustainable Tropical Fisheries and Aquaculture, James Cook University: Townsville, Australia, 2014; p. 70.

144. Tobin, A.; Maplestone, A. *Exploitation Dynamics and Biological Characteristics of the Queensland East Coast Spanish Mackerel (Scomberomorus commerson) Fishery*; CRC Reef Research Centre Technical Report No 51; CRC Reef Research Centre: Townsville, Australia, 2004; p. 61.

145. Champion, C.; Brodie, S.; Coleman, M.A. Climate-driven range shifts are rapid yet variable among recreationally important coastal-pelagic fisheries. *Front. Mar. Sci.* **2021**, *8*, 159. [CrossRef]

146. Espinoza, M.; Matley, J.; Heupel, M.R.; Tobin, A.J.; Fisk, A.T.; Simpfendorfer, C.A. Multi-tissue stable isotope analysis reveals resource partitioning and trophic relationships of large reef-associated predators. *Mar. Ecol. Prog. Ser.* **2019**, *615*, 159–176. [CrossRef]

147. Wortmann, J.; O'Neill, M.F.; Sumpton, W.; Campbell, M.J.; Stewart, J. *Stock Assessment of Australian east Coast Snapper, Chrysophrys auratus: Predictions of Stock Status and Reference Points for 2016*; Queensland Government Report: Brisbane, Australia, 2018; p. 122.

148. Leigh, G.M.; Yang, W.-H.; O'Neill, M.F.; McGilvray, J.G.; Wortmann, J. *Stock Assessments of Bream, Whiting and Flathead (Acanthopagrus australis, Sillago ciliata and Platycephalus fuscus) in South East Queensland*; Fisheries Queensland, Department of Agriculture and Fisheries: Brisbane, Australia, 2019; p. 197.

149. Roelfsema, C.M.; Kovacs, E.M.; Ortiz, J.C.; Callaghan, D.P.; Hock, K.; Mongin, M.; Johansen, K.; Mumby, P.J.; Wettle, M.; Ronan, M.; et al. Habitat maps to enhance monitoring and management of the Great Barrier Reef. *Coral Reefs* **2020**, *39*, 1039–1054. [CrossRef]

150. Yang, Z.; Luo, Q.; Liang, Y.; Mazumder, A. Processes and pathways of ciguatoxin in aquatic food webs and fish poisoning of seafood consumers. *Environ. Rev.* **2016**, *150*, 144–150. [CrossRef]

151. Halstead, B.W.; Bunker, N.C. A survey of the poisonous fishes of the Phoenix Islands. *Copeia* **1954**, *1*, 1–11. [CrossRef]

152. Banner, A.H.; Helfrich, P.; Piyakarnchana, T. Retention of ciguatera toxin by the red snapper, *Lutjanus bohar*. *Copeia* **1966**, *2*, 297–301.

153. Cooper, M.J. 1964 Ciguatera and other marine poisoning in the Gilbert Islands. *Pac. Sci.* **1964**, *4*, 411–440.

154. USFDA. 2021 Natural Toxins. In *Fish and Fishery Product Hazards and Controls Guidance*, 4th ed.; 2021; Table A-5; p. A5-11. Available online: http://www.fda.gov/media/80637/download (accessed on 20 July 2021).

155. Ferreira, B.P.; Russ, G.R. Age validation and estimation of growth rate of the coral trout, *Plectropomus leopardus*, (Lacepede 1802) from Lizard Island, norther Great Barrier Reef. *Fish. Bull.* **1994**, *92*, 46–57.

156. Begg, G.A.; Chen, C.C.-M.; O'Neill, M.F.O.; Rose, D.B. *Stock Assessment of the Torres Strait Spanish Mackerel Fishery*; CRC Reef Research Centre Technical Report No. 66; CRC Reef Research Centre: Townsville, Australia, 2006; p. 81.

157. Campbell, A.B.; O'Neill, M.F.; Staunton-Smith, J.; Atfield, J.; Kirkwood, J. *Stock Assessment of the Australian East Coast Spanish Mackerel (Scomberomorus commerson) Fishery*; Department of Employment, Economic Development and Innovation, Queensland Government: Brisbane, Australia, 2012; p. 138.

158. Oshiro, N.; Nagaswa, H.; Kuniyoshi, K.; Kobayashi, N.; Sugita-Konishi, Y.; Asakura, H.; Yasumoto, T. Characteristic distribution of ciguatoxins in the edible parts of a grouper, *Variola louti*. *Toxins* **2021**, *13*, 218. [CrossRef]

159. Vernoux, J.-P.; Gaign, M.; Riyeche, N.; Tagmouti, F.; Magras, L.P.; Nolen, J. Mise en evidence d'une toxine liposoluble de type ciguatérique chez *Caranx bartholomaei* pêché aux Antilles françaises. *Biochimie* **1982**, *64*, 933–939. [CrossRef]
160. Helfrich, P.; Piyakarnchana, T.; Miles, P. Ciguatera fish poisoning I: The ecology of ciguateric reef fishes in the Line Islands. *Occ. Pap. Bernice P. Bishop Mus.* **1968**, *23*, 305–370.
161. Banner, A.H. Ciguatera: A Disease from Coral Reef Fish. In *Biology and Geology of Coral Reefs*; Jones, A.O., Endean, R., Eds.; Academic Press: London, UK, 1976; Volume 3, pp. 177–212.
162. Mackie, M.; Gaughan, D.J.; Buckworth, R.C. Stock assessment of narrow-barred Spanish mackerel (*Scomberomorus commerson*) in Western Australia. *FRDC Proj.* **2003**, *151*, 242.
163. O'Toole, A.C.; Bottein, M.-Y.D.; Danylchuk, A.J.; Ramsdell, J.S.; Cooke, S.J. Linking ciguatera poisoning to spatial ecology of fish: A novel approach to examining the distribution of biotoxin levels in the great barracuda by combining non-lethal blood sampling and biotelemetry. *Sci. Total Environ.* **2012**, *427–428*, 98–105. [CrossRef] [PubMed]
164. Helfrich, P.; Banner, A.H. Experimental induction of ciguatera toxicity in fish through diet. *Nature* **1963**, *197*, 1025–1026. [CrossRef]
165. Chan, T.Y.K. Regional variations in the risk and severity of ciguatera caused by eating moray eels. *Toxins* **2017**, *9*, 201. [CrossRef]
166. Hahn, S.T.; Capra, M.F.; Walsh, T.P. Ciguatoxin-protein association in skeletal muscle of Spanish mackerel (*Scomberomorus commersoni*). *Toxicon* **1992**, *30*, 843–852. [CrossRef]
167. Jiang, X.-W.; Li, X.; Lam, P.K.S.; Cheng, S.H.; Schlenk, D.; de Mitcheson, Y.S.; Li, Y.; Gu, J.-D.; Chan, L.L. Proteomic analysis of hepatic tissue of ciguatoxin (CTX) contaminated coral reef fish *Cephalopholis argus* and the moray eel *Gymnothorax undulatus*. *Harmful Algae* **2012**, *13*, 65–71. [CrossRef]
168. Hossen, V.; Soliño, L.; Leroy, P.; David, E.; Velge, P.; Dragacci, S.; Krys, S.; Quintana, H.F.; Diogène, J. Contribution to the risk characterization of ciguatoxins: LOAEL estimated from eight ciguatera fish poisoning events in Guadeloupe (French West Indies). *Environ. Res.* **2015**, *143*, 100–108. [CrossRef]
169. Dalzell, P. Management of ciguatera fish poisoning in the South Pacific. *Mem. Qld. Mus.* **1994**, *34*, 471–479.
170. Tosteson, T.R.; Ballantine, D.L.; Durst, H.D. Seasonal frequency of ciguatoxic barracuda in southwest Puerto Rico. *Toxicon* **1988**, *26*, 795–801. [CrossRef]
171. Ginsberg, G.; Hattis, D.; Sonawane, B.; Russ, A.; Banati, P.; Kozlak, M.; Smolenski, S.; Goble, R. Evaluation of child/adult pharmacokinetic differences from a database derived from the therapeutic drug literature. *Toxicol. Sci.* **2002**, *66*, 185–200. [CrossRef] [PubMed]
172. Lovett, R.A.; Prosser, A.; Leigh, G.M.; O'Neill, M.F.; Stewart, J. *Stock Assessment of the Australian East Coast Sea Mullet (Mugil cephalus) Fishery*; Queensland Department of Agriculture and Fisheries Report: Brisbane, Australia, 2018; p. 80.
173. Bell, P.A.; O'Neill, M.F.; Leigh, G.M.; Courtney, A.J.; Peel, S.L. *Stock Assessment of the Queensland-New South Wales Sea Mullet Fishery (Mugil Cephalus)*; Queensland Department of Primary Industries and Fisheries: Brisbane, Australia, 2005; p. 93.
174. Yogi, K.; Sakugawa, S.; Oshiro, N.; Ikehara, T.; Sugiyama, K.; Yasumoto, T. Determination of toxins involved in ciguatera fish poisoning in the Pacific by LC/MS. *J. AOAC* **2014**, *97*, 398–402. [CrossRef] [PubMed]
175. Lewis, R.J.; Sellin, M.; Gillespie, N.C.; Holmes, M.J.; Keys, A.; Street, R.; Smythe, H.; Thaggard, H.; Bryce, S. Ciguatera and herbivores: Uptake and accumulation of ciguatoxins in *Ctenochaetus striatus* on the Great Barrier Reef. *Mem. Qld Mus.* **1994**, *34*, 565–570.
176. Colman, J.R.; Bottein, M.-Y.D.; Dickey, R.W.; Ramsdell, J.S. Characterization of the developmental toxicity of Caribbean ciguatoxins in finfish embryos. *Toxicon* **2004**, *44*, 59–66. [CrossRef]
177. Yan, M.; Mak, M.Y.L.; Cheng, J.; Li, J.; Gu, J.R.; Leung, P.T.Y.; Lam, P.K.S. Effects of dietary exposure to ciguatoxin P-CTX-1 on the reproductive performance in marine medaka (*Oryzias melastigma*). *Mar. Poll. Bull.* **2020**, *152*, 110837. [CrossRef]
178. Mak, Y.L.; Li, J.; Liu, C.-N.; Cheng, S.H.; Lam, P.K.S.; Cheng, J.; Chan, L.L. Physiological and behavioural impacts of Pacific ciguatoxin-1 (P-CTX-1) in marine medaka (*Oryzias melastigma*). *J. Hazard. Mater.* **2017**, *321*, 782–790. [CrossRef]
179. Yan, M.; Leung, P.T.Y.; Ip, J.C.H.; Cheng, J.-P.; Wu, J.-J.; Gu, J.-R.; Lam, P.K.S. Developmental toxicity and molecular responses of marine medaka (*Oryzias melastigma*) embryos to ciguatoxin P-CTX-1 -exposure. *Aquat. Toxicol.* **2017**, *185*, 149–159. [CrossRef]
180. Emslie, M.J.; Cheal, A.J.; Logan, M. The distribution and abundance of reef-associated predatory fishes on the Great Barrier Reef. *Coral Reefs* **2017**, *36*, 829–846. [CrossRef]
181. Gillespie, N.C. Possible origins of ciguatera. In *Toxic Plants and Animals: A Guide for Australia*; Covacevich, J., Davie, P., Pearn, J., Eds.; Queensland Museum: Brisbane, Austrilia, 1987; pp. 170–179.
182. Bagnis, R. Naissance et développment d'une flambée de ciguatera dans un aoll des Tuamotu. *Revue Corps Santé* **1969**, *10*, 783–795.
183. Cheal, A.J.; MacNeill, M.A.; Cripps, E.; Emslie, M.J.; Jonker, M.; Schaffelke, B.; Sweatman, H. Coral-macroalgal phase shifts or reef resilience: Links with diversity and functional roles of herbivorous fishes on the Great Barrier Reef. *Coral Reefs* **2010**, *29*, 1005–1015. [CrossRef]
184. Webley, J.; McInnes, K.; Teixeira, D.; Lawson, A.; Quinn, R. *Statewide Recreational Fishing Survey 2013–2014*; Queensland Government Report: Brisbane, Australia, 2015; p. 127.
185. Skinner, M.P.; Lewis, R.J.; Morton, S. Ecology of the ciguatera causing dinoflagellates from the Northern Great Barrier Reef: Changes in community distribution and coastal eutrophication. *Mar. Poll. Bull.* **2013**, *77*, 210–219. [CrossRef]
186. Davies, C.H.; Coughlan, A.; Hallegraeff, G.; Ajani, P.; Armbrecht, L.; Atikins, N.; Bonham, P.; Brett, S.; Brinkman, R.; Burford, M.; et al. A database of marine phytoplankton abundance, biomass and species composition in Australian waters. *Sci. Data* **2016**, *3*, 160043. [CrossRef]

187. Larsson, M.E.; Harwood, T.D.; Lewis, R.J.; Himaya, S.W.A.; Doblin, M.A. Toxicological characterization of *Fukuyoa paulensis* (Dinophyceae) from temperate Australia. *Phycol. Res.* **2019**, *67*, 65–71. [CrossRef]

188. Catania, D.; Richlen, M.L.; Mak, Y.L.; Morton, S.L.; Laban, E.H.; Xu, Y.; Anderson, D.M.; Chan, L.L.; Berumen, M.L. The prevalence of benthic dinoflagellates associated with ciguatera fish poisoning in the central Red Sea. *Harmful Algae* **2017**, *68*, 206–216. [CrossRef]

189. Gaiani, G.; Leonardo, S.; Tudó, À.; Toldrà, A.; Rey, M.; Andree, K.B.; Tsumuraya, T.; Hirama, M.; Diogène, J.; O'Sullivan, C.K.; et al. Rapid detection of ciguatoxins in *Gambierdiscus* and *Fukuyoa* with immunosensing tools. *Ecotoxicol. Environ. Saf.* **2020**, *204*, 111004. [CrossRef]

190. Kretzschmar, A.L.; Verma, A.; Harwood, D.T.; Hoppenrath, M.; Murray, S. Characterization of *Gambierdiscus lapillus* sp. nov. (Gonyaucales, Dinophyceae): A new toxic dinoflagellate from the Great Barrier Reef (Australia). *J. Phycol.* **2017**, *53*, 283–297. [CrossRef]

191. Kretzschmar, A.L.; Larsson, M.E.; Hoppenrath, M.; Doblin, M.A.; Murray, S.A. Characterisation of two toxic *Gambierdiscus* spp. (Gonyaulacales, Dinophyceae) from the Great Barrier Reef (Australia): *G. lewisii* sp. nov. and *G. holmesii* sp. nov. *Protist* **2019**, *170*, 125699. [CrossRef]

192. Yasumoto, T.; Inoue, A.; Bagnis, R.; Garcon, M. Ecological survey on a dinoflagellate possibly responsible for the induction of ciguatera. *Bull. Jpn. Soc. Sci. Fish.* **1979**, *45*, 395–399. [CrossRef]

193. Yasumoto, T.; Inoue, A.; Ochi, T.; Fujimoto, K.; Oshima, Y.; Fukuyo, Y.; Adachi, R.; Bagnis, R. Environmental studies on a toxic dinoflagellate responsible for ciguatera. *Bull. Jpn. Soc. Sci. Fish.* **1980**, *46*, 1397–1404. [CrossRef]

194. Lobel, P.S.; Anderson, D.M.; Durand-Clement, M. Assessment of ciguatera dinoflagellate populations: Sample variability and algal substrate selection. *Biol. Bull.* **1988**, *175*, 95–101. [CrossRef]

195. Caire, J.F.; Raymond, A.; Bagnis, R. Ciguatera: Study of Setting Up and the Evolution of Gambiediscus Toxicus Population on an Artificial Substrate Introduced in an Atoll Lagoon with Follow up of Associated Environmental Factors. In Proceedings of the Fifth International Coral Reef Congress: French Polynesian Coral Reefs, Tahiti, France, 27 May–1 June 1985; Delsalle, B., Galzin, R., Salvat, B., Eds.; Antenne Museum-Ephe: Moorea, France, 1985; Volume 1, pp. 429–435.

196. Tester, P.A.; Kibler, S.R.; Holland, W.C.; Usup, G.; Vandersea, M.W.; Leaw, C.P.; Teen, L.P.; Larsen, J.; Mohammad-Noor, N.; Faust, M.A.; et al. Sampling harmful benthic dinoflagellates: Comparison of artificial and natural substrate methods. *Harmful Algae* **2014**, *39*, 8–25. [CrossRef]

197. Yong, H.L.; Mustapa, N.I.; Lee, L.K.; Lim, Z.F.; Tan, T.H.; Usup, G.; Gu, H.; Litaker, R.W.; Tester, P.A.; Lim, P.T.; et al. Habitat complexity affects benthic harmful dinoflagellate assemblages in the fringing reef of Rawa Island, Malaysia. *Harmful Algae* **2018**, *78*, 56–68. [CrossRef] [PubMed]

198. Lee, L.K.; Lim, Z.F.; Gu, H.; Chan, L.L.; Litaker, R.W.; Tester, P.A.; Leaw, C.P.; Lim, P.T. Effects of substratum and depth on benthic harmful dinofagellate assemblages. *Sci. Rep.* **2020**, *10*, 1–14. [CrossRef]

199. Jauzein, C.; Açaf, L.; Accoroni, S.; Asnaghi, V.; Fricke, A.; Hachani, M.A.; abboud-Abi Saab, M.; Chiantore, M.; Mangialajo, L.; Totti, C.; et al. Optimization of sampling, cell collection and counting for the monitoring of benthic harmful algal blooms: Application to *Ostreopsis* spp. blooms in the Mediterranean Sea. *Ecol. Indic.* **2018**, *91*, 116–127. [CrossRef]

200. Parsons, M.L.; Brandt, A.L.; Ellsworth, A.; Leyne, A.L.; Rains, L.K.; Anderson, D.M. Assessing the use of artificial substrates to monitor *Gambierdiscus* populations in the Florida Keys. *Harmful Algae* **2017**, *68*, 52–66. [CrossRef]

201. Parsons, M.L.; Richlen, M.L.; Smith, T.B.; Solow, A.R.; Anderson, D.M. Evaluation of 24-h screen deployments as a standardized platform to monitor *Gambierdiscus* populations in the Florida Keys and U.S. Virgin Islands. *Harmful Algae* **2021**, *103*, 101998. [CrossRef]

202. Diaz-Pulido, G.; McCook, L.J.; Larkum, A.W.D.; Lotze, H.K.; Raven, J.A.; Schaffelke, B.; Smith, J.E.; Steneck, R.S. Chapter 7: Vulnerability of Macroalgae of the Great Barrier Reef to Climate Change. In *Climate Change and the Great Barrier Reef: A Vulnerability Assessment*; Johnson, J.E., Marshall, P.A., Eds.; Great Barrier Reef Marine Park Authority: Townsville, Australia, 2007; pp. 153–192.

203. Cruz-Rivera, E.; Villareal, T.A. Macroalgal palatability and the flux of ciguatera toxins through marine food webs. *Harmful Algae* **2006**, *5*, 497–525. [CrossRef]

204. Madin, E.M.P.; Gaines, S.D.; Warner, R.R. Field evidence for pervasive indirect effects of fishing on prey foraging behaviour. *Ecology* **2010**, *91*, 3563–3571. [CrossRef]

205. Madin, E.M.P.; Madin, J.S.; Harmer, A.M.T.; Barrett, N.S.; Booth, D.J.; Caley, M.J.; Cheal, A.J.; Edgar, G.J.; Emslie, M.J.; Gaines, S.D.; et al. Latitude and protection affect decadal trends in reef trophic structure over a continental scale. *Ecol. Evol.* **2020**, *10*, 6954–6966. [CrossRef]

206. Rasher, D.B.; Hoey, A.S.; Hay, M.E. Cascading predator effects in a Fijian coral reef ecosystem. *Nat. Sci. Rep.* **2017**, *7*, 15684. [CrossRef]

207. Rongo, T.; van Woesik, R. The effects of natural disturbances, reef state, and herbivorous fish densities on ciguatera poisoning in Rarotonga, southern Cook Islands. *Toxicon* **2013**, *64*, 87–95. [CrossRef]

208. Kelly, E.L.A.; Eynaud, Y.; Williams, I.D.; Sparks, R.T.; Dailer, M.L.; Sandin, S.A.; Smith, J.E. A budget of algal production and consumption by herbivore fisheries management area, Maui Hawaii. *Ecosphere* **2017**, *8*, e01899. [CrossRef]

209. GBRMPA. *Great Barrier Reef Outlook Report 2019*; Great Barrier Reef Marine Park Authority: Townsville, Australia, 2019; Available online: http://hdl.handle.net/11017/3474 (accessed on 20 July 2021).

210. Yasumoto, T.; Hashimoto, Y.; Bagnis, R.; Randall, J.E.; Banner, A.H. Toxicity of the surgeonfishes. *Bull. Jpn. Soc. Sci. Fish.* **1971**, *37*, 724–734. [CrossRef]
211. Caillaud, A.; de la Iglesia, P.; Barber, E.; Eixarch, H.; Mohammad-Noor, N.; Yasumoto, T.; Diogène, J. Monitoring of dissolved ciguatoxin and maitotoxin using solid-phase adsorption toxin tracking devices: Application to Gambierdiscus pacificus in culture. *Harmful Algae* **2011**, *10*, 433–446. [CrossRef]
212. Roué, M.; Darius, H.T.; Viallon, J.; Ung, A.; Gatti, C.; Harwood, D.T.; Chinain, M. Application of solid phase adsorption toxin tracking (SPATT) devices for the field detection of *Gambierdiscus* toxins. *Harmful Algae* **2018**, *71*, 40–49. [CrossRef] [PubMed]
213. Roué, M.; Smith, K.F.; Sibat, M.; Viallon, J.; Henry, K.; Ung, A.; Biessy, L.; Hess, P.; Darius, H.T.; Chinain, M. Assessment of ciguatera and other phycotoxin-related risks in Anaho Bay (Nuku Hiva Island, French Polynesia): Molecular, toxicological, and chemical analyses of passive samplers. *Toxins* **2020**, *12*, 321. [CrossRef] [PubMed]
214. Furnas, M.; Alongi, A.; McKinnon, D.; Trott, L.; Skuza, M. Regional-scale nitrogen and phosphorus budgets for the northern (14° S) and central (17° S) Great Barrier Reef shelf ecosystem. *Cont. Shelf Res.* **2011**, *31*, 1967–1990. [CrossRef]
215. Brodie, J.E.; Lewis, S.E.; Collier, C.J.; Woolridge, S.; Bainbridge, Z.T.; Waterhouse, J.; Rasheed, M.A.; Honchin, C.; Holmes, G.; Fabricus, K. Setting ecologically relevant targets for river pollutant loads to meet marine water quality requirements for the Great Barrier Reef, Australia: A preliminary methodology and analysis. *Ocean Coast. Manag.* **2017**, *143*, 136–147. [CrossRef]
216. Devlin, M.J.; McKinna, L.W.; Álvarez-Romero, J.G.; Petus, C.; Abott, B.; Harkenss, P.; Brodie, J. Mapping the pollutants in surface riverine flood plume waters in the Great Barrier Reef, Australia. *Mar. Poll. Bull.* **2012**, *65*, 224–235. [CrossRef]
217. Williams, S.L.; Carpenter, R.C. Nitrogen-limited primary productivity of coral reef algal turfs: Potential contribution of ammonium excreted by *Diadema antillarum*. *Mar. Ecol. Prog. Ser.* **1988**, *47*, 145–152. [CrossRef]
218. Karcher, D.B.; Roth, F.; Carvalho, S.; El-Khaled, Y.C.; Tilstra, A.; Kürten, B.; Struck, U.; Jones, B.H.; Wild, C. Nitrogen eutrophication particularly promotes turf algae in coral reefs of the central Red Sea. *PeerJ* **2020**, *8*, e8737. [CrossRef]
219. Faust, M.A. Mixotrophy in tropical benthic dinoflagellates. In *Harmful Algae, Proceedings of the VIII International Conference on Harmful Algae, Vigo, Spain, 25–29 June 1997*; Reguera, B., Blanco, J., Fernández, M.L., Wyatt, T., Eds.; Xunta de Galicia and International Oceanographic Commission of UNESCO: Santiago de Compostela, Spain, 1998; pp. 390–393.
220. Price, D.C.; Farinholt, N.; Gates, C.; Shumaker, A.; Wagner, N.E.; Bienfang, P.; Bhattacharya, D. Analysis of *Gambierdiscus* transcriptome data supports ancient origins of mixotrophic pathways in dinoflagellates. *Environ. Microbiol.* **2016**, *18*, 4501–4510. [CrossRef]
221. Clausing, R.J.; Annunziata, C.; Baker, G.; Lee, C.; Bittick, S.J.; Fong, P. Effects of sediment depth on algal turf height are mediated by interactions with fish herbivory on a fringing reef. *Mar. Ecol. Prog. Ser.* **2014**, *517*, 121–129. [CrossRef]
222. Tebbett, S.B.; Goatley, C.H.R.; Bellwood, D.R. Fine sediments suppress detritivory on coral reefs. *Mar. Poll. Bull.* **2017**, *114*, 934–940. [CrossRef]
223. Tebbett, S.B.; Bellwood, D.R.; Purcell, S.W. Sediment addition drives declines in algal turf yield to herbivorous coral reef fishes: Implications for reefs and reef fisheries. *Coral Reefs* **2018**, *37*, 929–937. [CrossRef]
224. Tebbett, S.B.; Goatley, C.H.R.; Streit, R.P.; Bellwood, D.R. Algal turf sediments limit the spatial extent of function delivery on coral reefs. *Sci. Total Environ.* **2020**, *734*, 139422. [CrossRef]
225. Tebbett, S.B.; Goatley, C.H.R.; Bellwood, D.R. The effects of algal turf sediments and organic loads on feeding by coral reef surgeonfishes. *PLoS ONE* **2017**, *12*, e0169479. [CrossRef]
226. Tebbett, S.B.; Bellwood, D.R. Sediments ratchet-down coral reef algal turf productivity. *Sci. Total Environ.* **2020**, *713*, 136709. [CrossRef]
227. Tebbett, S.B.; Goatley, C.H.R.; Bellwood, D.R. Algal turf sediments across the Great Barrier Reef: Putting coastal reefs in perspective. *Mar. Poll. Bull.* **2018**, *137*, 518–525. [CrossRef]
228. Gordon, S.E.; Goatley, C.H.R.; Bellwood, D.R. Low-quality sediments deter grazing by the parrotfish *Scarus rivulatus* on inner-shelf reefs. *Coral Reefs* **2016**, *35*, 285–291. [CrossRef]
229. Lewis, S.E.; Schaffelke, B.; Shaw, M.; Bainbridge, Z.T.; Rohde, K.W.; Kennedy, K.; Davis, A.M.; Masters, B.L.; Devlin, M.J.; Mueller, J.F.; et al. Assessing the additive risks of PSII herbicide exposure to the Great Barrier Reef. *Mar. Poll. Bull.* **2012**, *65*, 280–291. [CrossRef] [PubMed]
230. Shaw, C.M.; Brodie, J.; Mueller, J.F. Phytotoxicity induced in isolated zooxanthellae by herbicides extracted from Great Barrier Reef flood waters. *Mar. Poll. Bull.* **2012**, *65*, 335–362. [CrossRef]
231. McNeil, M.A.; Webster, J.M.; Beaman, R.J.; Graham, T.L. New constraints on the spatial distribution and morphology of the *Halimeda* bioherms of the Great Barrier Reef, Australia. *Coral Reefs* **2016**, *35*, 1343–1355. [CrossRef]
232. Sih, T.L.; Daniell, J.J.; Bridge, T.C.L.; Beaman, R.J.; Cappo, M.; Kingsford, M.J. Deep-reef communities of the Great Barrier Reef shelf-break: Trophic structure and habitat associations. *Diversity* **2019**, *11*, 26. [CrossRef]
233. Wiese, M.; D'Agostino, P.M.; Mihali, T.K.; Moffitt, M.C.; Neilan, B.A. Neurotoxic alkaloids: Saxitoxin and its analogs. *Mar. Drugs* **2010**, *8*, 2185–2211. [CrossRef] [PubMed]
234. Hahn, S.T.; Capra, M.F. The cyanobacterium *Oscillatoria erythraea*—A potential source of the toxin in the ciguatera food-chain. *Food Addit. Contam.* **1992**, *9*, 351–355. [CrossRef]
235. Kerbrat, A.-S.; Darius, H.T.; Pauillac, S.; Chinain, M.; Laurent, D. Detection of ciguatoxin-like and paralysing toxins in *Trichodesmium* spp. from New Caledonia lagoon. *Mar. Pol. Bull.* **2010**, *61*, 360–366. [CrossRef]

236. Clements, K.D.; German, D.P.; Piché, J.; Tribollet, A.; Choat, J.H. Integrating ecological roles and trophic diversification on coral reefs: Multiple lines of evidence identify parrotfishes as microphages. *Biol. J. Linn. Soc.* **2017**, *120*, 729–751. [CrossRef]
237. Nicholson, G.M.; Clments, K.D. Resolving resource partitioning in parrotfishes (Scarini) using microhistology of feeding substrata. *Coral Reefs* **2020**, *39*, 1313–1327. [CrossRef]
238. Laurent, D.; Kerbrat, A.-S.; Darius, H.T.; Cirarad, E.; Golubic, S.; Benoit, E.; Sauviat, M.-P.; Chinain, M.; Molgo, J.; Pauillac, S. Are cyanobacteria involved in ciguatera fish poisoning-like outbreaks in New Caledonia. *Harmful Algae* **2008**, *7*, 827–838. [CrossRef]
239. Bakus, G.J. Chemical defense mechanisms on the Great Barrier Reef, Australia. *Science* **1981**, *211*, 497–499. [CrossRef] [PubMed]
240. Holmes, M.J.; Gillespie, N.C.; Lewis, R.J. Toxicity and morphology of *Ostreopsis* cf. *siamensis* cultured from a ciguatera endemic region of Queensland, Australia. In Proceedings of the 6th International Coral Reef Symposium Executive Committee, Townsville, Australia, 8–12 August 1988; Choat, J.H., Barnes, D., Borowitzka, M.A., Coll, J.C., Davies, P.J., Flood, P., Hatcher, B.G., Hopley, D., Hutchings, P.A., Kinsey, D., et al., Eds.; Townsville, Australia, 1988; Volume 3, pp. 49–54.
241. Murakami, M.; Oshima, Y.; Yasumoto, T. Identification of okadaic acid as a toxic component of a marine dinoflagellate *Prorocentrum lima*. *Bull. Jpn. Soc. Sci. Fish.* **1982**, *48*, 69–72. [CrossRef]
242. Reguera, B.; Velo-Suárez, L.; Raine, R.; Park, M.G. Harmful *Dinophysis* species. *Harmful Algae* **2012**, *14*, 87–106. [CrossRef]
243. Murata, M.; Shimatani, M.; Sugitani, H.; Oshima, Y.; Yasumoto, T. Isolation and structural elucidation of the causative toxin of the diarrhetic shellfish poisoning. *Bull. Jpn. Soc. Sci. Fish.* **1982**, *48*, 549–552. [CrossRef]
244. Gamboa, P.M.; Park, D.L.; Fremy, J.-M. Extraction and Purification of Toxic Fractions from Barracuda (Sphyraena barracuda) Implicated in Ciguatera Poisoning. In Proceedings of the Third International Conference on Ciguatera Fish Poisoning, La Parguera, Puerto Rico, 30 April–5 May 1990; Tosteson, T.R., Ed.; Polyscience Publications: Quebec, QC, Canada, 1992; pp. 13–24.
245. Corriere, M.; Soliño, L.; Costa, P.R. Effects of the marine biotoxins okadaic acid and dinophysistoxins on fish. *J. Mar. Sci. Eng.* **2021**, *9*, 293. [CrossRef]
246. Usami, M.; Satake, M.; Ishida, S.; Inoue, A.; Kan, Y.; Yasumoto, T. Palytoxin analogs from the dinoflagellate *Ostreopsis siamensis*. *J. Am. Chem. Soc.* **1995**, *117*, 5389–5390. [CrossRef]
247. Parsons, M.L.; Aligizaki, K.; Bottein, M.-Y.D.; Fraga, S.; Morton, S.L.; Penna, A.; Rhodes, L. *Gambierdiscus* and *Ostreopsis*: Reassessment of the state of knowledge of their taxonomy, geography, ecophysiology, and toxicology. *Harmful Algae* **2012**, *14*, 107–129. [CrossRef]
248. Katikou, P.; Vlamis, A. Palytoxin and Analogs: Ecobiology and Origin, Chemistry, and Chemical Analysis. In *Seafood and freshwater Toxins. Pharmacology, Physiology and Detection*; Botana, L., Ed.; CRC Press: Boca Raton, FL, USA, 2014; pp. 695–740.
249. Kerbrat, A.S.; Amzil, Z.; Pawlowiez, R.; Golubic, S.; Sibat, M.; Darius, H.T.; Chinain, M.; Laurent, D. First evidence of palytoxin and 42-hydroxy-palytoxin in the marine cyanobacterium *Trichodesmium*. *Mar. Drugs* **2011**, *9*, 543–560. [CrossRef] [PubMed]
250. Onuma, Y.; Satake, M.; Ukena, T.; Roux, J.; Chanteau, S.; Rasolofonirina, N.; Ratsimaloto, M.; Naoki, H.; Yasumoto, T. Identification of putative palytoxin as the cause of clupeotoxism. *Toxicon* **1999**, *37*, 55–65. [CrossRef]
251. Tubaro, A.; Durando, P.; Favero, D.G.; Ansaldi, F.; Icardi, G.; Deeds, J.R.; Sosa, S. Case definitions for human poisonings postulated to palytoxins exposure. *Harmful Algae* **2011**, *57*, 478–495. [CrossRef]
252. Verma, A.; Hoppenrath, M.; Harwood, T.; Brett, S.; Rhodes, L.; Murray, S. Molecular phylogeny, morphology and toxicgenecity of *Ostreopsis* c.f. *siamensis* (Dinophyceae) from temperate south-east Australia. *Phycol. Res.* **2016**, *64*, 146–159. [CrossRef]
253. Verma, A.; Hughes, D.J.; Harwood, D.T.; Suggett, D.J.; Ralph, P.J.; Murray, S.A. Functional significance of phylogeographic structure in a toxic benthic marine microbial eukaryote over a latitudinal gradient along the East Australian Current. *Ecol. Evol.* **2020**, *10*, 6257–6273. [CrossRef]
254. Cvitanovic, C.; Fox, R.J.; Bellwood, D.R. *Herbivory by Fishes on the Great Barrier Reef: A Review of Knowledge and Understanding*; Unpublished Report to the Marine and Tropical Sciences Research Facility; Reef and Rainforest Research Centre Limited: Cairns, Australia, 2007; p. 33.
255. Cheal, A.; Emslie, M.; Miller, I.; Sweatman, H. The distribution of herbivorous fishes on the Great Barrier Reef. *Mar. Biol.* **2012**, *159*, 1143–1154. [CrossRef]
256. Purcell, S.W.; Bellwood, D.R. A functional analysis of food procurement in two surgeonfish species, *Acanthurus nigrofuscus* and *Ctenochaetus striatus* (Acanthuridae). *Env. Biol. Fish.* **1993**, *37*, 139–159. [CrossRef]
257. Tebbett, S.B.; Goatley, C.H.R.; Bellwood, D.R. Clarifying functional roles: Algal removal by the surgeonfishes *Ctenochaetus striatus* and *Acanthurus nigrofuscus*. *Coral Reefs* **2017**, *36*, 803–813. [CrossRef]
258. Bagnis, R. Natural versus anthropogenic disturbances to coral reefs: Comparison in epidemiological patterns of ciguatera. *Mem. Qld Mus.* **1994**, *34*, 455–460.
259. Yasumoto, T. Chemistry, etiology, and food chain dynamics of marine toxins. *Proc. Jpn. Acad.* **2005**, *81*, 43–51. [CrossRef]
260. Tebbett, S.B.; Goatley, C.H.R.; Huertas, V.; Mihalitsis, M.; Bellwood, D.R. A functional evaluation of feeding in the surgeonfish *Ctenochaetus striatus*: The role of soft tissues. *R. Soc. Open Sci.* **2018**, *5*, 171111. [CrossRef] [PubMed]
261. Campbell, B.; Nakagawa, L.K.; Kobayashi, M.N.; Hokama, Y. *Gambierdiscus toxicus* in gut content of the surgeonfish *Ctenochaetus strigosus* (herbivore) and its relationship to toxicity. *Toxicon* **1987**, *25*, 1125–1127.
262. Magnelia, S.J.; Kohler, C.C.; Tindall, D.R. Acanthurids do not avoid consuming cultured toxic dinoflagellates yet do not become ciguatoxic. *Trans. Am. Fish. Soc.* **1992**, *121*, 737–745. [CrossRef]
263. Marshell, A.; Mumby, P.J. The role of surgeonfish (Acanthuridae) in maintaining algal turf biomass on coral reefs. *J. Exp. Mar. Biol. Ecol.* **2015**, *473*, 152–160. [CrossRef]

264. Russ, G.R. Grazer biomass correlates more strongly with production than with biomass of algal turfs on a coral reef. *Coral Reefs* **2003**, *22*, 63–67. [CrossRef]
265. Vacarizas, J.; Benico, G.; Austero, N.; Azanza, R. Taxonomy and toxin production of *Gambierdiscus carpenteri* (Dinophyceae) in a tropical marine ecosystem: The first record from the Philippines. *Mar. Pol. Bull.* **2018**, *137*, 430–443. [CrossRef]
266. Bomber, J.W.; Guillard, R.R.L.; Nelson, W.G. Rôles of temperature, salinity, and light in seasonality, growth, and toxicity of ciguatera-causing *Gambierdiscus toxicus* Adachi et Fukuyo (Dinophyceae). *J. Exp. Mar. Biol. Ecol.* **1988**, *115*, 53–65. [CrossRef]
267. Kibler, S.R.; Litaker, R.W.; Holland, W.C.; Vandersea, M.W.; Tester, P.A. Growth of eight *Gambierdiscus* (Dinophyceae) species: Effects of temperature, salinity and irradiance. *Harmful Algae* **2012**, *19*, 1–14. [CrossRef]
268. Xu, Y.; Richlen, M.L.; Liefer, J.D.; Robertson, A.; Kullis, D.; Smith, T.B.; Parsons, M.L.; Andersen, D.M. Influence of environmental variables on *Gambierdiscus* sp. (Dinophyceae) growth and distribution. *PLoS ONE* **2016**, *11*, e0153197. [CrossRef]
269. Sparrow, L.; Momigliano, P.; Russ, G.R.; Heimann, K. Effects of temperature, salinity and composition of the dinoflagellate assemblage on the growth of *Gambierdiscus carpenteri* isolated from the Great Barrier Reef. *Harmful Algae* **2017**, *65*, 52–60. [CrossRef] [PubMed]
270. Mustapa, N.I.; Yong, H.L.; Lee, L.K.; Lim, Z.F.; Lim, H.C.; Teng, S.T.; Luo, Z.; Gu, H.; Leaw, C.P.; Lim, P.T. Growth and epiphytic behaviour of three *Gambierdiscus* species (Dinophyceae) associated with various macroalgal substrates. *Harmful Algae* **2019**, *89*, 101671. [CrossRef] [PubMed]
271. Russ, G.R.; Cheal, A.J.; Dolman, A.M.; Emslie, M.J.; Evans, R.D.; Miller, I.; Sweatman, H.; Williamson, D.H. Rapid increase in fish numbers follows creation of world's largest marine reserve network. *Curr. Biol.* **2008**, *18*, R514–R515. [CrossRef]
272. McCook, L.J.; Ayling, T.; Cappo, M.; Choat, J.H.; Evans, R.D.; De Freitas, D.M.; Heupel, M.; Hughes, T.; Jones, G.P.; Mapstone, B.; et al. Adaptive management of the Great Barrier Reef: A globally significant demonstration of the benefits of networks of marine reserves. *Proc. Natl. Acad. Sci. USA* **2010**, *107*, 18278–18285. [CrossRef] [PubMed]
273. Rizzari, J.R.; Bergseth, B.J.; Frisch, A.J. Impact of conservation areas on trophic interactions between apex predators and herbivores on coral reefs. *Conserv. Biol.* **2015**, *29*, 418–429. [CrossRef]
274. Casey, J.M.; Baird, A.H.; Brandl, S.J.; Hoogenboom, M.O.; Rizzari, J.R.; Frisch, A.J.; Mirbach, C.E.; Connolly, S.R. A test of trophic cascade theory: Fish and benthic assemblages across a predator density gradient on coral reefs. *Oceologia* **2017**, *183*, 161–175. [CrossRef] [PubMed]
275. Madin, E.M.P.; Madin, J.S.; Booth, D.J. Landscape of fear visible from space. *Nature Sci. Rep.* **2011**, *1*, 14. [CrossRef] [PubMed]
276. Johnson, G.B.; Taylor, B.M.; Robbins, W.D.; Franklin, E.C.; Toonen, R.; Bowen, B.; Choat, J.H. Diversity and structure of parrotfish assemblages across the northern Great Barrier Reef. *Diversity* **2019**, *11*, 14. [CrossRef]
277. Mellin, C.; McNeil, M.A.; Cheal, A.J.; Emslie, M.J.; Caley, M.J. Marine protected areas increase resilience among coral reef communities. *Ecol. Lett.* **2016**, *19*, 629–637. [CrossRef]
278. Newton, K.; Côté, I.M.; Pilling, G.M.; Jennings, S.; Dulvy, N.K. Current and future sustainability of island coral reef fisheries. *Curr. Biol.* **2007**, *17*, 655–658. [CrossRef] [PubMed]
279. Hughes, T.P.; Rodrigues, M.J.; Bellwood, D.R.; Ceccarelli, D.; Hoegh-Guldberg, O.; McCook, L.; Moltschaniwskyj, N.; Pratchett, M.S.; Steneck, R.S.; Willis, B. Phase shifts, herbivory, and the resilience of coral reefs to climate change. *Curr. Biol.* **2007**, *17*, 360–365. [CrossRef]
280. Steneck, R.S.; Arnold, S.N.; Mumby, P.J. Experiment mimics fishing on parrotfish: Insights on coral reef recovery and alternative attractors. *Mar. Ecol. Prog. Ser.* **2014**, *506*, 115–127. [CrossRef]
281. Russ, G.R.; Questel, S.-L.A.; Rizzari, J.R.; Alcala, A.C. The parrotfish-coral relationship: Refuting the ubiquity of a prevailing paradigm. *Mar. Biol.* **2015**, *162*, 2029–2045. [CrossRef]
282. Bellwood, D.R.; Hughes, T.P.; Hoey, A.S. Sleeping functional group drives coral-reef recovery. *Curr. Biol.* **2006**, *16*, 2434–2439. [CrossRef]
283. Fox, R.J.; Bellwood, D.R. Remote video bioassays reveal the potential feeding impact of the rabbitfish *Siganus canaliculatus* (f: Siganidae) on an inner-shelf reef of the Great Barrier Reef. *Coral Reefs* **2008**, *27*, 605–615. [CrossRef]
284. Tebbett, S.B.; Hoey, A.S.; Depczynski, M.; Wismer, S.; Bellwood, D.R. Macroalgae removal on coral reefs: Realised ecosystem functions transcend biogeographic locations. *Coral Reefs* **2020**, *39*, 203–214. [CrossRef]
285. Morais, R.A.; Depczynski, M.; Fulton, C.; Marnane, M.; Narvaez, P.; Huertas, V.; Brandl, S.J.; Bellwood, D. Severe coral loss shifts energetic dynamics on a coral reef. *Funct. Ecol.* **2020**, *34*, 1507–1518. [CrossRef]
286. Leigh, G.M.; Campbell, A.B.; Lunow, C.P.; O'Neill, M.F. *Stock Assessment of the Queensland East Coast Common Coral Trout (Plectropomus leopardus) Fishery*; Queensland Department of Agriculture, Fisheries and Forestry: Brisbane, Australia, 2014; p. 113.
287. Heaven, C. *Queensland Fisheries Summary, October 2018*; Queensland Department of Agriculture and Fisheries: Brisbane, Australia, 2018; p. 62.
288. Vermeij, M.J.A.; van der Heijden, R.A.; Olthuis, J.G.; Marhaver, K.L.; Smith, J.E.; Visser, P.M. Survival and dispersal of turf algae and macroalage consumed by herbivorous coral reef fishes. *Oecologia* **2013**, *171*, 417–425. [CrossRef]
289. Goatley, C.H.R.; Bellwood, D.R. Biologically mediated sediment fluxes on coral reefs: Sediment removal and off-reef transportation by the surgeonfish *Ctenochaetus striatus*. *Mar. Ecol. Prog. Ser.* **2010**, *415*, 237–245. [CrossRef]
290. Kramer, M.J.; Bellwood, O.; Bellwood, D.R. The trophic importance of algal turfs for coral reef fishes: The crustacean link. *Coral Reefs* **2013**, *32*, 575–583. [CrossRef]

291. Brandl, S.J.; Goatley, C.H.R.; Bellwood, D.R.; Tornabene, L. The hidden half: Ecology and evolution of cryptobenthic fishes on coral reefs. *Biol. Rev.* **2018**, *93*, 1846–1873. [CrossRef] [PubMed]

292. Brandl, S.J.; Tornabene, L.; Goatley, C.H.R.; Casey, J.M.; Morais, R.A.; Côte, I.M.; Baldwin, C.C.; Parravicini, V.; Schiettekatte, N.M.D.; Bellwood, D.R. Demographic dynamics of the smallest marine vertebrates fuel coral-reef ecosystem functioning. *Science* **2019**, *364*, 1189–1192. [CrossRef]

293. Gillespie, N.C. Ciguatera poisoning. In *Toxic Plants and Animals: A Guide for Australia*; Covacevich, J., Davie, P., Pearn, J., Eds.; Queensland Museum: Brisbane, Australia, 1987; pp. 160–169.

294. Sumpton, W.; Mayer, D.; Brown, I.; Sawynok, B.; McLennan, M.; Butcher, A.; Kirkwood, J. Investigation of movement and factors influencing post-release survival of line-caught coral reef fish using recreational tag-recapture data. *Fish. Res.* **2008**, *92*, 189–195. [CrossRef]

295. Currey, L.M.; Heupel, M.R.; Simpfendorfer, C.A.; Williams, A.J. Sedentary or mobile? Variability in space and depth use of an exploited coral reed fish. *Mar Biol.* **2014**, *161*, 2155–2166. [CrossRef]

296. Frisch, A.J.; Ireland, M.; Baker, R. Trophic ecology of large predatory reef fishes: Energy pathways, trophic level, and implications for fisheries in a changing climate. *Mar. Biol.* **2014**, *161*, 61–73. [CrossRef]

297. Holmes, B.; Leslie, M.; Keag, M.; Roelofs, A.; Winning, M.; Zeller, B. *Stock Status of Queensland's Fisheries Resources 2012*; Queensland Department of Agriculture, Fisheries and Forestry: Brisbane, Australia, 2013; p. 120.

298. Campbell, A.; Leigh, G.; Bessell-Browne, P.; Lovett, R. *Stock Assessment of the Queensland East Coast Common Coral Trout (Plectropomus leopardus) Fishery*; Department of Employment, Economic Development and Innovation, Queensland Government: Brisbane, Australia, 2019; p. 61.

299. Campbell, A.B.; Northrop, A.R. *Stock Assessment of the Common Coral Trout (Plectropomus leopardus) in Queensland, Australia*; Department of Employment, Economic Development and Innovation, Queensland Government: Brisbane, Australia, 2020; p. 45.

300. Kingsford, M.J. Spatial and temporal variation in predation on reef fishes by coral trout (*Plectropomus leopardus*, Serranidae). *Coral Reefs* **1992**, *11*, 193–198. [CrossRef]

301. St John, J.; Russ, G.R.; Brown, I.W.; Squire, L.C. The diet of the large coral reef serranid *Plectropomus leopardus* in two fishing zones on the Great Barrier Reef, Australia. *Fish. Bull.* **2001**, *99*, 180–192.

302. Matley, J.K.; Maes, G.E.; Devloo-Delva, F.; Huerlimann, R.; Chua, G.; Tobin, A.J.; Fisk, A.T.; Simpfendorfer, C.A.; Heupel, M.R. Integrating complementary methods to improve diet analysis in fishery-targeted species. *Ecol. Evol.* **2018**, *8*, 9503–9515. [CrossRef]

303. St John, J. Temporal variation in the diet of a coral reef piscivore (Pisces: Serranidae) was not seasonal. *Coral Reefs* **2001**, *20*, 163–170. [CrossRef]

304. Streit, R.P.; Cumming, G.S.; Bellwood, D.R. Patchy delivery of functions undermines functional redundancy in a high diversity system. *Funct. Ecol.* **2019**, *33*, 1144–1155. [CrossRef]

305. Tebbett, S.B.; Chase, T.J.; Bellwood, D.R. Farming damselfishes shape algal turf sediment dynamics on coral reefs. *Mar. Environ. Res.* **2020**, *160*, 104988. [CrossRef]

306. Wilson, S.; Bellwood, D.R. Cryptic dietary components of territorial damselfishes (Pomacentridae, Labroidei). *Mar. Ecol. Prog. Ser.* **1997**, *153*, 299–310. [CrossRef]

307. Edwards, C.B.; Friedlander, A.M.; Green, A.G.; Hardt, M.J.; Sala, E.; Sweatman, H.P.; Williams, I.D.; Zglicynski, B.; Sandin, S.A.; Smith, J.E. Global assessment of the status of coral reef herbivorous fishes: Evidence for fishing effects. *Proc. Royal Soc. B* **2013**, *281*, 20131835. [CrossRef]

308. Andersen, C.; Clarke, K.; Higgs, J.; Ryan, S. *Ecological Assessment of the Queensland Coral Reef Fin Fish Fishery*; Queensland Department of Primary Industries: Brisbane, Australia, 2005; p. 149.

309. Marriott, R.J.; Mapstone, B.D.; Begg, G.A. Age-specific demographic parameters, and their implication for management of the red bass, *Lutjanus bohar* (Forsskál 1775): A large, long-lived fish. *Fish. Res.* **2007**, *83*, 204–215. [CrossRef]

310. Choat, J.H.; Axe, L.M. Growth and longevity in acanthurid fishes; an analysis of otolith increments. *Mar. Ecol. Prog. Ser.* **1996**, *134*, 15–26. [CrossRef]

311. Kraa, E.; Campbell, B. Ciguatera outbreak, NSW, New South Wales. *Public Health Bull.* **1994**, *5*, 69. Available online: http://www.health.nsw.gov.au/phb/Documents/1994-06.pdf (accessed on 20 July 2021).

312. Lewis, R.J.; Ruff, T.A. Ciguatera: Ecological, clinical and socioeconomic perspectives. *Crit. Rev. Environ. Sci. Technol.* **1993**, *23*, 137–156. [CrossRef]

313. Wong, C.-K.; Hung, P.; Lo, J.Y.C. Ciguatera fish poisoning in Hong Kong-A 10-year perspective on the class of ciguatoxins. *Toxicon* **2014**, *86*, 96–106. [CrossRef]

314. Leigh, G.M.; Williams, A.J.; Begg, G.A.; Gribble, N.A.; Whybird, O.J. *Stock Assessment of the Queensland East Coast Red Throat Emperor (Lethrinius miniatus) Fishery*; Queensland Department of Primary Industries and Fisheries: Brisbane, Australia, 2006; p. 127.

315. Hessel, D.W.; Halstead, B.W.; Peckham, N.H. Marine biotoxins. 1. Ciguatera poison: Some biological and chemical aspects. *Ann. N. Y. Acad. Sci.* **1960**, *90*, 788–797. [CrossRef] [PubMed]

316. Banner, A.H.; Helfrich, P.; Scheuer, P.J.; Yoshida, T. Research on ciguatera in the tropical Pacific. In Proceedings of the 16th Gulf and Caribbean Fisheries Insititute, Coral Gables, FL, USA, May 1964; pp. 84–98.

317. Vernoux, J.-P. La ciguatera dans l'île de Saint-Barthélémy: Aspects épidémiologiques, toxicologiques et préventifs. *Oceanol. Acta* **1988**, *11*, 37–46.

318. Bravo, J.; Cabrera Suárez Ramírez, A.S.; Acosta, F. Ciguatera, an emerging human poisoning in Europe. *J. Aqua. Mar. Biol.* **2015**, *3*, 53. [CrossRef]
319. Tsumuraya, T.; Hirama, M. Rationally designed synthetic haptens to generate anti-ciguatoxin monoclonal antibodies, and development of a practical sandwich ELISA to detect ciguatoxins. *Toxins* **2019**, *11*, 533. [CrossRef] [PubMed]
320. Leonardo, S.; Gaiani, G.; Tsumuraya, T.; Hirama, M.; Turquet, J.; Sagristá, N.; Rambla-Alegre, M.; Flores, C.; Caixach, J.; Diogène, J.; et al. Addressing the analytical challenges for the detection of ciguatoxins using an electrochemical biosensor. *Anal. Chem.* **2020**, *92*, 4858–4865. [CrossRef] [PubMed]
321. Bagnis, R.; Bennett, J.; Barsinas, M.; Drollet, J.H.; Jacquet, G.; Legrand, A.M.; Cruchet, P.H.; Pascal, H. Correlation between Ciguateric Fish and Damage to Reefs in the Gambier Islands (French Polynesia). In Proceedings of the 6th International Coral Reef Symposium Executive Committee, Townsville, Australia, 1 January 1988; Volume 2, pp. 195–200.
322. Belliveau, S.A.; Paul, V.J. Effects of herbivory and nutrients on the early colonization of crustose coralline and fleshy algae. *Mar. Ecol. Prog. Ser.* **2002**, *232*, 105–114. [CrossRef]
323. Stuart-Smith, R.D.; Brown, C.J.; Ceccarelli, D.M.; Edgar, G.J. Ecosystem restructuring along the Great Barrier Reef following mass coral bleaching. *Nature* **2018**, *560*, 92–96. [CrossRef]
324. Bellwood, D.R.; Pratchett, M.S.; Morrison, T.H.; Gurney, G.G.; Highes, T.P.; Álvarez-Romero, J.G.; Day, J.C.; Grantham, R.; Grech, A.; Hoey, A.S.; et al. Coral reef conservation in the Anthropocene: Confronting spatial mismatches and prioritizing functions. *Biol. Cons.* **2019**, *236*, 604–615. [CrossRef]
325. Cheal, A.J.; Emslie, M.; MacNeil, A.; Miller, I.; Sweatman, H. Spatial variation in the functional characteristics of herbivorous fish communities and the resilience of coral reefs. *Ecol. Appl.* **2013**, *23*, 174–188. [CrossRef]
326. Turner, M.G.; Calder, W.J.; Cumming, G.S.; Hughes, T.P.; Jentsch, A.; LaDeau, S.L.; Lenton, T.M.; Shuman, B.N.; Turetsky, M.R.; Ratajczak, Z.; et al. Climate change, ecosystems and abrupt change: Science priorities. *Phil. Trans. R. Soc. B* **2020**, *375*, 20190105. [CrossRef]
327. Kaly, U.L.; Jones, G.P. Test of the effect of disturbance on ciguatera in Tuvalu. *Mem. Qld Mus.* **1994**, *34*, 523–532.
328. Robinson, J.P.W.; McDevitt-Irwin, J.; Dajka, J.-C.; Hadj-Hammou, J.; Howlett, S.; Graba-Landry, A.; Hoey, A.S.; Nash, K.L.; Wilson, S.K.; Graham, N.A.J. Habitat and fishing control grazing potential on coral reefs. *Funct. Ecol.* **2019**, *34*, 240–251. [CrossRef]
329. McClure, E.C.; Richardson, L.E.; Graba-Landry, A.; Loffler, Z.; Russ, G.R.; Hoey, A.S. Cross-shelf differences in the response of herbivorous fish assemblage to severe environmental disturbances. *Diversity* **2019**, *11*, 23. [CrossRef]
330. Hales, S.; Weinstein, P.; Woodward, A. Ciguatera (fish poisoning), El Niño, and Pacific sea surface temperatures. *Ecosyst. Health* **1999**, *5*, 20–25. [CrossRef]
331. Chateau-Degat, M.-L.; Chinain, M.; Cerf, N.; Gingras, S.; Hubert, B.; Dewailly, É. Seawater temperature, *Gambierdiscus* spp. variability and incidence of ciguatera in French Polynesia. *Harmful Algae* **2005**, *4*, 1053–1062. [CrossRef]
332. Llewellyn, L.E. Revisiting the association between sea surface temperature and the epidemiology of fish poisoning in the South Pacific: Reassessing the link between ciguatera and climate change. *Toxicon* **2010**, *56*, 691–697. [CrossRef]
333. Tester, P.A.; Feldman, R.L.; Nau, A.W.; Kibler, S.R.; Litaker, R.W. Ciguatera fish poisoning and sea surface temperatures in the Caribbean Sea and the West Indies. *Toxicon* **2010**, *56*, 698–710. [CrossRef] [PubMed]
334. Heimann, K.; Capper, A.; Sparrow, L. Ocean surface warming: Impact on toxic benthic dinoflagellates causing ciguatera. In *The Encyclopaedia of Life Sciences A23373*; Wiley Publishing: New York, NY, USA, 2011.
335. Wells, M.L.; Karlson, B.; Wulff, A.; Kudela, R.; Trick, C.; Asnaghi, V.; Berdalet, E.; Cochlan, W.; Davidson, K.; De Rijcke, M.; et al. Future HAB science: Directions and challenges in a changing climate. *Harmful Algae* **2020**, *91*, 101632. [CrossRef]
336. Hallegraeff, G.M.; Fraga, S. Bloom dynamics of the toxic dinoflagellate *Gymnodinium catenatum*, with emphasis on Tasmanian and Spanish coastal waters. In *Physiological Ecology of Harmful Algal Blooms*; Anderson, D.M., Cembella, A.D., Hallegraeff, G.M., Eds.; Springer: Berlin/Heidelberg, Germany, 1998; pp. 59–80.
337. Holmes, M.J.; Bolch, C.J.S.; Green, D.H.; Cembella, A.D.; Teo, S.L.M. Singapore isolates of the dinoflagellate *Gymnodinium catenatum* (Dinophyceae) produce a unique profile of paralytic shellfish poisoning toxins. *J. Phycol.* **2002**, *38*, 96–106. [CrossRef]
338. Lewis, R.J. Socioeconomic impacts and management ciguatera in the Pacific. *Bull. Path. Soc. Exot.* **1992**, *85*, 427–434.
339. Zheng, L.; Gatti, C.M.; Gamarro, E.G.; Suzuki, A.; Teah, H.Y. Modeling the time-lag effect of sea surface temperatures on ciguatera poisoning in the South Pacific: Implications for surveillance and response. *Toxicon* **2020**, *182*, 21–29. [CrossRef] [PubMed]

Article

First Report of Domoic Acid Production from *Pseudo-nitzschia multistriata* in Paracas Bay (Peru)

Cecil Tenorio [1,2,*], Gonzalo Álvarez [2,3,4,*], Sonia Quijano-Scheggia [5], Melissa Perez-Alania [6], Natalia Arakaki [1], Michael Araya [4], Francisco Álvarez [3], Juan Blanco [7] and Eduardo Uribe [2,3]

1   Banco de Germoplasma de Organismos Acuáticos, Instituto del Mar del Perú, Callao 07021, Peru; narakaki@imarpe.gob.pe

2   Doctorado en Acuicultura, Programa Cooperativo Universidad de Chile, Pontificia Universidad Católica de Valparaíso, Coquimbo 17811421, Chile; euribe@ucn.cl

3   Facultad de Ciencias del Mar, Departamento de Acuicultura, Universidad Católica del Norte, Coquimbo 1281, Chile; falvarezsego@gmail.com

4   Centro de Investigación y Desarrollo Tecnológico en Algas (CIDTA), Facultad de Ciencias del Mar, Universidad Católica del Norte, Coquimbo 1281, Chile; mmaraya@ucn.cl

5   Centro Universitario de Investigaciones Oceanológicas, Universidad de Colima, Manzanillo 28218, Mexico; quijano@ucol.mx

6   Facultad de Ciencias, Universidad Nacional Agraria La Molina, Avenida La Universidad s/n, La Molina, Lima 15024, Peru; mlperal17@gmail.com

7   Centro de Investigacións Mariñas (Xunta de Galicia), 36620 Vilanova de Arousa, Pontevedra, Spain; juan.carlos.blanco.perez@xunta.gal

*   Correspondence: ltenorio@imarpe.gob.pe (C.T.); gmalvarez@ucn.cl (G.Á.); Tel.: +511-2088650 (C.T.); +56-2-209766 (G.Á.)

Citation: Tenorio, C.; Álvarez, G.; Quijano-Scheggia, S.; Perez-Alania, M.; Arakaki, N.; Araya, M.; Álvarez, F.; Blanco, J.; Uribe, E. First Report of Domoic Acid Production from *Pseudo-nitzschia multistriata* in Paracas Bay (Peru). *Toxins* **2021**, *13*, 408. https://doi.org/10.3390/toxins13060408

Received: 12 May 2021
Accepted: 7 June 2021
Published: 9 June 2021

Publisher's Note: MDPI stays neutral with regard to jurisdictional claims in published maps and institutional affiliations.

Abstract: The Peruvian sea is one of the most productive ecosystems in the world. Phytoplankton production provides food for fish, mammals, mollusks and birds. This trophic network is affected by the presence of toxic phytoplankton species. In July 2017, samples of phytoplankton were obtained from Paracas Bay, an important zone for scallop (*Argopecten purpuratus*) aquaculture in Peru. Morphological analysis revealed the presence of the genus *Pseudo-nitzschia*, which was isolated and cultivated in laboratory conditions. Subsequently, the monoclonal cultures were observed by scanning electron microscopy (SEM), and identified as *P. multistriata*, based on both the morphological characteristics, and internal transcribed spacers region (ITS2) sequence phylogenetic analysis. Toxin analysis using liquid chromatography (LC) with high-resolution mass spectrometry (HRMS) revealed the presence of domoic acid (DA) with an estimated amount of 0.004 to 0.010 pg cell^{-1}. This is the first report of DA from the coastal waters of Peru and its detection in *P. multistriata* indicates that it is a potential risk. Based on our results, routine monitoring of this genus should be considered in order to ensure public health.

Keywords: harmful algae; amnesic shellfish poisoning; ITS2; *Argopecten purpuratus*; scallop

Key Contribution: The presence of *P. multistriata*, domoic acid-producing species in the phytoplankton communities of Paracas Bay. Relevant information on the monitoring of harmful phytoplankton species along the Peruvian coast.

1. Introduction

Marine planktonic diatoms of the genus *Pseudo-nitzschia* are found in polar, warm and tropical regions; most of their species are cosmopolitan [1–3]. Currently, 56 species [4,5] of this genus have been reported, 26 of them producing the neurotoxic compound domoic acid (DA) [6]. Examination of the morphology by optical microscopy is frequently inconclusive, and for this reason scanning and/or transmission electron microscopy is also required. Additionally, molecular analysis of ITS2 [7] of the nuclear encoded ribosomal DNA can identify the species at the molecular level and differentiate between cryptic and pseudo-cryptic species [2,8–12].

The first intoxication in humans by DA was reported at Prince Edward Island, Canada, in 1987. More than 100 people reported becoming ill and at least three people died after eating blue mussels (*Mytilus edulis*) [13]. Digestive problems and short-term memory loss were the main symptoms of this intoxication, these led to the syndrome being named amnesic shellfish poisoning (ASP) [13–15]. Since the report of the first outbreak detected in Canada produced by the diatom *Pseudo-nitzschia multiseries*, episodes of DA have been recorded in many areas around the world [2,3]. Besides its effects on humans, DA also has severe effects on the trophic transfer between harmful microalgae, filter feeders and mollusks [16–20], spreading to fish [21] and causing mortality of birds and marine mammals [2,22–24].

Blooms of *Pseudo-nitzschia* spp., which produce DA, generate large economic losses due to long periods of closures for recreational and commercial fisheries or marketing of aquaculture products. Recently, these toxic outbreaks have been reported in the United States from 2015 to 2016, affecting the fisheries of the razor clam (*Siliqua patula*), Dungeness crab (*Metacarcinus magister*) and rock crabs (*Cancer productus*, *Metacarcinus anthonyi* and *Romaleon antennarium*) [3,24–26]. Likewise, in Europe, the DA-producing diatoms of genus *Pseudo-nitzschia* have led to bans on harvesting the natural populations of the king scallop *Pecten maximus*, and to the discouragement of aquaculture efforts for this species [27,28], due to its high capability to retain the toxin [29,30].

The coastal upwelling system of Peru constitutes a large part of the Humboldt Current system and is considered one of the most productive regions in the world, fixing 3000–4000 mg C m^{-2} d^{-1} [31–34]. Due to this high productivity, the area is susceptible to harmful algal blooms [2,35,36]. For Peruvian oceanic and coastal areas, the first report is from Hasle [37], who described the presence of *Pseudo-nitzschia pungens* (Grunow ex Cleve) Hasle and *P. australis* Frenguelli (as *P. pseudo seriata* G.R. Hasle) [2,37]. More recently, Tenorio, et al. [38] reported the presence of a non-toxic strain of *P. subpacifica* (Hasle) Hasle on the central coast of Peru, between San Lorenzo Island, Callao (12°03′S), and Paracas Bay, Ica (13°49′S). In northern Chile, within the framework of the Molluscan Shellfish Safety Program of the National Fisheries and Aquaculture Service (SERNAPESCA), elevated levels of DA have been detected in shellfish from many of the primary scallop *Argopecten purpuratus* aquaculture sites [39]. The blooms associated with those events have been dominated by diatom *Pseudo-nitzschia australis* with densities around 1.6×10^6 cell L^{-1} [40].

In Peru, during the period 2011–2012, intoxication of fur seals (*Arctocephalus australis*) and sea lions (*Otaria byronia*) was reported in San Juan de Marcona, Ica (15°20′S), associated with the detection of DA in feces of these marine mammals. During this episode, *Pseudo-nitzschia* spp. were detected with a maximum of 88,580 cell L^{-1} in Paracas Bay sampling station (distance of ~155 km to the north) [41]. Unfortunately, there is no additional information about the species that formed the *Pseudo-nitzschia* assemblage.

Paracas Bay is a traditional aquaculture area of the scallop *A. purpuratus*, the most important bivalve species in Peru [42]. To date, within the framework of the Molluscan Shellfish Control Program run by the National Fisheries Health Organization of Peru (SANIPES), there is no information about the presence of DA in this bivalve. Nevertheless, the detection of DA in marine mammals in nearby areas indicates that the toxin is a potential risk to aquaculture and suggests that more research is necessary in order to identify different *Pseudo-nitzschia* species and their capability to produce DA on the coast of Peru.

In 2017, phytoplankton were collected in Paracas Bay to establish monoclonal cultures of *Pseudo-nitzschia* spp., for their morphological, molecular and toxicological characterization, in order to understand the potentially toxic species of the genus *Pseudo-nitzschia* in Peruvian waters.

2. Results

2.1. Morphological Analysis

The isolated strain was morphologically identified as *Pseudo-nitzschia multistriata* (H. Takano) H. Takano. The cells of strain IMP-BG 440 (Figure 1) had a sigmoid shape in girdle view, and formed stepped chains, with up to four in length and an overlap of 1/6 of the total cell length. The cells were symmetrical and broad lanceolate in valve view. The apical axis ranged from 40 to 48 μm, while the transapical axis of the valves ranged from 3.20 to 4.30 μm. A large central interspace and a central nodule were absent. Valves contained 24 to 28 fibulae and 40 to 42 striae per 10 μm. Striae were formed by 2–3 rows of poroids with a density between 8 to 13 per 1 μm.

Figure 1. *Pseudo-nitzschia multistriata* (MEB), (**A**) whole valve; (**B**) valve end (2 μm); (**C**) mid valve, no central interspace (5 μm).

2.2. Molecular Analysis

Final alignment yielded 414 characters and comprised 128 ITS2 sequences, including a short sequence (MZ312514, 132 pb) of the strain IMP-BG 440. Phylogenetic analysis of ITS2 sequences using maximum likelihood (ML) and Bayesian inference (BI) showed congruent topologies (Figure 2). Within the genus *Pseudo-nitzschia*, a well-supported monophyletic clade (BI/ML, 1/100) corresponded to *P. multistriata*, and 11 strains from Australia, Malaysia, Taiwan, Japan, South Korea, China, Greece, Italy, Spain and Portugal. The clade containing *P. multistriata* also contained species of *Pseudo-nitzschia* from France (*P. americana*) and Malaysia (*P. braziliana*) with low support (BI, 0.73). This *Pseudo-nitzschia* clade was positioned within a larger unsupported clade (BI, 0.70) containing *Pseudo-nitzschia* species from Malaysia (genera type *P. pungens*), Japan and the USA (*P. multiseries*). Additionally, the phylogenetic tree shows that *P. multistriata* is grouped with other species of *Pseudo-nitzschia* with different levels of support. However, five sequences of *Fragilariopsis* from Artic, Antarctica and the USA form a supported clade within species of *Pseudo-nitzschia* (BI/ML, 0.99/64).

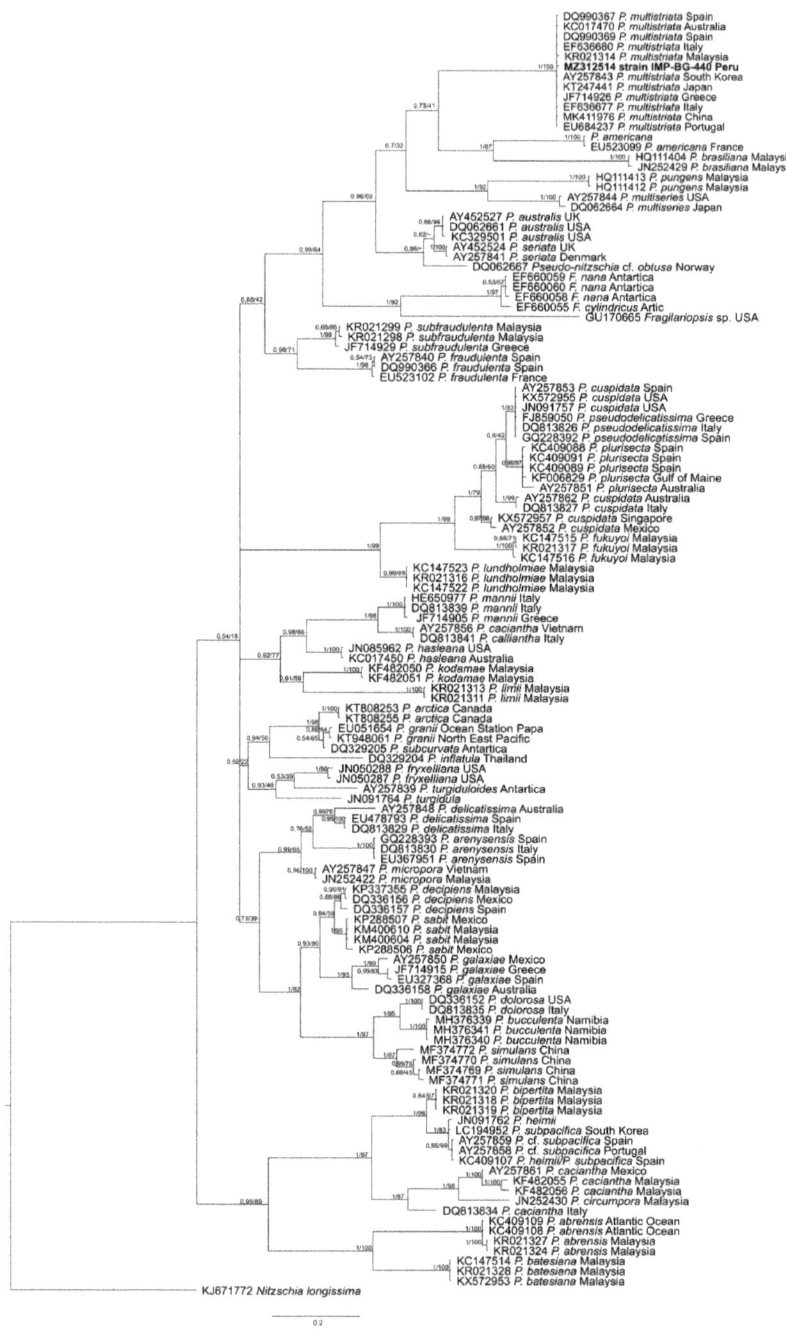

Figure 2. *Pseudo-nitzschia* Bayesian tree based on ITS2 sequences. Numbers above lines represent BI posterior probabilities/ML bootstrap values. "-" indicates a different phylogeny structure for ML analysis. Boldface indicates the studied strain as *P. multistriata*. Phylogenetic ITS2 trees (BI and ML) showed six general groups. The taxon *P. multistriata* is included in one of these groups comprising also *P. americana* + *P. brasiliana*, *P. pungens* + *P. multiseries*, *P. australis* + *P.* seriata + *P.* cf. *obtusa*, *Fragilariopsis nana* + *F. cylindricus* + *Fragilariopsis* sp. and *P. subfraudulenta* + *P. fraudulenta*.

2.3. Toxin Analysis

Analysis of extracts (*n* = 3) of the strain IMP-BG 440 showed that it contained domoic acid (Figure 3). Toxin analysis by LC-HRMS showed a chromatographic peak with a retention time of 6.60 min corresponding to the ion [M+H]$^+$ 312.1449 *m/z* (mass deviation: 0.64 ppm). The fragmentation mass spectrum of the ion [M+H]$^+$ 312.1449 *m/z* confirmed the identification of domoic acid (DA) because of its characteristic fragment MS/MS at 294.1332, 266.1384, 248.1277 and 220.1329 *m/z* (mass deviation in Table S1; Supplementary Materials). The estimated amount was between 0.004 to 0.010 pg cell^{-1}.

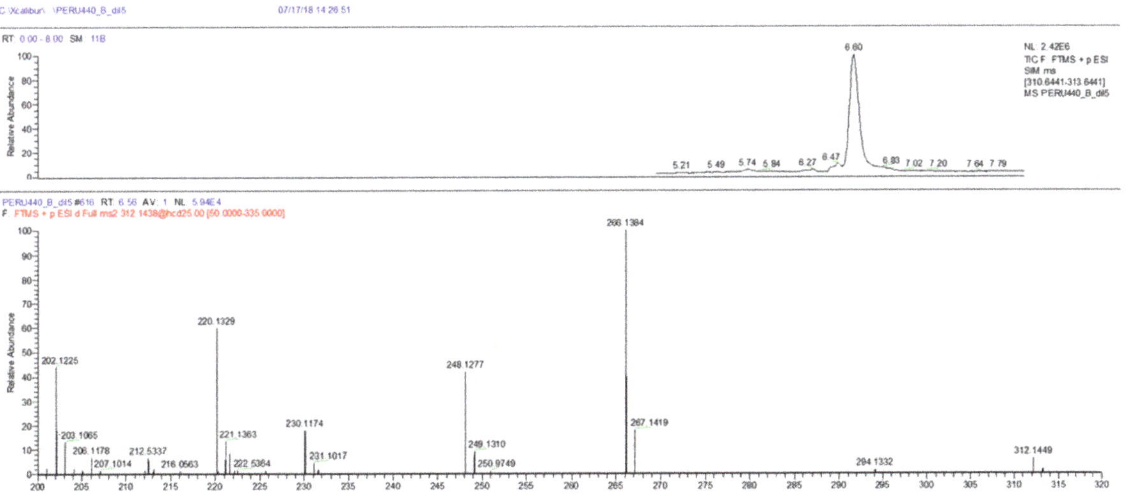

Figure 3. SIM chromatogram (upper panel) and mass spectrum (lower panel) of DA in *Pseudo-nitzschia multistriata* culture.

3. Discussion

The species of the genus *Pseudo-nitzschia* are distributed throughout all the coasts of the world [1–3]. The present study provides the first report of *Pseudo-nitzschia multistriata* in Peruvian coastal waters and, as far as we know, in the southeastern Pacific area. The presence of this species has been reported from different geographical locations around the world (Table S2, Supplementary Materials).

The morphological examination of *Pseudo-nitzschia* cells (Table S2, Supplementary Materials) from the obtained cultures agrees in length of apical axis with the description of strains of *P. multistriata* from China [43], Tunisia [44], Catalan Coast [45,46], Gulf of Naples [47,48] and the Western Adriatic Sea [49]. Similarly, the width of the transapical axis corresponds to descriptions of cells from Ria de Aveiro, Portugal [50], Tokyo bay [51], New Zealand [52], Mexico [53] and Uruguay [54]. However, the length of the apical axis and the width of the transapical axis of the cell do not match the first description made by Takano [55], given that those were smaller.

The number of fibulae of Paracas strains was close to descriptions of cells from Ria de Aveiro, Portugal [50], Greek coastal waters [56], Gulf Naples, Italy [57] and Gulf of Trieste, Northern Adriatic Sea, Italy [58]. Finally, the striae were formed by 2–3 rows of poroids and their density were similar to those described in cells from Fukukoka Bay, Japan [55] and the Sea of Japan [59].

Phylogenetic analysis of ITS2 sequences support the morphological identification of the strain isolated from Paracas Bay as *P. multistriata*. This strain from Peru is situated with *P. multistriata* strains from Australia, Malaysia, South Korea, Taiwan, Japan, China and Europe, forming a well-supported monophyletic clade (BI/ML, 1/100). The phylogenetic tree also shows that *P. multistriata* is grouped with *P. brasiliana* and *P. americana*, within a clade comprising *P. pungens* (genera type), *P. multiseries*, *P. australis*, *P. seriata* and *P. obtuse*. Previous phylogenetic analyses of ITS2 sequences had pointed to different relationships of *P. multistriata* with *P. americana*, *P. brasiliana*, and *P. pungens*. Huang, et al. [60] showed that *P. americana* is placed on the base of the clade, not clustered with *P. multistriata* and *P. brasiliana*. On the contrary, Lim, et al. [61] showed that *P. multiseries* and *P. pungens* are placed at the base of the tree. Morphological characteristics have been included by Lim, et al. [61] as representative of some species of the *Pseudo-nitzschia* clade. Thus, morphological characters, 2–4 rows of poroids, the absence of a central nodule and the lower number of fibulae versus striae in 10 μm, were observed in *P. multistriata* from Japan [62], matching the description of the strain of *P. multistriata* (IMP-BG 440) from Peru.

This study confirms *P. multistriata* as an unequivocal source of domoic acid (DA) on the coast of Peru. The strain IMP-BG 440 tested was able to produce the toxin in culture with a concentration between 0.004 and 0.010 pg cell^{-1}, which is comparable to those reported by Pistocchi, et al. [49] (0.003 pg cell^{-1}) in a strain obtained from the Adriatic Sea that was cultured under similar conditions (16–18 °C; 60–100 μmol photons m^{-2} s^{-1}). These concentrations were lower than the values reported in Australian strains by Ajani, et al. [63] (1–11 pg cell^{-1}), Rhodes, et al. [64] (1.5 pg cell^{-1}) and in Italian strains registered by Amato, et al. [65] (0.28 pg cell^{-1}), Orsini, et al. [47] (0.69 pg cell^{-1}) and Sarno [48] (0.65 pg cell^{-1}).

The Humboldt Current system (HCS) is considered one of the most productive fishery regions in the world oceans [33,34,66,67]. As mentioned above, due to its high productivity, this upwelling area is susceptible to harmful algal blooms (HABs) [35,36]. In this context, other toxic *Pseudo-nitzschia* species have been reported in the HCS, specifically on the northern Chilean coast [68,69]. In some cases, DA concentrations have exceeded the regulatory limit (20 mg·kg^{-1}) and the harvesting of scallops (*A. purpuratus*), from aquaculture sites, has therefore been banned [40]. The DA content in *P. multistriata* (strain IMP-BG 440) was substantially lower than those reported in *P. australis* (1.74 pg cell^{-1}) for the southeastern Pacific; however, it was close to the value of *P. calliantha* (0.01 pg cell^{-1}) [39]. The low content of DA in *P. multistriata* in Peruvian waters could be one of the reasons that there has not been any detection of this toxin in scallops cultivated in Paracas Bay in the framework of the Molluscan Shellfish Control Program run by SANIPES. A second reason could be the rapid DA depuration of this bivalve in the natural environment as has been demonstrated by Álvarez, et al. [70] in scallops cultivated in Tongoy Bay, Chile. However, the information provided by this work should be taken into consideration in the development of the Molluscan Shellfish Control Program ran by SANIPES [71].

Regarding the intoxication of marine mammals with low levels of DA on the Peruvian coast [41], it is clear that *P. multistriata* could be involved. However, with the available information we cannot discard the possibility that other species of *Pseudo-nitzschia* or more toxic strains than the one found in this study could be the principal cause of pinniped intoxication. Finally, more research is needed to find other toxic species, as well as the roll of different environmental variables in the production of DA in different strains of *P. multistriata* obtained along the Peruvian coast.

4. Conclusions

Pseudo-nitzschia multistriata has been identified from the Peruvian coast based on morphological, phylogenetic and molecular evidence. This is the first report of this species for the Southeast Pacific. The species is confirmed to be a producer of DA which makes it the first known DA producer from Peruvian waters. The presence of toxic *P. multistriata* is a potential risk for mammals, making it necessary to routinely monitor this species in order to protect public health, as well as the ecosystem of Paracas Bay.

5. Materials and Methods

5.1. Biological Samples and Establishment of Cultures

Phytoplankton samples were obtained periodically in August 2017 in Paracas Bay (13°49′S, 76°17′O) (Figure 4) with temperatures of around 15 to 17 °C and salinity of 35.

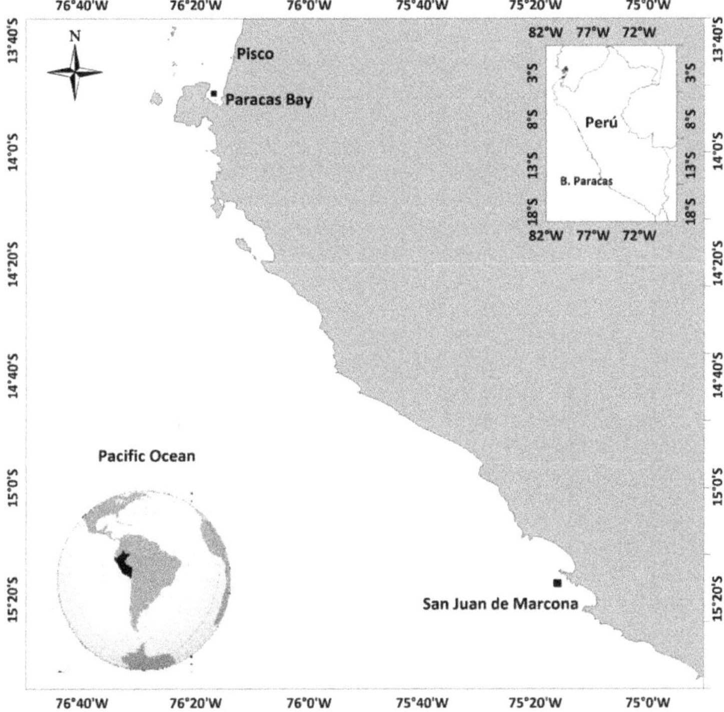

Figure 4. Location of the sampling station Paracas Bay, Peru.

Samples were collected using vertical net hauls (20 μm mesh), stored in 250 mL glass bottles and transported to the laboratory in the dark and chilled on ice (10 °C). To establish cultures of the *Pseudo-nitzschia* species, single chains of *Pseudo-nitzschia* cells were picked by micropipette and transferred to multi-well culture plates (hydrobios, Germany) filled with 2 mL of L1 culture medium [72] with a salinity of 30. The plates were maintained at 15 °C in a 12:12-h light: dark cycle, with a photon flux of 60 μmol photons m^{-2} s^{-1}. Established cultures were transferred to borosilicate Erlenmeyer flasks with 150 mL of f/2 medium and grown at 15 °C in a 12:12-h light: dark cycle, with a photon flux of 80 μmol photons m^{-2} s^{-1}. Mass cultures were grown in borosilicate bottles with 1 L of f/2 medium in triplicate under the above conditions. Two milliliter aliquots of the cultures were preserved with Lugol's solution for the direct count of the cells. The cell densities in the samples were quantified by the Utermöhl method described by Hasle [73].

5.2. Morphological Analysis

Scanning electron microscopy (SEM) was used to perform detailed morphological analyses of the *Pseudo-nitzschia* cells. Organic matter was removed from the frustules following the methodology described by Lundholm, et al. [74]. The clean material was retained on a 5.0 μm membrane filter (Isopore Merck KGaA, Darmstadt, Germany), and washed with distilled water to remove salts and preservatives. After being airdried overnight, specimens were gold-coated in a JEOL JFC-1100 (JEOL Ltd., Tokyo, Japan) and observed with a Hitachi SU3500 scanning electron microscope (Hitachi High-Technologies Corporation,

Tokyo, Japan). *Pseudo-nitzschia* cells were examined for morphometric characteristics that included width and length of the valve, density of striae, fibulae and poroids.

5.3. Molecular and Phylogenetic Analysis

Molecular identification of the strain of the *Pseudo-nitzschia* genus was performed by analyzing sequences of the internal transcribed spacer two (ITS2) region. When initial cultures of the strain reached the exponential growth phase, cells were concentrated by successive centrifugations and frozen at $-80\ ^\circ$C prior to DNA extraction (24 h). Total genomic DNA was extracted following the cetyltrimethylammonium bromide (CTAB) method [75]. Part of the internal transcribed spacer (ITS) region was amplified by PCR, using ITS1/ITS4 primers (BIOSEARCH TECHNOLOGIES, Petaluma, CA, USA) [76]. The polymerase chain reaction (PCR) conditions for ITS include pre denaturation at 95 $^\circ$C for 3 min, followed by 39 cycles of 95 $^\circ$C for 30 s, 45 $^\circ$C for 30 s and 72 $^\circ$C for 50 s; and finally, 71 $^\circ$C for 7 min. The amplicons were visualized on agarose gel (1.2%) (Invitrogen, Carlsbad, USA), purified and sequenced for one strand by Macrogen Inc. (South Korea).

Sequences obtained were checked in BIOEDIT v.7.0.5.3 (Raleigh, NC, USA., 2005) [77] and compared in the GenBank public database using the Basic Local Alignment Search Tool BLASTn. The data block used in the molecular analyses consisted of 128 ITS sequences (Table S3), including the one sequence of *Pseudo-nitzschia* obtained in this study, sequences of *Pseudo-nitzschia* and 5 sequences of *Fragilariopsis* available in the public database and a sequence of *Nitzschia longissima* as an outgroup. The alignment was constructed using the Muscle algorithm in MEGA7v.7.0.26 (Philadelphia, PA, USA., 2016) [78], checked visually, corrected and trimmed using MEGA7 so that it only contained sequences of the ITS2 region. Final alignment was independently analyzed using maximum likelihood (ML) and Bayesian inference (BI). The best evolutionary models for ML and BI were calculated in jModelTest 2 (Spain, 2012) [79] using the Akaike information criterion (AIC) and the Bayesian information criterion (BIC), respectively. ML analysis was carried out in RAxML v.8.2.X (Karlsruhe, Germany, 2014) [80] using the graphic user interface raxmlGUI v.1.5 5b1 (Frankfurt, Germany, 2012) [81] with the selected model (GTR+I+G) and 1000 bootstrap replications. BI was carried out in MrBayes v.3.2.6 [82] with the selected model (HKY+I+G), two runs of 10 million Markov chain Monte Carlo generations each with 1 cold chain and 3 heated chains, sampling and printing every 1000 generations. The convergence of the runs was checked using Tracer v.1.6.0 (Edinburgh, UK, 2014). A consensus tree was constructed after a burn-in of 25%, and posterior probabilities were estimated.

5.4. Sample Preparation and Toxin Analysis

A 1 L sample from the culture of *Pseudo-nitzschia* spp. (densities ranged from 267,170 to 305,691 cell mL^{-1}) was taken in the stationary phase of growth. The sample was concentrated by centrifugation at 4000 g for 10 min with a centrifuge (Hettich Rotina 420R, Germany). The obtained pellets were mixed with 10 mL of aqueous methanol (Merck KGaA, Darmstadt, Germany) (50%, *v/v*) and the cells disrupted with a Branson Ultrasonic 250 (Danbury, CT, USA). The extract was clarified by centrifugation at 10,000 g for 20 min (Centurion K2015R, Centurion Scientific Ltd., Stoughton, West Sussex, UK). A one-milliliter aliquot was filtered through 0.22 μm Clarinert nylon syringe filters (13 mm diameter) (Bonna-Agela technologies, Torrance, CA, USA) and stored in an autosampler vial at $-20\ ^\circ$C until analysis. The presence of DA (cellular content) in the extracts was determined following the method described by de la Iglesia, et al. [83] with modifications. The instrumental analysis was developed using a Dionex Ultimate 3000 UHPLC system (Thermo Fisher Scientific, Sunnyvale, CA, USA). A reversed-phase HPLC column Kinetex C18 (50 mm × 2.1 mm; 2.6 μm) with an Ultra Guard column C18 from Phenomenex (Torrance, CA, USA) was used. The flow rate was set to 0.28 mL min^{-1}, and the injection volume was 10 μL. Mobile phases A and B were water (Milli-Q) and MeOH, respectively, both containing 0.1% formic acid. The following gradient was used to achieve the chromatographic separation: 100% phase A, held for the first 0.5 min. Afterwards, separation was carried out at 12.5% B up to 3 min, decreased to 3% B over 7 min and then returned to the initial conditions over 2 min. The total analysis run time was 12 min.

The detection of DA was carried out by a high-resolution mass spectrometer Q Exactive Focus equipped with an electrospray interphase HESI II (Thermo Fisher Scientific, Sunnyvale, CA, USA). The interface was operated in positive ionization mode with a spray voltage of 3.5 kV. The temperature of the ion transfer tube and the HESI II vaporizer were set at 250 °C. Nitrogen (>99.98%) was employed as a sheath gas and auxiliary gas at pressures of 20 and 10 arbitrary units, respectively. Data were acquired in selected ion monitoring (SIM) and data-dependent (ddMS^2) acquisition modes (for quantification and confirmation, respectively). In SIM mode, the mass was set to 312.1404 m/z, the scan mass range was set at m/z 100–1000 with a mass resolution of 70,000, the automatic gain control (AGC) was set at 5×10^4 and the maximum injection time (IT) 3000 ms. For dds^2 the mass resolution was set at 70,000, AGC at 5×10^4 and IT 3000 ms. In both cases, the isolation windows were 2 m/z. DA was quantified by external calibration, using a DA-certified reference solution (CRM-DA-g) (NRC, CNRC, Canada). Limits of detection were calculated as the average concentration of DA producing a signal-to-noise ratio (S/N) of 3 and corresponding to 0.5 ng mL^{-1}, while the limit of quantification of the method was 2 ng mL^{-1}.

Supplementary Materials: The following are available online at https://www.mdpi.com/article/10.3390/toxins13060408/s1, Table S1: Accurate mass and mass deviation (ppm) of domoic acid and its main fragments. Table S2: Comparison of morphometric data between Peruvian strain of *Pseudo-nitzschia multistriata* with strains obtained from different locations around the world, Table S3: List of sequences of *Pseudo-nitzschia*, *Fragilariopsis* and *Nitzschia* strains included in the molecular analysis. Species, locality, strain code and GenBank accession numbers for the ITS2 gene marker.

Author Contributions: Conceptualization, C.T., G.Á., E.U., S.Q.-S. and J.B.; methodology, C.T., G.Á., E.U., S.Q.-S., M.A. and N.A.; software, C.T., S.Q.-S., N.A., M.P.-A., M.A. and G.Á.; investigation, C.T., G.Á., S.Q.-S., M.P.-A., N.A., M.A., F.Á., J.B. and E.U.; formal analysis, C.T., G.Á., S.Q.-S., E.U. and J.B.; visualization, G.Á., S.Q.-S., E.U. and J.B.; writing—original draft preparation, C.T., G.Á., M.A., E.U., S.Q.-S., E.U. and J.B.; writing—review and editing, G.Á., S.Q.-S. and J.B.; visualization, C.T., G.Á., M.A. and J.B.; supervision, G.Á., S.Q.-S., E.U. and J.B.; funding acquisition, G.Á. All authors have read and agreed to the published version of the manuscript.

Funding: This research was funded by the "ANID + FONDEF/PRIMER CONCURSO INVESTIGACIÓN TECNOLÓGICA TEMATICO ENSISTEMAS PESQUERO ACUICOLAS FRENTE A FLORECIMIENTOS ALGALES NOCIVOS FANS IDeA DEL FONDO DE FOMENTO AL DESARROLLO CIENTÍFICO Y TECNOLÓGICO, FONDEF/ANID 2017, IT17F10002" developed within the framework of a cooperation agreement between the Consellería do Mar, Xunta de Galicia, Spain and the Universidad Católica del Norte, Chile and the Budget National Program N°094 (PpR094): "Management and Development of Aquaculture"—PRODUCE, PERU, through the project "Technological Development in Aquaculture".

Institutional Review Board Statement: Not applicable.

Informed Consent Statement: Not applicable.

Data Availability Statement: Not applicable.

Acknowledgments: We acknowledge the project the FONDEQUIP150109-Chile.

Conflicts of Interest: The authors declare no conflict of interest.

References

1. Hasle, G.R. Are most of the domoic acid-producing species of the diatom genus Pseudo-nitzschia cosmopolites? *Harmful Algae* **2002**, *1*, 137–146. [CrossRef]
2. Trainer, V.L.; Bates, S.S.; Lundholm, N.; Thessen, A.E.; Cochlan, W.P.; Adams, N.G.; Trick, C.G. Pseudo-nitzschia physiological ecology, phylogeny, toxicity, monitoring and impacts on ecosystem health. *Harmful Algae* **2012**, *14*, 271–300. [CrossRef]
3. Bates, S.S.; Hubbard, K.A.; Lundholm, N.; Montresor, M.; Leaw, C.P. Pseudo-nitzschia, Nitzschia, and domoic acid: New research since 2011. *Harmful Algae* **2018**, *79*, 3–43. [CrossRef]

4. Chen, X.M.; Pang, J.X.; Huang, C.X.; Lundholm, N.; Teng, S.T.; Li, A.; Li, Y. Two New and Nontoxigenic Pseudo-nitzschia species (Bacillariophyceae) from Chinese Southeast Coastal Waters. *J. Phycol.* **2021**, *57*, 335–344. [CrossRef]

5. Guiry, M.D.; Guiry, G.M. AlgaeBase. Available online: http://www.algaebase.org (accessed on 1 March 2021).

6. Lundholm, N. Bacillariophyceae, in IOC-UNESCO Taxonomic Reference List of Harmful Micro Algae. Available online: http://www.marinespecies.org/hab (accessed on 25 February 2021).

7. Coleman, A.W. ITS2 is a double-edged tool for eukaryote evolutionary comparisons. *Trends Genet.* **2003**, *19*, 370–375. [CrossRef]

8. Quijano-Scheggia, S.I.; Garcés, E.; Lundholm, N.; Moestrup, Ø.; Andree, K.; Camp, J. Morphology, physiology, molecular phylogeny and sexual compatibility of the cryptic Pseudo-nitzschia delicatissima complex (Bacillariophyta), including the description of P. arenysensis sp. nov. *Phycologia* **2009**, *48*, 492–509. [CrossRef]

9. Quijano-Scheggia, S.I.; Olivos-Ortiz, A.; Garcia-Mendoza, E.; Sánchez-Bravo, Y.; Sosa-Avalos, R.; Salas Marias, N.; Lim, H.C. Phylogenetic relationships of Pseudo-nitzschia subpacifica (Bacillariophyceae) from the Mexican Pacific, and its production of domoic acid in culture. *PLoS ONE* **2020**, *15*, e0231902. [CrossRef] [PubMed]

10. Lundholm, N.; Moestrup, Ø.; Hasle, G.R.; Hoef-Emden, K. A study of the Pseudonitzschia pseudodelicatissima/cuspidata complex (Bacillariophyceae): What is P. pseudodelicatissima? *J. Phycol.* **2003**, *39*, 797–813. [CrossRef]

11. Lundholm, N.; Bates, S.S.; Baugh, K.A.; Bill, B.D.; Connell, L.B.; Léger, C.; Trainer, V.L. Cryptic and pseudo-cryptic diversity in diatoms—with descriptions of pseudo-nitzschia hasleana sp. nov. and p. fryxelliana sp. nov. 1. *J. Phycol.* **2012**, *48*, 436–454. [CrossRef]

12. Amato, A.; Kooistra, W.H.; Ghiron, J.H.L.; Mann, D.G.; Pröschold, T.; Montresor, M. Reproductive isolation among sympatric cryptic species in marine diatoms. *Protist* **2007**, *158*, 193–207. [CrossRef]

13. Bates, S.; Bird, C.J.; Freitas, A.d.; Foxall, R.; Gilgan, M.; Hanic, L.A.; Johnson, G.R.; Mc Culloch, A.; Odense, P.; Pocklington, R.; et al. Pennate diatom Nitzschia pungens as the primary source of domoic acid, a toxin in shellfish from eastern Prince Edward Island, Canada. *Can. J. Fish. Aquat. Sci.* **1989**, *46*, 1203–1215. [CrossRef]

14. Perl, T.M.; Bédard, L.; Kosatsky, T.; Hockin, J.C.; Todd, E.C.; Remis, R.S. An outbreak of toxic encephalopathy caused by eating mussels contaminated with domoic acid. *N Engl. J. Med.* **1990**, *322*, 1775–1780. [CrossRef]

15. Wright, J.; Boyd, R.; Freitas, A.d.; Falk, M.; Foxall, R.; Jamieson, W.; Laycock, M.; McCulloch, A.; McInnes, A.; Odense, P. Identification of domoic acid, a neuroexcitatory amino acid, in toxic mussels from eastern Prince Edward Island. *Can. J. Chem.* **1989**, *67*, 481–490. [CrossRef]

16. Bates, S. Toxic phytoplankton on the Canadian east coast: Implications for aquaculture. *Bull. Aquacult. Assoc. Can.* **1997**, *97*, 9–18.

17. La Barre, S.; Bates, S.S.; Quilliam, M.A. Domoic acid. In *Outstanding Marine Molecules: Chemistry, Biology, Analysis*; La Barre, S., Kornprobst, J.M., Eds.; Wiley-VCH Verlag GmbH & KgaA: Weinheim, Germany, 2014; pp. 189–216.

18. Goldberg, J.D. Domoic Acid in the Benthic Food Web of Monterey Bay, California. Master's Thesis, University Monterey Bay, Seaside, CA, USA, 2003.

19. Zabaglo, K.; Chrapusta, E.; Bober, B.; Kaminski, A.; Adamski, M.; Bialczyk, J. Environmental roles and biological activity of domoic acid: A review. *Algal Res.* **2016**, *13*, 94–101. [CrossRef]

20. Bates, S.S.; Garrison, D.L.; Horner, R.A. Bloom dynamics and physiology of domoic-acid-producing Pseudo-nitzschia species. *Nato Asi Ser. G Ecol. Sci.* **1998**, *41*, 267–292.

21. Lefebvre, K.A.; Frame, E.R.; Kendrick, P.S. Domoic acid and fish behavior: A review. *Harmful Algae* **2012**, *13*, 126–130. [CrossRef]

22. D'Agostino, V.C.; Degrati, M.; Sastre, V.; Santinelli, N.; Krock, B.; Krohn, T.; Dans, S.L.; Hoffmeyer, M.S. Domoic acid in a marine pelagic food web: Exposure of southern right whales Eubalaena australis to domoic acid on the Peninsula Valdes calving ground, Argentina. *Harmful Algae* **2017**, *68*, 248–257. [CrossRef]

23. Di Liberto, T. This summer's West Coast algal bloom was unusual. *What would Usual Look Like?* 30 September 2015.

24. McCabe, R.M.; Hickey, B.M.; Kudela, R.M.; Lefebvre, K.A.; Adams, N.G.; Bill, B.D.; Gulland, F.M.; Thomson, R.E.; Cochlan, W.P.; Trainer, V.L. An unprecedented coastwide toxic algal bloom linked to anomalous ocean conditions. *Geophys. Res. Lett.* **2016**, *43*, 10366–310376. [CrossRef]

25. Ritzman, J.; Brodbeck, A.; Brostrom, S.; McGrew, S.; Dreyer, S.; Klinger, T.; Moore, S.K. Economic and sociocultural impacts of fisheries closures in two fishing-dependent communities following the massive 2015 US West Coast harmful algal bloom. *Harmful Algae* **2018**, *80*, 35–45. [CrossRef]

26. Du, X.; Peterson, W.; Fisher, J.; Hunter, M.; Peterson, J. Initiation and development of a toxic and persistent Pseudo-nitzschia bloom off the Oregon coast in spring/summer 2015. *PLoS ONE* **2016**, *11*, e0163977. [CrossRef]

27. Arévalo, F.; Bermúdez de la Puente, M.; Salgado, C. Seguimiento de biotoxinas marinas en las Rías Gallegas: Control y evolución durante los años 1995–1996. *V Reunión Ibérica de Fitoplancton Tóxico y Biotoxinas.* ANFACO-CECOPESCA, Vigo **1997**, 90–101.

28. Blanco, J.; Acosta, C.; De La Puente, M.B.; Salgado, C. Depuration and anatomical distribution of the amnesic shellfish poisoning (ASP) toxin domoic acid in the king scallop Pecten maximus. *Aquat. Toxicol.* **2002**, *60*, 111–121. [CrossRef]

29. Mauriz, A.; Blanco, J. Distribution and linkage of domoic acid (amnesic shellfish poisoning toxins) in subcellular fractions of the digestive gland of the scallop Pecten maximus. *Toxicon* **2010**, *55*, 606–611. [CrossRef] [PubMed]

30. Blanco, J.; Mauríz, A.; Álvarez, G. Distribution of Domoic Acid in the Digestive Gland of the King Scallop Pecten maximus. *Toxins* **2020**, *12*, 371. [CrossRef] [PubMed]

31. Calienes, R.; Guillén, O.; Lostaunau, N. Variabilidad espacio-temporal de clorofila, producción primaria y nutrientes frente a la costa peruana. *Boletin Instituto del Mar del Peru* **1985**, *10*, 1–44.

32. Graco, M.; Ledesma, J.; Flores, G.; Giron, M. Nutrients, oxygen and biogeochemical processes in the Humboldt upwelling current system off Peru. *Rev. Peru. Biol* **2007**, *14*, 117–128.

33. Echevin, V.; Gévaudan, M.; Espinoza-Morriberón, D.; Tam, J.; Aumont, O.; Gutierrez, D.; Colas, F. Physical and biogeochemical impacts of RCP8. 5 scenario in the Peru upwelling system. *Biogeosciences* **2020**, *17*, 3317–3341. [CrossRef]

34. Bakun, A.; Weeks, S.J. The marine ecosystem off Peru: What are the secrets of its fishery productivity and what might its future hold? *Prog. Oceanogr.* **2008**, *79*, 290–299. [CrossRef]

35. Pitcher, G.C.; Jiménez, A.B.; Kudela, R.M.; Reguera, B. Harmful algal blooms in eastern boundary upwelling systems. *Oceanography* **2010**, *30*, 22. [CrossRef]

36. Trainer, V.L.; Pitcher, G.C.; Reguera, B.; Smayda, T.J. The distribution and impacts of harmful algal bloom species in eastern boundary upwelling systems. *Prog. Oceanogr.* **2010**, *85*, 33–52. [CrossRef]

37. Hasle, G.R. Nitzschia and Fragilariopsis species studied in the light and electron microscopes. II. The group Pseudonitzschia. *Skr. Nor. Vidensk-Akad. I. Mat.-Nat. Kl. Ny Ser.* **1965**, *18*, 1–45.

38. Tenorio, C.; Uribe, E.; Gil-Kodaka, P.; Blanco, J.; Álvarez, G. Morphological and toxicological studies ofPseudo-nitzschiaspecies from the central coast of Peru. *Diatom Res.* **2016**, *31*, 331–338. [CrossRef]

39. Álvarez, G.; Uribe, E.; Quijano-Scheggia, S.; López-Rivera, A.; Mariño, C.; Blanco, J. Domoic acid production by Pseudo-nitzschia australis and Pseudo-nitzschia calliantha isolated from North Chile. *Harmful Algae* **2009**, *8*, 938–945. [CrossRef]

40. Díaz, P.A.; Álvarez, A.; Varela, D.; Pérez-Santos, I.; Díaz, M.; Molinet, C.; Seguel, M.; Aguilera-Belmonte, A.; Guzmán, L.; Uribe, E. Impacts of harmful algal blooms on the aquaculture industry: Chile as a case study. *Perspect. Phycol* **2019**. [CrossRef]

41. Fire, S.E.; Adkesson, M.J.; Wang, Z.; Jankowski, G.; Cárdenas-Alayza, S.; Broadwater, M. Peruvian fur seals (Arctocephalus australis ssp.) and South American sea lions (Otaria byronia) in Peru are exposed to the harmful algal toxins domoic acid and okadaic acid. *Mar. Mammal Sci.* **2017**, *33*, 630–644. [CrossRef]

42. Kluger, L.C.; Taylor, M.H.; Wolff, M.; Stotz, W.; Mendo, J. From an open-access fishery to a regulated aquaculture business: The case of the most important Latin American bay scallop (Argopecten purpuratus). *Rev. Aquac.* **2019**, *11*, 187–203. [CrossRef]

43. Lü, S.; Li, Y.; Lundholm, N.; Ma, Y.; Ho, K. Diversity, taxonomy and biogeographical distribution of the genus Pseudo-nitzschia (Bacillariophyceae) in Guangdong coastal waters, South China Sea. *Nova Hedwig.* **2012**, *95*, 123–152. [CrossRef]

44. Sahraoui, I.; Grami, B.; Bates, S.S.; Bouchouicha, D.; Chikhaoui, M.A.; Mabrouk, H.H.; Hlaili, A.S. Response of potentially toxic Pseudo-nitzschia (Bacillariophyceae) populations and domoic acid to environmental conditions in a eutrophied, SW Mediterranean coastal lagoon (Tunisia). *Estuar. Coast. Shelf Sci.* **2012**, *102*, 95–104. [CrossRef]

45. Quijano-Scheggia, S.; Garcés, E.; Andree, K.B.; De la Iglesia, P.; Diogène, J.; Fortuño, J.M.; Camp, J. Pseudo-nitzschia species on the Catalan coast: Characterization and contribution to the current knowledge of the distribution of this genus in the Mediterranean Sea. *Sci. Mar.* **2010**, *74*, 395–410. [CrossRef]

46. Quijano-Sheggia, S.; Garcés, E.; Sampedro, N.; Van Lenning, K.; Flo Arcas, E.; Andree, K.; Fortuño Alós, J.M.; Camp, J. Identification and characterisation of the dominant Pseudo-nitzschia species (Bacillariophyceae) along the NE Spanish coast (Catalonia, NW Mediterranean). *Sci. Mar.* **2008**, *72*, 343–359.

47. Orsini, L.; Sarno, D.; Procaccini, G.; Poletti, R.; Dahlmann, J.; Montresor, M. Toxic Pseudo-nitzschia multistriata (Bacillariophyceae) from the Gulf of Naples: Morphology, toxin analysis and phylogenetic relationships with other Pseudo-nitzschia species. *Eur. J. Phycol.* **2002**, *37*, 247–257. [CrossRef]

48. Sarno, D. Production of domoic acid in another species of Pseudo-nitzschia: P. multistriata in the Gulf of Naples (Mediterranean Sea). *Harmful Algal News* **2000**, *21*, 5.

49. Pistocchi, R.; Guerrini, F.; Pezzolesi, L.; Riccardi, M.; Vanucci, S.; Ciminiello, P.; Dell'Aversano, C.; Forino, M.; Fattorusso, E.; Tartaglione, L. Toxin levels and profiles in microalgae from the North-Western Adriatic Sea—15 years of studies on cultured species. *Mar. Drugs* **2012**, *10*, 140–162. [CrossRef]

50. Churro, C.I.; Carreira, C.C.; Rodrigues, F.J.; Craveiro, S.C.; Calado, A.J.; Casteleyn, G.; Lundholm, N. Diversity and abundance of potentially toxic Pseudo-nitzschia Peragallo in Aveiro coastal lagoon, Portugal and description of a new variety, P. pungens var. aveirensis var. nov. *Diatom Res.* **2009**, *24*, 35–62. [CrossRef]

51. Yap-Dejeto, L.G.; Omura, T.; Nagahama, Y.; Fukuyo, Y. Observations of eleven Pseudo nitzschia species in Tokyo Bay, Japan. *La mer* **2010**, *48*, 1–16.

52. Rhodes, L.L.; Adamson, J.; Scholin, C. Pseudo-nitzschia multistriata(Bacillariophyceae) in New Zealand. *New Zealand J. Mar. Freshw. Res.* **2000**, *34*, 463–467. [CrossRef]

53. Rivera-Vilarelle, M.; Quijano-Scheggia, S.; Olivos-Ortiz, A.; Gaviño-Rodríguez, J.H.; Castro-Ochoa, F.; Reyes-Herrera, A. The genus Pseudo-nitzschia (Bacillariophyceae) in Manzanillo and Santiago Bays, Colima, Mexico. *Bot. Mar.* **2013**, *56*, 357–373. [CrossRef]

54. Méndez, S.M.; Ferrario, M.; Cefarelli, A.O. Description of toxigenic species of the genus Pseudo-nitzschia in coastal waters of Uruguay: Morphology and distribution. *Harmful Algae* **2012**, *19*, 53–60. [CrossRef]

55. Takano, H. Marine diatom Nitzschia multistriata sp. nov. common at inlets of southern Japan. *Diatom* **1993**, *8*, 39–41.

56. Moschandreou, K.K.; Baxevanis, A.D.; Katikou, P.; Papaefthimiou, D.; Nikolaidis, G.; Abatzopoulos, T.J. Inter- and intra-specific diversity of Pseudo-nitzschia (Bacillariophyceae) in the northeastern Mediterranean. *Eur. J. Phycol.* **2012**, *47*, 321–339. [CrossRef]

57. D'Alelio, D.; Amato, A.; Kooistra, W.H.; Procaccini, G.; Casotti, R.; Montresor, M. Internal transcribed spacer polymorphism in Pseudo-nitzschia multistriata (Bacillariophyceae) in the Gulf of Naples: Recent divergence or intraspecific hybridization? *Protist* **2009**, *160*, 9–20. [CrossRef]

58. Dermastia, T.T.; Cerino, F.; Stanković, D.; France, J.; Ramšak, A.; Tušek, M.Ž.; Beran, A.; Natali, V.; Cabrini, M.; Mozetič, P. Ecological time series and integrative taxonomy unveil seasonality and diversity of the toxic diatom Pseudo-nitzschia H. Peragallo in the northern Adriatic Sea. *Harmful Algae* **2020**, *93*, 101773. [CrossRef] [PubMed]

59. Stonik, I.V.; Orlova, T.Y.; Lundholm, N. Diversity of Pseudo-nitzschia H. Peragallo from the western North Pacific. *Diatom Res.* **2011**, *26*, 121–134. [CrossRef]

60. Huang, C.X.; Dong, H.C.; Lundholm, N.; Teng, S.T.; Zheng, G.C.; Tan, Z.J.; Lim, P.T.; Li, Y. Species composition and toxicity of the genus Pseudo-nitzschia in Taiwan Strait, including P. chiniana sp. nov. and P. qiana sp. nov. *Harmful Algae* **2019**, *84*, 195–209. [CrossRef]

61. Lim, H.C.; Tan, S.N.; Teng, S.T.; Lundholm, N.; Orive, E.; David, H.; Quijano-Scheggia, S.; Leong, S.C.Y.; Wolf, M.; Bates, S.S.; et al. Phylogeny and species delineation in the marine diatom Pseudo-nitzschia (Bacillariophyta) using cox1, LSU, and ITS2 rRNA genes: A perspective in character evolution. *J. Phycol.* **2018**, *54*, 234–248. [CrossRef] [PubMed]

62. Stonik, I.; Orlova, T.Y.; Propp, L.; Demchenko, N.; Skriptsova, A. An autumn bloom of diatoms of the genus Pseudo-nitzschia H. Peragallo, 1900 in Amursky Bay, the Sea of Japan. *Russ. J. Mar. Biol.* **2012**, *38*, 211–217. [CrossRef]

63. Ajani, P.; Murray, S.; Hallegraeff, G.; Lundholm, N.; Gillings, M.; Brett, S.; Armand, L. The diatom genus Pseudo-nitzschia (Bacillariophyceae) in New South Wales, A ustralia: Morphotaxonomy, molecular phylogeny, toxicity, and distribution. *J. Phycol.* **2013**, *49*, 765–785. [CrossRef]

64. Rhodes, L.; Jiang, W.; Knight, B.; Adamson, J.; Smith, K.; Langi, V.; Edgar, M. The genus Pseudo-nitzschia (Bacillariophyceae) in New Zealand: Analysis of the last decade's monitoring data. *New Zealand J. Mar. Freshw. Res.* **2013**, *47*, 490–503. [CrossRef]

65. Amato, A.; Lüdeking, A.; Kooistra, W.H. Intracellular domoic acid production in Pseudo-nitzschia multistriata isolated from the Gulf of Naples (Tyrrhenian Sea, Italy). *Toxicon* **2010**, *55*, 157–161. [CrossRef]

66. Oyarzún, D.; Brierley, C.M. The future of coastal upwelling in the Humboldt current from model projections. *Clim. Dyn.* **2019**, *52*, 599–615. [CrossRef]

67. Brink, K.; Halpern, D.; Huyer, A.; Smith, R. The physical environment of the Peruvian upwelling system. *Prog. Oceanogr.* **1983**, *12*, 285–305. [CrossRef]

68. Suárez-Isla, B.A.; López, A.; Hernández, C.; Clement, A.; Guzmán, L. Impacto económico de las floraciones de microalgas nocivas en Chile y datos recientes sobre la ocurrencia de veneno amnésico de los mariscos. Floraciones Algales Nocivas en el Cono Sur Americano, Inst. Esp. Oceanogr. Madrid, España. In *Floraciones Algales Nocivas en el Cono Sur Americano*; Sar, E.A., Ferrario, M.E., Reguera, B., Eds.; Instituto Español de Oceanografía: Madrid, Spain, 2002; pp. 257–268.

69. Lopez-Rivera, A.; Pinto, M.; Insinilla, A.; Suarez Isla, B.; Uribe, E.; Alvarez, G.; Lehane, M.; Furey, A.; James, K.J. The occurrence of domoic acid linked to a toxic diatom bloom in a new potential vector: The tunicate Pyura chilensis (piure). *Toxicon* **2009**, *54*, 754–762. [CrossRef]

70. Álvarez, G.; Rengel, J.; Araya, M.; Álvarez, F.; Pino, R.; Uribe, E.; Díaz, P.A.; Rossignoli, A.E.; López-Rivera, A.; Blanco, J. Rapid Domoic Acid Depuration in the Scallop Argopecten purpuratus and Its Transfer from the Digestive Gland to Other Organs. *Toxins* **2020**, *12*, 698. [CrossRef]

71. National. Fisheries Health Organization of Perú-Organismo Nacional de Sanidad Pesquera(SANIPES). Available online: https://www.sanipes.gob.pe/web/index.php/es/fitoplancton (accessed on 15 April 2021).

72. Guillard, R.R.L.; Hargraves, P.E. Stichochrysis immobilis is a diatom, not a chrysophyte. *Phycologia* **1993**, *32*, 234–236. [CrossRef]

73. Hasle, G.R. *The Inverted-Microscope Method*; Sournia, A., Ed.; Phytoplankton Manual. Monographs on Oceanic Methodology; 1978; pp. 88–96.

74. Lundholm, N.; Hasle, G.R.; Fryxell, G.A.; Hargraves, P.E. Morphology, phylogeny and taxonomy of species within the Pseudo-nitzschia americana complex (Bacillariophyceae) with descriptions of two new species, Pseudo-nitzschia brasiliana and Pseudo-nitzschia linea. *Phycologia* **2002**, *41*, 480–497. [CrossRef]

75. Doyle, J.J.; Doyle, J.L. A rapid DNA isolation procedure for small quantities of fresh leaf tissue. *Phytochem. Bull.* **1987**, *19*, 11–15.

76. White, T.J.; Bruns, T.; Lee, S.; Taylor, J. Amplification and direct sequencing of fungal ribosomal RNA genes for phylogenetics. *Pcr Protoc. A Guide Methods Appl.* **1990**, *18*, 315–322.

77. Hall, T.A. BioEdit: A user-friendly biological sequence alignment editor and analysis program for Windows 95/98/NT. *Nucleic Acids Symp. Ser.* **1999**, *41*, 95–98.

78. Kumar, S.; Stecher, G.; Tamura, K. MEGA7: Molecular evolutionary genetics analysis version 7.0 for bigger datasets. *Mol. Biol. Evol.* **2016**, *33*, 1870–1874. [CrossRef] [PubMed]

79. Darriba, D.; Taboada, G.L.; Doallo, R.; Posada, D. jModelTest 2: More models, new heuristics and parallel computing. *Nat. Methods* **2012**, *9*, 772. [CrossRef]

80. Stamatakis, A. RAxML version 8: A tool for phylogenetic analysis and post-analysis of large phylogenies. *Bioinformatics* **2014**, *30*, 1312–1313. [CrossRef] [PubMed]

81. Silvestro, D.; Michalak, I. raxmlGUI: A graphical front-end for RAxML. *Org. Divers. Evol.* **2012**, *12*, 335–337. [CrossRef]

82. Ronquist, F.; Teslenko, M.; Van Der Mark, P.; Ayres, D.L.; Darling, A.; Höhna, S.; Larget, B.; Liu, L.; Suchard, M.A.; Huelsenbeck, J.P. MrBayes 3.2: Efficient Bayesian phylogenetic inference and model choice across a large model space. *Syst. Biol.* **2012**, *61*, 539–542. [CrossRef] [PubMed]
83. de la Iglesia, P.; Gimenez, G.; Diogene, J. Determination of dissolved domoic acid in seawater with reversed-phase extraction disks and rapid resolution liquid chromatography tandem mass spectrometry with head-column trapping. *J. Chromatogr. A* **2008**, *1215*, 116–124. [CrossRef]

Article

Acute Toxicity of Gambierone and Quantitative Analysis of Gambierones Produced by Cohabitating Benthic Dinoflagellates

J. Sam Murray [1,2,3,*], Sarah C. Finch [4], Jonathan Puddick [1], Lesley L. Rhodes [1], D. Tim Harwood [1,2], Roel van Ginkel [1] and Michèle R. Prinsep [3]

1 Cawthron Institute, Private Bag 2, Nelson 7042, New Zealand; Jonathan.puddick@cawthron.org.nz (J.P.); Lesley.rhodes@cawthron.org.nz (L.L.R.); tim.harwood@cawthron.org.nz (D.T.H.); roel.vanginkel@cawthron.org.nz (R.v.G.)
2 New Zealand Food Safety Science and Research Centre, Massey University, Private Bag 11 222, Palmerston North 4442, New Zealand
3 School of Science, University of Waikato, Private Bag 3105, Hamilton 3240, New Zealand; michele.prinsep@waikato.ac.nz
4 AgResearch, Ruakura Research Centre, Private Bag 3123, Hamilton 3240, New Zealand; sarah.finch@agresearch.co.nz
* Correspondence: sam.murray@cawthron.org.nz

Abstract: Understanding the toxicity and production rates of the various secondary metabolites produced by *Gambierdiscus* and cohabitating benthic dinoflagellates is essential to unravelling the complexities associated with ciguatera poisoning. In the present study, a sulphated cyclic polyether, gambierone, was purified from *Gambierdiscus cheloniae* CAWD232 and its acute toxicity was determined using intraperitoneal injection into mice. It was shown to be of low toxicity with an LD_{50} of 2.4 mg/kg, 9600 times less toxic than the commonly implicated Pacific ciguatoxin-1B, indicating it is unlikely to play a role in ciguatera poisoning. In addition, the production of gambierone and 44-methylgambierone was assessed from 20 isolates of ten *Gambierdiscus*, two *Coolia* and two *Fukuyoa* species using quantitative liquid chromatography–tandem mass spectrometry. Gambierone was produced by seven *Gambierdiscus* species, ranging from 1 to 87 pg/cell, and one species from each of the genera *Coolia* and *Fukuyoa*, ranging from 2 to 17 pg/cell. The production of 44-methylgambierone ranged from 5 to 270 pg/cell and was ubiquitous to all *Gambierdiscus* species tested, as well as both species of *Coolia* and *Fukuyoa*. The relative production ratio of these two secondary metabolites revealed that only two species produced more gambierone, *G. carpenteri* CAWD237 and *G. cheloniae* CAWD232. This represents the first report of gambierone acute toxicity and production by these cohabitating benthic dinoflagellate species. While these results demonstrate that gambierones are unlikely to pose a risk to human health, further research is required to understand if they bioaccumulate in the marine food web.

Keywords: ciguatera poisoning; liquid chromatography–tandem mass spectrometry; nuclear magnetic resonance spectroscopy; LD_{50}; *Gambierdiscus*; *Coolia*; *Fukuyoa*

Key Contribution: This paper is the first to report both the acute toxicity of gambierone and the quantitative assessment of gambierone and 44-methylgambierone production by cohabitating benthic dinoflagellates from the genera *Gambierdiscus*; *Coolia* and *Fukuyoa*.

Citation: Murray, J.S.; Finch, S.C.; Puddick, J.; Rhodes, L.L.; Harwood, D.T.; van Ginkel, R.; Prinsep, M.R. Acute Toxicity of Gambierone and Quantitative Analysis of Gambierones Produced by Cohabitating Benthic Dinoflagellates. *Toxins* **2021**, *13*, 333. https://doi.org/10.3390/toxins13050333

Received: 10 April 2021
Accepted: 3 May 2021
Published: 5 May 2021

1. Introduction

Ciguatera fish poisoning (CFP) is the most common non-microbial foodborne illness related to finfish consumption in the world [1,2]. It is generally associated with the bioaccumulation of ciguatoxins (CTXs) in the flesh and viscera of fish from all trophic levels. While published estimates of CFP vectors range between 60 [3,4] and 400 species [1], it is the carnivorous fish species that are the most commonly implicated in CFP cases as they are often targeted by commercial and recreational fishermen. In recent years, however, a variety of marine invertebrates including echinoderms (e.g., urchins and starfish), gastropods (e.g., cone snails) and bivalve molluscs (e.g., giant clams) [5–9] have also been identified as CFP vectors. Based on these new findings, the Food and Agriculture Organisation of the United Nations and World Health Organisation (FAO and WHO) expert panel have reclassified this illness as ciguatera poisoning (CP) [10].

The poisoning syndrome is prevalent in all circumtropical regions of the world [11,12], and is particularly prolific throughout the tropical and subtropical waters of the South Pacific Ocean, affecting both populated and remote indigenous island communities [13–15]. These communities are intrinsically linked to the reef system for subsistence and trade, which leaves them vulnerable to both the direct and indirect effects of CP [16].

While the existence of CP has been known for centuries [17–19], with the first historical event being reported in 1521 [20], the true impact of this illness is not known. It is estimated that 25,000–50,000 people are affected annually, with epidemiological studies indicating that <20% of actual cases are reported [19]. Intoxications manifest as a wide array of symptoms including gastrointestinal discomfort (e.g., vomiting, diarrhoea, nausea), neurological impairment (e.g., inversion of hot and cold, dysaesthesia and paraethesia) and/or cardiovascular complications (e.g., hypotension and bradycardia) [14,21]. Interestingly, differences in symptoms and intrinsic potencies can be geographically assigned, for example, in the Pacific region, neurological symptoms are commonly associated with intoxication events [14,22].

Current thinking is that the causative compounds of CP are produced by the epiphytic, benthic dinoflagellate genus *Gambierdiscus*. This genus of microalgae is found attached to various substrates and the toxins they produce enter the marine food web by herbivorous reef fish grazing on the macroalgae (e.g., in the Pacific region *Gambierdiscus* favours filamentous red and calcareous green species), coralline turfs, dead corals and volcanic sands [23–25]. *Gambierdiscus* is regarded as an opportunistic dinoflagellate that proliferates following damage to the reef system from tropical hurricanes, crown of thorn starfish outbreaks or coral bleaching events [26,27]. In addition, the complexity of CP is heightened as *Gambierdiscus* is regularly found in microalgal assemblages with other toxin-producing benthic dinoflagellates from the genera *Amphidinium*, *Coolia*, *Fukuyoa*, *Ostreopsis* and *Prorocentrum* [28].

To date, 18 species of *Gambierdiscus* have been described: *G. australes*, *G. balechii*, *G. belizeanus*, *G. caribaeus*, *G. carolinianus*, *G. carpenteri*, *G. cheloniae*, *G. excentricus*, *G. holmesii*, *G. honu*, *G. jejuensis*, *G. lapillus*, *G. lewisii*, *G. pacificus*, *G. polynesiensis*, *G. scabrosus*, *G. silvae* and *G. toxicus*; although new species are regularly being discovered [29–32]. This has heightened as global awareness of CP is increasing and research efforts are being focused on this neglected tropical disease. With 18 species, the genus *Gambierdiscus* is one of the largest genera of marine benthic dinoflagellates, and 16 of the reported species have been isolated from the western South Pacific.

The genus *Gambierdiscus* has been shown to produce a complex array of bioactive, ladder-shaped polyether secondary metabolites which have varying levels of toxicity. These include CTXs, which have been demonstrated to biomagnify and biotransform into more toxic analogues as they move up the marine food web, maitotoxins (MTXs), which are some of the most potent non-peptide toxins known, gambieric acids [33], gambierol [34], gambieroxide [35] and gambierones (Figure 1) [36]. While CTXs are thought to be the causative compounds of CP, with the FDA having established a guidance level for finfish of 0.01 µg/kg P-CTX-1B [37], it is currently unclear whether any of the other compounds produced by *Gambierdiscus* play a role in intoxication events.

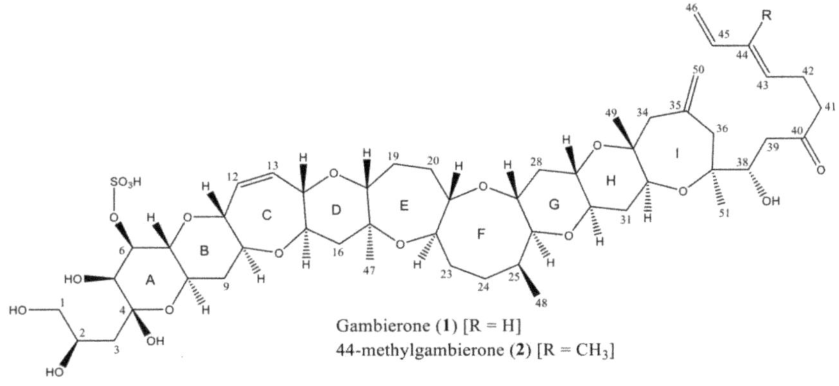

Figure 1. Structures of gambierone and 44-methylgambierone (monoisotopic masses 1024.5 and 1038.5 g/mol, respectively).

Determining the toxicity of these secondary metabolites and their abundance in *Gambierdiscus* and cohabitating genera is a critical step in understanding whether they have the potential to cause human illness. Our recent study assessed one of these secondary metabolites, 44-methylgambierone, and determined it had low acute toxicity via intraperitoneal (i.p) injection to mice (LD_{50} between 20 and 38 mg/kg). Its production was also qualitatively assessed in commonly found cohabitating genera of toxin-producing benthic dinoflagellates. The results demonstrated that three genera produced this secondary metabolite: *Gambierdiscus*, *Coolia* and *Fukuyoa* [28].

The work presented in this manuscript expands on this knowledge, where another structurally related secondary metabolite, gambierone, was purified in sufficient quantities to allow for the first determination of its acute toxicity and ascertain if it could play a role in CP events. In addition, quantitative analysis of both gambierone and 44-methylgambierone production by 20 isolates of cohabitating dinoflagellates from the genera *Gambierdisucs*, *Coolia* and *Fukuyoa* was performed, as reference material had been generated.

2. Results

2.1. Purification and Identification of Gambierone from Gambierdiscus cheloniae CAWD232

Gambierone was purified from *G. cheloniae* CAWD232 monoclonal cultures using a combination of liquid–liquid partitioning, solid-phase clean-up, flash-chromatography and preparative high-performance liquid chromatography.

Analysis of the purified gambierone material by electrospray ionization (ESI) mass spectrometry (positive and negative ion modes; m/z 850–1150), revealed a $[M-H]^-$ ion at m/z 1023.3 and a $[M+H]^+$ ion at m/z 1025.3 (respectively; Figure 2). Additional ions observed in the +ESI spectrum represented water loss (m/z 1007.3, 989.3) and a $[M-SO_3+H]^+$ ion (m/z 945.3) plus sequential water-loss ions (m/z 927.3, 909.3, 891.3, 873.3). These results aligned with those published by Rodriguez et al., 2015 [38] (Supplementary Figure S1). Fragmentation via collision induced dissociation in –ESI mode, revealed a single dominant fragment ion representing a bisulphate anion (m/z 96.8), and in +ESI there were a variety of unassigned fragment ions (Figure 3).

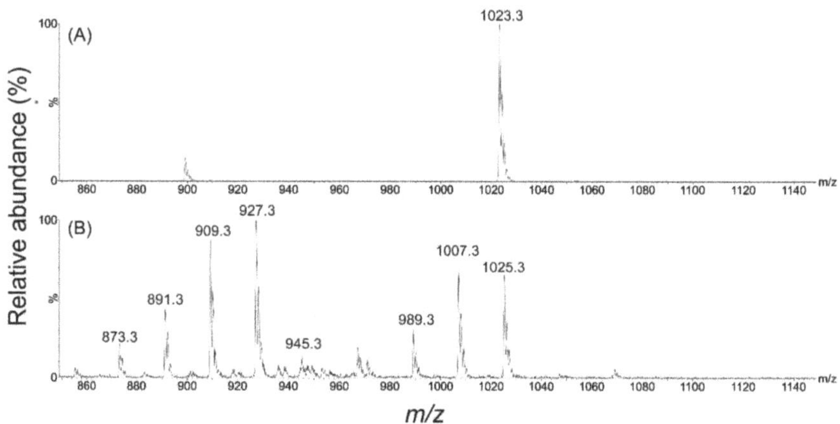

Figure 2. Full scan mass spectra of gambierone (*m/z* 850–1150) in (**A**) −ESI mode showing the [M–H]⁻ ion (*m/z* 1023.3) and (**B**) +ESI mode showing the [M+H]⁺ (*m/z* 1025.3), [M–H₂O+H]⁺ (*m/z* 1007.3), [M–2H₂O +H]⁺ (*m/z* 989.3) and [M–SO₃+H]⁺ (*m/z* 945.3) ions, as well as a series of sulphite plus sequential water-loss ions (*m/z* 927.3, 909.3, 891.3, 873.3).

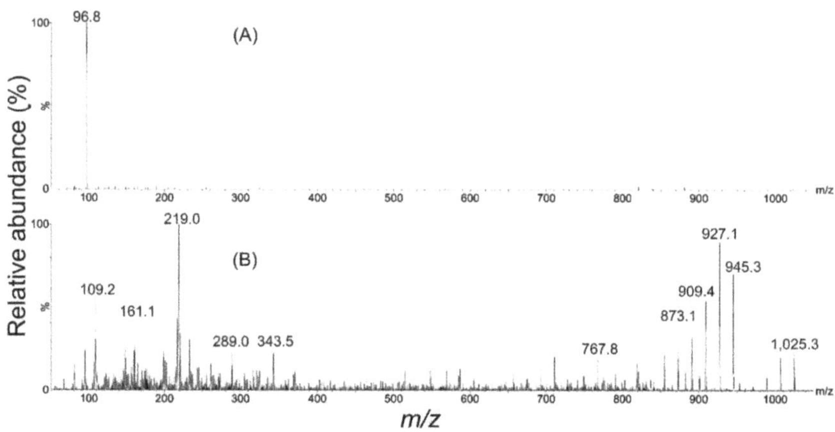

Figure 3. Collision-induced dissociation tandem MS spectra (*m/z* 50–1050) of purified gambierone generated from (**A**) the [M–H]⁻ parent ion (*m/z* 1023.3) in –ESI mode, collision energy 70 eV, showing a single dominant fragment ion representing the bisulphate anion (*m/z* 96.8) and (**B**) the [M+H]⁺ parent ion (*m/z* 1025.3) in +ESI mode, collision energy 25 eV, showing a variety of unassigned fragment ions.

Analysis of the ¹H NMR spectrum (Table 1; Supplementary Figures S2–S6) revealed an analogous spectrum, close to that published by Rodriguez et al. [38]. Key signals included those of: the 1,3-diene (H-43 at 5.70 ppm (dt, 15.1, 7.2), H-44 at 6.08 ppm (dd, 15.1, 10.4), H-45 at 6.28 ppm (dt, 16.9, 10.3), H-46a at 5.07 ppm (dd, 17.1, 1.7) and H-46b at 4.94 (dd, 10.3, 1.7)), the connection point of the monosulphate on ring A (H-6 at 4.70 ppm (d, 3.2)) and the terminal diol (H-2 at 4.10 (m)). Other signals observed included those of the alkene protons in ring C (H-12 at 5.64 ppm (dd, 12.5, 2.4) and H-13 at 5.74 ppm (dd, 12.4, 2.5)), the terminal methylene (H-50 at 4.98 (s) and 4.85 ppm (s)) and the methyl groups (H-47 at 1.21 ppm (3H, s), H-48 at 1.00 ppm (3H, s), H-49 at 1.19 ppm (3H, s) and H-51 at 1.13 ppm (3H, s)). For the full comparison of the ¹H NMR chemical shifts (ppm), multiplicity and coupling constants (Hz) refer to Supplementary Table S1.

Table 1. Comparison of the ^{1}H (500 MHz) NMR chemical shifts (ppm), multiplicity and coupling constants (Hz) of the key structural signals of gambierone purified from *Gambierdiscus cheloniae* CAWD232 and those published by Rodriguez et al., 2015 [38].

Structural Feature	Atom	Purified Gambierone	Published Gambierone [38]
Terminal diol	2	4.10 (m)	4.11 (m)
Monosulphate	6	4.70 (dd, 10.0, 3.2)	4.70 (dd, 10.0, 3.2)
Alkene in ring C	12	5.64 (dd, 12.5, 2.4)	5.64 (dd, 12.5, 2.5)
	13	5.74 (dd, 12.4, 2.5)	5.75 (dd, 12.5, 2.5)
1,3-diene	43	5.70 (dt, 15.1, 7.2)	5.70 (dt, 15.0, 7.0)
	44	6.08 (dd, 15.1, 10.4)	6.08 (dd, 15.0, 10.5)
	45	6.28 (dt 16.9, 10.3)	6.29 (dt, 17.0, 10.4)
	46a	5.07 (dd, 17.1, 1.7)	5.08 (dd, 17.0, 1.8)
	46b	4.94 (dd, 10.3, 1.7)	4.94 (dd, 10.3, 1.8)
Methyl group	47	1.21 (3H, s)	1.20 (3H, s)
	48	1.00 (3H, d, 7.3)	1.00 (3H, d, 7.3)
	49	1.19 (3H, s)	1.19 (3H, s)
Terminal methylene	50	4.98 (br s), 4.85 (br s)	4.98 (br s), 4.86 (br s)
Methyl group	51	1.13 (3H, s)	1.13 (3H, s)

In addition to the ^{1}H NMR spectrum, COSY (Supplementary Figure S7) and HSQC spectra (Supplementary Figure S8) were acquired. The ^{1}H–^{1}H and ^{1}H–^{13}C correlations were also closely aligned to those published by Rodriguez et al. [38] (Supplementary Figures S9–S12). Collectively, the MS and NMR data provided confirmation that the isolated compound was gambierone.

Quantification of the gambierone purified from *G. cheloniae* CAWD232 against a qNMR-calibrated 44-methlygambierone standard using LC–MS/MS, and a relative response factor of 1, determined the yield to be 1.84 mg.

2.2. Acute Toxicity of Gambierone by Intraperitoneal Injection

On the basis of the low i.p toxicity previously determined for 44-methylgambierone (between 20 and 38 mg/kg) [28], it was anticipated that gambierone would have a similarly low toxicity. Single mice were dosed at 10, 5, 3.2 and 2.54 mg/kg, all of which died within 4−9 h post-dosing (Table 2). Symptoms of toxicity included abdominal stretching, which was evident 10 min post-dosing. Mice also showed classic signs of discomfort such as pulling their ears back and orbital tightening [39]. As the toxicity progressed the mice became reluctant to move and any movement was jerky. Further deterioration resulted in laboured breathing, and at this point, mice were euthanised to prevent long-term suffering in accordance with the requirements of the Organisation for Economic Cooperation and Development (OECD) Human Endpoints Guidance Document [40]. A single mouse dosed at 2.04 mg/kg gambierone survived, and having established an appropriate dosing range within the OECD guideline 425 [41], the up-and-down method was used to determine the LD$_{50}$ of gambierone as 2.4 mg/kg with 95% confidence intervals of 2.04 and 2.54 mg/kg. All mice dosed at 2.54 mg/kg died in an analogous manner to that described above (Table 2). At necropsy, the small intestine and caecum of each mouse that was administered a lethal dose of gambierone, were found to contain a pale green fluid. All mice dosed at less than 2.04 mg/kg survived. These mice showed abdominal stretching as well as the signs of discomfort mentioned above. All mice appeared normal within 3.5 and 7 h post-dosing. A low food intake and subsequent weight loss over the first 24-h period post-dosing was observed in mice dosed at 2.04 and 1.58 mg/kg. However, for the remainder of the 14-day observation period, food intake and weight gain were normal. This effect on food intake was not noted in mice dosed at rates lower than 1.58 mg/kg. At necropsy, none of the surviving mice showed any abnormalities and the organ weights, as expressed as % of bodyweight, were all within the normal range (data not provided).

Table 2. Lethality, death time, recovery time and symptoms of toxicity following acute i.p. injection of gambierone to mice.

Dose (mg/kg)	Lethality	Death (h)	Recovery (h)	Symptoms of Toxicity
1.00	0/1		$4\frac{1}{2}$	Abdominal stretching
1.26	0/1		7	Abdominal stretching
1.58	0/1		$3\frac{1}{2}$	Abdominal stretching, ears back, low food intake for 1 day
2.04	0/3		3–$3\frac{1}{2}$–$4\frac{1}{2}$	Abdominal stretching, ears back, orbital tightening, low food intake for 1 day
2.54	2/2	$5^{\,a}$–$6^{\,a}$		Abdominal stretching, ears back, orbital tightening, prostration, jerky movement, laboured breathing
5.00	1/1	$9^{\,a}$		Abdominal stretching, ears back, orbital tightening, prostration, jerky movement, laboured breathing
10.0	1/1	4		Abdominal stretching, ears back, orbital tightening, prostration, jerky movement, laboured breathing

[a] Mice were euthanised when breathing became laboured to prevent long-term suffering.

2.3. Gambierone and 44-Methylgambierone Production by Cohabitating Benthic Dinoflagellates

Dinoflagellate isolates (n = 20) were analysed for the production of gambierone and 44-methylgambierone by LC–MS/MS to determine pg/cell quotas. Quantitative analysis was now possible due to the generation of well-characterised reference material for both secondary metabolites. 44-Methylgambierone production was ubiquitous in the ten *Gambierdiscus* species tested, with levels ranging from 26 to 270 pg/cell (Table 3). The highest levels were produced by *G. lapillus* CAWD338 (270 pg/cell) and *G. australes* CAWD149 (259 pg/cell), while the *G. lapillus* CAWD336 isolate produced considerably less (46 pg/cell). Compared to the *Gambierdiscus* species, 44-methylgambierone cell quotas from the *Coolia* and *Fukuyoa* isolates were generally lower (5–65 pg/cell).

The production of gambierone was more varied, with only seven of the *Gambierdiscus* species and one each of the *Coolia* and *Fukuyoa* species producing detectable levels (LoQ = 0.01 pg/cell). The gambierone cell quotas for the *Gambierdiscus* isolates ranged from 1 to 87 pg/cell (Table 3), with the highest levels observed in *G. carpenteri* CAWD237 (87 pg/cell). When *G. carpenteri* CAWD237 was grown in K media instead of f/2 media, a higher gambierone quota was observed (87 compared to 65 pg/cell). Similar to what was observed for 44-methylgambierone production, the range and level of gambierone cell quotas by the *Coolia* and *Fukuyoa* isolates was lower (2–17 pg/cell). The largest variation was observed between the two *C. malayensis* isolates CAWD154 and CAWD175 (2 and 17 pg/cell, respectively). Gambierone was not detected in the two *C. tropicalis* or the two *F. paulensis* isolates.

The relative production ratio, calculated as pmol/cell, between gambierone and 44-methylgambierone was varied and ranged from 0.01 to 2.15. Of the 14 microalgal species tested, only two produced more gambierone than 44-methylgambierone; *G. carpenteri* CAWD237 (1.19 in K media and 1.46 in f/2 media) and *G. cheloniae* (2.15). Of the isolates where gambierone was detected, the lowest levels of relative production were observed for *G. lewisii* CAWD369 and *G. pacificus* CAWD337, both of which had a ratio of 0.01.

Table 3. Production of gambierone and 44-methylgambierone in isolates of *Gambierdiscus*, *Coolia* and *Fukuyoa*.

Scientific Name	Culture ID [a]	44-Methylgambierone (pg/Cell)	Gambierone (pg/Cell)	Production Ratio [b]
G. australes	CAWD149	259	<0.01	–
G. caribaeus	CAWD301	44	<0.01	–
G. carpenteri (K media)	CAWD237	74	87	1.19
G. carpenteri	CAWD237	45	65	1.46
G. cheloniae	CAWD232	26	55	2.15
G. holmesii	CAWD368	97	20	0.20
G. honu	CAWD242	182	38	0.21
G. lapillus	CAWD336	46	<0.01	–
G. lapillus	CAWD338	270	<0.01	–
G. lewisii	CAWD369	68	1	0.01
G. pacificus	CAWD337	100	1	0.01
G. polynesiensis	CAWD212	29	13	0.45
G. polynesiensis	CAWD267	44	13	0.30
C. malayensis	CAWD154	9	2	0.29
C. malayensis	CAWD175	24	17	0.72
C. tropicalis	UTS2	14	<0.01	–
C. tropicalis	UTS3	15	<0.01	–
F. paulensis	CAWD238	5	<0.01	–
F. paulensis	CAWD306	65	<0.01	–
F. ruetzeri	S044	12	8	0.62
F. ruetzeri	S051	13	6	0.47

[a] Unique laboratory identifier, [b] Production ratio of gambierone/44-methylgambierone calculated as pmol/cell (monoisotopic masses 1024.5 and 1038.5, respectively).

3. Discussion

In the present study, *G. cheloniae* CAWD232 was grown (100 L) to purify enough gambierone to determine its acute toxicity and ascertain if it could play a role in CP events. Fractionation of a methanolic extract and subsequent purification was tracked using LC–MS/MS. The compound was identified as gambierone through its mass spectral properties (the deprotonated and protonated molecular ions; the observed sulphate-/water-loss ions characteristic of cyclic polyethers) and comparison of NMR spectroscopic data with those previously published by Rodriguez et al. in 2015 [38]. The chemical shifts, coupling constants and overlaying of the ^{1}H, COSY and HSQC NMR spectra provided unequivocal evidence that the isolated compound was gambierone. As the HSQC spectrum was used to assign the ^{13}C chemical shifts, quaternary carbons were unable to be assigned using the current dataset. Small differences in the peak resolution, chemical shifts and coupling constants between the two studies were observed and are a result of the ^{1}H NMR data in the current study being acquired at 500 MHz, and the Rodriguez et al., 2015 [38] data being acquired at 750 MHz.

Due to the structural similarities between gambierone and 44-methylgambierone (i.e., the presence of a monosulphate, terminal diol and 1,3-diene) the quantity of gambierone generated was determined by LC–MS/MS against a qNMR reference standard of 44-methylgambierone [28,36]. A relative response factor of 1 was used as the only structural difference between these two compounds is an additional methyl group on C-44, which is unlikely to grossly affect its ionisation properties. Analysis was performed in –ESI mode for two reasons: (1) it afforded a clean spectrum with good sensitivity and (2) the charge is on the sulphate group, which is at the opposite end of the molecule to the additional methyl group, hence, it would not yield a noticeable effect on the ionisation efficiency of the two compounds.

Based on the low acute i.p toxicity of 44-methylgambierone (between 20 and 38 mg/kg) [28], it was anticipated that gambierone would also have a low acute toxicity. The LD$_{50}$ of gambierone was found to be 2.4 mg/kg, indicating that it is 10–15 times more toxic than

44-methylgambierone. Given that the only structural difference between these two compounds is the addition of a methyl group on C-44, this difference in toxicity is surprising. It also highlights that toxicity cannot be predicted on the basis of structure, and that small structural changes have the potential to greatly affect toxicity. This may be due to the structural change affecting the three-dimensional shape of the molecule such that its access to the active site responsible for the toxic effect is altered. While the i.p toxicity data of gambierone is new and important information, oral dosing is the most relevant administration route to assess the risk posed to humans. Previous studies on other shellfish toxins have shown that, although toxicity is reduced when given orally rather than by i.p, there is no correlation between the two. For example, among the paralytic shellfish toxins, the difference between oral and i.p toxicity ranges from 30-fold for decarbamoyl neosaxitoxin to 400-fold for gonyautoxin 5 [42]. These differences are likely due to differences in metabolism or absorption of the metabolites, parameters which are bypassed when the compounds are administered by i.p. Unfortunately, it was not possible to isolate the quantities of gambierone that would be required to determine an oral LD_{50}.

Although the i.p LD_{50} of gambierone showed it to be more toxic than 44-methylgambierone, an LD_{50} of 2.4 mg/kg is still indicative of low toxicity. This is consistent with results published by Rodriguez et al. [38], who showed that gambierone had low activity in sodium channels and little impact on cytosolic Ca^{2+} levels in the neuroblastoma SH-SY5Y cell line. In comparison, other toxins implicated in CP are of far greater toxicity. Algal CTXs have i.p LD_{50} values of $2-10$ μg/kg and fish metabolite CTXs have i.p LD_{50} values of $0.25-0.45$ μg/kg, thereby ranging from being 240 (for 10 μg/kg) to 9600 (for 0.25 μg/kg) times more toxic than gambierone [43]. Although further work is required to assess the bioaccumulation of this secondary metabolite in the marine food web, the hydrophilic nature of gambierone along with the vast difference in toxicity compared to CTXs mean that it is highly unlikely that gambierone contributes to CP. The symptoms of gambierone intoxication are also significantly different to those observed for the CTXs. CTX toxicity is characterised by diarrhoea and hypersalivation, neither of which were observed in this study. A comparison of the symptoms induced by gambierone and 44-methylgambierone is limited, since only one mouse dosed with 44-methylgambierone (38 mg/kg, a lethal dose) showed toxicity. However, it is noted that this mouse did not show the abdominal stretching that was consistently seen in mice dosed with gambierone, although both toxins affected food intake. In addition, this effect was limited to $1-2$ days for gambierone, whereas a much more pronounced anorexic effect was seen in the one mouse given a lethal dose of 44-methylgambierone [28]. Although there are some apparent differences in symptomology between gambierone and 44-methylgambierone, additional toxicity work with the latter compound is required for confirmation.

Previous research on 44-methylgambierone production showed that it was ubiquitous to all *Gambierdiscus* species tested [28]. The current work has expanded on this by demonstrating that *G. holmesii* CAWD368 and *G. lewisii* CAWD369 also produce 44-methylgambierone, and by assessing the gambierone production in a range of *Gambierdiscus*, *Coolia* and *Fukuyoa* isolates ($n = 20$). These isolates were sourced from the Cawthron Institute Culture Collection of Microalgae (CICCM) or were donated from research collaborators in Hong Kong and Australia. The *Gambierdiscus* isolates were collected in Australia, the Cook Islands, French Polynesia, the Kermadec Islands and the Federated States of Micronesia. Representative isolates from the ten species held in culture—*G. australes*, *G. caribaeus*, *G. carpenteri*, *G. cheloniae*, *G. holmesii*, *G. honu*, *G. lapillus*, *G. lewisii*, *G. pacificus* and *G. polynesiensis*—were selected. Two isolates were selected for both *G. lapillus* and *G. polynesiensis*, and for *G. carpenteri* the selected isolate was grown in two different growth media to investigate if differences in the production of secondary metabolites would be observed. The four *Coolia* isolates from two species, *C. malayensis* and *C. tropicalis*, and four *Fukuyoa* isolates from two species, *F. paulensis* and *F. ruetzeri*, were selected as they had previously been shown to produce 44-methylgambierone [28].

A quantitative analysis of 44-methylgambierone and gambierone was also undertaken in the current study to evaluate cell quotas. Production of 44-methylgambierone varied between the dinoflagellate isolates, with *G. lapillus* CAWD338 producing the most. Interestingly, the second highest producer was *G. australes* CAWD149, which is the only known MTX-1 producer [36]. Our previous work with *G. carpenteri* isolates from around the Pacific region reported this species to produce 44-methylgambierone, except in the case of *G. carpenteri* CAWD237 collected in Australia [28]. In contrast, the present study demonstrated that the isolate does produce this secondary metabolite. However, because the original analysis was performed on a cell pellet grown at a different laboratory, it is possible that the different seawater/growth medium source, along with different growth cabinets (light intensity, temperature controls), affected the production of this secondary metabolite. This suggests that culturing conditions play a critical role in the production of these compounds, but further investigations are required to properly understand this.

The production of gambierone by *Gambierdiscus* species was varied, with only seven species producing detectable levels (>0.01 pg/cell; *G. carpenteri*, *G. cheloniae*, *G. holmesii*, *G. honu*, *G. lewisii*, *G. pacificus* and *G. polynesiensis*). *G. carpenteri* CAWD237 was the highest producer followed by *G. cheloniae* CAWD232, which was the isolate used for the purification of gambierone in the present study. To date, *G. belizeanus* has been the only species reported to produce gambierone, thereby making this the first report of gambierone production by an additional seven *Gambierdiscus* species.

An interesting observation was made with the *G. carpenteri* CAWD237 isolate, in that it produced more gambierone and 44-methylgambierone when grown in K media compared to f/2 media. This is unusual, as it is well known that the preferred growth medium for *Gambierdiscus* is f/2 media [44]. K media has been specifically designed for growing marine dinoflagellates that cannot survive in higher levels of trace metals. One potential explanation is that this culture had reduced metal availability resulting in a stressed state [45], which in turn increased the production rates of these secondary metabolites. Further research with the remaining isolates is required to test this hypothesis, as well as the effects of the presence/absence of inorganic phosphates on growth and toxin production [45].

Both strains of the two *Coolia* species, *C. malayensis* CAWD154 and CAWD175 and *C. tropicalis* UTS2 and UTS3, produced 44-methylgambierone, whilst only the *C. malayensis* isolates produced gambierone (2 and 17 pg/cell). A similar result was observed with both strains of the two *Fukuyoa* species, *F. paulensis* CAWD238 and CAWD306 and *F. ruetzeri* S044 and S051, producing 44-methylgambierone, while only the *F. ruetzeri* isolates produced gambierone (6 and 8 pg/cell). This is the first report of gambierone production by the genera *Coolia* and *Fukuyoa*.

Investigation of the production ratio between gambierone and 44-methylgambierone across all species revealed that only two species produced more gambierone than 44-methylgambierone. The first was *G. cheloniae* CAWD232, which produced more than twice the amount of gambierone compared to 44-methylgambierone (ratio of 2.12), followed by *G. carpenteri* CAWD237 (ratio of 1.17 and 1.44 for K and f/2 media, respectively). The largest inverse variation was observed for *G. lewisii* CAWD369 and *G. pacificus* CAWD337 (both with a ratio of 0.01).

While this is the first study to accurately quantify gambierone and 44-methylgambierone production by *Gambierdiscus* species, little research has been conducted to accurately quantify cells quotas using LC–MS/MS for the additional cyclic polyethers produced by this genus. The few that have been reported show that *G. australes* isolates can produce MTX-1 at 2−9 pg/cell, and *G. polynesiensis* CAWD212 produces total CTXs (sum of P-CTX-3B, P-CTX-3C, P-CTX-4A and P-CTX-4B) at 0.44 pg/cell [46]. One research group demonstrated that *G. polynesiensis* TB92 produced 11.9 ± 0.4 pg P-CTX-3C equivalence/cell, as determined using the receptor-binding assay. Subsequent analysis using LC–MS/MS revealed that five CTX analogues were present (P-CTX-3B, P-CTX-3C, P-CTX-4A, P-CTX-4B and M-seco-CTX-3C); however, they were not quantified [13].

4. Conclusions

Gambierdiscus produces a complex array of bioactive ladder-shaped polyether secondary metabolites. Understanding the toxicity of these compounds is essential in determining if they contribute to CP. One of these compounds, gambierone, was purified from *G. cheloniae* CAWD232 and found to have low acute toxicity by i.p injection in mice (LD$_{50}$ 2.4 mg/kg). The hydrophilic nature of this secondary metabolite and the low acute toxicity compared to CTXs, 9600 times less toxic than P-CTX-1B, indicate that gambierone is unlikely to play a role in CP. However, to confirm this prediction, further research is required to assess the bioaccumulation of this secondary metabolite in the marine food web. In addition, cell quotas were determined and seven of the ten *Gambierdiscus* species tested produced gambierone, along with one species from the genera *Coolia* and *Fukuyoa*. The cell quotas of another structurally related analogue, 44-methylgambierone, were also determined and ubiquitous production by all *Gambierdiscus* species, as well as both the *Coolia* and *Fukuyoa* species tested, was demonstrated. This is the first report of the acute toxicity of gambierone and the quantitative analysis of gambierones produced by these genera of cohabitating benthic dinoflagellates.

5. Materials and Methods

5.1. Purification of Gambierone

5.1.1. Microalgal Culturing

Gambierdiscus cheloniae CAWD232 collected from Rarotonga, The Cook Islands, in 2014, was cultured at 25 °C ($\pm$2 °C), 40–70 µmol m^{-2} s^{-1} photon irradiance (12:12 h light:dark cycle) [46]. The isolate was grown in f2/seawater (1:3; UV treated and filtered down to 0.22 µm using a Millipore filtration system; Millipore, Toronto, Canada) [47]. Consecutive 5 L monoclonal cultures (total of 100 L), equating to 1.6×10^8 cells, were harvested during the stationary phase of the growth cycle by centrifugation ($3200\times g$, 10 °C, 10 min; Eppendorf 5810 R, Hamburg, Germany).

5.1.2. Extraction and Isolation

The cell pellets from Section 5.1.1. were frozen (-20 °C) before being extracted three times with 90% aq. methanol (MeOH; Fisher-Optima), at a ratio of 1 mL per 2×10^5 cells, and ultrasonication (10 min at 59 kHz; model 160HT, Soniclean Pty, Adelaide, Australia). Cellular debris was pelleted by centrifugation ($3200\times g$, 4 °C, 5 min) between extractions and the supernatant combined in a Schott bottle. To remove lipids, the 90% aq. MeOH extract was subjected to a liquid–liquid partition with *n*-hexane (1:1, *v/v*; Thermo-Fisher). The 90% aq. MeOH layer was collected and frozen (-20 °C) to precipitate extracellular co-extractives, followed by centrifugation ($3200\times g$, 4 °C, 10 min) and sequential membrane and glass-fibre filtration (8, 2 and 1.6 µm) to remove any fine particulates.

The extract was then diluted to 60% aq. MeOH before completing a second liquid–liquid partition with dichloromethane (DCM; 1:1, *v/v*; HiPerSolv Chromanorm, VWR International) to remove any remaining lipophilic compounds. The 60% aq. MeOH containing gambierone was collected, dried down using rotary-evaporation (50 mBar and 50 °C), resuspended in 30 mL Milli-Q water (18.2 MΩ; Millipore, Toronto, ON, Canada) and loaded onto a Strata-X prepacked solid-phase cartridge (10 g; Phenomenex; Torrance, CA, USA). The column was conditioned with ethanol, MeOH and then Milli-Q water (200 mL of each) and washed with 40% aq. MeOH (200 mL). The gambierone was eluted with 100% MeOH (200 mL).

Fractionation was then performed on a Reveleris Flash Chromatography system (Büchi, Flawil, Switzerland) fitted with an Agilent Superflash C$_{18}$ SF 25–75 g column (four injections; Santa Clara, CA, USA). The column was eluted at 20 mL/min with (A) Milli-Q water and (B) acetonitrile (MeCN; Fisher-Optima) mobile phases. The initial solvent composition was 20% B for 5 min before a linear gradient to 95% B from 5–30 min, and then held at 95% B for 10 min. Fractions were collected every 30 s (10 mL). A final fractionation was performed on a Shimadzu preparative HPLC system (Kyoto, Japan) using isocratic elution, 35% aq. MeCN with 0.2% (*v/v*) of a 25% ammonium hydroxide (NH_4OH) solution, on a Phenomenex Gemini C18 column (5 μm, 150 × 21 mm; five injections) with a flow rate of 25 mL/min. The fractions containing gambierone were combined, dried down using rotary-evaporation (50 mBar and 50 °C) and resuspended in 100% MeOH (1 mL; for a schematic of the purification scheme refer Supplementary Figure S13).

5.1.3. Nuclear Magnetic Resonance Spectroscopy Evaluation of Gambierone

1- and 2-D NMR spectroscopy experiments were conducted on a Bruker Avance III 500 MHz instrument with a 5 mm BBOF smart probe (Billerica, MA, USA). The instrument was operated at 500 MHz for ^1H and 125 MHz for ^{13}C, with the chemical shifts being determined at 303 K and referenced to the MeOH signal at 3.31 ppm. The gambierone material was taken to dryness under a stream of N_2 gas at 40 °C, resuspended in d4-MeOH (600 μL, >99.8% deuterium; Sigma-Aldrich, St. Louis, MO, USA) and transferred to a Wilmad® (Vineland, NJ, USA) 5 mm high-precision NMR tube prior to analysis.

5.1.4. Liquid Chromatography–Mass Spectrometry Evaluation of Gambierone

The purified gambierone (from Section 5.1.2) was assessed using a Waters Xevo TQ-S triple quadrupole mass spectrometer coupled to a Waters Acquity UPLC i-Class with a flow-through needle sample manager (Waters; Milford, MA, USA). The mass spectrometer utilized electrospray ionization (positive and negative ion modes) with a scan range of *m/z* 850–1150. Chromatographic separation used a Waters Acquity UPLC BEH phenyl column (1.7 μm, 100 × 2.1 mm) held at 50 °C. The column was eluted at 0.55 mL/min using mobile phases containing 0.2% (*v/v*) of a 25% NH_4OH solution in (A) Milli-Q water and (B) MeCN. Initial solvent conditions were 5% B for 1 min with a linear gradient to 95% B from 1.0 to 7.5 min, held at 95% B for 1 min, followed by a linear gradient back to 5% B from 8.5 to 9 min. The column was re-equilibrated with 5% B for 1 min before the next injection. The autosampler chamber was maintained at 10 °C and the injection volume was 1 μL.

5.2. Acute Toxicity of Gambierone by Intraperitoneal Injection
5.2.1. Animals

Female Swiss albino mice (18–22 g) were bred at AgResearch, Ruakura, New Zealand. The mice were housed individually during the experiments and were allowed unrestricted access to food (Rat and Mouse Cubes, Specialty Feeds Ltd., Glen Forrest, Western Australia) and water. All experiments were approved by the Ruakura Animal Ethics Committee established under the Animal Protection (code of ethical conduct) Regulations Act, 1987 (New Zealand), Project Number 14988 (approval date 5 March 2020) and Project Number 15296 (approval date 4 March 2021).

5.2.2. Toxicity Assessment

Acute toxicity was determined using the principles of OECD guideline 425 [41]. This guideline employs an up-and-down procedure, whereby one animal is dosed and if it survives, the next mouse receives an increased dose, whereas if it dies, the next mouse is dosed with a reduced amount of the test material. To determine the LD_{50}, dosing is continued until four live-death reversals have been achieved.

Toxicity was determined by i.p injection. Each mouse was weighed prior to dosing and the required amount of gambierone calculated to yield the chosen dose on a mg/kg basis. The dose was prepared by taking the appropriate volume of stock solution (pure gambierone in 90% aq. MeOH), drying it down under nitrogen and immediately redis-solving in 1% Tween 60 in normal saline (1 mL). This freshly prepared solution was then immediately injected into mice. All dosing was conducted between 8 and 9.30 a.m. to avoid any diurnal variations in response. Mice were monitored closely during the day of dosing and those that survived were monitored for a 14 day observation period, which included a daily measurement of food consumption and bodyweight. After 14 days, the mice were euthanised by carbon dioxide inhalation and necropsied. The weights of the liver, kidneys, spleen, heart, lungs, stomach (full and empty) and the whole gut were measured and calculated as a percentage of bodyweight (data not provided).

5.3. Quantitive Analysis of Gambierone and 44-Methylgambierone Production

5.3.1. Microalgal Culturing and Sample Extraction

The microalgal isolates studied (20 in total) consisted of 14 species from three genera, *Gambierdiscus*, *Coolia* and *Fukuyoa*. All isolates were grown in f2/seawater (1:3) except for the *G. carpenteri* isolate, which was also grown in K media. The growth chamber was set at 25 °C ($\pm$2 °C), 40–70 μmol m^{-2} s^{-1} photon irradiance (12:12 h light:dark cycle). Isolates were either sourced from the CICCM; or donated by researchers from Hong Kong and Australia. Cultures were harvested in the late exponential or stationary phase and contained at least 5 $\times$ 10^5 cells. The cells were harvested by centrifugation (3200$\times$ g, 10 °C, 10 min), the growth medium was decanted and the resulting cell pellets were frozen (-20 °C).

Each cell pellet was extracted twice with 90% aq. MeOH, at a ratio of 1 mL per 2 $\times$ 10^5 cells, and ultrasonication (10 min at 59 kHz). Cellular debris was pelleted by centrifugation (3200$\times$ g, 4 °C, 5 min) between extractions and the supernatant transferred to another vial. The resulting supernatants were pooled to give a final extract concentration equivalent to 1 $\times$ 10^5 cells/mL. The combined extracts were stored at -20 °C for 24–48 h to precipitate insoluble matrix co-extractives, which were removed using centrifugation (3200$\times$ g, 4 °C, 5 min) prior to analysis. An aliquot of the clarified extract was transferred into a 2 mL glass autosampler vial and analysed using a modification of the LC–MS/MS method described in Murray et al., 2018 [48].

5.3.2. Liquid Chromatography–Tandem Mass Spectrometry Analysis

Quantitative analysis of gambierone and 44-methylgambierone (monoisotopic masses 1024.5 and 1038.5 g/mol, respectively) was performed on a Waters Xevo TQ-S triple quadrupole mass spectrometer coupled to a Waters Acquity UPLC i-Class with a flow-through needle sample manager. Chromatographic separation used a Waters Acquity UPLC BEH phenyl column (1.7 μm, 100 $\times$ 2.1 mm) held at 50 °C. The column was eluted at 0.55 mL/min with (A) Milli-Q water and (B) MeCN mobile phases, each containing 0.2% (v/v) of a 25% NH$_4$OH solution. Fresh mobile phases were prepared daily to ensure optimal sensitivity and stable retention times. The initial solvent composition was 5% B with a linear gradient to 50% B from 0 to 2.5 min, ramped up to 95% B by 3 min and held at 95% B until 3.2 min, followed by a linear gradient back to 5% B at 3.5 min. The column was then re-equilibrated with 5% B until 4 min. The autosampler chamber was maintained at 10 °C and the injection volume was 1 μL. The mass spectrometer used an electrospray ionization source operated in negative ion mode. Other settings were capillary voltage 3.0 kV, cone voltage 40 V, source temperature 150 °C, nitrogen gas desolvation flow rate 1000 L/h at 600 °C, cone gas 150 L/h and the collision cell was operated with 0.15 mL/min argon. Multiple reaction monitoring (MRM) transitions for gambierone were m/z 1023.3 > 96.8 (Channel 1) and m/z 899.6 > 96.8 (Channel 2), with collision energies of 50 eV. 44-Methylgambierone was monitored using the m/z 1037.6 > 96.8 (Channel 1) and 899.6 > 96.8 (Channel 2), with collision energies of 70 and 48 eV, respectively. All transitions had a dwell time of 30 ms. Channels 1 and 2 were used for quantitation and confirmation, respectively.

Data acquisition and processing were performed with TargetLynx software (Waters, Milford, MA, USA). Gambierone and 44-methylgambierone were identified in sample extracts based on the retention time (2.54 and 2.58 min, respectively) and a fragment ion ratio of 8:1 (Channel 1/Channel 2; as determined using reference material). Quantitative analysis was performed using a five-point linear regression calibration (0–1000 ng/mL) prepared in 90% aq. MeOH and a relative response factor of 1 to 44-methylgambierone. The LoQ of the analytical method was 1 ng/mL, which equates to 0.01 pg/cell in an extract generated from a cell pellet of 1×10^5 cells/mL.

5.3.3. Quantitation of Gambierone Using Liquid Chromatography–Tandem Mass Spectrometry

The purified gambierone material (Section 5.1.2) was quantified against a qNMR-calibrated 44-methylgambierone reference standard [28] using LC–MS/MS and a relative response factor of 1. The instrument parameters and chromatographic conditions outlined above (Section 5.3.2) were used. Triplicate injections of a 100 ng/mL standard were compared, followed by calibration of the gambierone material using a five-point linear regression calibration (0–1000 ng/mL) of 44-methylgambierone.

Supplementary Materials: The following are available online at https://www.mdpi.com/article/10.3390/toxins13050333/s1. Table S1. Comparison of the ^{13}C (125 MHz) and ^{1}H (500 MHz) NMR chemical shifts (ppm), multiplicity and coupling constants (Hz) for gambierone purified from *Gambierdiscus cheloniae* CAWD232, generated in d4-MeOH, and those published by Rodriguez et al., 2015 [38]. Figure S1. Comparison of the mass spectra of gambierone in positive electrospray ionization mode (**A**) purified from *Gambierdiscus cheloniae* CAWD232 and (**B**) published by Rodriguez et al., 2015 [38]. Figure S2. ^{1}H NMR spectrum of gambierone purified from *Gambierdiscus cheloniae* CAWD232 acquired on a Bruker Advance III 500 MHz instrument in CD$_3$OH ($\geq$99.8% atom D). Figure S3. Expansion of the ^{1}H NMR spectrum (0.6–2.7 ppm) of gambierone in CD$_3$OD ($\geq$99.8% atom D) at 500 MHz. Figure S4. Expansion of the ^{1}H NMR spectrum (2.8–5.2 ppm) of gambierone in CD$_3$OD ($\geq$99.8% atom D) at 500 MHz. Figure S5. Expansion of the ^{1}H NMR spectrum (5.3–7.3 ppm) of gambierone in CD$_3$OD ($\geq$99.8% atom D) at 500 MHz. Figure S6. Comparison of the ^{1}H NMR spectrum of (**A**) gambierone purified from *Gambierdiscus cheloniae* CAWD232 acquired on a Bruker Advance III 500 MHz instrument and (**B**) the published spectrum from Rodriguez et al., acquired on a Varian Inova 750 MHz instrument [38]. Figure S7. COSY NMR spectrum of gambierone purified from *Gambierdiscus cheloniae* CAWD232 in CD$_3$OD ($\geq$99.8% atom D) at 500 MHz. Figure S8. HSQC NMR spectrum of gambierone purified from *Gambierdiscus cheloniae* CAWD232 in CD$_3$OD ($\geq$99.8% atom D) at 500 MHz. Figure S9. Comparison of the COSY NMR spectrum (1.0–6.5 ppm) of gambierone purified from *Gambierdiscus cheloniae* CAWD232 (black) and that published by Rodriguez et al., 2015 (red) [38]. Long range (~2 Hz) couplings not displayed. Figure S10. Expansion (^{1}H: 1.0–2.7 ppm; ^{13}C: 10–55 ppm) of the HSQC spectrum of gambierone purified from *Gambierdiscus cheloniae* CAWD232 (black), and the published spectrum from Rodriguez et al., 2015 (red) [38]. Figure S11. Expansion (^{1}H: 2.5–4.7 ppm; ^{13}C: 62–85 ppm) of the HSQC spectrum of gambierone purified from *Gambierdiscus cheloniae* CAWD232 (black), and the published spectrum from Rodriguez *et al.*, 2015 (red) [38]. Figure S12. Expansion (^{1}H: 4.7–6.5 ppm; ^{13}C: 110–140 ppm) of the HSQC spectrum of gambierone purified from *Gambierdiscus cheloniae* CAWD232 (black), and the published spectrum from Rodriguez et al., 2015 (red) [38]. Figure S13. Purification scheme for the isolation of gambierone from *Gambierdiscus cheloniae* CAWD232.

Author Contributions: Experimental design was performed by J.S.M., J.P., D.T.H. and M.R.P.; the research was carried out by J.S.M.; isolations and microalgal culturing was performed by L.L.R.; compound purification was conducted by J.S.M. and R.v.G.; toxicity assessment was performed by S.C.F.; funding was acquired by J.S.M. and D.T.H.; All authors have read and agreed to the published version of the manuscript.

Funding: This research was conducted with funding from a doctoral scholarship from the New Zealand Food Safety Science and Research Centre awarded to J.S.M, the Cawthron Institute Capability Investment Fund and the Seafood Safety research programme (contract CAWX1801).

Acknowledgments: The authors would like to acknowledge Kirsty Smith from Cawthron for collecting the microalgal samples in the Cook Islands, Lucy Thompson from Cawthron for her technical assistance, Tom Trnski from the Auckland War Memorial Museum for collecting microalgal samples from the Kermadec Islands, Meng Yan and Priscilla Leung for the *Fukuyoa* isolates from Hong Kong, Michaela Larsson for the *Coolia* isolates from Australia, and Andrew Lewis from Callaghan Innovation for the NMR analysis.

Conflicts of Interest: The authors declare no conflict of interest.

References

1. FAO. *Application of Risk Assessment in the Fish Industry*; Food and Agriculture Organisation of the United Nations—FAO Fisheries Technical Paper; FAO: Rome, Italy, 2014; Volume 442.
2. Friedman, M.A.; Fleming, L.E.; Fernandez, M.; Bienfang, P.; Schrank, K.; Dickey, R.; Bottein, M.-Y.; Backer, L.; Ayyar, R.; Weisman, R.; et al. Ciguatera fish poisoning: Treatment, prevention and management. *Mar. Drugs* **2008**, *6*, 456–479. [CrossRef] [PubMed]
3. Gaboriau, M.; Ponton, D.; Darius, H.T.; Chinain, M. Ciguatera fish toxicity in French Polynesia: Size does not always matter. *Toxicon* **2014**, *84*, 41–50. [CrossRef] [PubMed]
4. Kohli, G.S.; Farrell, H.; Murray, S.A. Gambierdiscus, the cause of ciguatera fish poisoning: An increased human health threat influenced by climate change. In *Climate Change and Marine and Freshwater Toxins*; De Gruyter: Berlin, Germany, 2015; pp. 273–312.
5. Roué, M.; Darius, H.T.; Picot, S.; Ung, A.; Viallon, J.; Gaertner-Mazouni, N.; Sibat, M.; Amzil, Z.; Chinain, M. Evidence of the bioaccumulation of ciguatoxins in giant clams (*Tridacna maxima*) exposed to *Gambierdiscus* spp. cells. *Harmful Algae* **2016**, *57*, 78–87. [CrossRef] [PubMed]
6. Kelly, A.M.; Kohler, C.C.; Tindall, D.R. Are crustaceans linked to the ciguatera food chain? *Environ. Biol. Fishes* **1992**, *33*, 275–286. [CrossRef]
7. Silva, M.; Rodriguez, I.; Barreiro, A.; Kaufmann, M.; Neto, A.I.; Hassouani, M.; Sabour, B.; Alfonso, A.; Botana, L.M.; Vasconcelos, V. First Report of Ciguatoxins in Two Starfish Species: *Ophidiaster ophidianus* and *Marthasterias glacialis*. *Toxins* **2015**, *7*, 3740–3757. [CrossRef]
8. Darius, H.T.; Roué, M.; Sibat, M.; Viallon, J.; Gatti, C.M.; Vandersea, M.W.; Tester, P.A.; Litaker, R.W.; Amzil, Z.; Hess, P.; et al. Toxicological Investigations on the Sea Urchin *Tripneustes gratilla* (Toxopneustidae, Echinoid) from Anaho Bay (Nuku Hiva, French Polynesia): Evidence for the Presence of Pacific Ciguatoxins. *Mar. Drugs* **2018**, *16*, 122. [CrossRef]
9. Gatti, C.M.; Lonati, D.; Darius, H.T.; Zancan, A.; Roué, M.; Schicchi, A.; Locatelli, C.A.; Chinain, M. *Tectus niloticus* (Tegulidae, Gastropod) as a Novel Vector of Ciguatera Poisoning: Clinical Characterization and Follow-Up of a Mass Poisoning Event in Nuku Hiva Island (French Polynesia). *Toxins* **2018**, *10*, 102. [CrossRef]
10. FAO; WHO. *Report of the Expert Meeting on Ciguatera Poisoning: Rome, 19–23 November 2018*; FAO: Rome, Italy, 2020; p. 156.
11. Bagnis, R.; Chanteau, S.; Chungue, E.; Hurtel, J.M.; Yasumoto, T.; Inoue, A. Origins of ciguatera fish poisoning: A new dinoflagellate, Gambierdiscus toxicus Adachi and Fukuyo, definitively involved as a causal agent. *Toxicon* **1980**, *18*, 199–208. [CrossRef]
12. Botana, L.M. Toxicology of seafood toxins: A critical review. In *Seafood and Freshwater Toxins: Pharmacology, Physiology, and Detection*; Botana, L.M., Ed.; CRC Press: Boca Raton, FL, USA, 2014.
13. Chinain, M.; Darius, H.T.; Ung, A.; Cruchet, P.; Wang, Z.; Ponton, D.; Laurent, D.; Pauillac, S. Growth and toxin production in the ciguatera-causing dinoflagellate *Gambierdiscus polynesiensis* (Dinophyceae) in culture. *Toxicon* **2010**, *56*, 739–750. [CrossRef]
14. Friedman, M.A.; Fernandez, M.; Backer, L.C.; Dickey, R.W.; Bernstein, J.; Schrank, K.; Kibler, S.; Stephan, W.; Gribble, M.O.; Bienfang, P.; et al. An updated review of ciguatera fish poisoning: Clinical, epidemiological, environmental, and public health management. *Mar. Drugs* **2017**, *15*, 72. [CrossRef]
15. Rhodes, L.; Smith, K.; Harwood, T.; Selwood, A.; Argyle, P.; Bedford, C.; Munday, R. *Gambierdiscus* and *Ostreopsis* from New Zealand, the Kermadec Islands and the Cook Islands and the risk of ciguatera fish poisoning in New Zealand. In Proceedings of the 16th International Conference on Harmful Algae, Wellington, NZ, USA, 27–31 October 2014; pp. 180–183.
16. Skinner, M.P.; Brewer, T.D.; Johnstone, R.; Fleming, L.E.; Lewis, R.J. Ciguatera fish poisoning in the Pacific Islands (1998 to 2008). *PLoS ONE Negl. Trop. Dis.* **2011**, *5*, e1416. [CrossRef]
17. FAO. *Marine Biotoxins*; Food and Agriculture Organisation of the United Nations—FAO Food and Nutritional Paper; FAO: Rome, Italy, 2004; p. 80.
18. Swift, A.E.; Swift, T.R. Ciguatera. *J. Toxicol. Clin. Toxicol.* **1993**, *31*, 1–29. [CrossRef]
19. ILM. Institute Loius Malarde—Ciguatera Online. 2014. Available online: http://www.ciguatera-online.com/index.php/en/ (accessed on 1 July 2020).
20. Urdaneta, A. Relación de los sucesos de la armada del Comendador Loaiza a Las Islas de la Especiería o Molucas en 1525 y sucesos acaecidos en ellas hasta el 1536. *Real Bibl.* **1580**, *II*, 1465.
21. Caillaud, A.; de la Iglesia, P.; Darius, H.T.; Pauillac, S.; Aligizaki, K.; Fraga, S.; Chinain, M.; Diogene, J. Update on methodologies available for ciguatoxin determination: Perspectives to confront the onset of ciguatera fish poisoning in Europe. *Mar. Drugs* **2010**, *8*, 1838–1907. [CrossRef]
22. Zhang, F.; Xu, X.; Li, T.; Liu, Z. Shellfish toxins targeting voltage-gated sodium channels. *Mar. Drugs* **2013**, *11*, 4698–4723. [CrossRef]
23. Rhodes, L.; Smith, K.; Murray, S.; Harwood, D.; Trnski, T.; Munday, R. The epiphytic genus *Gambierdiscus* (Dinophyceae) in the Kermadec Islands and Zealandia regions of the southwestern Pacific and the associated risk of ciguatera fish poisoning. *Mar. Drugs* **2017**, *15*, 219. [CrossRef]
24. Ledreux, A.; Brand, H.; Chinain, M.; Bottein, M.-Y.D.; Ramsdell, J.S. Dynamics of ciguatoxins from *Gambierdiscus polynesiensis* in the benthic herbivore *Mugil cephalus*: Trophic transfer implications. *Harmful Algae* **2014**, *39*, 165–174. [CrossRef]
25. Yasumoto, T.; Hashimoto, Y.; Bagnis, R.; Randall, J.E.; Banner, A.H. Toxicity of the surgeonfishes. *Bull. Jpn. Soc. Sci. Fish.* **1971**, *37*, 724–734. [CrossRef]

26. Xu, Y.; Richlen, M.L.; Liefer, J.D.; Robertson, A.; Kulis, D.; Smith, T.B.; Parsons, M.L.; Anderson, D.M. Influence of environmental variables on *Gambierdiscus* spp. (Dinophyceae) growth and distribution. *PLoS ONE* **2016**, *11*, e0153197. [CrossRef]

27. Rongo, T.; van Woesik, R. The effects of natural disturbances, reef state, and herbivorous fish densities on ciguatera poisoning in Rarotonga, southern Cook Islands. *Toxicon* **2013**, *64*, 87–95. [CrossRef]

28. Murray, J.S.; Nishimura, T.; Finch, S.C.; Rhodes, L.L.; Puddick, J.; Harwood, D.T.; Larsson, M.E.; Doblin, M.A.; Leung, P.; Yan, M.; et al. The role of 44-methylgambierone in ciguatera fish poisoning: Acute toxicity, production by marine microalgae and its potential as a biomarker for *Gambierdiscus* spp. *Harmful Algae* **2020**, *97*, 101853. [CrossRef] [PubMed]

29. Smith, K.F.; Rhodes, L.; Verma, A.; Curley, B.G.; Harwood, D.T.; Kohli, G.S.; Solomona, D.; Rongo, T.; Munday, R.; Murray, S.A. A new *Gambierdiscus* species (Dinophyceae) from Rarotonga, Cook Islands: *Gambierdiscus cheloniae* sp. nov. *Harmful Algae* **2016**, *60*, 45–56. [CrossRef]

30. Kretzschmar, A.L.; Verma, A.; Harwood, T.; Hoppenrath, M.; Murray, S. Characterization of *Gambierdiscus lapillus* sp. nov. (Gonyaulacales, Dinophyceae): A new toxic dinoflagellate from the Great Barrier Reef (Australia). *J. Phycol.* **2017**, *53*, 283–297. [CrossRef] [PubMed]

31. Fraga, S.; Rodríguez, F.; Riobó, P.; Bravo, I. *Gambierdiscus balechii* sp. nov (Dinophyceae), a new benthic toxic dinoflagellate from the Celebes Sea (SW Pacific Ocean). *Harmful Algae* **2016**, *58*, 93–105. [CrossRef] [PubMed]

32. Kretzschmar, A.L.; Larsson, M.E.; Hoppenrath, M.; Doblin, M.A.; Murray, S.A. Characterisation of Two Toxic *Gambierdiscus* spp. (Gonyaulacales, Dinophyceae) from the Great Barrier Reef (Australia): *G. lewisii* sp. nov. and *G. holmesii* sp. nov. *Protist* **2019**, *170*, 125699. [CrossRef] [PubMed]

33. Morohashi, A.; Satake, M.; Nagai, H.; Oshima, Y.; Yasumoto, T. The absolute configuration of gambieric acids A-D, potent antifungal polyethers, isolated from the marine dinoflagellate *Gambierdiscus toxicus*. *Tetrahedron* **2000**, *56*, 8995–9001. [CrossRef]

34. Morohashi, A.; Satake, M.; Yasumoto, T. The absolute configuration of gambierol, a toxic marine polyether from the dinoflagellate, *Gambierdiscus toxicus*. *Tetrahedron Lett.* **1999**, *40*, 97–100. [CrossRef]

35. Watanabe, R.; Uchida, H.; Suzuki, T.; Matsushima, R.; Nagae, M.; Toyohara, Y.; Satake, M.; Oshima, Y.; Inoue, A.; Yasumoto, T. Gambieroxide, a novel epoxy polyether compound from the dinoflagellate *Gambierdiscus toxicus* GTP2 strain. *Tetrahedron Lett.* **2013**, *69*, 10299–10303. [CrossRef]

36. Murray, J.S.; Selwood, A.I.; Harwood, D.T.; van Ginkel, R.; Puddick, J.; Rhodes, L.L.; Rise, F.; Wilkins, A.L. 44-Methylgambierone, a new gambierone analogue isolated from *Gambierdiscus australes*. *Tetrahedron Lett.* **2019**, *60*, 621–625. [CrossRef]

37. USFDA (U.S. Food and Drug Administration). *Fish and Fishery Products—Hazards and Controls Guidance*, 4th ed.; Chapter 6: Natural Toxins; Center for Food Safety and Applied Nutrition: College Park, MD, USA, 2011.

38. Rodríguez, I.S.; Genta-Jouve, G.G.; Alfonso, C.; Calabro, K.; Alonso, E.; Sánchez, J.A.; Alfonso, A.; Thomas, O.P.; Botana, L.M. Gambierone, a ladder-shaped polyether from the dinoflagellate *Gambierdiscus belizeanus*. *Org. Lett.* **2015**, *17*, 2392–2395. [CrossRef]

39. Langford, D.J.; Bailey, A.L.; Chanda, M.L.; Clarke, S.E.; Drummond, T.E.; Echols, S.; Glick, S.; Ingrao, J.; Klassen-ross, T.; Lacroix-fralish, M.L.; et al. Coding of facial expressions of pain in the laboratory mouse. *Nat. Methods* **2010**, *7*, 447–449. [CrossRef]

40. OECD. OECD Guidance Document on the Recognition, Assessment and Use of Clinical Signs as Humane Endpoints for Experimental Animals Used in Safety Evaluation. Available online: https://www.aaalac.org/accreditation/RefResources/RR_HumaneEndpoints.pdf (accessed on 20 December 2000).

41. OECD. OECD Guideline for Testing of Chemicals 425. Acute Oral Toxicity-Up-and-Down-Procedure (UDP). Available online: http://www.epa.gov/oppfead1/harmonization/docs/E425guideline.pdf (accessed on 2 November 2007).

42. Selwood, A.I.; Waugh, C.; Harwood, D.T.; Rhodes, L.L.; Reeve, J.; Sim, J.; Munday, R. Acute toxicities of the saxitoxin congeners gonyautoxin 5, gonyautoxin 6, decarbamoyl gonyautoxin 2&3, decarbamoyl neosaxitoxin, C-1&2 and C-3&4 to mice by various routes of administration. *Toxins* **2017**, *9*, 73.

43. Rhodes, L.L.; Smith, K.F.; Murray, J.S.; Nishimura, T.; Finch, S.C. Ciguatera fish poisoning: The risk from and Aotearoa/New Zealand perspective. *Toxins* **2020**, *12*, 50. [CrossRef]

44. Guillard, R.R.L.; Ryther, J.H. Studies of marine planktonic diatoms: I. *Cyclotella nana* Hustedt and *Detonula confervacea*. *Can. J. Microbiol.* **1962**, *8*, 229–239. [CrossRef]

45. Huerlimann, R.; de Nys, R.; Heimann, K. Growth, lipid content, productivity, and fatty acid composition of tropical microalgae for scale-up production. *Biotechnol. Bioeng.* **2010**, *107*, 245–257. [CrossRef]

46. Munday, R.; Murray, J.S.; Rhodes, L.L.; Larsson, M.E.; Harwood, D.T. Ciguatoxins and maitotoxins in extracts of sixteen *Gambierdiscus* isolates and one *Fukuyoa* isolate from the south Pacific and their toxicity to mice by intraperitoneal and oral administration. *Mar. Drugs* **2017**, *15*, 208. [CrossRef]

47. Guillard, R.R.L. Culture of Phytoplankton for Feeding Marine Invertebrates. In *Culture of Marine Invertebrate Animals: Proceedings—1st Conference on Culture of Marine Invertebrate Animals Greenport*; Smith, W.L., Chanley, M.H., Eds.; Springer: Boston, MA, USA, 1975; pp. 29–60. [CrossRef]

48. Murray, J.S.; Boundy, M.J.; Selwood, A.I.; Harwood, D.T. Development of an LC–MS/MS method to simultaneously monitor maitotoxins and selected ciguatoxins in algal cultures and P-CTX-1B in fish. *Harmful Algae* **2018**, *80*, 80–87. [CrossRef]

Article

Shellfish Toxin Uptake and Depuration in Multiple Atlantic Canadian Molluscan Species: Application to Selection of Sentinel Species in Monitoring Programs

Wade A. Rourke [1],[*], Andrew Justason [2], Jennifer L. Martin [3] and Cory J. Murphy [1]

[1] Dartmouth Laboratory, Canadian Food Inspection Agency, 1992 Agency Drive, Dartmouth, NS B3B 1Y9, Canada; Cory.Murphy@canada.ca

[2] New Brunswick Operations, Canadian Food Inspection Agency, 99 Mount Pleasant Road, P.O. Box 1036, St. George, NB E5C 3S9, Canada; Andrew.Justason@canada.ca

[3] St. Andrews Biological Station, Fisheries and Oceans Canada, 125 Marine Science Drive, St. Andrews, NB E5B 0E4, Canada; Jennifer.Martin@dfo-mpo.gc.ca

[*] Correspondence: Wade.Rourke@canada.ca

Abstract: Shellfish toxin monitoring programs often use mussels as the sentinel species to represent risk in other bivalve shellfish species. Studies have examined accumulation and depuration rates in various species, but little information is available to compare multiple species from the same harvest area. A 2-year research project was performed to validate the use of mussels as the sentinel species to represent other relevant eastern Canadian shellfish species (clams, scallops, and oysters). Samples were collected simultaneously from Deadmans Harbour, NB, and were tested for paralytic shellfish toxins (PSTs) and amnesic shellfish toxin (AST). Phytoplankton was also monitored at this site. Scallops accumulated PSTs and AST sooner, at higher concentrations, and retained toxins longer than mussels. Data from monitoring program samples in Mahone Bay, NS, are presented as a real-world validation of findings. Simultaneous sampling of mussels and scallops showed significant differences between shellfish toxin results in these species. These data suggest more consideration should be given to situations where multiple species are present, especially scallops.

Keywords: shellfish; marine toxins; monitoring; phytoplankton; sentinel species

Key Contribution: Data from simultaneous sampling of multiple shellfish species suggest that the monitoring of additional shellfish species may be necessary, in addition to mussels as a sentinel species, to represent the risk in all species as part of a toxin monitoring program.

Citation: Rourke, W.A.; Justason, A.; Martin, J.L.; Murphy, C.J. Shellfish Toxin Uptake and Depuration in Multiple Atlantic Canadian Molluscan Species: Application to Selection of Sentinel Species in Monitoring Programs. *Toxins* **2021**, *13*, 168. https://doi.org/10.3390/toxins13020168

Received: 7 January 2021
Accepted: 15 February 2021
Published: 22 February 2021

1. Introduction

Shellfish toxins have been present and monitored for many decades on the Canadian Atlantic coast [1]. The first North American shellfish sanitation regulations came into effect in 1925 [2], and a Canada/USA bilateral agreement on shellfish sanitation was enacted in 1948 [3]. This agreement is still in place, and the key principles are delivered through the National Shellfish Sanitation Program (NSSP) in the USA and the Canadian Shellfish Sanitation Program (CSSP) in Canada. The CSSP is delivered by three government departments: the Canadian Food Inspection Agency (CFIA), Environment and Climate Change Canada (ECCC) and Fisheries and Oceans Canada (DFO). The CFIA performs marine toxin monitoring for paralytic shellfish toxins (PSTs), amnesic shellfish toxin (AST), and lipophilic shellfish toxins (LSTs). Health Canada has established maximum limits (MLs) for these toxins in bivalve shellfish edible tissue [4]: 0.8 mg saxitoxin (STX) equivalents/kg for PSTs, 20 mg/kg domoic acid (DA) for AST, 0.2 mg okadaic acid (OA) equivalents/kg and 0.2 mg pectenotoxin (PTX)/kg for LSTs. These toxins have all been responsible for Canadian harvest area closures from time to time when shellfish concentrations have exceeded an ML [5]. A thorough review of occurrence, modes of action and chemical properties for these toxins is included in Daneshian et al. [6].

Laboratory methods available to detect and quantify these toxins have changed considerably over time. Monitoring of PSTs was originally completed using a mouse bioassay method [7], but now there are multiple chemistry-based analytical methods [8,9] and a receptor-binding assay [10] that have been validated and approved as AOAC Official Methods of Analyses. Additionally, LC-MS/MS methods have now been validated for PST analysis [11,12] and offer even greater selectivity and confirmation ability. Monitoring of AST has been performed consistently with chemistry-based analytical methods [13], with improvements as technology has advanced [14]. Monitoring of LSTs, like PSTs, was previously widely performed using a mouse bioassay [15], but has now advanced to analytical methods using LC-MS/MS [16]. The use of chemistry-based analytical methods requires purified standards for each individual toxin, and known toxic equivalence factors (TEF) in order to calculate results. Despite these additional needs, the lower detection limits and toxin profile information that these methods provide are invaluable in modern monitoring programs [17,18].

Phytoplankton monitoring results have the potential to be used as an early warning for elevated toxin levels in shellfish, although there are many variables that are not well understood with regard to phytoplankton population dynamics and toxin production [19, 20]. Many different species of phytoplankton can cause toxin outbreaks, for example, *Alexandrium catenella* has been responsible for producing PSTs in Deadmans Harbour, NB, (Figure 1a) in the Bay of Fundy [19,21] and *Pseudo-nitzschia pseudodelicatissima* has been responsible for producing AST in the Bay of Fundy [22]. Some countries monitor phytoplankton counts as part of routine shellfish monitoring programs [23,24], although it is not required as part of the CSSP.

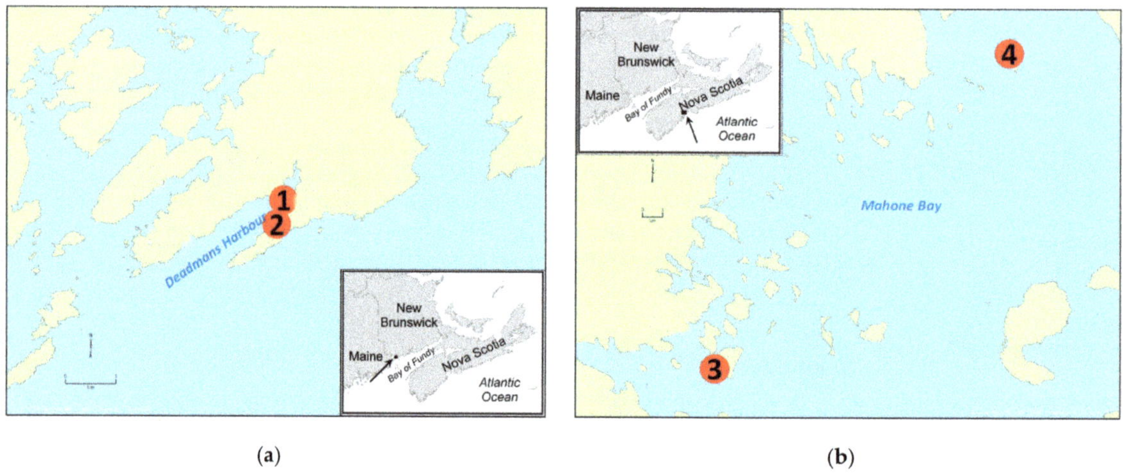

(a)   (b)

Figure 1. Maps of sampling areas: (**a**) Deadmans Harbour, NB; (**b**) Mahone Bay, NS. Sampling sites are labelled as follows: 1. Deadmans Harbour natural clam bed, 2. Deadmans Harbour experimental site for submerged cages, 3. Indian Point, 4. Snake Island.

There have been studies examining toxin accumulation and depuration of PSTs, AST, and LSTs in shellfish; some of these studies have involved opportunistic sampling during toxin blooms [25–29], while others have been controlled laboratory studies where shellfish were fed toxic phytoplankton [28,30–37]. These studies have described observations in single species, and in some cases, included comparisons between species. The diversity and complexity of shellfish environments mean that these studies cannot fully describe the processes of toxin accumulation and depuration in the natural environment. Other studies have suggested ways to mitigate the impact of toxic episodes, such as methods for decreasing toxin accumulation [38,39] and increasing depuration rates [40]. These techniques may not be practical on a large scale, are generally only applicable to aquaculture settings, and have not negated the need for routine monitoring programs.

The CSSP has been very effective, with only a single documented outbreak of shellfish-toxin related illnesses associated with legally harvested shellfish in recent history [41]; however, efforts are always being made to improve monitoring. It is important to ensure that the risk of elevated toxin levels is adequately assessed in each harvesting area, and mussels (*Mytilus spp.*) are the most common species used for this purpose in Canada and other areas [23]. Monitoring toxin levels in multiple shellfish species simultaneously poses a difficulty in assessing risk in shellfish harvest areas. Furthermore, the unpredictable nature of toxin-producing algal blooms makes it difficult to plan experiments to gather information to develop strategies to address these issues.

This paper describes results from a combination of (1) a designed research project and (2) opportunistic sampling. The objective of the designed research project was to validate the use of mussels as the sentinel species for monitoring PSTs (and represent the highest PST risk in various bivalve species). The study was completed in Deadmans Harbour, NB, where historic monitoring results demonstrated the annual presence of both *A. catenella* cells and PSTs in shellfish; PSTs are the most prevalent shellfish toxins detected in eastern Canada, and there are few sites with this level of predictability. Although *P. pseudodelicatissima* was observed annually at the test site, the shellfish rarely accumulated AST. This experiment was later expanded when AST was detected in routine monitoring of shellfish samples. Blue mussels (*Mytilus edulis*), soft-shell clams (*Mya arenaria*), Atlantic sea scallops (*Placopecten magellanicus*), and eastern oysters (*Crassostrea virginica*) from other harvest areas were stored in, and subsequently simultaneously harvested from, submerged cages 150–250 m distance from the natural clam bed (a routine CSSP monitoring station). All shellfish were harvested from other areas, tested to ensure that they contained no toxins when transferred, and then allowed to acclimate to conditions in Deadmans Harbour for at least 3 weeks before being sampled. PST results were used to compare accumulation and depuration rates between species. Water samples were also collected in close proximity to the suspended cages and the total phytoplankton community was analyzed. Species enumeration included cell counts for *A. catenella*, which were compared with PST levels in shellfish to assess the application of phytoplankton monitoring as a predictor of PSTs in shellfish.

Additional data were obtained from opportunistic sampling at two harvest sites in Mahone Bay, NS, (Figure 1b) when mussels and scallops were sampled simultaneously due to elevated toxin levels noted at these sites. This sampling was a combination of planned monitoring samples and targeted sampling in response to increasing toxin levels. As the generation of these data was not specifically designed to support this paper, not all species were sampled at each time point. These data are presented as validation of the results from the research study. Table 1 shows the number of samples collected at each site and which toxin groups were analyzed.

Table 1. Sampling locations and numbers of samples included in this study for paralytic shellfish toxins (PST), amnesic shellfish toxin (AST), and lipophilic shellfish toxins (LST).

Toxin	Harvest Site	Mussels	Scallops	Oysters
PST	Deadmans Harbour, NB [1]	63	63	62
AST	Deadmans Harbour, NB [1]	20	20	19
AST	Indian Point, NS [2]	26	25	− [4]
AST	Snake Island, NS [2]	16	21	− [4]
LST	Indian Point, NS [3]	22	22	− [4]
LST	Snake Island, NS [3]	13	17	− [4]

[1] Sampled January 2013–May 2015; [2] Sampled April–December 2017; [3] Sampled April–October 2017; [4] No oysters at these sites.

Despite the length of time that routine monitoring has been in place around the world, much remains to be understood about the toxin accumulation and depuration rates of various species [28,42–45]. This paper contributes information about the toxin uptake of multiple species, which can be used to improve the design and implementation of shellfish toxin monitoring programs.

2. Results

2.1. Research Project—Deadmans Harbour, NB

2.1.1. Toxin Monitoring in Shellfish

Results from PST analyses of clams held in submerged cages were inconsistent with the results from all other species in this study and did not demonstrate any peak in toxin level (Figure A1). This could be explained by a number of factors, including that the clams were not in their natural environment (tidal mud flats), experienced more turbulent oceanographic conditions, or had altered feeding rates due to stocking density or fouling of cages. Appendix A includes a detailed rational for excluding soft-shell clam results from this study, including analysis of long-term comparison of mussel and soft-shell clams at another CSSP monitoring site (Table A1). The original study design was to compare PST concentrations between samples from experimental cages, and to use the sample results from the natural clam bed as validation of the results; instead, the results from the natural clam bed were used to exclude soft-shell clam data.

Figure 2 shows PST results from multiple shellfish species. All species accumulated some PSTs during the project, but oysters were the only species in which PST concentrations never exceeded the ML. The PST concentrations in mussels changed rapidly and coincided with the rise and fall of adjacent *A. catenella* cell counts (Figure 2). Scallop samples exceeded the ML before mussel samples during the toxic episodes in 2013 and 2014, and provided the earliest warning of increasing PST levels. Analysis of shellfish before the project began confirmed no PSTs were present in any species, and eliminated the possibility of contamination from previous toxic episodes.

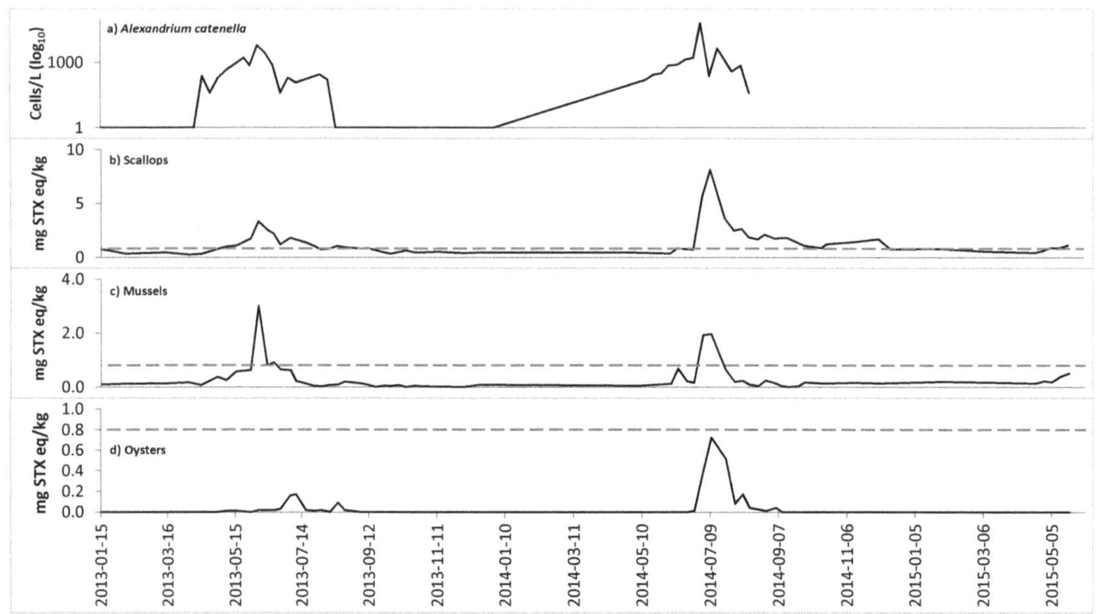

Figure 2. Cell counts (cells/L) for (**a**) *Alexandrium catenella* and total paralytic shellfish toxins (PST) concentrations (mg STX eq/kg) for species harvested from experimental cages in Deadmans Harbour, NB (**b**) scallops; (**c**) mussels; (**d**) oysters. The dashed red lines represent the Canadian PST maximum limit (ML).

Oyster PST concentrations remained significantly lower than mussels and scallops, and there was a delay in observable PST concentrations in oysters relative to mussels and scallops. Figure 3 shows that the PST concentrations in mussels began to increase as surface water temperatures increased in late April/early May.

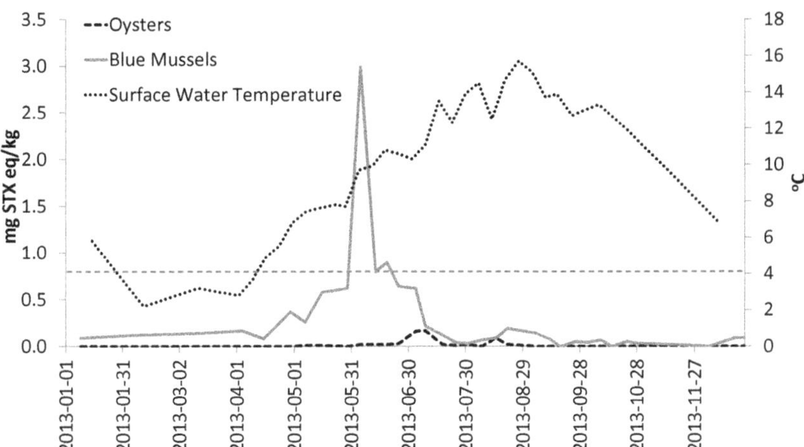

Figure 3. Total PST concentrations (mg STX eq/kg) in mussels and oysters harvested from experimental cages in Deadmans Harbour, NB, with accompanying surface water temperature (°C) for 2013. The dashed red line represents the Canadian PST ML.

The highest AST levels were detected in scallops, which were the only species to exceed the ML (Figure 4). Mussel samples also contained AST, but for a much shorter duration: 6 days in mussels vs. 175 days in scallops. Only a single oyster sample had detectable AST levels during the same time period.

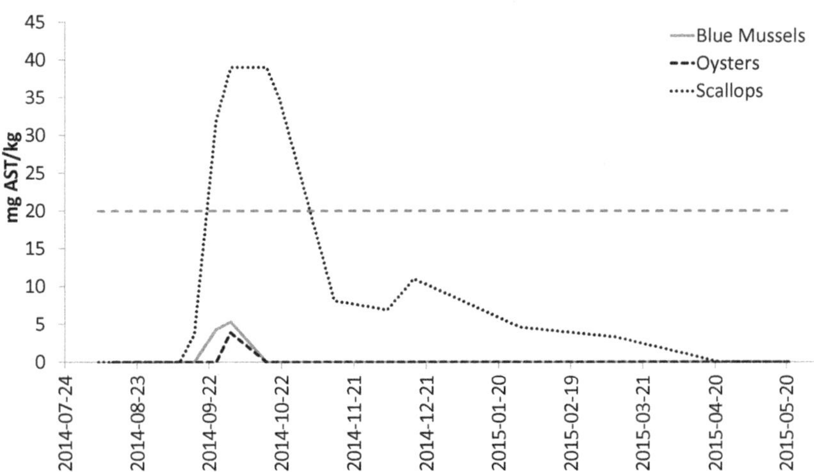

Figure 4. AST concentrations (mg AST/kg) for shellfish harvested from experimental cages in Deadmans Harbour, NB. The dashed red line represents the Canadian AST ML.

2.1.2. Phytoplankton Monitoring

The total phytoplankton community was analyzed as part of a long-term dataset initiated in 1988, which provided the opportunity to capture a weekly picture of species initiation, development, and decline [46]. The subset of *A. catenella* cell concentrations showed a strong temporal correlation with PST presence in shellfish (Figure 2), while cell counts showed no correlation with PST concentrations in shellfish, as has been observed in previous Bay of Fundy studies [19,47]. Low levels of PSP toxicity can be detected at very low concentrations of A. catenella (20–40 cells/L). The *A. catenella* cell counts changed more rapidly than the shellfish toxin levels. This may have been due to physical oceanography, bloom (duration, intensity and toxicity), very low numbers resulting in shellfish toxicity, patchiness of the cell distributions, the fact that *A. catenella* is often not the dominant phytoplankton species in the community and shellfish can selectively feed on other species, and/or retention and conversion of toxins in shellfish for extended periods. Weekly sampling for *A. catenella* indicates that this frequency of sampling is sufficient to provide an indication of increasing PSTs in tissues. Following the bloom, an absence of *A. catenella* cells within the water column indicates that the shellfish have the potential to depurate and PST concentrations can decrease. This absence of cells can act as a signal to increase PST analyses in order to measure the decline in toxins and determine the timing for the safe marketing of shellfish. *Pseudo-nitzschia pseudodelicatissima* cell counts were not available for 2014.

2.2. Validation of Research Findings with Routine Monitoring Samples—Mahone Bay, NS

Samples obtained from two sites in Mahone Bay, NS, highlighted significant differences in LST concentrations between scallops and mussels. At one site (Indian Point), LST levels were 10 × higher in mussels than in scallops, while at the other site (Snake Island), scallop LST levels were higher (Figure 5). It is also noteworthy that scallops did not retain LST for an extended time as they did for PST and AST.

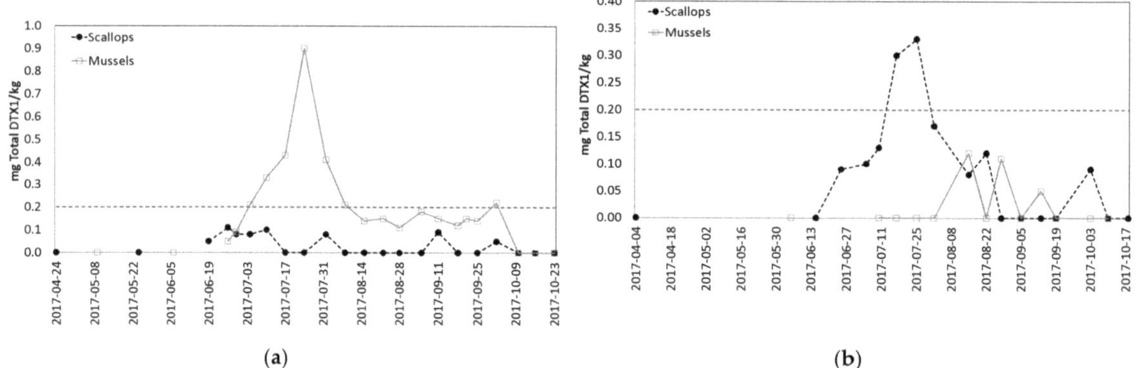

(a)   (b)

Figure 5. LST concentrations (mg total DTX1/kg) for shellfish harvested in Mahone Bay, NS: (a) Indian Point, NS; (b) Snake Island, NS. The dashed red lines represent the Canadian LST ML.

The only LST toxins detected were dinophysistoxin-1 (DTX1) and DTX1 esters; no OA or DTX2 were detected. Mussel samples were contaminated with both free and esterified forms of DTX1, with esterified forms contributing an average of 51% of the total toxicity (ranging from 32–100%) (Figure 6). No free DTX1 was detected in scallops; all toxins observed in scallops were present in the esterified form. The data in Figure 6 are displayed by species (combination of Indian Point and Snake Island samples); no differences in esterification rates were detected between those sites.

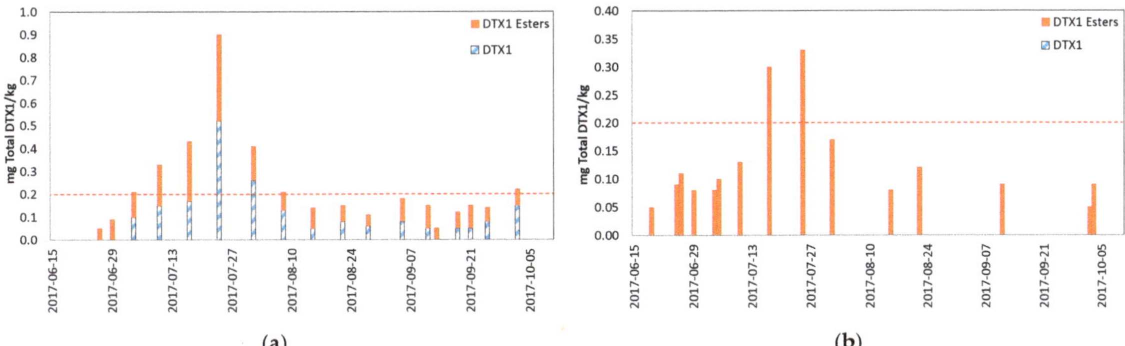

Figure 6. LST toxin profile concentrations (mg total DTX1/kg) for shellfish harvested in Mahone Bay, NS (Indian Point and Snake Island combined): (**a**) mussels; (**b**) scallops. The dashed red lines represent the Canadian LST ML.

The AST results from Mahone Bay, NS are presented in Figure 7. The onset of toxicity in scallops was not captured, because mussels and scallops were not sampled simultaneously until AST was detected. These data show that scallops accumulated higher AST concentrations than mussels, and scallops also retained the toxin over a much greater period of time than mussels. Scallops and mussels were both tested and found to have no toxins one month before AST was detected (markers visible on x-axis indicate sampling events when no toxins were detected); this is significant because it confirms that AST detected in shellfish were a result of a new contamination, not residual levels from previous exposures.

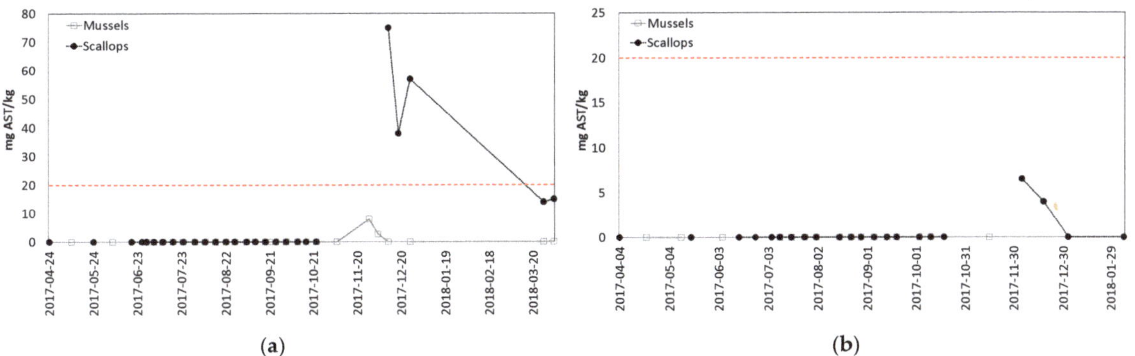

Figure 7. AST concentrations (mg AST/kg) for shellfish harvested in Mahone Bay, NS: (**a**) Indian Point, NS; (**b**) Snake Island, NS. The dashed red lines represent the Canadian AST ML.

Raw data from all figures is available in the Supplementary Materials.

3. Discussion

The PST concentrations in mussels changed rapidly in comparison to the other species in this study and coincided with the rise and fall of *A. catenella* cell counts. This rapid rise and subsequent fall of PST levels in mussels suggests that they tend to accumulate and depurate the toxins more rapidly than other species, which greatly increases the possibility of missing spikes in toxin levels, because the window of time with elevated mussel PST levels can be quite narrow; this is consistent with previous observations [17]. A spike in toxin levels could occur between PST sampling events and be missed if mussels were the only species sampled. However, weekly *A. catenella* sampling would indicate the presence and magnitude of *A. catenella* cell concentrations, thus providing warning of PST shellfish toxicity so that shellfish sampling and analysis could be increased. The peak PST concentration in mussels from this study was higher in 2013 than 2014, but the opposite trend was observed in scallops and oysters from the submerged cages and clams from the natural clam bed (Appendix A). Since mussel PST levels change rapidly, this may indicate that a mussel sample was collected in 2014 just before or after a rapid change in PST concentrations, but before the concentrations changed in the other species. The observation of extended PST retention in scallops compared to mussels suggests that scallops may need to be monitored, in addition to mussels, to reopen harvest areas when scallops are present and being harvested.

Scallops were the first species with detectable levels of AST, with 1-week and 2-week delays before AST levels were observed in mussels and oysters, respectively. Scallops also retained AST longer than mussels; this is consistent with published literature documenting retention of AST in scallops [20,25,30,31], as well as rapid depuration of AST from mussels [20,26,30]. These AST results further support the conclusion that scallops may need to be monitored, in addition to mussels, when scallops are being harvested.

A delay was observed in the rise of PST levels in oysters relative to mussels and scallops. This delay could be related to the temperature dependence of oyster feeding behavior [33], as this study confirmed that oysters did not accumulate PSTs when the surface water temperature was <10 °C, consistent with previous research [33]. Recent research [34,35,48,49] with other shellfish species suggests that PST accumulation and depuration rates could be significantly different (lower toxin accumulation rates and slower depuration rates) at warmer temperatures. The effect of increased surface water temperatures on shellfish accumulation and depuration rates will need to be explored with relevant shellfish species to determine if there could be a potential future impact on shellfish monitoring in eastern Canada.

The use of phytoplankton monitoring as an early PST warning was also considered. *Alexandrium catenella* cell counts are predictive of shellfish PST levels in some areas [50,51], but in other areas, this relationship is only qualitative and cannot be used to predict PST levels in shellfish tissue [19,52]. The current study confirmed that *A. catenella* cell counts are not predictive of shellfish PST levels in the Bay of Fundy, since there was no delay between the peak *A. catenella* cell counts and peak shellfish toxicity. This was not unexpected, as it is rare for peak *A. catenella* cell abundance to show a correlation with shellfish toxicity in the Bay of Fundy [19]. However, *A. catenella* monitoring can provide valuable complementary data for PST monitoring programs and industry in Atlantic Canada, but based on this limited data set and previous work in the Bay of Fundy [19], it is not in a position to replace or allow for reduced shellfish sampling at this particular monitoring site (and is not included in CSSP monitoring). Other work has reached a similar conclusion for LSTs [53]. In addition, higher cell counts increase the potential for toxic episodes, but they are not always predictive since some phytoplankton species do not produce toxins consistently, and higher toxin production can sometimes be observed with relatively low cell counts [19]. It is noteworthy that *A. catenella* from the Bay of Fundy always produce toxins; however, other factors impact shellfish uptake rates (such as position in the water column, weather conditions, physical oceanography of an area, etc.). These data do not diminish the value of phytoplankton monitoring to add context to PST results (especially when PST levels in shellfish are below the ML); elevated cell counts still indicate an increased potential for toxin production, which could be used as a trigger for targeted sampling (or other actions).

An opportunity to validate conclusions from the research project was presented when AST and LST concentrations rose in Mahone Bay, NS during 2017. Toxin levels were elevated at harvest sites with both mussels and scallops during the summer and fall, and resulted in sampling of both species. The simultaneous sampling of both species necessary to confirm the relative accumulation and depuration rates was not conducted because this was reactive sampling completed to ensure food safety, not a designed research project; only one species was sampled during some weeks. The conclusions based on AST results from the research project at Deadmans Harbour were confirmed by these real-life monitoring results; scallops accumulated higher AST levels and retained the toxin longer than mussels. Results from LST analyses also demonstrated that a single species cannot always represent the risk for other species, as mussels were the higher risk species at one site (Indian Point), while scallops were the higher risk species at the other site (Snake Island). Samples of both species were tested and found to contain no toxins within a month of onset of this toxic episode. These results make it particularly difficult to select a single species for routine monitoring. The fact that LST levels in scallops at Indian Point never exceeded the ML is noteworthy, as is the fact that the LST levels were present in mussels for longer than scallops. This is consistent with LST observations in different mussel and scallop species [27], although different than the toxin retention behavior observed in scallops for PSTs [1] and AST [25,30,31]. A difference was also observed in the LST toxin profiles for mussels and scallops. LSTs detected in mussels were both free and esterified forms of DTX1; the esterified forms of DTX1 contributed an average of 51% to the total toxicity (ranging from 32–100%). No free DTX1 was detected in scallops from either site; all toxins observed in scallops were present only in the esterified form. This highlights the differences in toxin profiles between species [45], as well as the importance of performing the alkaline hydrolysis step necessary to liberate the esterified forms of these toxins during analysis. There could be a significant under-estimation of risk if scallops were selected as the sentinel species and analyzed without looking for the esterified forms of LSTs.

These data all demonstrate that a single species cannot always represent the risk accurately when multiple shellfish species are present in a harvest area; Bresnan et al. reached the same conclusion [20]. The routine monitoring data from Mahone Bay presented cannot fully validate the conclusions of the research project because there were no PST levels detected in shellfish sampled for routine monitoring at these times; however, when all the available data are considered together, they highlight that different approaches may be needed to deal with different species and risks associated with different toxins. The risks in all situations included in this paper, and especially those in Mahone Bay, were managed through routine CSSP procedures; harvest areas were closed appropriately and no illnesses were linked with any of these results. Mussels are a hardy species, easily maintained in cages, present at many harvest areas, and easily sampled; all these factors support using mussels for toxin monitoring. The presented data do not suggest that mussels should not be used for monitoring, but that the appropriate context should be considered when interpreting toxin levels in mussels. This is especially true in areas with multiple shellfish species present. As an example, a sampling procedure employed in Mahone Bay, NS, was to sample both species regularly, alternating between mussels and scallops until toxins were detected, and then to sample both species simultaneously to ensure that the highest risk was identified. This is also consistent with the decision by CFIA in BC to sample geoducks and mussels when both species are present, because data have demonstrated that PST levels in mussels do not represent PST levels in geoducks [54].

4. Conclusions

Mussels have been used as the sentinel species in Atlantic Canadian shellfish toxin monitoring programs for many years. These samples and programs have generally protected consumers from illnesses related to shellfish toxins. Additional context may be necessary if there are multiple species in the same harvest area, where additional species may need to be sampled along with mussels to evaluate the food safety risk associated with each species.

5. Materials and Methods

All testing was completed using validated methods in an ISO 17025 accredited laboratory.

5.1. Samples

Shellfish (blue mussels, soft-shell clams, Atlantic sea scallops, and eastern oysters) were harvested from other areas, tested to ensure they contained no toxins when transferred, and then allowed to acclimate to conditions in Deadmans Harbour, NB for at least 3 weeks before being sampled. Shellfish were kept in submerged cages on the ocean floor for the research project in Deadmans Harbour, NB (2013–2015). Shellfish were sampled simultaneously, and sampling frequency ranged from weekly in summer to monthly in winter, according to PST risk and CSSP sampling frequency.

The retention of PSTs in various scallop tissues has been documented for a long time [1, 55,56]. Atlantic sea scallop (*Placopecten magellanicus*) adductor muscles are commonly consumed in Canada (not whole scallops) and lower (or no) PST levels are present in Atlantic sea scallop adductor muscle compared to whole scallops which contain digestive materials that accumulate toxins. Whole scallops were analyzed in this study to assess toxin concentrations, and because whole tissue is the easiest tissue to use for regulatory monitoring.

Samples of mussels and scallops were obtained from aquaculture operations in Mahone Bay, NS in support of CSSP monitoring.

Water samples (250 mL) were collected weekly (although not always on the same day as shellfish samples) for phytoplankton analysis from the surface in close proximity to the suspended cages and preserved with 2.5% formalin acetic acid (FAA) (Fisher Scientific, Nepean, ON, Canada). Later, 50 mL subsamples were settled in counting chambers for 16 h and the whole surface area was counted and enumerated for total phytoplankton community, including *A. catenella* and *P. pseudodelicatissima* concentrations (as cells/L or chains of cells/L) using a Nikon inverted microscope (Nikon Instruments, Melville, NY, USA). A vertical 20 μm mesh phytoplankton 30 cm net sample was collected, preserved with FAA for qualitative analysis of dominant phytoplankton (as well as harmful) species using a compound Nikon microscope (Nikon Instruments, Melville, NY, USA).

5.2. Reagents and Chemicals

Instrument solvents, test reagents, and chemicals for the analysis of all sample types were either HPLC- or LC-MS-grade, as appropriate for the assay. Certified reference materials (CRMs) used for preparing instrumental calibrants were all obtained from Biotoxin Metrology, NRCC, Halifax, Canada.

5.3. Sample Preparation

Samples of shellfish were shucked and extracted following internal laboratory protocols prior to analysis for marine toxins. Subsamples of the tissue homogenates were extracted as noted below. Whole shellfish tissue was analyzed in all cases.

5.4. PST

All PST testing was conducted on 5 ± 0.25 g subsamples of tissue homogenate by LC with post-column oxidation and fluorescence detection (LC-PCOX-FLD) following AOAC OMA 2011.02 [9] using single-point calibration [57,58]. Analyses were carried out with Agilent 1200 LC systems (Agilent Technologies, Kirkland, PQ, Canada) and Waters reagent manager pumps and post-column reaction modules (Waters Limited, Milford, MA, USA) fitted with 1.0 mL knitted teflon reaction coils (Sigma-Aldrich Canada, Oakville, ON, Canada). PST analogues included in the method were saxitoxin (STX), neosaxitoxin (NEO), decarbamoyl saxitoxin (dcSTX), gonyautoxins 1 to 5 (GTX1–5), decarbamoyl gonyautoxins 2 and 3 (dcGTX2&3), and N-sulfocarbamoyl gonyautoxins 2 and 3 (C1&2). PST method LOD estimates are shown in Table 2.

Table 2. LOD estimates for compounds in PST method (µg STX eq/kg).

PST Analogue	GTX4	GTX1	dcGTX3	GTX5	dcGTX2	GTX3	GTX2	NEO	dcSTX	STX	C1	C2
LOD (µg STX eq/kg)	10	25	1.1	5.1	3.5	1.5	7.5	19	11	11	0.3	1.4

5.5. AST

All AST testing was conducted on subsamples of tissue homogenate by LC-UV using an in-house method based on [13,14]. Analyses were carried out with an Agilent 1290 UHPLC system (Agilent Technologies, Kirkland, PQ, Canada) with UV-diode array detection. Tissue homogenate was weighed (5 ± 0.25 g) into a 50 mL polypropylene centrifuge tube. Then, 5.0 mL water was added, the mixture was vortexed before adding 10.0 mL methanol (Caledon Laboratories, Georgetown, ON, Canada). The mixture was vortexed again and then centrifuged at ≥1000 g for 10 min. Approximately 1.5 mL resulting supernatant was filtered through a 0.2 µm nylon syringe filter. Filtered sample extract (750 µL) was transferred to an autosampler vial and diluted with 750 µL water, and vortexed. Injections of 2 µL were performed on a Waters Acquity UPLC BEH C18, 1.7 µM, 2.1 × 50 mm column (Waters Limited, Taunton, MA, USA) at 50 °C, and a flow rate of 0.7 mL/min. Mobile phase A was water + 0.1% formic acid (Sigma-Aldrich Canada, Oakville, ON, Canada), and mobile phase B was acetonitrile (Caledon Laboratories, Georgetown, ON, Canada) + 0.1% formic acid. AST was eluted during a 1.2 min isocratic hold at 8% mobile phase B, and this was followed by a 0.5 min isocratic hold at 50% mobile phase B to flush the column, and 0.3 min isocratic hold at 8% mobile phase B to re-equilibrate at starting conditions for the next injection. AST peaks were measured at 242 nm and confirmed by spectral comparison with external calibration standards. The method LOD was 0.7 mg/kg shellfish tissue. AST analogues included domoic acid and epi-domoic acid.

5.6. LST

All LST testing was conducted on 2 ± 0.05 g subsamples of tissue homogenate by LC-MS/MS [59] with no SPE cleanup and separation on a Waters Acquity UPLC BEH Shield RP18, 1.7 µm, 2.1 × 100 mm column (Waters Limited, Taunton, MA, USA) with acidic mobile phase [16]. Analyses were carried out with either an Agilent 1290 UHPLC system (Agilent Technologies, Kirkland, PQ, Canada) coupled to an AB Sciex 5500 QTrap MS/MS (AB Sciex, Concorde, ON, Canada) or a Waters I-class UPLC coupled to a Waters Xevo TQ-S Micro MS/MS (Waters Limited, Millford, MA, USA). LST analogues included in the method were gymnodimine (GYM), pinnatoxins A, E, F, and G (PnTX-A, PnTX-E, PnTX-F, PnTX-G), PnTX esters, azaspiracids 1 to 3 (AZA1-3), okadaic acid (OA), OA esters, dinophysistoxins 1 to 2 (DTX1-2), DTX1-2 esters, yessotoxin (YTX), homoYTX, 45-OH YTX, and 45-OH homoYTX. LST method LOD estimates are shown in Table 3.

Table 3. LOD estimates for regulated compounds in LST method (ng/g).

LST Analogues	AZA1	AZA2	AZA3	DTX1	DTX2	OA	PTX2	YTX
LOD (ng/g)	0.5	0.5	0.4	49	17	23	0.8	50

Supplementary Materials: The following are available online at https://www.mdpi.com/2072-6651/13/2/168/s1, Excel: Sentinel species supplemental information.

Author Contributions: Conceptualization, W.A.R. and A.J.; methodology, W.A.R.; formal analysis, W.A.R. and C.J.M.; phytoplankton and temperature data, J.L.M.; writing—original draft preparation, W.A.R.; writing—review and editing, W.A.R. and C.J.M.; visualization, W.A.R.; supervision, W.A.R. and C.J.M. All authors have read and agreed to the published version of the manuscript.

Funding: This research received no external funding.

Institutional Review Board Statement: Not applicable.

Data Availability Statement: The data presented in this study are available in supplementary material.

Acknowledgments: Michael Doon is acknowledged for assisting with placement and maintenance of the submerged cages, as well as collecting shellfish samples. The Toxins Unit of the CFIA Dartmouth Laboratory is acknowledged for performing the PST, AST, and LST analyses on all shellfish samples.

Conflicts of Interest: The authors declare no conflict of interest.

Appendix A

Figure A1 shows total PST concentrations of clams from the natural bed and mussels and clams from the experimental cages 150–250 m away. The fact that mussel results consistently remained lower than soft-shell clam results (a vs b) throughout the 2-year project was unexpected. Mussels are sampled at a lower frequency than soft-shell clams, because it is illegal to harvest mussels in that area, and soft-shell clams are the primary species used to monitor PST concentrations. Mussel sampling is maintained at some sites to monitor differences between species, and as a potential early warning of toxic episodes. A 15-year dataset from routine monitoring at a nearby harvest site, Lepreau Basin, NB was analyzed for trends between PST levels in soft-shell clams and mussels. These data are summarized in Table A1. The fact that mussels had the higher PST concentration in 67% of simultaneous sampling events, and that the difference between PST concentrations in mussels and soft-shell clams was much larger when mussels had higher PST concentration both suggest that soft-shell clam PST concentrations are generally lower than mussel PST concentrations, and very rarely are soft-shell clam concentrations significantly higher than mussel PST concentrations. This called the validity of the current data into question.

Figure A1 (b vs. c) shows total PST concentrations of clams from the natural bed and clams from the experimental cages 150–250 m away. There is a large difference between results, with clams from the experimental cages never demonstrating a peak in total PST concentration; the concentration remained consistent throughout the project. These data support the conclusion that soft-shell clam samples in this study were not representative, and should be removed from further data analysis.

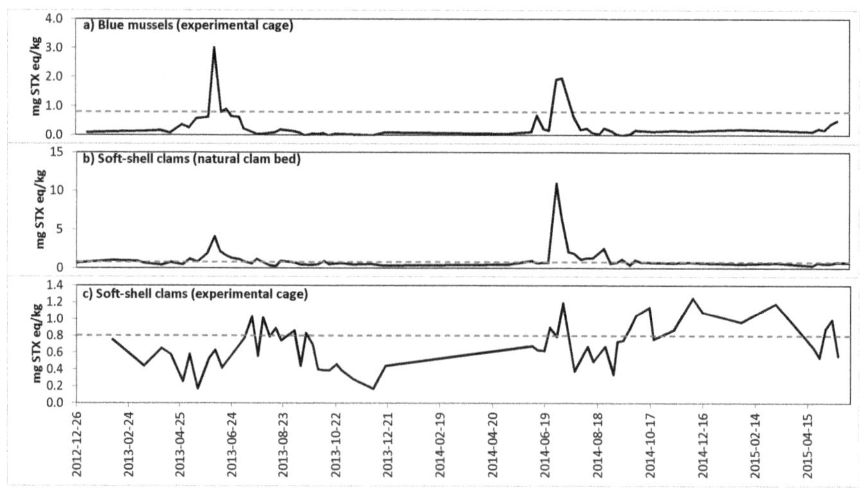

Figure A1. Total PST concentrations of mussels and clams from Deadmans Harbour (**a**) blue mussels (experimental cage), (**b**) soft-shell clams (natural clam bed) and (**c**) soft-shell clams (experimental cage). The dashed red lines represent the Canadian PST ML.

Table A1. Summary of mussel and clam sampling from Lepreau Basin, NB from 2000–2015.

Description	Mussels	Clams
Total samples	129	615
Sampling events when both species were collected simultaneously	89	
Correlation of PST concentration between mussels and clams	0.84	
Samples with highest PST concentration	60 [1]	15 [1]
PST range (mg STX equiv/kg)	0.04–37.2	0.12–9.29
Largest PST difference (mg STX equiv/kg)	33.9 [2]	0.75 [3]
Average PST difference (mg STX equiv/kg)	2.76 [2]	0.25 [3]

[1] 14 sampling events had equal PST concentrations for both species; [2] when mussels had highest PST concentration; [3] when clams had highest PST concentration.

References

1. Medcof, J.C.; Leim, A.H.; Needler, A.B.; Needler, A.W.H.; Gibbard, J.; Naubert, J. Paralytic shellfish poisoning on the Canadian Atlantic Coast. *Fish. Res. Board Can.* **1947**, *75*, 1–32.
2. World Health Organization. *Safe Management of Shellfish and Harvest Waters*, 1st ed.; Rees, G., Pond, K., Kay, D., Bartram, J., Santo Domingo, J., Eds.; IWA Publishing: London, UK, 2010; p. 360.
3. Government of Canada. Canada-US Bilateral Agreement on Shellfish Sanitation. Available online: https://www.canada.ca/en/environment-climate-change/corporate/international-affairs/partnerships-countries-regions/north-america/canada-united-states-agreement-shellfish-sanitation.html (accessed on 8 March 2018).
4. Health Canada. Health Canada's Maximum Levels for Chemical Contaminants in Foods. Available online: https://www.canada.ca/en/health-canada/services/food-nutrition/food-safety/chemical-contaminants/maximum-levels-chemical-contaminants-foods.html (accessed on 9 March 2018).
5. Bates, S.S.; Beach, D.G.; Comeau, L.A.; Haigh, N.; Lewis, N.I.; Locke, A.; Martin, J.L.; McCarron, P.; McKenzie, C.H.; Michel, C.; et al. Marine harmful algal blooms and phycotoxins of concern to Canada. *Can. Tech. Rep. Fish. Aquat. Sci.* **2020**, *3384*, 1–323.
6. Daneshian, M.; Botana, L.M.; Bottein, M.-Y.D.; Buckland, G.; Campas, M.; Dennison, N.; Dickey, R.W.; Diogene, J.; Fessard, V.; Hartung, T.; et al. A roadmap for hazard monitoring and risk assessment of marine biotoxins on the basis of chemical and biological test systems. *Altex* **2013**, *30*, 487–545. [CrossRef]
7. AOAC INTERNATIONAL. Method 959.08. In *Official Methods of Analysis*, 19th ed.; AOAC INTERNATIONAL: Gaithersburg, MD, USA, 2012; Volume 19.
8. AOAC INTERNATIONAL. Method 2005.06. In *Official Methods of Analysis*, 19th ed.; AOAC INTERNATIONAL: Gaithersburg, MD, USA, 2012; Volume 19.
9. AOAC INTERNATIONAL. Method 2011.02. In *Official Methods of Analysis*, 19th ed.; AOAC INTERNATIONAL: Gaithersburg, MD, USA, 2012; Volume 19.
10. AOAC INTERNATIONAL. Method 2011.27. In *Official Methods of Analysis*, 19th ed.; AOAC INTERNATIONAL: Gaithersburg, MD, USA, 2012; Volume 19.
11. Turner, A.D.; McNabb, P.S.; Harwood, D.T.; Selwood, A.I.; Boundy, M.J. Single-laboratory validation of a multitoxin ultra-performance LC-hydrophilic interaction LC-MS/MS method for quantitation of paralytic shellfish toxins in bivalve shellfish. *J. AOAC Int.* **2015**, *98*, 609–621. [CrossRef] [PubMed]
12. Turner, A.D.; Dhanji-Rapkova, M.; Fong, S.Y.T.; Hungerford, J.; McNabb, P.S.; Boundy, M.J.; Harwood, D.T. Ultra-high-performance hydrophilic interaction liquid chromatography with tandem mass spectrometry method for the determination of paralytic shellfish toxins and tetrodotoxin in mussels, oysters, clams, cockles, and scallops: Collaborative study. *J. AOAC Int.* **2020**, *103*, 533–562. [CrossRef]
13. Quilliam, M.A.; Xie, M.; Hardstaff, W.R. Rapid extraction and cleanup for liquid chromatographic determination of domoic acid in unsalted seafood. *J. AOAC Int.* **1995**, *78*, 543–554. [CrossRef]
14. De La Iglesia, P.; Barber, E.; Giménez, G.; Rodríguez-Velasco, M.L.; Villar-González, A.; Diogéne, J. High-throughput analysis of amnesic shellfish poisoning toxins in shellfish by ultra-performance rapid resolution LC-MS/MS. *J. AOAC Int.* **2011**, *94*, 555–564. [CrossRef] [PubMed]
15. EURLMB. *EU-Harmonised Standard Operating Procedure for Detection of Lipophilic Toxins by Mouse Bioassay*; Version 6; European Union Reference Laboratory for Marine Biotoxins: Vigo, Spain, 2013.
16. Villar Gonzalez, A.; Rodriguez-Velasco, M.L.; Gago Martinez, A. Determination of lipophilic toxins by LC/MS/MS: Single-laboratory validation. *J. AOAC Int.* **2011**, *94*, 909–922. [CrossRef]
17. Rourke, W.A.; Murphy, C.J. Animal-free paralytic shellfish toxin testing—the Canadian perspective to improved health protection. *J. AOAC Int.* **2014**, *97*, 334–338. [CrossRef]

18. Turner, A.D.; Hatfield, R.G.; Maskrey, B.H.; Algoet, M.; Lawrence, J.F. Evaluation of the new European Union reference method for paralytic shellfish toxins in shellfish: A review of twelve years regulatory monitoring using pre-column oxidation LC-FLD. *TrAC Trends Anal. Chem.* **2019**, *113*, 124–139. [CrossRef]

19. Martin, J.L.; LeGresley, M.M.; Hanke, A.R. Thirty years—*Alexandrium fundyense* cyst, bloom dynamics and shellfish toxicity in the Bay of Fundy, eastern Canada. *Deep. Sea Res. Part II* **2014**, *103*, 27–39. [CrossRef]

20. Bresnan, E.; Fryer, R.; Fraser, S.; Smith, N.; Stobo, L.; Brown, N.; Turrell, E. The relationship between Pseudo-nitzschia (Peragallo) and domoic acid in Scottish shellfish. *Harmful Algae* **2017**, *63*, 193–202. [CrossRef]

21. Prud'homme van Reine, W. Report of the Nomenclature Committee for Algae: 15. *Taxon* **2017**, *66*, 191–192. [CrossRef]

22. Martin, J.L.; Haya, K.; Burridge, L.E.; Wildish, D.J. *Nitzschia pseudodelicatissima*—A source of domoic acid in the Bay of Fundy, eastern Canada. *Mar. Ecol. Prog. Ser.* **1990**, *67*, 177–182.23. [CrossRef]

23. Trainer, V.L.; Hardy, F.J. Integrative monitoring of marine and freshwater harmful algae in Washington State for public health protection. *Toxins* **2015**, *7*, 1206–1234. [CrossRef] [PubMed]

24. European Union. Regulation (EC) No 854/2004 of the European Parliament and of the Council of 29th April 2004 laying down specific rules for the organisation of official controls on products of animal origin intended for human consumption. *Off. J. Eur. Union* **2004**, *L226*, 83–127.

25. Blanco, J.; Acosta, C.; Bermúdez de la Puente, M.; Salgado, C. Depuration and anatomical distribution of the amnesic shellfish poisoning (ASP) toxin domoic acid in the king scallop *Pecten maximus*. *Aquat. Toxicol.* **2002**, *60*, 111–121. [CrossRef]

26. Blanco, J.; Bermudez de la Puente, M.; Arevalo, F.; Salgado, C.; Moroño, A. Depuration of mussels (*Mytilus galloprovincialis*) contaminated with domoic acid. *Aquat. Living Resour.* **2002**, *15*, 53–60. [CrossRef]

27. Suzuki, T.; Mitsuya, T. Comparison of dinophysistoxin-1 and esterified dinophysistoxin-1 (dinophysistoxin-3) contents in the scallop *Patinopecten yessoensis* and the mussel *Mytilus galloprovincialis*. *Toxicon* **2001**, *39*, 905–908. [CrossRef]

28. Mafra, L.L.; Ribas, T.; Alves, T.P.; Proença, L.A.O.; Schramm, M.A.; Uchida, H.; Suzuki, T. Differential okadaic acid accumulation and detoxification by oysters and mussels during natural and simulated *Dinophysis* blooms. *Fish. Sci.* **2015**, *81*, 749–762. [CrossRef]

29. Matsushima, R.; Uchida, H.; Watanabe, R.; Oikawa, H.; Oogida, I.; Kosaka, Y.; Kanamori, M.; Akamine, T.; Suzuki, T. Anatomical distribution of diarrhetic shellfish toxins (DSTs) in the Japanese scallop *Patinopecten yessoensis* and individual variability in scallops and *Mytilus edulis* mussels: Statistical considerations. *Toxins* **2018**, *10*, 395. [CrossRef]

30. Wohlgeschaffen, G.D.; Mann, K.H.; Rao, D.V.S.; Pocklington, R. Dynamics of the phycotoxin domoic acid: Accumulation and excretion in two commercially important bivalves. *J. Appl. Phycol.* **1992**, *4*, 297–310. [CrossRef]

31. Douglas, D.J.; Kenchington, E.R.; Bird, C.J.; Pocklington, R.; Bradford, B.; Silvert, W. Accumulation of domoic acid by the sea scallop (*Placopecten magellanicus*) fed cultured cells of toxic *Pseudo-nitzschia multiseries*. *Can. J. Fish. Aquat. Sci.* **1997**, *54*, 907–913. [CrossRef]

32. Nielsen, L.T.; Hansen, P.J.; Krock, B.; Vismann, B. Accumulation, transformation and breakdown of DSP toxins from the toxic dinoflagellate *Dinophysis acuta* in blue mussels, *Mytilus edulis*. *Toxicon* **2016**, *117*, 84–93. [CrossRef] [PubMed]

33. Mafra, L.L.; Bricelj, V.M.; Ouellette, C.; Bates, S.S. Feeding mechanics as the basis for differential uptake of the neurotoxin domoic acid by oysters, *Crassostrea virginica*, and mussels, *Mytilus edulis*. *Aquat. Toxicol.* **2010**, *97*, 160–171. [CrossRef]

34. Farrell, H.; Seebacher, F.; O'Connor, W.; Zammit, A.; Harwood, D.T.; Murray, S. Warm temperature acclimation impacts metabolism of paralytic shellfish toxins from *Alexandrium minutum* in commercial oysters. *Glob. Chang. Biol.* **2015**, *21*, 3402–3413. [CrossRef]

35. Turner, L.M.; Alsterberg, C.; Turner, A.D.; Girisha, S.K.; Rai, A.; Havenhand, J.N.; Venugopal, M.N.; Karunasagar, I.; Godhe, A. Pathogenic marine microbes influence the effects of climate change on a commercially important tropical bivalve. *Sci. Rep.* **2016**, *6*, srep32413. [CrossRef]

36. Qiu, J.; Ji, Y.; Fang, Y.; Zhao, M.; Wang, S.; Ai, Q.; Li, A. Response of fatty acids and lipid metabolism enzymes during accumulation, depuration and esterification of diarrhetic shellfish toxins in mussels (*Mytilus galloprovincialis*). *Ecotoxicol. Environ. Saf.* **2020**, *206*, 111223. [CrossRef]

37. Liu, Y.; Kong, F.-Z.; Xun, X.-G.; Dai, L.; Geng, H.-X.; Hu, X.-L.; Yu, R.-C.; Bao, Z.-M.; Zhou, M.-J. Biokinetics and biotransformation of paralytic shellfish toxins in different tissues of Yesso scallops, *Patinopecten yessoensis*. *Chemosphere* **2020**, *261*, 128063. [CrossRef]

38. Blanco, J.; Martín, H.; Mariño, C. Reduction of diarrhetic shellfish poisoning (DSP) toxins accumulation in cultured mussels by means of rope clustering and hydrodynamic barriers. *Aquaculture* **2017**, *479*, 120–124. [CrossRef]

39. Lu, G.; Song, X.; Yu, Z.; Cao, X.; Yuan, Y. Environmental effects of modified clay flocculation on *Alexandrium tamarense* and paralytic shellfish poisoning toxins (PSTs). *Chemosphere* **2015**, *127*, 188–194. [CrossRef]

40. Peña-Llopis, S.; Serrano, R.; Pitarch, S.R.; Beltrán, E.; Ibáñez, M.V.; Hernández, F.H.; Peña, J.B. N-Acetylcysteine boosts xenobiotic detoxification in shellfish. *Aquat. Toxicol.* **2014**, *154*, 131–140. [CrossRef] [PubMed]

41. Taylor, M.; McIntyre, L.; Ritson, M.; Stone, J.; Bronson, R.; Bitzikos, O.; Rourke, W.; Galanis, E.; Outbreak Investigation Team. Outbreak of diarrhetic shellfish poisoning associated with mussels, British Columbia, Canada. *Mar. Drugs* **2013**, *11*, 1669–1676. [CrossRef]

42. Ajani, P.; Harwood, D.T.; Murray, S.A. Recent trends in marine phycotoxins from Australian coastal waters. *Mar. Drugs* **2017**, *15*, 33. [CrossRef]

43. Oyaneder Terrazas, J.; Contreras, H.R.; García, C. Prevalence, variability and bioconcentration of saxitoxin-group in different marine species present in the food chain. *Toxins* **2017**, *9*, 190. [CrossRef] [PubMed]

44. Costa, P.R.; Costa, S.T.; Braga, A.C.; Rodrigues, S.M.; Vale, P. Relevance and challenges in monitoring marine biotoxins in non-bivalve vectors. *Food Control* **2017**, *76*, 24–33. [CrossRef]

45. Blanco, J. Accumulation of Dinophysis toxins in bivalve molluscs. *Toxins* **2018**, *10*, 453. [CrossRef] [PubMed]

46. Martin, J.L.; LeGresley, M.M.; Gidney, M.E. Phytoplankton monitoring in the western isles region of the Bay of Fundy during 2007–2013. *Can. Tech. Rep. Fish. Aquat. Sci.* **2014**, *3105*, 262.

47. Martin, J.L.; LeGresley, M.M.; Richard, D.J.A. Toxic phytoplankton, PSP and ASP toxicity data during the years 1988–1996 from the southwest Bay of Fundy, eastern Canada. In *Harmful Algae*; Reguera, B., Blanco, J., Fernandez, M.L., Wyatt, T., Eds.; Xunta de Galicia Intergovernmental Oceanographic Commission of UNESCO: Paris, France, 1998; pp. 233–234.

48. Braga, A.C.; Camacho, C.; Marques, A.; Gago-Martínez, A.; Pacheco, M.; Costa, P.R. Combined effects of warming and acidification on accumulation and elimination dynamics of paralytic shellfish toxins in mussels *Mytilus galloprovincialis*. *Environ. Res.* **2018**, *164*, 647–654. [CrossRef]

49. Trainer, V.L.; Moore, S.K.; Hallegraeff, G.; Kudela, R.M.; Clement, A.; Mardones, J.I.; Cochlan, W.P. Pelagic harmful algal blooms and climate change: Lessons from nature's experiments with extremes. *Harmful Algae* **2020**, *91*, 101591. [CrossRef]

50. Vandersea, M.W.; Kibler, S.R.; Tester, P.A.; Holderied, K.; Hondolero, D.E.; Powell, K.; Baird, S.; Doroff, A.; Dugan, D.; Litaker, R.W. Environmental factors influencing the distribution and abundance of *Alexandrium catenella* in Kachemak Bay and lower cook inlet, Alaska. *Harmful Algae* **2018**, *77*, 81–92. [CrossRef]

51. Davidson, K.; Anderson, D.M.; Mateus, M.; Reguera, B.; Silke, J.; Sourisseau, M.; Maguire, J. Forecasting the risk of harmful algal blooms. *Harmful Algae* **2016**, *53*, 1–7. [CrossRef] [PubMed]

52. Crespo, B.G.; Keafer, B.A.; Ralston, D.K.; Lind, H.; Farber, D.; Anderson, D.M. Dynamics of *Alexandrium fundyense* blooms and shellfish toxicity in the Nauset Marsh System of Cape Cod (Massachusetts, USA). *Harmful Algae* **2011**, *12*, 26–38. [CrossRef] [PubMed]

53. Reguera, B.; Riobó, P.; Rodríguez, F.; Díaz, P.A.; Pizarro, G.; Paz, B.; Franco, J.M.; Blanco, J. *Dinophysis* toxins: Causative organisms, distribution and fate in shellfish. *Mar. Drugs* **2014**, *12*, 394–461. [CrossRef]

54. Canadian Food Inspection Agency. Monitoring for Marine Toxins in Geoduck in British Columbia. Available online: https://inspection.gc.ca/food-safety-for-industry/compliance-continuum/guidance-for-inspectors/sip/monitoring-for-marine-biotoxins-in-geoduck/eng/1541104736952/1541104737206 (accessed on 5 March 2020).

55. Cembella, A.D.; Shumway, S.E.; Larocque, R. Sequestering and putative biotransformation of paralytic shellfish toxins by the sea scallop *Placopecten magellanicus*. *J. Exp. Mar. Biol. Ecol.* **1994**, *180*, 1–22. [CrossRef]

56. Shumway, S.E.; Cembella, A.D. The impact of toxic algae on scallop culture and fisheries. *Rev. Fish. Sci.* **1993**, *1*, 121–150. [CrossRef]

57. Turner, A.D.; Lewis, A.M.; Rourke, W.A.; Higman, W.A. Interlaboratory comparison of two AOAC liquid chromatographic fluorescence detection methods for paralytic shellfish toxin analysis through characterization of an oyster reference material. *J. AOAC Int.* **2014**, *97*, 380–390. [CrossRef]

58. Van De Riet, J.M.; Gibbs, R.S.; Chou, F.W.; Muggah, P.M.; Rourke, W.A.; Burns, G.; Thomas, K.; Quilliam, M.A. Liquid chromatographic post-column oxidation method for analysis of paralytic shellfish toxins in mussels, clams, scallops, and oysters: Single-laboratory validation. *J. AOAC Int.* **2009**, *92*, 1690–1704.

59. EURLMB. *EU-Harmonized Standard Operating Procedure for Determination of Lipophilic Marine Biotoxins in Molluscs by LC-MS/MS*; Version 5; European Union Reference Laboratory for Marine Biotoxins: Vigo, Spain, 2015.

Article

Lobster Supply Chains Are Not at Risk from Paralytic Shellfish Toxin Accumulation during Wet Storage

Alison Turnbull [1,2,*], Andreas Seger [1,2], Jessica Jolley [1], Gustaaf Hallegraeff [2], Graeme Knowles [3] and Quinn Fitzgibbon [2]

[1] South Australian Research and Development Institute, GPO Box 397, Adelaide, SA 5001, Australia; andreas.seger@utas.edu.au (A.S.); Jessica.jolley@sa.gov.au (J.J.)

[2] Institute for Marine and Antarctic Studies, University of Tasmania, Private Bag 129, Hobart, TAS 7001, Australia; gustaaf.hallegraeff@utas.edu.au (G.H.); quinn.fitzgibbon@utas.edu.au (Q.F.)

[3] Animal Health Laboratory DPIPWE Tasmania, 165 Westbury Rd, Prospect, TAS 7250, Australia; graeme.knowles@dpipwe.tas.gov.au

* Correspondence: Alison.Turnbull@utas.edu.au; Tel.: +61-418-348450

Abstract: Lobster species can accumulate paralytic shellfish toxins (PST) in their hepatopancreas following the consumption of toxic prey. The Southern Rock Lobster (SRL), *Jasus edwardsii*, industry in Tasmania, Australia, and New Zealand, collectively valued at AUD 365 M, actively manages PST risk based on toxin monitoring of lobsters in coastal waters. The SRL supply chain predominantly provides live lobsters, which includes wet holding in fishing vessels, sea-cages, or processing facilities for periods of up to several months. Survival, quality, and safety of this largely exported high-value product is a major consideration for the industry. In a controlled experiment, SRL were exposed to highly toxic cultures of *Alexandrium catenella* at field relevant concentrations (2×10^5 cells L^{-1}) in an experimental aquaculture facility over a period of 21 days. While significant PST accumulation in the lobster hepatopancreas has been reported in parallel experiments feeding lobsters with toxic mussels, no PST toxin accumulated in this experiment from exposure to toxic algal cells, and no negative impact on lobster health was observed as assessed via a wide range of behavioural, immunological, and physiological measures. We conclude that there is no risk of PST accumulation, nor risk to survival or quality at the point of consumption through exposure to toxic algal cells.

Keywords: lobster health; toxic algae; *Alexandrium*; *Jasus edwardsii*

Key Contribution: Lobsters exposed to toxic algae during wet storage in long supply chains do not take up paralytic shellfish toxins. Furthermore, exposure does not cause health issues for lobsters and will therefore not have an impact on their survival and health in the supply chain.

Citation: Turnbull, A.; Seger, A.; Jolley, J.; Hallegraeff, G.; Knowles, G.; Fitzgibbon, Q. Lobster Supply Chains Are Not at Risk from Paralytic Shellfish Toxin Accumulation during Wet Storage. *Toxins* **2021**, *13*, 129. https://doi.org/10.3390/toxins13020129

Received: 12 January 2021
Accepted: 5 February 2021
Published: 9 February 2021

1. Introduction

The Southern Rock Lobster (*Jasus edwardsii* Hutton) is sold in high value live export fisheries in Tasmania, Australia, and New Zealand worth AUD 97 M and AUD 268 M, respectively [1,2]. Lobsters are known to accumulate paralytic shellfish toxins (PST) during blooms of PST-producing algal species in Tasmanian and New Zealand coastal waters [3,4]. The causative alga in Tasmania is *Alexandrium catenella* (Whedon and Kofoid) Balech, whilst New Zealand blooms may be *A. minutum* Halim, *A. pacificum* Litaker (previously identified as *A. catenella*) and *Gymnodinium catenatum* Graham [5,6]. The toxins accumulate in the lobster hepatopancreas via the consumption of contaminated prey but are not found in the tail meat [7,8].

Whilst there is no Australian or New Zealand food standard for PST in lobster, several key export markets such as China and Hong Kong stipulate a maximum level of 0.8 mg saxitoxin (STX) equivalents kg^{-1}. Furthermore, human health risk assessment has shown a risk of illness for consumers if consuming large quantities of lobster hepatopancreas. This risk is significantly reduced if the bivalve regulatory level of 0.8 mg STX equivalents kg^{-1} is applied [9]. In both Tasmania and New Zealand, the public health and market access risks associated with PST in *J. edwardsii* are managed during high-risk periods through weekly or fortnightly biotoxin monitoring of bivalve sentinel species in coastal waters, followed by direct monitoring of lobster hepatopancreases when bivalves indicate risk.

The *J. edwardsii* supply chain is focused on live seafood markets in Asia. Wet storage is employed to maintain animal quality and facilitate maximum price return during market fluctuations. Animals are moved into wet storage immediately after capture and remain in specialised holding facilities as they move through the supply chain [10], as depicted in Figure 1. The seawater used for wet storage is sourced from local coastal waters.

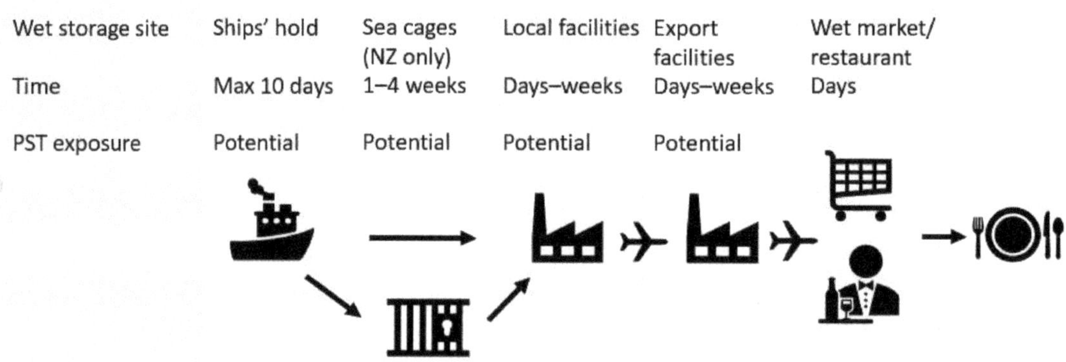

Figure 1. Supply chain for *J. edwardsii* from Tasmania and New Zealand to Asian markets, showing wet storage and potential paralytic shellfish toxin (PST) exposure sites. Biotoxin risk monitoring occurs in coastal waters prior to entry into the supply chain.

Wet storage times may range from a few days to several weeks depending on the time of year and market demands. If PST-producing algal blooms are present in coastal waters, these algae will be inadvertently pumped into holding compartments on-vessel, in sea cages (New Zealand only), or in local/export holding facilities. As a result, *J. edwardsii* may be exposed to PST in the supply chain, post regulatory monitoring programs.

Crustacean gills have multiple functions, such as ionic transport mediating haemolymph osmoregulation, acid–base balance, and ammonia excretion. Heavy metal accumulation in crustaceans also occurs via the gills [11]. No studies have examined the potential for PST uptake in lobsters when directly exposed to toxic algae. Furthermore, lobsters may be in the supply chain for significant periods of time and subjected to more than one period of emersion during transport. To maximise value in the market, lobsters need to survive the rigours of international transport and thus it is integral that they start their journey in strong health [12]. Whilst a recent study showed no impact of PST feeding-related accumulation on *J. edwardsii* health [13], no studies have so far examined the sensitivity of lobster gill cells to the superoxide radicals, exudate phycotoxins, and fatty acids that are known to be produced by toxic *Alexandrium* spp. and to have a deleterious impact on fish gill cells [14,15].

A range of indicators have been used to assess stress and predict mortality in commercially-important crustacean species [12]. In this study, we have taken a holistic approach to determine the impact of PST, examining both whole organism indicators (survival, nutritional condition, reflex, behaviour, and health) and cellular indicators (haemolymph immunity, biochemical parameters, and gill histopathology).

Commercial operators assess *J. edwardsii* health during grading using a subjective vitality scale which is based on lobster reflex and behavioural responses [16]. Reflex actions are consistent, involuntary, nearly instantaneous responses to stimuli which can reliably indicate crustacean whole-body health status independently of animal size, strength, motivation and gender [12]. Crustacean reflex scores have been previously used to provide an accurate indicator of crustacean performance in supply chain studies [12,17–20]. Other commonly used methods to assess lobster gross or whole-body performance include survival and nutritional condition [16,19,21,22]

Physiological indicators for the assessment of the health and vitality of crustaceans commonly include immune responses (raised bacteraemia levels and changes in haemocyte counts [21,23–26]; and haemolymph concentration of ions (e.g., potassium, sodium, magnesium, calcium, bicarbonate, pH), metabolites (e.g., ammonia, urea, glucose, lactate) and hormones (e.g., crustacean hyperglycaemic hormone) [21,25,27,28].

The present study aimed to determine if *J. edwardsii* could accumulate PST through exposure to PST-producing algae, and to ascertain whether direct exposure to these algae could impact lobster health and vitality. In a biosecure experimental aquaculture facility, lobsters were exposed to field-relevant concentrations of toxin-producing *A. catenella* algae for 21 days. PST concentrations in the hepatopancreas were measured, as well as a range of measures commonly employed to assess survivability and lobster health.

2. Results

2.1. Stocking Animals

Lobsters arrived in good condition, with vitalities on receipt ranging from 4–5 (maximum score possible is 5). There was no significant difference in the lobster harvest wet weights (514 ± 34 g; p value = 0.59) or carapace lengths (104 ± 3 mm; p value = 0.37) between each treatment group. No lobsters moulted during the experiment, but two lobsters in the high exposure treatment groups died; one on day 10, and one on day 11.

2.2. Specific Feed Intake

Lobsters from all treatment groups ate well during the experiment, with no significant difference in the specific feed intake (SFI) between treatment groups. Feed consumption decreased during the experiment, with SFI during weeks 2 and 3 being significantly less than that at week 1 (Figure 2; p value = 0.0009).

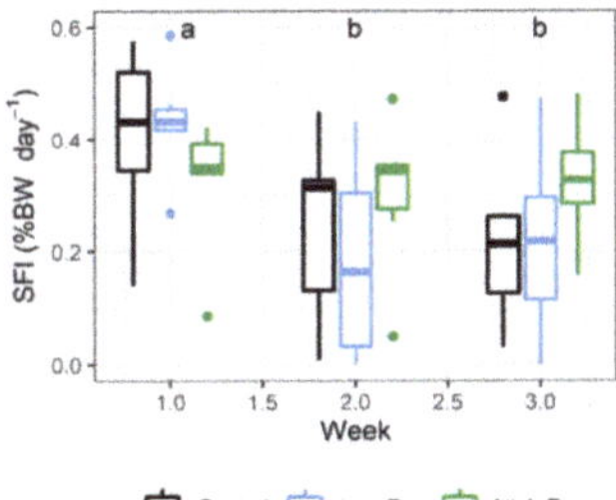

Figure 2. Weekly specific feed intake (SFI) of *J. edwardsii* lobsters exposed to 0, 1 × 10^5 or 2 × 10^5 cells of *A. catenella* per litre of tank water (control, low, or high exposure (exp) groups, respectively) across three weeks of exposure. Weeks where SFI is not significantly different share the same letter. BW: body weight.

2.3. PST Accumulation

Lobsters had low concentrations of PST in the hepatopancreas on receipt (mean 0.03 ± 0.01 mg STX.2HCl equiv. kg^{-1}). Lobsters harvested on day 21 had significantly lower PST than those harvested on day 7 (p value = 0.001), but not on day 0 (p value = 0.53; Figure 3; Supplementary Table S1). There was no significant difference between PST levels among treatment groups (p value = 0.64).

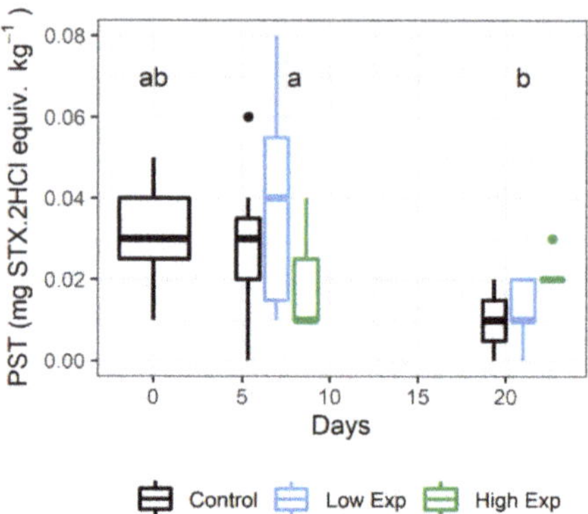

Figure 3. PST concentrations in the hepatopancreas of *J. edwardsii* lobsters harvested on days 0, 7 and 21 after exposure to 0, 1×10^5 or 2×10^5 cells of *A. catenella* per litre of tank water (control, low, or high exposure groups respectively). Days where the PST concentration is not significantly different share the same letter. STX: saxitoxin.

2.4. Lobster Health Responses

There was minimal difference between treatment groups across a wide range of behavioural, immunological, and physiological parameters measured, as summarised in Table 1. The means and standard deviations of continuous variables are provided in Supplementary Table S2.

2.5. Behavioural Responses

No significant difference between treatments or across days in the experimental system was seen in any of the behaviour measures tested (other than SFI, as discussed above). Lobster vitality remained high throughout the experiment, with 36 animals scoring the maximum vitality score (5), seven scoring a vitality of 4, and three scoring a vitality of 3. All lobsters responded quickly to being placed ventrum-up, righting within 28 s. Impairment of reflexes was low, with 32 lobsters showing impairment of three or less reflexes (Supplementary Table S2). Reflex impairment scores were significantly related to vitality scores (p value = 0.0003; Supplementary Table S3).

2.6. Immune Health Response

There was no change to bacteraemia concentrations or the prevalence of necrosis between treatment groups or days, however haemocyte counts did increase significantly across all treatment groups at day 21 compared to days 0 and 7 (Table 1, Figure 4; p values = 0.0007 and 0.002, respectively).

2.7. Nutritional Response

There were no significant differences between the nutritional indicators of Brix and hepatopancreas index across treatments or across days (Table 1, Supplementary Table S3).

Table 1. Summary of differences in behavioural, reflex, immunological, and blood chemistry parameters between control and exposed *J. edwardsii* (Treatment); across days; and for the interaction between treatment and days, as measured by ANOVA, ordinal logistic regression (OLR) or logistic regression (LR). Significant differences are marked with asterisks (* p = 0.05–0.01, ** p = 0.01–0.001, *** $p < 0.001$). ZMA: Zobell's marine agar, TCBS: thiosulphate-citrate-bile salts agar.

Variable	Treatment	Days	Treatment: Days	Two-Way ANOVA, OLR, or LR
Behaviour				
Vitality				NS
Time to right				NS
Reflex Impairment Score (RIS)				NS, vitality and RIS sig. related ($p < 0.001$)
Specific feed intake		***		F = 12.3, week 1 higher than weeks 2 and 3
Immune Response				
Haemocyte count		***		F = 18.0, weeks 1 and 2 lower than week 3
Bacteraemia on ZMA media				NS
Bacteraemia on TCBS media				NS
Necrosis				NS
Nutritional				
Hepatopancreas Index				NS
Brix				NS
Hemolymph/Biochemical				
pH	*			F = 4.0, control higher than low exp treatment
Sodium				NS
Potassium				NS
Sodium:potassium				NS
Chloride				NS
Magnesium				NS
Bicarbonate				NS
Calcium				NS
Phosphorus				NS
Glucose				NS
Lactate				NS
Cholesterol		***		F = 8.8, week 1 lower than week 3
Triglyceride		*	*	F = 7.2 (days), week 1 lower than week 3F = 3.5 (treatment:days)
Total protein		*		F = 4.9, week 1 lower than week 3
Albumin				NS
Globulin				NS
Albumin:globulin				NS
Uric acid				NS
Lipase	**			F = 7.4, control is lower than both exposed treatments
Glutamate dehydrogenase				NS
Measured osmolality				NS

2.8. Haemolymph Biochemical Response

Of the range of electrolytes, minerals, metabolites, and enzymes examined, only pH and lipase showed any significant difference between treatments (Figure 4). The low exposure treatment group showed significantly lower pH than the control group (p value = 0.02), whilst both low and high exposure groups showed significantly higher lipase concentrations than the control group (p value = 0.03 and 0.02, respectively).

No difference was observed in electrolyte, mineral, or enzyme levels across the course of the experiment, but the metabolites cholesterol, total protein, and triglycerides all increased significantly across the duration of the experiment (Figure 5).

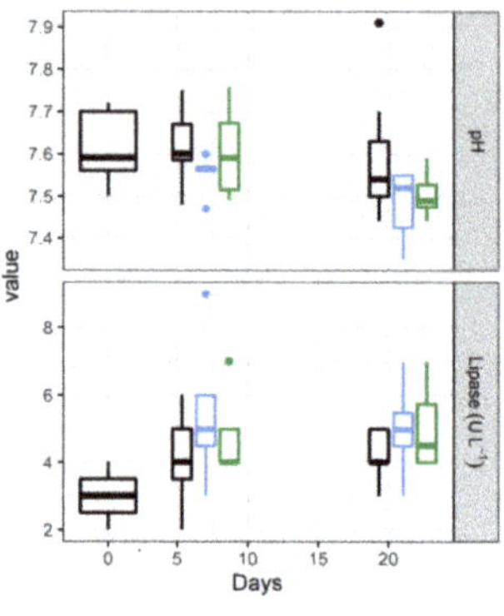

Figure 4. Haemolymph biochemical parameters that showed significant differences between *J. edwardsii* treatment groups. Low exposed (dark grey) lobsters had significantly lower pH than the control group (*p* value = 0.02). Both high and low exposed lobster groups had significantly higher lipase than the control group (*p* values = 0.03 and 0.02, respectively).

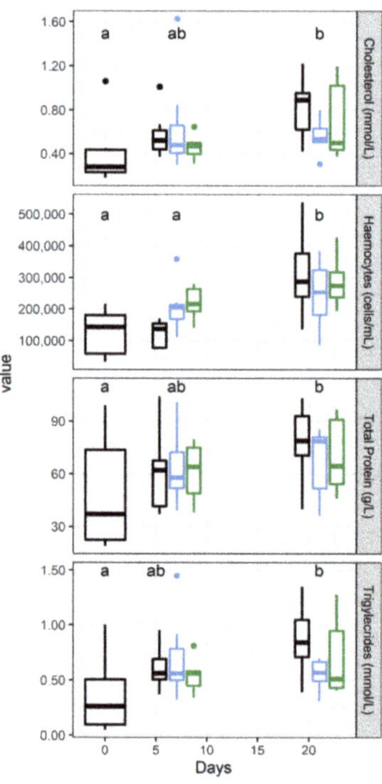

Figure 5. *J. edwardsii* haemolymph parameters that changed significantly over the course of the experiment in control (black), low (blue), and high (red) algal exposure treatments (0, 1×10^5, 2×10^5 *A. catenella* cells/L, respectively). Days which are not significantly different from each other share the same letter.

2.9. Histopathological Findings in Gills

Initial examination of gill filaments from the six high exposed and three of the control animals showed no differences in histopathological findings, so no further examinations of other lobster gills occurred. Diffuse pooling of haemolymph was observed in the lamellae and central axis of all gills examined, consistent with agonal change. All gills examined also showed multifocal deposits of rod and/or filamentous bacteria and low to moderate numbers of ciliated protozoa. Low numbers of free-living larval nematodes were found between lamellae on three of the six samples of gill tissues examined from exposed lobsters, and one of the three gill tissues from control lobsters. The gills of two of the high exposure lobsters and one control lobster showed microvesicles in reserve inclusion cells (Figure 6), likely storing lipids [29].

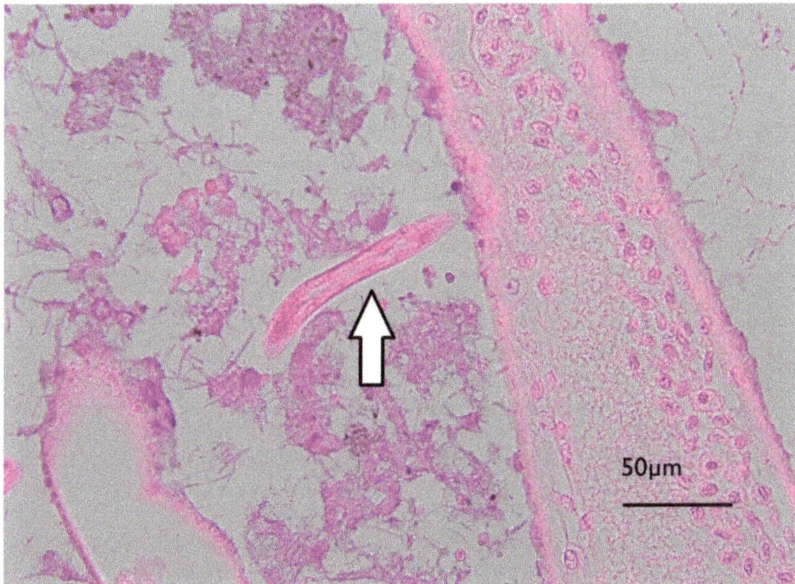

Figure 6. Micrograph of *J. edwardsii* gill filament with biofouling (mixed bacterial colonies, including Leucothrix-like organisms, adhered to filaments) and a free-living nematode between filaments as indicated by the arrow.

3. Discussion

No uptake of PST was detected in *J. edwardsii* exposed to high but field-relevant concentrations of *A. catenella* in an experimental setting over a three-week period. This was in striking contrast to significant PST accumulation in the lobster hepatopancreas (reaching a maximum of 9.0 mg STX.2HCl equiv. kg^{-1}) observed in parallel experiments involving feeding lobsters with toxic mussels [8], and uptake of PST by abalone when exposed to toxic algal cells in a similar experiment to the current study [30]. In the latter experiment, abalone were exposed to the same highly toxic strain of *A. catenella* (AT. TR/F) at the same level of the high exposure group in this experiment (2×10^5 cells L^{-1}). Abalone were able to accumulate up to 128 μg STX.2HCl equiv. kg^{-1} in this experiment, although it is unknown if this accumulation occurred across the gills, epipodium, via the viscera, or a combination of these routes.

Furthermore, the minimal impact on lobster health, demonstrated in this experiment across a range of organismal and cellular levels, indicates that there is no detrimental effect on the survivability and vitality of these animals as a result of exposure to toxic cells. Minimal whole organism and cellular responses were observed in *J. edwardsii* following the accumulation of high levels of PST in their hepatopancreas [13] and when exposed to toxic cells (current experiment), indicating that lobsters are relatively resistant to the action of PST. However, this response is in contrast to significant histopathology and mortality experienced by Blue mussel and Pacific oyster larvae when exposed to extracellular exudates of the same Tasmanian *A. catenella* strain at equivalent cell concentrations of 100 to 1000 cells mL^{-1} [31].

The present study exposed lobsters to aliquots of cultured algae and replicated environmental conditions where animals would be exposed to both cells and cell exudates. The toxic cells were presented to the algae at the highest level recorded from the Tasmanian blooms [3]. It is likely that toxins in wet storage would be equal to or less than those found in the field, as wet storage areas either draw directly from coastal waters in a continuous flow through systems or recirculate sea water through filtration and sedimentation systems to maintain water conditions.

The cultured *A. catenella* strain (AT. TR/F) was originally isolated from a bloom on the east coast of Tasmania and contained up to 21.2 pg STX.2HCl equiv. cell^{-1}, a relatively high PST cell quota from cultured algae [32–35]. The toxin profiles of the *A. catenella* cells were predominantly di-sulfated carbomoyl (C) 1,2 and gonyautoxin (GTX) 1,4, with minor levels of C3,4, neosaxitoxin (Neo), GTX2,3, STX, decarbamyolated gonyautoxin (dcGTX) 2,3, and GTX5,6 (see Section 4.2 below). These analogues are the same as those found in toxic shellfish from the east coast of Tasmania [36,37] and are thus considered representative of the Tasmanian *A. catenella* blooms. The same PST analogues are also found in New Zealand *A. pacificum*, *A. minutum* and *G. catenatum* isolates, although toxin proportions vary [5,38]. In particular, the proportion of C3,4 toxins tends to be higher in *A. pacificum* isolates, Neo and STX higher in *A. minutum* isolates, and C3,4 and GTX2,3 higher in *G. catenatum* isolates. Absolute concentrations present at any time will vary with the cell number and toxin content of the cells, and the amount of cell exudate.

Bioactive exudates from ichthyotoxic species have been demonstrated to have harmful impacts on the gills of adult Pacific oysters [39] and a range of fish species [15,40,41]. Compromised gills show necrotising degeneration of the epithelium of the secondary lamellae and sloughing and swelling of the primary lamellar epithelium with congestion of branchial vessels [42–44]. None of these effects were seen in the lobster gills exposed to high concentrations of *A. catenella* in this experiment.

A wide range of indicators used to predict the health of lobsters throughout the supply chain were assessed in this study. The only characteristics that demonstrated significant differences between treatment groups was haemolymph pH and lipase concentration. A difference in pH was found between control and low exposure groups, with the control group showing higher pH. A decrease in pH is a common stress response in lobsters caused by respiratory and metabolic acidosis. Other studies found a decrease in pH associated with emersion and high temperature [25,45–47]. Given that the high exposure group was not significantly different to the control and low exposure group, it is unlikely this difference was caused by the exposure to *A. catenella*.

Lipase plays an important role in the digestion of fats. The increase seen between the control and exposed groups of lobsters was influenced by the relatively low level of lipase measured in the control group at the start of the experiment. No significant differences were seen between control and the two treatment groups if the lobsters on day 0 were excluded from the analysis (p value = 0.21). This observed difference could be related to the improved nutritional condition of the lobsters across the experiment, as shown by significant increases in cholesterol, protein, and triglyceride levels. This increase in nutritional status was also detected in similar experiments involving feeding *J. edwardsii* mussels to excess daily [13] and was also associated with a similar decrease in feed intake during the course of the experiment as seen here.

The only immune response indicator that showed variation across the experiment was haemocyte levels, which increased with time in the system in all treatment groups. Other studies looking at stress in lobster supply chains have found varying results; some have found an increase in haemocyte levels with starvation, capture, emersion, storage and transport [21,25,26]; and others have found a decrease [18,24,48–50]. No difference in haemocyte levels was seen in a similar experimental study looking at the impact of PST accumulation on *J. edwardsii* health [13]. It is possible that the increase over time in all treatment groups may be related to the static experimental system.

In conclusion, we have conducted the first reported experiment to examine the uptake of PST by lobsters during exposure to PST-producing algal cells, as would potentially occur in lobster supply chains. The algal culture used was highly toxic, produced a range of commonly found PST analogues, and was presented at relatively high concentrations. The lobsters were exposed to toxic cells for three weeks, longer than would normally be experienced in the *J. edwardsii* supply chain in Tasmania and New Zealand. From the lack of uptake of PST in lobsters during this study and the lack of impact on animal health, we conclude that the wet storage of lobsters in coastal waters contaminated with the PST-producing algae typically found in Tasmania and New Zealand, as occurs in the *J. edwardsii* supply chain, does not pose a human health risk, nor an animal health risk. Therefore, no market access or risk to commercial returns through ill health exists from this practice.

4. Materials and Methods

4.1. Experimental System

The experimental system used is described in detail in Turnbull et al. [8], with the exception that in this case, a static system was employed with total daily water exchange. Briefly, 450–600 g adult *J. edwarsdii*, (n = 48 male and 1 female) were sourced directly from South Australian fishing vessels from a mix of shallow and deep habitats with no known bloom activity. The lobsters were transported to the South Australian Aquatic Biosecurity Centre at Roseworthy, where they were held in individual 30 L tanks maintained between 13.1 and 16.3 °C, salinity of less than 37 ppt, and pH between 7.7–8.2 (supplementing the seawater supply with bicarbonate soda as necessary). Water quality was maintained using pre-conditioned sponge biofilters. Dissolved oxygen was >90% saturation and 10 lumens of light was provided on a 12:12 h light:dark cycle. Lobsters were acclimated for 7 days prior to exposure.

4.2. Algal Cultures

Batch cultures of *A. catenella* strain AT.TR/F (previously known as *A. tamarense* group 1; isolated from Triabunna, Tasmania at the Institute for Marine and Antarctic Studies, Hobart, Australia) were cultivated in 15 L carboys following the method of Seger et al. [30]. Cultures were maintained in sterile filtered seawater (0.22 μm) supplemented with modified GSe nutrient concentrations (final media = 3/4 GSe nutrients, 5 mM sodium bicarbonate and 7.5 pM H_2SeO_3 to replace the soil extract in the basal recipe). Cultures were grown at 18 ± 1 °C under 120 μmol photons m^{-2} s^{-1} of light supplied by low temperature light emitting diodes on a 12:12 h light:dark cycle. During the dark period, the carboys were gently aerated (0.15 L min^{-1}) with ambient air, which was enriched with 1.5–2.5% (*v/v*) CO_2 in the light.

Cultures used for the exposure experiments were in late exponential/early stationary phase (>2 × 10^7 cells L^{-1}, 2.5% CO_2 aeration) and contained 3.5–21.2 pg STX.2HCl equiv. $cells^{-1}$ [30]. Cell PST quotas were determined in parallel experiments by Seger et al. [27] to be 3.5–21.2 pg STX.2HCl equiv. $cell^{-1}$. Briefly, suspensions of toxic cells were concentrated from four different batches of the same monoclonal source culture in late exponential/early stationary phase through centrifugation. Extracts of these suspensions were produced by lysis and further centrifugation to remove cell fragments. The toxin content of the extracts was determined via LCMS-MS analysis by the Cawthron Institute, New Zealand, as described below. The average toxin profile on a molar basis was 55% C1, 2, 36% GTX1,4, 3% C3,4, 2% Neo, 2% GTX2,3, 2% dcGTX2,3, and small percentages (<2% each) of STX, GTX5, GTX6, and doSTX.

4.3. Lobster Treatments

Lobsters were fed to excess (3 in-shell blue mussels, *Mytilus galloprovincialis* Lamarck) at the same time each day during the 7 day acclimation period and for the course of the experiment. The mussels were sourced from Coffin Bay, South Australia and were confirmed to be free of toxins via LCMS-MS [51,52] at the Cawthron Institute, New Zealand.

Lobsters were randomly allocated to three treatment groups: control (n = 21), low exposure (n = 14) and high exposure (n = 14). Each treatment group was further divided into harvest groups of seven replicates each. One control group was harvested on day 0, and the control, low, and high exposure groups were each harvested on days 7 and 21. Seven replicates were used in each group to minimise the number of experimental animals for ethical reasons whilst still allowing statistical rigour. The cell density of the *A. catenella* culture was determined daily via haemocytometer counts, and aliquots of culture were added to the low and high exposed lobsters at final concentrations of 1×10^5 cells/L and 2×10^5 cells/L, respectively, immediately after the morning water exchange each day.

4.4. Specific Feed Intake

The apparent feed intake (AFI) of lobsters were measured each week following the method of Fitzgibbon et al. [53]. Feed control tanks for the control, low, and high exposure groups were included in the random allocation, with no lobsters placed in these tanks. The feed control tanks each received 3 mussels at the same time as the lobster tanks each afternoon. Uneaten mussel meat from each tank (control and exposed animals, and feed control tanks) was collected at the beginning of each day. The shucked meat was frozen cumulatively over the period of a week. Subsequently, the uneaten mussel meat was dried at 105 °C for 24 h and weighed. The AFI of each lobster was calculated (dry weight of the uneaten food in the treatment tank subtracted from that of the respective control tank, divided by 7) and converted to SFI by dividing by the wet weight of the lobster.

4.5. Lobster Harvest Protocols

The harvest protocols and tests are described in detail in Turnbull et al. [13]. Briefly, lobster behavioural responses and tissue collection were conducted in the same order by the same researchers on each harvest day. Following behavioural measurements, lobsters were euthanised in an ice slurry, and then haemolymph samples (5–15 mL) were taken from the sinus under the right fifth leg joint. The animals were kept on ice overnight, then weighed, and the carapace length was measured prior to tissue dissection and collection. Gill tissue samples from each animal were immediately placed in Davidson's fixative for 24 h, which was then replaced with 70% ethanol. Hepatopancreases were stored at −80 °C prior to PST analysis.

4.6. Behavioural Responses

Lobsters were first assessed for 7 reflex responses following Turnbull et al. [13]: primary and secondary pereopod lift, antennae and secondary antennal lift, and tail arch via photography whilst emersed; rapid (<1 sec) eye stalk return to normal after gently squeezing together; and rapid antennal touch of hand placed directly in front of immersed animal. Two behavioural responses were then assessed (righting response time measured by placing each animal ventrum-up in a tank of saltwater and recording the time taken to return to dorsum-up [54]; and vitality visually assessed on a lobster commercial operator 1–5 scale similar to that described by Spanoghe and Bourne [16]; 1 = dead; 2 = limp tail, no escape response, no response to handling; 3 = limp tail, some response to handling, i.e., leg movement; 4 = mostly alert, tail held erect; 5 = alert with vigorous escape behaviour). Each reflex response was scored (positive response = 0, negative response = 1) and summed into a reflex impairment score for each animal, as described by Stoner et al. [19]. Potential scores ranged from 0–7, with 0 indicating maximum vigour.

4.7. Immune Health Response

Haemolymph samples were preserved immediately after extraction by adding 200–300 µL chilled anticoagulant Lillie's formol calcium (1.3 M formalin, 126 mM calcium acetate) and haemocytes were counted in an Improved Neubauer haemocytometer at 40× magnification (Olympus CX41 RF) within 48 h. To assess bacteraemia levels, 100 µL of haemolymph was sterilely plated onto each of Zobell's marine and thiosulphate-citrate-bile salts agars (ZMA and TCBS, Thermofisher), which were incubated at 26 °C for 48 h prior to colonies being counted. Shell necrosis was visually noted as present/absent during dissection.

4.8. Nutritional Health Response

Nutritional responses were assessed via Brix index (Hanna Refractometer H196801) and hepatopancreas index (the ratio of hepatopancreas wet weight to lobster wet weight [55,56]).

4.9. Haemolymph Biochemical Response

Haemolymph pH was measured using a Radiometer Analytical pH meter PHM210 with micro-electrode B10C162, following which the haemolymph was spun at 10,000× g for 5 min (Sigma Microcentrifuge 1–14). The supernatant was snap frozen in liquid nitrogen, stored at −80 °C and sent to Crustipath Laboratories, Canada, for analysis using a Cobas c501 automated biochemistry analyser (Roche Diagnostics Corporation, Indianopolis, IN, USA) as described by Day et al. [21] and Fitzgibbon et al. [22]. Sodium (Na^+), chloride (Cl^-), and potassium (K^+) were measured using an Ion-Selective Electrode, whilst Mg and bicarbonate (bicarb) were measured photometrically. The minerals calcium (Ca) and phosphorous (P); metabolites glucose (Gluc), lactate (Lact), cholesterol (Chol), triglycerides (Trig), total protein (TP), albumin (Alb), globulin (Glob), urea, and uric acid (UA); and enzymes lipase (Lip), amylase (Amy), alanine aminotransferase (ALT), aspartate aminotransferase (AST), alkaline phosphatase (ALP), sorbital (SDH), glutamate dehydrogenases (GD) and gamma-glutamyl transferase (GGT) were measured photometrically. Osmolality was measured on a Micro-Osmette (Precision Systems Inc., Natick, MA, USA) via freezing point depression.

4.10. PST Analysis

Paralytic shellfish toxins in the hepatopancreas were analysed at the Cawthron Institute, New Zealand by LC-MS/MS (Waters Acquity UPLC i-Class system coupled to a Waters Xevo TQ-S triple quadrupole mass spectrometer with electrospray ionization), following the methods of Boundy et al. [51] and Turner et al. [52], with minor variations as detailed in Turnbull et al. [8]. Results were calculated using Food and Agriculture Organisation of the United Nations (FAO) toxicity equivalency factors [57]. Results reported as part of this study were corrected based on spike recoveries observed for the different sample matrices analysed. The limit of reporting for each PST analogue differed for each matrix tested.

4.11. Gill Histology

Histopathological analysis of gill tissues was conducted at the Animal Health Laboratory, Department of Primary Industries, Parks, Water and Environment in Tasmania. Gills stored in ethanol were embedded in paraffin, cut at 5 µm thickness, mounted and stained with haematoxylin and eosin using standard techniques. All slides were read by the same American College of Veterinary Pathologists (ACVP) board-certified veterinary pathologist.

4.12. Data Analysis

Statistical analyses were performed using R Software (R Core Development Team version 3.6, April 2019). Continuous datasets were checked for normality and homoscedasticity using the Shapiro–Wilk test and Levene's test respectively, with appropriate transformations if necessary (no transformation: SFC, haemocyte count, brix, pH, Na, Cl, TP, Glob,

Alb:Glob (A:G), and GD; log transformations: time to right, Ca, Gluc, Chol, and Alb; square root transformations: P, bicarb, Trig, and UA). Analysis of variance was used to test for significant differences between groups for data with normal distributions, followed by post-hoc analysis using Tukey honestly significant difference (HSD) tests. Prior to transformations, P, Lact, bicarb, Trig, UA concentrations that were reported as less than the level of detection (LOD) were replaced with 0.5* LOD (n = 1, 12, 1, 2, 1 respectively). Two-way random permutation tests were used to test for significant differences in continuous data that could not be transformed to a normal distribution (bacterial counts on ZMA and TCBS, K, Na:K, Mg, Lact and measured osmolality).

Ordinal logistic regression was used to test for significant differences between discrete and ordinal datasets (reflex impairment score and vitality respectively), with p values calculated by comparing the t-value against the standard normal distribution. Ordinal chi-squared analysis was used to test for association between vitality and RIS. Significant differences between groups in binary datasets (necrosis and gill parasites) were tested using logistic regression.

Analytes where most of the data were below the LOD were not tested for significant differences between groups (creatinine, urea, ALT, ALP, AMY, AST, GGT, and SDH). Differences were considered statistically significant when p values < 0.05.

Supplementary Materials: The following are available online at https://www.mdpi.com/2072-6651/13/2/129/s1, Table S1: The median and standard deviation of behavioural, immunological and biochemical parameters measured in control and PST exposed *J. edwardsii* over different time periods, Table S2: Frequency table of reflex impairment scores (RIS) for *J. edwardsii* control low and high exposure treatment groups (0, 1×10^5 and 2×10^5 cells *A. catenella* respectively) on days 0, 7 and 21, Table S3: Reflex impairment scores in relation to vitality in *J. edwardsii*. All treatment groups combined.

Author Contributions: Conceptualization, A.T., G.H. and Q.F.; methodology, A.T., J.J., A.S., G.K. and Q.F.; formal analysis, A.T.; investigation, A.T., A.S., G.K.; data curation, A.T.; writing—original draft preparation, A.T.; writing—review and editing, J.J., A.S., G.H., G.K. and Q.F.; project administration, A.T.; funding acquisition, A.T. and G.H. All authors have read and agreed to the published version of the manuscript.

Funding: This research was funded by Fisheries Research and Development Corporation, grant numbers 2017-051 and 2017-086.

Institutional Review Board Statement: Ethical review and approval for studies on crustacea are not required in South Australia. The study used the minimum number of animals required to ensure statistical rigour and used ethical procedures to euthanise all animals.

Informed Consent Statement: Not applicable.

Data Availability Statement: The data presented in this study are available in the supplementary material.

Acknowledgments: This research was conducted at the South Australian Aquatic Animal Biosecurity Centre. We thank Navreet Malhi, Geoff Holds, Jan Lee and Elliot Brown for technical support at the South Australian Aquatic Biosecurity Centre; Dane Hayes and Wendy Allen from the Department of Primary Industries Parks Water and Environment Animal Health Laboratory for histological preparation of slides.

Conflicts of Interest: The authors declare no conflict of interest. The funders had no role in the design of the study; in the collection, analyses, or interpretation of data; in the writing of the manuscript, or in the decision to publish the results.

References

1. Patterson, H.; Williams, A.; Woodhams, J.; Curtotti, R. *Fishery Status Reports 2019*; Australian Bureau of Agricultural and Resource Economics: Canberra, Australia, 2019.
2. Williams, J.; Stokes, F.; Dixon, H.; Hurren, K. *The Economic Contribution of Commercial Fishing to the New Zealand Economy*; BERL: Wellington, New Zealand, 2017; pp. 1–53.

3.   Condie, S.A.; Oliver, E.C.J.; Hallegraeff, G.M. Environmental drivers of unprecedented *Alexandrium catenella* dinoflagellate blooms off eastern Tasmania, 2012–2018. *Harmful Algae* **2019**, *87*, 101628. [CrossRef] [PubMed]
4.   Hallegraeff, G.; Bolch, C.; Condie, S.; Dorantes-Aranda, J.J.; Murray, S.; Quinlan, R.; Ruvindy, R.; Turnbull, A.; Ugalde, S.; Wilson, K. Unprecedented Alexandrium blooms in a previously low biotoxin risk area of Tasmania, Australia. In *Proceedings of the 17th International Conference on Harmful Algae 2016*; International Society for the Study of Harmful Algae and Intergovernmental Oceanographic Commission of UNESCO: Paris, France, 2017; pp. 38–41.
5.   MacKenzie, A.L. The risk to New Zealand shellfish aquaculture from paralytic shellfish poisoning (PSP) toxins. *New Zealand J. Mar. Freshw. Res.* **2014**, *48*, 430–465. [CrossRef]
6.   Rhodes, L.; Smith, K.F.; MacKenzie, L.; Wood, S.A.; Ponikla, K.; Harwood, D.T.; Packer, M.; Munday, R. The Cawthron Institute Culture Collection of Micro-algae: A significant national collection. *New Zealand J. Mar. Freshw. Res.* **2016**, *50*, 291–316. [CrossRef]
7.   Madigan, T.; Malhi, N.; Tan, J.; McLeod, C.; Stewart, I.; Harwood, T.; Mann, G.; Turnbull, A. Experimental uptake and depuration of paralytic shellfish toxins in Southern Rock Lobster, *Jasus edwardsii. Toxicon* **2018**, *143*, 44–50. [CrossRef] [PubMed]
8.   Turnbull, A.; Malhi, N.; Seger, A.; Harwood, T.; Jolley, J.; Fitzgibbon, Q.; Hallegraeff, G. Paralytic shellfish toxin uptake, tissue distribution, and depuration in the Southern Rock Lobster *Jasus edwardsii* Hutton. *Harmful Algae* **2020**, *95*, 101818. [CrossRef]
9.   McLeod, C.; Kiermeier, A.; Stewart, I.; Tan, J.; Turnbull, A.; Madigan, T. Paralytic shellfish toxins in Australian Southern Rock Lobster (Jasus edwardsii): Acute human exposure from consumption of hepatopancreas. *Hum. Ecol. Risk Assess.* **2018**, *24*, 1872. [CrossRef]
10. Patel, K.K.; Fitzgibbon, Q.; Caraguel, C.G.B. Investigation of risk factors associated with sub-optimal holding survival in southern rock lobster (*Jasus edwardsii*) in Australia. *Prev. Vet. Med.* **2020**, *183*, 105122. [CrossRef]
11. Henry, R.P.; Lucu, C.; Onken, H.; Weihrauch, D. Multiple functions of the crustacean gill: Osmotic/ionic regulation, acid-base balance, ammonia excretion, and bioaccumulation of toxic metals. *Front. Physiol.* **2012**, *3*, 431. [CrossRef]
12. Stoner, A.W. Assessing stress and predicting mortality in economically significant crustaceans. *Rev. Fish. Sci.* **2012**, *20*, 111–135. [CrossRef]
13. Turnbull, A.; Malhi, N.; Seger, A.; Jolley, J.; Hallegraeff, G.; Fitzgibbon, Q. Accumulation of paralytic shellfish toxins by Southern Rock Lobster *Jasus edwardsii* causes minimal impact on lobster health. *Aquat. Toxicol.* **2020**. [CrossRef]
14. Dorantes-Aranda, J.J.; Seger, A.; Mardones, J.I.; Nichols, P.D.; Hallegraeff, G.M. Progress in understanding algal bloom-mediated fish kills: The role of superoxide radicals, phycotoxins and fatty acids. *PLoS ONE* **2015**, *10*, e0133549. [CrossRef]
15. Mardones, J.I.; Dorantes-Aranda, J.J.; Nichols, P.D.; Hallegraeff, G.M. Fish gill damage by the dinoflagellate *Alexandrium catenella* from Chilean fjords: Synergistic action of ROS and PUFA. *Harmful Algae* **2015**, *49*, 40–49. [CrossRef]
16. Spanoghe, P.T.; Bourne, P.K. Relative influence of environmental factors and processing techniques on *Panulirus cygnus* morbidity and mortality during simulated live shipments. *Mar. Freshw. Res.* **1997**, *48*, 839–844. [CrossRef]
17. James, P.; Izquierdo-Gómez, D.; Siikavuopio, S.I. Use of reflex indicators for measuring vitality and mortality of the snow crab (*Chionoecetes opilio*) in captivity. *Aquac. Res.* **2019**, *50*, 1321–1328. [CrossRef]
18. Paterson, B.D.; Spanoghe, P.T.; Davidson, G.W.; Hosking, W.; Nottingham, S.; Jussila, J.; Evans, L.H. Predicting survival of western rock lobsters *Panulirus cygnus* using discriminant analysis of haemolymph parameters taken immediately following simulated handling treatments. *New Zealand J. Mar. Freshw. Res.* **2005**, *39*, 1129–1143. [CrossRef]
19. Stoner, A.W.; Rose, C.S.; Munk, J.E.; Hammond, C.F.; Davis, M.W. An assessment of discard mortality for two Alaskan crab species, Tanner crab (*Chionoecetes bairdi*) and snow crab (*C. opilio*), based on reflex impairment. *Fish. Bull.* **2008**, *106*, 337.
20. Stoner, A.W. Prediction of discard mortality for Alaskan crabs after exposure to freezing temperatures, based on a reflex impairment index. *Fish. Bull.* **2009**, *107*, 451.
21. Day, R.D.; Fitzgibbon, Q.P.; Gardner, C. The impact of holding stressors on the immune function and haemolymph biochemistry of Southern Rock Lobsters (*Jasus edwardsii*). *Fish Shellfish Immunol.* **2019**, *89*, 660–671. [CrossRef] [PubMed]
22. Fitzgibbon, Q.P.; Day, R.D.; McCauley, R.D.; Simon, C.J.; Semmens, J.M. The impact of seismic air gun exposure on the haemolymph physiology and nutritional condition of spiny lobster, *Jasus edwardsii. Mar. Pollut. Bull.* **2017**, *125*, 146–156. [CrossRef] [PubMed]
23. Fotedar, S.; Evans, L.; Jones, B. Effect of holding duration on the immune system of western rock lobster, *Panulirus cygnus. Comp. Biochem. Physiol. Part A* **2006**, *143*, 479–487. [CrossRef]
24. Fotedar, S.; Tsvetnenko, E.; Evans, L. Effect of air exposure on the immune system of the rock lobster *Panulirus cygnus. Mar. Freshw. Res.* **2001**, *52*, 1351–1355. [CrossRef]
25. Simon, C.J.; Mendo, T.C.; Green, B.S.; Gardner, C. Predicting transport survival of brindle and red rock lobsters *Jasus edwardsii* using haemolymph biochemistry and behaviour traits. *Comp. Biochem. Physiol. Part A* **2016**, *201*, 101–109. [CrossRef]
26. Tsvetnenko, E.; Fotedar, S.; Evans, L. Antibacterial activity in the haemolymph of western rock lobster, *Panulirus cygnus. Mar. Freshw. Res.* **2001**, *52*, 1407–1412. [CrossRef]
27. Chandrapavan, A.; Gardner, C.; Green, B.S. Haemolymph condition of deep-water southern rock lobsters (*Jasus edwardsii*) translocated to inshore reefs. *Mar. Freshw. Behav. Physiol.* **2011**, *44*, 21–32. [CrossRef]
28. Morris, S.; Oliver, S. Circulatory, respiratory and metabolic response to emersion and low temperature of *Jasus edwardsii*: Simulation studies of commercial shipping methods. *Comp. Biochem. Physiol. A Mol. Integr. Physiol.* **1999**, *122*, 299–308. [CrossRef]
29. Travis, D.F. The moulting cycle of the spiny lobster, *Panulirus argus* Latreille. II. Pre-ecdysial histological and histochemical changes in the hepatopancreas and integumental tissues. *Biol. Bull.* **1954**, *108*, 88–112. [CrossRef]

30. Seger, A.; Hallegraeff, G.; Stone, D.; Bansemer, M.; Harwood, T.D.; Turnbull, A. Uptake of paralytic shellfish toxins by blacklip abalone (*Haliotis rubra rubra* leach) from direct exposure to *Alexandrium catenella* microalgal cells and toxic aquaculture feed. *Harmful Algae* **2020**, *99*. [CrossRef] [PubMed]
31. Supono, S.; Knowles, G.; Bolch, C. Toxicity and histopathological effects of toxic dinoflagellate, *Alexandrium catenella* exudates on larvae of blue mussel, *Mytilus galloprovincialis*, and Pacific oyster, *Crassostrea gigas*. *J. Ilm. Perikan. Dan Kelaut.* **2020**, *12*, 188–198. [CrossRef]
32. Jester, R.; Rhodes, L.; Beuzenberg, V. Uptake of paralytic shellfish poisoning and spirolide toxins by paddle crabs (*Ovalipes catharus*) via a bivalve vector. *Harmful Algae* **2009**, *8*, 369–376. [CrossRef]
33. Lilly, E.L.; Kulis, D.M.; Gentien, P.; Anderson, D.M. Paralytic shellfish poisoning toxins in France linked to a human-introduced strain of *Alexandrium catenella* from the western Pacific: Evidence from DNA and toxin analysis. *J. Plankton Res.* **2002**, *24*, 443–452. [CrossRef]
34. Sekiguchi, K.; Ogata, T.; Kaga, S.; Yoshida, M.; Fukuyo, Y.; Kodama, M. Accumulation of paralytic shellfish toxins in the scallop *Patinopecten yessoensis* caused by the dinoflagellate *Alexandrium catenella* in Otsuchi Bay, Iwate Prefecture, northern Pacific coast of Japan. *Fish. Sci.* **2001**, *67*, 1157–1162. [CrossRef]
35. Varela, D.; Paredes, J.; Alves-de-Souza, C.; Seguel, M.; Sfeir, A.; Frangópulos, M. Intraregional variation among *Alexandrium catenella* (Dinophyceae) strains from southern Chile: Morphological, toxicological and genetic diversity. *Harmful Algae* **2012**, *15*, 8–18. [CrossRef]
36. Dorantes-Aranda, J.J.; Tan, J.Y.C.; Hallegraeff, G.M.; Campbell, K.; Ugalde, S.C.; Harwood, D.T.; Bartlett, J.K.; Campàs, M.; Crooks, S.; Gerssen, A.; et al. Detection of paralytic shellfish toxins in mussels and oysters using the qualitative Neogen Lateral-Flow Immunoassay: An interlaboratory study. *J. AOAC Int.* **2018**, *101*, 468–479. [CrossRef] [PubMed]
37. Turnbull, A.R.; Tan, J.Y.C.; Ugalde, S.C.; Hallegraeff, G.M.; Campbell, K.; Harwood, D.T.; Dorantes-Aranda, J.J. Single-laboratory validation of the Neogen Qualitative Lateral Flow Immunoassay for the detection of paralytic shellfish toxins in mussels and oysters. *J. AOAC Int.* **2018**, *101*, 480–489. [CrossRef] [PubMed]
38. Harwood, D.T.; Boundy, M.; Selwood, A.I.; van Ginkel, R.; MacKenzie, L.; McNabb, P.S. Refinement and implementation of the Lawrence method (AOAC 2005.06) in a commercial laboratory: Assay performance during an *Alexandrium catenella* bloom event. *Harmful Algae* **2013**, *24*, 20–31. [CrossRef]
39. Castrec, J.; Soudant, P.; Payton, L.; Tran, D.; Miner, P.; Lambert, C.; Le Goïc, N.; Huvet, A.; Quillien, V.; Boullot, F.; et al. Bioactive extracellular compounds produced by the dinoflagellate *Alexandrium minutum* are highly detrimental for oysters. *Aquat. Toxicol.* **2018**, *199*, 188–198. [CrossRef] [PubMed]
40. Chen, C.-Y.; Chou, H.-N. Ichthyotoxicity studies of milkfish *Chanos chanos* fingerlings exposed to a harmful dinoflagellate *Alexandrium minutum*. *J. Exp. Mar. Biol. Ecol.* **2001**, *262*, 211–219. [CrossRef]
41. Ma, H.; Krock, B.; Tillmann, U.; Bickmeyer, U.; Graeve, M.; Cembella, A. Mode of action of membrane-disruptive lytic compounds from the marine dinoflagellate *Alexandrium tamarense*. *Toxicon* **2011**, *58*, 247–258. [CrossRef]
42. Shimada, M.; Imahayashi, T.; Ozaki, H.S.; Murakami, T.H.; Toyoshima, T.; Okaichi, T. Effects of sea bloom, *Chattonella antiqual*, on gill primary lamellae of the young yellowtail, *Seriola quinqueradiata*. *Acta Histochem. Cytochem.* **1983**, *16*, 232–244. [CrossRef]
43. Hallegraeff, G.; Dorantes-Aranda, J.; Mardones, J.I.; Seger, A. Review of progress in our understanding of fish-killing microalgae: Implications for management and mitigation. In *Proceedings of the 17th International Conference on Harmful Algae 2016*; International Society for the Study of Harmful Algae and Intergovernmental Oceanographic Commission of UNESCO: Paris, France, 2017; pp. 150–155.
44. Roberts, R.J.; Bullock, A.M.; Turners, M.; Jones, K.; Tett, P. Mortalities of *Salmo gairdneri* exposed to cultures of *Gyrodinium aureolum*. *J. Mar. Biol. Assoc. U.K.* **1983**, *63*, 741–743. [CrossRef]
45. Dove, A.D.M.; Allam, B.; Powers, J.J.; Sokolowski, M.S. A prolonged thermal stress experiment on the American Lobster *Homarus americanus*. *J. Shellfish Res.* **2005**, *24*, 761–765.
46. Lorenzon, S.; Giulianini, P.G.; Martinis, M.; Ferrero, E.A. Stress effect of different temperatures and air exposure during transport on physiological profiles in the American lobster *Homarus americanus*. *Comp. Biochem. Physiol. Part A Mol. Integr. Physiol.* **2007**, *147*, 94–102. [CrossRef] [PubMed]
47. Whiteley, E.; Taylor, E. Oxygen and acid-base disturbances in the hemolymph of the lobster *Homarus gammarus* during commercial transport and storage. *J. Crustacean Biol.* **1992**, *12*, 19–30. [CrossRef]
48. Stewart, J.E.; Cornick, J.W.; Foley, D.M.; Li, M.F.; Bishop, C.M. Muscle weight relationship to serum proteins, hemocytes and hepatopancreas in the lobster, *Homarus americanus*. *Fish. Res. Board Can.* **1967**, *24*, 2339–2354. [CrossRef]
49. Jussila, J.; Jago, J.; Tsvetnenko, E.; Dunstan, B.; Evans, L.H. Total and differential haemocyte counts in western rock lobsters (Panulirus cygnus George) under post-harvest stress. *Mar. Freshw. Res.* **1997**, *48*, 863–868. [CrossRef]
50. Basti, D.; Hoyt, K.; Bouchard, D.; Bricknell, I.; Chang, E.S.; Halteman, W. Factors affecting post-capture survivability of lobster *Homarus americanus*. *Dis. Aquat. Org.* **2010**, *90*, 153–166. [CrossRef]
51. Boundy, M.J.; Selwood, A.I.; Harwood, T.D.; McNabb, P.S.; Turner, A.D. Development of a sensitive and selective liquid chromatography–mass spectrometry method for high throughput analysis of paralytic shellfish toxins using graphitised carbon solid phase extraction. *J. Chromatogr. A* **2015**, *1387*, 1–12. [CrossRef]

52. Turner, A.D.; McNabb, P.S.; Harwood, T.; Selwood, A.I.; Boundy, M.J. Single-laboratory validation of a multitoxin ultra-performance LC-hydrophilic interaction LC-MS/MS method for quantitation of paralytic shellfish toxins in bivalve shellfish. *J. AOAC Int.* **2015**, *98*, 609–621. [CrossRef]
53. Fitzgibbon, Q.P.; Simon, C.J.; Smith, G.G.; Carter, C.G.; Battaglene, S.C. Temperature dependent growth, feeding, nutritional condition and aerobic metabolism of juvenile spiny lobster, *Sagmariasus verreauxi*. *Comp. Biochem. Physiol. Part A Mol. Integr. Physiol.* **2017**, *207*, 13–20. [CrossRef]
54. Day, R.D.; McCauley, R.D.; Fitzgibbon, Q.P.; Hartmann, K.; Semmens, J.M. Seismic air guns damage rock lobster mechanosensory organs and impair righting reflex. *Proc. R. Soc. B Biol. Sci.* **2019**, *286*. [CrossRef]
55. Simon, C.J.; Fitzgibbon, Q.P.; Battison, A.; Carter, C.G.; Battaglene, S.C. Bioenergetics of nutrient reserves and metabolism in spiny lobster juveniles *Sagmariasus verreauxi*: Predicting nutritional condition from hemolymph biochemistry. *Physiol. Biochem. Zool. Ecol. Evol. Approaches* **2015**, *88*, 266–283. [CrossRef] [PubMed]
56. Landman, M.J.; Fitzgibbon, Q.P.; Wirtz, A.; Codabaccus, B.M.; Ventura, T.; Smith, G.G.; Carter, C.G. Physiological status and nutritional condition of cultured juvenile *Thenus australiensis* over the moult cycle. *Comp. Biochem. Physiol. Part B Biochem. Mol. Biol.* **2020**, *250*, 110504. [CrossRef] [PubMed]
57. Food and Agriculture Organization of the United Nations and World Health Organisation. *Technical Paper on Toxicity Equivalency Factors for Marine Biotoxins Associated with Bivalve Molluscs*; FAO/WHO: Rome, Italy, 2016.

Article

Brevisulcenals-A1 and A2, Sulfate Esters of Brevisulcenals, Isolated from the Red Tide Dinoflagellate *Karenia brevisulcata*

Masayuki Satake [1,*], Raku Irie [1,2], Patrick T. Holland [3], D Tim Harwood [3], Feng Shi [3], Yoshiyuki Itoh [4], Fumiaki Hayashi [5] and Huiping Zhang [5]

[1] Department of Chemistry, School of Science, The University of Tokyo, Hongo, Bunkyo-ku, Tokyo 113-0033, Japan; irie@yokohama-cu.ac.jp
[2] Graduate School of Nanobioscience, Yokohama City University, Seto 22-2, Kanazawa-ku, Yokohama 236-0027, Japan
[3] Cawthron Institute, Private Bag 2, Nelson 7010, New Zealand; Patrick.Holland@cawthron.org.nz (P.T.H.); Tim.Harwood@cawthron.org.nz (D.T.H.); FShi@aaapl.com.au (F.S.)
[4] MS Business Unit, JEOL Ltd., Musashino, Akishima, Tokyo 196-8558, Japan; yoitol@jeol.co.jp
[5] NMR Science and Development Division, RIKEN SPring-8 Center, Suehiro-cho, Tsurumi-ku, Yokohama 230-0045, Japan; fumiaki.hayashi@riken.jp (F.H.); huiping.zhang@riken.jp (H.Z.)
* Correspondence: msatake@chem.s.u-tokyo.ac.jp; Tel.: +81-3-5841-4357

Abstract: Two different types of polycyclic ether toxins, namely brevisulcenals (KBTs) and brevisulcatic acids (BSXs), produced by the red tide dinoflagellate *Karenia brevisulcata*, were the cause of a toxic incident that occurred in New Zealand in 1998. Four major components, KBT-F, -G, -H, and -I, shown to be cytotoxic and lethal in mice, were isolated from cultured *K. brevisulcata* cells, and their structures were elucidated by spectroscopic analyses. New analogues, brevisulcenal-A1 (KBT-A1) and brevisulcenal-A2 (KBT-A2), toxins of higher polarity than that of known KBTs, were isolated from neutral lipophilic extracts of bulk dinoflagellate culture extracts. The structures of KBT-A1 and KBT-A2 were elucidated as sulfated analogues of KBT-F and KBT-G, respectively, by NMR and matrix-assisted laser desorption/ionization tandem mass spectrometry (MALDI TOF/TOF), and by comparison with the spectra of KBT-F and KBT-G. The cytotoxicities of the sulfate analogues were lower than those of KBT-F and KBT-G.

Keywords: marine polyether; dinoflagellate; red tide incident; harmful algal bloom

Key Contribution: The structures of new brevisulcenal analogues, brevisulcenal-A1, and brevisulcenal-A2 were elucidated by NMR and spiral MALDI TOF/TOF spectroscopy. Brevisulcenal-A1 and brevisulcenal-A2 are sulfate esters of brevisulcenal-F and brevisulcenal-G, respectively. The cytotoxicity of sulfated brevisulcenals is lower than that of known brevisulcenals.

Citation: Satake, M.; Irie, R.; Holland, P.T.; Harwood, DT.; Shi, F.; Itoh, Y.; Hayashi, F.; Zhang, H. Brevisulcenals-A1 and A2, Sulfate Esters of Brevisulcenals, Isolated from the Red Tide Dinoflagellate *Karenia brevisulcata*. *Toxins* **2021**, *13*, 82. https://doi.org/10.3390/toxins13020082

Received: 29 December 2020
Accepted: 20 January 2021
Published: 22 January 2021

1. Introduction

A widespread bloom of the dinoflagellate *Karenia brevisulcata* [1] occurred in the central and south-east coast of the North Island of New Zealand in early 1998 with deadly and devastating consequences to fish and other marine organisms in Wellington Harbour [2,3]. In addition to the devastating damages to marine animals, more than 500 patients were reported during the red tide incident. Characteristic symptoms were respiratory distress, inflammation of skin and eyes, severe headaches, and facial sunburn sensations, which were similar to symptoms caused by *Karenia brevis* [4–6]. Further toxicological studies are needed to elucidate the correlation between human illness and the toxins of *K. brevisulcata*. *Karenia brevisulcata* is morphologically very similar to *Karenia brevis* and *K. mikimotoi*, which produce ladder-frame polyether toxins, brevetoxins [7,8], and gymnocins, respectively [9,10]; however, the cells of *K. brevisulcata* are smaller, and brevetoxins and gymnocins were not detected in the cell extracts. Therefore, the toxins produced by *K. brevisulcata* that caused the toxic event in New Zealand in 1998 were presumably unknown.

Cell extracts of *K. brevisulcata* exhibited potent mouse lethality and cytotoxicity. The extracts were partitioned between chloroform and aqueous methanol under neutral conditions during initial investigations. Two different types of toxins were detected in the extracts [3]. The aqueous methanol layer contained brevisulcatic acids (BSXs), brevetoxin-like ladder-frame polyether compounds possessing side-chain carboxylic acids. BSXs displayed potent cytotoxicity against neuro2A cells with veratridine and ouabain [11,12]. The lipophilic layer contained brevisulcenals (KBTs), which exhibited potent mouse lethality and cytotoxicity against the P388 (murine leukemia) cell line.

NMR and matrix-assisted laser desorption/ionization (MALDI) tandem mass spectrometry (MS) with high-energy collision-induced dissociation (HE-CID) led to the elucidation of the four KBT structures, brevisulcenal-F (KBT-F), KBT-G, KBT-H, and KBT-I, as shown in Figure 1 [13,14]. KBT structural features include molecular weights above 2000 and extensive ladder-frame polyether compounds comprising 24 ether rings, including a dihydrofuran, decorated with hydroxy groups, methyl groups, and conjugated unsaturated aldehyde side chains. The longest contiguous ether ring chain of KBTs comprises 17 units (A–Q). Apart from a single oxidation, the skeletal structure of KBT-F and KBT-H is the same, as is that of KBT-G and KBT-I. KBT-H and KBT-I possess branched primary alcohols on C-2, generated by oxidation of olefinic methyls in the side chains of KBT-F and KBT-G, respectively. Such branched oxidized structures are very rare among polycyclic ethers produced by dinoflagellates. The cytotoxicity of KBT-H and KBT-I was more potent than that of KBT-F and KBT-G, indicating that oxidation of terminal olefinic methyls enhances toxicity. The long contiguous ether ring assembly with an unsaturated aldehyde terminus is analogous to the structures of gymnocins A and B [9,10], but KBTs displayed more potent mouse lethality and cytotoxicity. Structural elucidation of KBT analogues is important for the development and improvement of detection methods [15] and for investigating the mode of action and structure-activity relationships of these toxins, which may be beneficial for developing preventative strategies to mitigate the effects of KBT red tide events, should they occur again. Our ongoing efforts in this area led to the isolation of new KBT analogues, brevisulcenal-A1 (KBT-A1), and brevisulcenal-A2 (KBT-A2), which are sulfate esters of KBT-F and KBT-G, respectively. KBT-A1 and KBT-A2 eluted before KBT-F and KBT-G when using reversed-phase chromatography. In this study, we elucidated the structures of KBT-A1 and KBT-A2 by detailed analyses of their NMR and spiral MALDI TOF/TOF spectra.

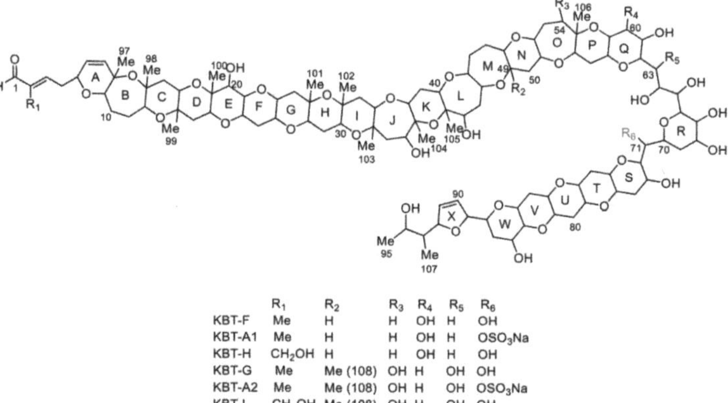

	R_1	R_2	R_3	R_4	R_5	R_6
KBT-F	Me	H	H	OH	H	OH
KBT-A1	Me	H	H	OH	H	OSO$_3$Na
KBT-H	CH$_2$OH	H	H	OH	H	OH
KBT-G	Me	Me (108)	OH	H	OH	OH
KBT-A2	Me	Me (108)	OH	H	OH	OSO$_3$Na
KBT-I	CH$_2$OH	Me (108)	OH	H	OH	OH

Figure 1. Structures of brevisulcenals.

2. Results

2.1. Extraction and Isolation of Brevisulcenals

Brevisulcenals (KBTs) were extracted from mature bulk cultures of *Karenia brevisulcata* employing resin-based isolation. Cells were lysed with acetone and after stirring for 1 h, the cultures were diluted with water and passed through HP20 resin. Brevisulcenals were recovered by washing the resin with acetone. The lyophilized extract was dissolved in MeOH and diluted to 55% MeOH with pH 7.2 phosphate buffer, and then partitioned with chloroform ($CHCl_3$). The $CHCl_3$ extract was chromatographed on a diol-silica column by stepwise elution with ethyl acetate (EtOAc) and MeOH, EtOAc:MeOH (9:1, 8:2, 7:3, 6:4), and fraction collection was guided by the cytotoxicity assay against mouse leukemia P388 cells and UV absorption. Further purification of KBTs was performed using reversed phase chromatography with a linear gradient elution from 80% to 100% MeOH. The final purification on a reversed phase column with isocratic elution of 80% MeOH led to the isolation of 0.2 mg of KBT-A2 from 150 L of *K. brevisulcata* culture. Isolation and accumulation of KBT-A2 was difficult because of its low abundance and short retention time on the reversed phase column. From 690 L of ^{13}C labeled cultures, 0.8 mg of KBT-A2 was generated for NMR studies. On the other hand, only 0.2 mg of KBT-A1 was isolated from non-labeled cultures due to its early elution on reversed-phase columns.

2.2. Structural Elucidation of Brevisulcenal-A2

In the positive ion MALDI mass spectrum of KBT-A2, high-intensity signals were observed at *m/z* 2207.8 and 2105.5, and the difference between the two ions was 102 Da, corresponding to desulfonation. Therefore, these ions were identified as a sodium adduct ion $[M+Na]^+$ and a desulfonated ion $[M-SO_3Na+H+Na]^+$, respectively. A signal for $[M-Na]^-$ ion at *m/z* 2161.7 in the negative-ion MALDI mass spectrum confirmed that KBT-A2 possessed a sulfate ester as a sodium salt. Accurate mass measurement using MALDI-Spiral TOF (Figure S1) [16] indicated that the molecular formula of KBT-A2 was $C_{108}H_{161}O_{42}SNa$ (sodium salt) ($[M+Na]^+$ 2207.9971, calcd. 2207.9973). The molecular formula indicated that KBT-A2 was a sulfate ester of KBT-G ($C_{108}H_{162}O_{39}$).

The UV maximum of KBT-A2 was observed at 227 nm (ε 1.2 × 10^4). The UV absorption suggested that KBT-A2 has an enal side chain analogous with KBT-G. The ^{1}H NMR spectra of KBT-A2 (Figure S2), obtained at 800 and 500 MHz, resembled those of KBT-G. Comparison of the KBT-G and KBT-A2 HSQC spectra (Figure S3) revealed that the ^{1}H and ^{13}C chemical shifts of KBT-A2 arising from CH-1 to CH-59 were close to those of KBT-G (Table 1).

Table 1. NMR spectroscopic data (800 MHz, pyridine-d_5) for brevisulcenal-A1 (KBT-A1) and KBT-A2.

Pos.	KBT-A1[a] δ_H	KBT-A1[a] δ_C	KBT-A2[a] δ_H	KBT-A2[a] δ_C	Pos.	KBT-A1[a] δ_H	KBT-A1[a] δ_C	KBT-A2[a] δ_H	KBT-A2[a] δ_C	Pos.	KBT-A1[a] δ_H	KBT-A1[a] δ_C	KBT-A2[a] δ_H	KBT-A2[a] δ_C
1	9.51	197.1	9.54	197.1	36	4.11	78.7	4.11	78.3	70	4.92	70.7	4.92	72.6
2		143.3		143.3	37		82.7		82.7	71	5.72	71.1	5.70	73.3
3	6.61	151.9	6.61	151.9	38	4.31	80.0	4.30	80.0	72	4.20	81.7	4.06	83.4
4a	2.45	37.0	2.46	36.6	39a	2.11	33.3	2.12	32.8	73	4.77	64.7	4.67	67.1
4b	2.55		2.56		39b	2.22		2.24		74a	2.39	38.9	2.44	39.9
5	4.37	76.8	4.35	76.2	40	4.13	78.5	4.10	77.8	74b	2.69		2.83	
6	5.45	127.3	5.45	127.0	41		82.7		82.7	75	3.35	79.8	3.19	79.7
7	5.97	140.6	6.01	139.9	42	4.00	75.7	4.02	75.9	76	3.71	76.4	3.48	79.1
8		78.3		78.3	43a	2.22	42.5	2.20	42.5	77a	2.16	37.7	1.82	37.9
9	3.85	80.8	3.85	80.0	43b	2.64		2.43		77b	2.51		2.49	
10a	1.96	26.9	1.97	26.4	44	3.82	84.5	4.14	73.3	78	3.38	79.4	3.26	79.5
10b	2.03		2.03		45	4.13	85.6	4.24	85.0	79	3.45	80.1	3.45	80.1

Table 1. *Cont.*

Pos.	KBT-A1[a] δ_H	KBT-A1[a] δ_C	KBT-A2[a] δ_H	KBT-A2[a] δ_C	Pos.	KBT-A1[a] δ_H	KBT-A1[a] δ_C	KBT-A2[a] δ_H	KBT-A2[a] δ_C	Pos.	KBT-A1[a] δ_H	KBT-A1[a] δ_C	KBT-A2[a] δ_H	KBT-A2[a] δ_C
11a	1.88	27.0	1.90	26.9	46a	2.11	33.7	2.14	35.2	80a	2.29	38.7	2.20	37.5
11b	2.04		2.05		46b	2.25		2.20		80b	2.56		2.55	
12	4.04	76.7	4.07	75.4	47a	1.98	30.7	1.72	27.2	81	3.22	80.7	3.28	79.7
13		81.4		81.4	47b	2.04		2.02		82	3.38	80.6	3.38	79.2
14a	2.03	44.7	2.05	43.7	48	3.09	85.6	3.17	88.5	83a	1.72	39.1	1.74	38.4
14b	2.08		2.08		49	3.21	85.3		78.3	83b	2.50		2.52	
15	3.91	73.0	3.90	72.7	50a	1.86	42.5	1.89	49.1	84	4.19	66.9	4.26	66.4
16		76.4		76.4	50b	2.50		2.44		85	3.22	83.7	3.27	83.3
17a	1.99	41.9	1.98	41.6	51	3.41	82.6	4.39	79.7	86	4.37	67.8	4.40	67.4
17b	2.24		2.22		52	3.21	83.1	3.52	83.0	87a	1.95	34.6	2.01	34.0
18	4.52	74.7	4.53	74.2	53a	1.67	32.0	2.14	39.0	87b	2.55		2.58	
19		79.6		79.6	53b	1.97		2.48		88	3.67	79.1	3.69	78.4
20	4.39	75.3	4.39	75.1	54	1.76	40.5	4.06	76.8	89	5.87	87.5	5.86	86.8
21	3.67	82.8	3.67	82.4		1.97				90	6.09	131.8	6.10	132.0
22	4.20	75.3	4.20	75.4	55		80.1		82.0	91	6.31	132.8	6.31	132.3
23a	1.89	38.5	1.90	38.0	56	3.31	83.7	4.26	77.1	92	5.13	92.3	5.01	90.7
23b	2.72		2.72		57a	2.08	32.9	2.08	34.5	93	1.58	48.9	1.49	48.4
24	3.42	81.8	3.41	81.8	57b	2.31		2.33		94	4.38	68.8	4.4	67.4
25	3.59	80.6	3.59	80.6	58	4.04	71.6	3.25	80.9	95	1.36	23.8	1.37	23.1
26a	1.73	45.9	1.74	46.13	59	3.63	72.8	3.76	71.0	96	1.74	11.8	1.74	11.4
26b	2.32		2.39		60a	4.43	70.8	1.67	41.9	97	1.40	24.1	1.47	24.3
27		76.3		76.3	60b			2.41		98	1.56	23.3	1.61	22.9
28	3.44	86.5	3.46	85.9	61	4.07	73.1	4.46	73.6	99	1.48	17.6	1.48	17.3
29a	2.02	30.2	2.02	30.2	62	4.78	74.0	4.07	82.4	100	1.49	19.8	1.50	17.6
29b	2.07		2.07		63a	2.59	35.7	4.72	77.2	101	1.45	24.9	1.47	24.3
30	3.67	76.6	3.67	76.2	63b	2.71				102	1.63	23.3	1.69	23.2
31		77.2		77.2	64	5.18	70.3	5.26	72.7	103	1.57	24.2	1.58	22.8
32a	1.89	45.6	1.91	44.8	65	5.17	72.8	4.89	73.6	104	1.39	21.9	1.40	21.4
32b	2.33		2.41		66	4.11	75.6	4.39	76.5	105	1.39	24.9	1.40	23.2
33	4.85	81.5	4.89	80.9	67	4.47	72.6	4.50	75.4	106	1.39	19.5	1.46	18.2
34		81.7		81.7	68	4.87	68.2	5.11	72.4	107	1.02	12.6	1.04	11.5
35a	2.39	48.9	2.29	48.4	69a	2.11	33.2	2.09	35.1	108			1.38	17.0
35b	2.59		2.60		69b	3.52		3.46						

Pos.: Position, δ_H: ^{1}H chemical shifts, δ_C: ^{13}C chemical shifts. [a] ^{13}C chemical shifts were assigned based on HSQC and HMBC spectra.

However, the ^{1}H and ^{13}C chemical shifts corresponding to the CH-71 region differed significantly. The $\Delta\delta_H$ ($\delta_{H\text{-}KBTG} - \delta_{H\text{-}KBTA2}$) and $\Delta\delta_C$ ($\delta_{C\text{-}KBTG} - \delta_{C\text{-}KBTA2}$) values in the vicinity of CH-71 are summarized in Table 2. The ^{1}H chemical shift of H-71 shifted downfield from δ_H 5.05 in KBT-G to δ_H 5.70 in KBT-A2, thus $\Delta\delta_H$ of H-71 was -0.65. Similarly, the ^{13}C chemical shift of C-71 shifted downfield from δ_C 67.1 in KBT-G to δ_C 73.3 in KBT-A2, thus $\Delta\delta_C$ of C-71 was -6.2. The $\Delta\delta_H$ and $\Delta\delta_C$ of CH-71 were greater than those of the remaining protons and carbons in that region. In addition, in the ^{1}H NMR spectrum of KBT-G, proton coupling between H-71 and a hydroxy proton was observed, while this was not the case for KBT-A2, due to the substitution of OH with OSO$_3$Na. Therefore, it was deduced that the sulfate ester was bonded to C-71 (Figure 1).

Table 2. $\Delta\delta$ values ($\delta_{KBTG} - \delta_{KBTA2}$) for the CH-71 region.

Pos.	$\Delta\delta_H$	$\Delta\delta_C$	Pos.	$\Delta\delta_H$	$\Delta\delta_C$	Pos.	$\Delta\delta_H$	$\Delta\delta_C$
65	+0.08	0.0	69a	+0.10	+0.4	72	+0.15	−0.8
66	+0.05	0.0	69b	−0.30		73	−0.10	−0.3
67	0.00	0.0	70	−0.19	+1.6	74a	0.00	+1.7
68	−0.61	−0.3	71	−0.65	−6.2	74b	0.00	

As MALDI tandem MS with HE-CID using a MALDI-SpiralTOF-TOF [17] instrument proved useful for the structural determination of KBTs, structural confirmation of KBT-A2 was also conducted with this experimental setup. The sulfate ester in KBT-A2 is a suitable charge site for negative ion tandem MS measurements. A $[M-Na]^-$ ion at m/z 2162 was selected as the precursor ion for the HE-CID tandem MS experiments. Product ions generated by bond cleavage from both the aldehyde (C-1) and the methyl (C-95) terminals were observed because the position of the charge site, the sulfate ester, resided in the middle of the molecule. Product ion assignments are explained in Figure 2. A product ion at m/z 2078 was generated by cleavage of a 2-methylbut-2-enal side-chain (Figure 2a). Product ions arising from ring A to ring Q were clearly observed (Figure 2a). The ions observed at m/z 1940, 1870, 1800, 1714, 1658, 1602, 1532, and 1462 confirmed the presence of contiguous six-membered ether rings, C–I, methyl positions on B/C, C/D, D/E, G/H, and H/I junctions, and hydroxy-substitution on ring E (Figure 2a). Prominent product ions at m/z 1191, 1121, 1051, and 965 were observed in the spectrum of KBT-A2, and the mass differences between those product ions were 70, 70, and 86 Da, respectively. The presence of these ions strongly supported that rings M to P comprised a 7-membered ether ring, a 6-membered ether ring with a methyl, a 7-membered ether ring with a hydroxy group, and a 6-membered ether ring with a methyl, respectively, analogous to KBT-G. The product ions at m/z 895, 825, 689, and 617 confirmed the absence of a hydroxy group at C60 on ring Q, the presence of a 1,2,3-trihydroxypropyl linker, a 6-membered ether ring with two hydroxy groups (ring R), and the location of the sulfate ester (Figure 2a). Below m/z 600, only sulfate-related ions at m/z 96.8 and 79.8 were observed. Product ions generated by cleavage from the C-95 terminus are assigned in Figure 2b. The product ions at m/z 2088 and 2020 were generated by cleavage of the C93–C95 side chain and 2,5-dihydrofuran, respectively. The product ions at m/z 1948, 1892, 1836, 1780, and 1710 confirmed the ring arrangement from S to W of KBT-A2 as 6/6/6/6/6/ where the NMR signals overlapped, and the position of the sulfate ester. Thus, these product ions supported the structure of KBT-A2 deduced from NMR spectral analyses.

2.3. Structural Elucidation of Brevisulcenal-A1

The UV absorption of brevisulcenal-A1 (KBT-A1) was at 225 nm (ε 0.6 $\times$ 10^4), indicating that KBT-A1 had an enal side chain. MALDI-SpiralTOF (Figure S4) confirmed that the molecular formula of KBT-A1 was $C_{107}H_{159}O_{41}SNa$ (sodium salt) ($[M+Na]^+$ 2177.9867, calcd. 2177.9867). This indicated that KBT-A1 is a sulfate ester of KBT-F. Although the very small amount (0.2 mg) of KBT-A1 isolated from non-labeled cultures made it difficult to acquire NMR spectra (Figures S5 and S6), the HSQC spectra (Figure S7) of KBT-A1 enabled us to assign the ^{1}H and ^{13}C chemical shifts by comparison to the HSQC spectra of KBT-F (Table 1). The proton chemical shift for H-71 in KBT-A1 was observed at δ_H 5.72, while that of KBT-F was observed at δ_H 5.13. The $\Delta\delta_H$ value of -0.59 suggested that a sulfate ester resided on C-71, analogous to KBT-A2.

Structural validation of KBT-A1 was also accomplished by a MALDI-SpiralTOF-TOF experiment (see Figures S8 and S9 in the Supplemental Material). A $[M-Na]^-$ ion at m/z 2132 was selected as the precursor ion for HE-CID tandem MS experiments. The product ions from ring A to ring Q were clearly observed (Figure S7). Similar to KBT-A2, the product ions observed at m/z 1909, 1839, 1769, 1684, 1628, 1572, 1502, 1432, 1331, 1261, and 1161 confirmed the existence of the contiguous six-membered rings C–I and the J–K ring sequence. These product ions were smaller by 30 Da than the corresponding ions of KBT-A2. Prominent product ions at m/z 1091, 1035, and 965 were observed in the spectrum of KBT-A1. The mass differences between the product ions were 56 and 70 Da, indicating that rings N and O comprised a 6- and 7-membered ether ring, respectively. The ions at m/z 895, 823, 809, 765, and 735 confirmed the presence of a hydroxy group at C60, and the presence of a 2,3-dihydroxypropyl linker. The product ions generated by cleavage of rings S-W, 2,5-dihydrofuran, and the C93–C95 side chain were clearly observed at m/z 2058, 1917, 1861, 1805, and 1749 (Figure S8). Therefore, the structure of KBT-A1 is shown in Figure 1.

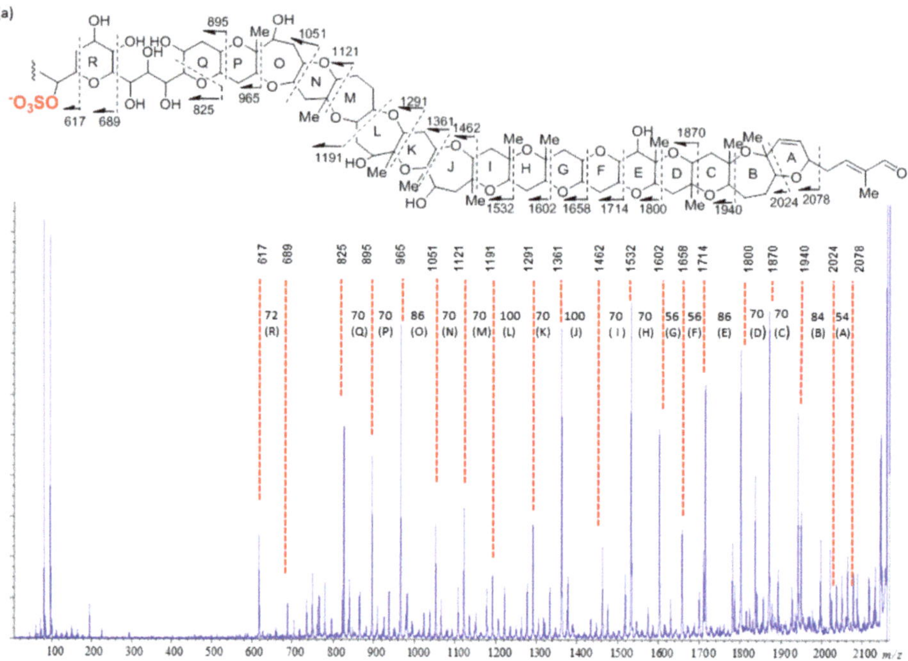

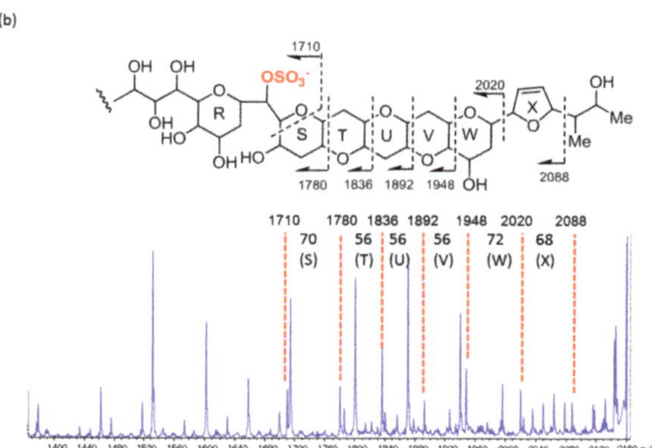

Figure 2. TOF-TOF spectra of brevisulcenal-A2 and assignments of prominent product ions in the partial structures: (**a**) Whole spectrum and assignments from the C-1 terminus, (**b**) expanded spectrum *m/z* 1400–2000 and assignments from the C-95 terminus.

3. Discussion

The cytotoxicity of KBT-A1 and KBT-A2 against mouse leukemia P388 cells was evaluated, obtaining LC_{50}s of 10.9 and 3.5 nM, respectively. That of KBT-F and KBT-G was 2.7 and 0.7 nM, respectively This indicated that the presence of a sulfate ester on brevisulcenal reduces its cytotoxicity. KBTs possess rigid polyether assemblies linked by flexible linear alkyl chains and comprise hydrophobic (rings A–Q) and hydrophilic (linkers and rings R–X) portions. The sulfate ester is positioned on the linker in the hydrophilic portion, increasing its polarity. Structural modifications, such as oxidation of the terminal olefinic methyls and substitution by a sulfate ester, influence the activity of KBTs. Marine polycyclic ethers, maitotoxin (MTX) [18], and yessotoxin (YTX) [19] possess two sulfate esters in their framework and show potent mouse lethality and cytotoxicity. Contrary to KBTs, the cytotoxicity of desulfonated derivatives of MTX and YTX are lower than those of MTX and YTX [20–22]. Therefore, sulfate esters of marine polycyclic ether toxins play an important role in their biological activities, as they affect target protein interactions.

4. Materials and Methods

4.1. General Methods

All purchased solvents (FUJIFILM Wako Pure Chemicals, Osaka, Japan) were of the highest commercial grade and used as received, unless otherwise noted. UV-visible absorption spectra were collected on a JASCO V-550 UV spectrometer (JASCO Co., Tokyo, Japan). A JASCO PU-98 pump and a JASCO UV-970 UV detector were used for liquid chromatography. NMR spectra were recorded on three NMR instruments (Bruker CO., Bremen, Germany and JEOL, Tokyo, Japan): At 800 MHz (200 MHz for ^{13}C) or 500 MHz. Chemical shift values are reported in ppm (δ) referenced to internal signals of residual protons (^{1}H NMR; C_5HD_4N (7.21); ^{13}C NMR, C_5D_5N (125.8)).

4.2. Culture Growth and Harvesting

Karenia brevisulcata (CAWD82) was collected from the Wellington Harbor in 1998 and was kept at Cawthron Institute Culture Collection of Microalgae (CICCM), Cawthron Institute, Nelson. Bulk cultures (150–250 L batches) were grown in 12 L carboys using 100% GP + Se media under a 12/12 h day/night-timed cool white fluorescent lighting regime, with 25 min of aeration at 30 min intervals. Starter culture (14–21 days old) was added to 100% GP + Se media at a ratio from 1:10 to 1:15. Cultures were maintained for up to 21 days. Aliquots of culture were assessed for cell numbers by inverted microscopy. For ^{13}C-enrichment, cultures were augmented at 0 and 10 days with $NaH^{13}CO_3$ (0.25 g per 12 L). Toxin production was assessed by liquid chromatography-mass spectrometry (LC-MS), following SPE using a 50 mL aliquot of culture extracted with Strata-X (60 mg, Phenomenex Inc., CA), washing with Milli-Q water and 20% methanol, and eluting with methanol or methanol followed by acetone (3 mL each).

Toxins were extracted from mature cultures using Diaion HP20 resin. Briefly, the pre-washed resin was packed in a polypropylene column. *K. brevisulcata* cultures were transferred to a 200 L barrel, and cells were lysed by the addition of acetone to 7% *v/v*. The cultures were allowed to settle for 1 h and then diluted with reversed osmosis purified water (RO water) to 5% *v/v* acetone before pumping at 0.3 L/min first through a filter system then through a HP20 resin column. The column was then washed with water, and the HP20 resin was transferred to a 2 L flask. Toxins were recovered by soaking the resin in AR acetone (1 L) and decanting ($\times$3). The combined acetone extracts were evaporated in vacuo to produce a dried crude extract.

4.3. Isolation of KBTs

The crude HP20 extract was dissolved in methanol and diluted to 55% *v/v* with pH 7.2 phosphate buffer. The solution was partitioned with $CHCl_3$ ($\times$2), and the combined chloroform fraction containing neutral toxins was evaporated. Brevisulcenals were isolated from the neutral fractions of 1450 L of bulk cultures by column chromatography using a diol cartridge (500 mg) with stepwise elution of ethyl acetate (EtOAc), EtOAc:MeOH (9:1 and 6:4), and MeOH, and guided by the P388 cytotoxicity assay. The EtOAc:MeOH (6:4) and MeOH fractions were combined. Further purification was conducted by preparative HPLC (250 mm × 4.6 mm i.d. Develosil C30-UG-5, Nomura Chemical Co., Japan) with linear gradient elution from 80% MeOH/H_2O to 100% MeOH and guided by UV absorbance at 230 nm. Final chromatographic purification was accomplished on a Develosil C30-UG-5 column with isocratic elution of 80% MeOH at 1 mL/min. The retention time of KBT-A1 and KBT-A2 was 9–10 min.

4.4. MALDI MS and MALDI Tandem MS Measurements

MALDI mass spectra were recorded on a JEOL JMS-S3000 SpiralTOF instrument (JEOL, Tokyo, Japan) using α-cyano-p-hydroxycinnamic acid (CHCA, FUJIFILM Wako Pure Chemicals, Osaka, Japan) as a matrix for the positive ion mode. KBT-A1 and KBT-A2 were dissolved in MeOH:CHCl$_3$, mixed with a norharmane matrix, and subjected to MALDI tandem MS measurements in negative ion mode. The product ion mass spectra were recorded with laser irradiation at 349 nm, a frequency of 250 Hz, and acceleration voltage of -20 kV in the first TOF stage. The collision energy was set at 20 keV to afford the HE-CID. Product ions formed by collision-induced dissociation were accelerated by 9 kV for analysis in the second TOFMS stage.

4.5. Chemical Properties of Brevisulcenal-A2

Brevisulcenal-A1: Isolated as a colorless amorphous solid: UV maxima (λ) 225 (ε 6000) nm. The high-resolution MALDI-TOF mass spectrum gave [M+Na]$^+$ at *m/z* 2177.9867 for C$_{107}$H$_{159}$O$_{41}$SNa$_2$ ([M+Na]$^+$ calcd. 2177.9867). ^{1}H and ^{13}C NMR data are presented in Table 1.

Brevisulcenal-A2: Isolated as a colorless amorphous solid: UV maxima (λ) 227 (ε 12,000) nm. The high-resolution MALDI-TOF mass spectrum gave [M+Na]$^+$ at *m/z* 2207.9971 for C$_{108}$H$_{161}$O$_{42}$SNa$_2$ ([M+Na]$^+$, calcd. 2207.9973). ^{1}H and ^{13}C NMR data are presented in Table 1.

4.6. Cytotoxicity

Mouse leukemia cells, P388, were cultured in RPMI 1640 supplemented with 5% penicillin/streptomycin solution and 10% inactivated fetal bovine serum at 37 °C under an atmosphere of 5% CO$_2$ in a CO$_2$ incubator. Each well of a 96-well microplate was filled with 100 μL of P388 cell suspension containing 1.0×10^4 cells/mL, followed by the addition of 100 μL of KBT solution dissolved in RPMI 1640 medium. The plate was incubated in a CO$_2$ incubator at 37 °C for 72 h. After 72 h of incubation, the plate was incubated for another 4 h with 50 μL of 1.0 mg/mL aqueous MTT solution under the same conditions. The obtained precipitate was dissolved in DMSO, and absorbance at 570 nm was measured with a multiwavelength spectrometer. Cytotoxicity against P388 cells was determined using the MTS colorimetric reaction method (CellTiter 96®AQueous One Solution Reagent, detected at 490 nm).

Supplementary Materials: The following are available online at https://www.mdpi.com/2072-665 1/13/2/82/s1, Figure S1: Spiral MALDI MS of KBT-A2, Figure S2: ^{1}H NMR spectrum of KBT-A2 (800 MHz, pyridine-d_5), Figure S3: HSQC of KBT-A2 (800 MHz, pyridine-d_5). Figure S4: Spiral MLDI MS of KBT-A1, Figure S5: ^{1}H NMR spectrum of KBT-A1 (800 MHz, pyridine-d_5), Figure S6: ^{1}H-^{1}H COSY of KBT-A1 (800 MHz, pyridine-d_5), Figure S7: HSQC of KBT-A1 (800 MHz, pyridine-d_5), Figure S8: Spiral MALDI TOF-TOF of KBT-A1, Figure S9: Spiral MALDI TOF-TOF of KBT-A1 (*m/z* 1600–2150).

Author Contributions: Culture and extraction of dinoflagellate, F.S.; isolation, R.I.; chemical analysis, M.S., R.I., P.T.H., D.T.H., and F.S.; MALDI-TOF MS and MS/MS analysis, Y.I.; NMR experiments, F.H. and H.Z.; writing—original draft preparation, M.S.; writing—review and editing, P.T.H. and D.T.H.; funding acquisition, M.S., P.T.H., and D.T.H. All authors have read and agreed to the published version of the manuscript.

Funding: This work was supported by KAKENHI (19H02720) and bilateral Japan-New Zealand programs funded by the Japan Society for the Promotion of Science (JSPS) and the New Zealand Ministry for Business, Innovation and Employment (contracts CAWX0703, CAWX0804, and CAWX1108).

Data Availability Statement: Not applicable. Data is provided in the manuscript and Supplementary Materials.

Acknowledgments: We are grateful to Yasukatsu Oshima (Tohoku University) for the initial isolation. The measurements with the 800 MHz NMR spectrometer were performed under the non-proprietary use program supported by RIKEN.

Conflicts of Interest: The authors declare no conflict of interest.

References

1. Hoe Chang, F.H. *Gymnodinium brevisulcatum* sp. nov. (Gymnodiniales, Dinophyceae), a new species isolated from the 1998 summer toxic bloom in Wellington Harbour, New Zealand. *Phycologia* **1999**, *38*, 377–384. [CrossRef]
2. Chang, F.H.; Chiswell, S.M.; Uddstrom, M.J. Occurrence and distribution of *Karenia brevisulcata* (Dinophyceae) during the 1998 summer toxic outbreaks on the central east coast of New Zealand. *Phycologia* **2001**, *40*, 215–221. [CrossRef]
3. Holland, P.T.; Shi, F.; Satake, M.; Hamamoto, Y.; Ito, E.; Beuzenberg, V.; McNabb, P.; Munday, R.; Briggs, L.; Truman, P.; et al. Novel toxins produced by the dinoflagellate *Karenia brevisulcata*. *Harmful Algae* **2012**, *13*, 47–57. [CrossRef]
4. Backer, L.C.; Fleming, L.E.; Rowan, A.; Cheng, Y.-S.; Benson, J.; Pierce, R.H.; Zaias, J.; Bean, J.; Bossart, G.D.; Jonson, D.; et al. Recreational exposure to aerosolized brevetoxins during Florida red tide events. *Harmful Algae* **2003**, *2*, 19–28. [CrossRef]
5. Fleming, L.E.; Kirkpatrick, B.; Backer, L.C.; Bean, J.A.; Wanner, A.; Reich, A.; Cheng, Y.S.; Pierce, R.; Naar, J.; Abraham, W.M.; et al. Aerosolide red-tide toxins (brevetoxins) and asthma. *CHEST* **2007**, *131*, 187–194. [CrossRef]
6. Bean, J.A.; Fleming, L.E.; Kirkpatrick, B.; Backer, L.C.; Nierenberg, K.; Reich, A.; Cheng, Y.S.; Wanner, A.; Benson, J.; Naar, J.; et al. Florida red tide toxins (brevetoxins) and longitudinal respiratory effects in asthmatics. *Harmful Algae* **2011**, *10*, 744–748. [CrossRef]
7. Lin, Y.-Y.; Risk, M.; Ray, S.M.; Van Engen, D.; Clardy, J.; Golik, J.; James, J.C.; Nakanishi, K. Isolation and structure of brevetoxin B from the red tide dinoflagellate *Ptychodiscus brevis* (*Gymnodinium breve*). *J. Am. Chem. Soc.* **1981**, *103*, 6773–6775. [CrossRef]
8. Shimizu, Y.; Chou, H.N.; Bando, H.; Van Duyne, G.; Clardy, J. Structure of brevetoxin A (GB-1 toxin), the most potent toxin in the Florida red tide organism *Gymnodinium breve* (*Ptychodiscus brevis*). *J. Am. Chem. Soc.* **1986**, *108*, 514–515. [CrossRef]
9. Satake, M.; Shoji, M.; Oshima, Y.; Naoki, H.; Fujita, T.; Yasumoto, T. Gymnocin-A, a cytotoxic polyether from the notorious red tide dinoflagellate, *Gymnodinium mikimotoi*. *Tetrahedron Lett.* **2002**, *43*, 5829–5832. [CrossRef]
10. Satake, M.; Tanaka, Y.; Ishikura, Y.; Oshima, Y.; Naoki, H.; Yasumoto, T.; Gymnocin, B. with the largest contiguous polyether rings from the red tide dinoflagellate *Karenia* (formerly *Gymnodinium*) *mikimotoi*. *Tetrahedron Lett.* **2005**, *46*, 3537–3540. [CrossRef]
11. Suzuki, R.; Irie, R.; Harntaweesup, Y.; Tachibana, K.; Holland, P.T.; Harwood, D.T.; Shi, F.; Beuzenberg, V.; Itoh, Y.; Pascal, S.; et al. Brevisulcatic acids, marine ladder-frame polyethers frm the red tide dinoflagellate *Karenia brevisulcata* in New Zealand. *Org. Lett.* **2014**, *16*, 5850–5853. [CrossRef]
12. Irie, R.; Suzuki, R.; Tachibana, K.; Holland, P.T.; Harwood, D.T.; Shi, F.; McNabb, P.; Beuzenberg, V.; Hayashi, F.; Zhang, H.; et al. Brevisulcatic acids from a marine microalgal species implicated in a toxic event in New Zealand. *Heterocycles* **2016**, *92*, 45–54.
13. Hamamoto, Y.; Tachibana, K.; Holland, P.T.; Shi, F.; Beuzenberg, V.; Itoh, Y.; Satake, M. Brevisulcenal-F: A polycyclic ether toxin associated with massive Fish-kills in New Zealand. *J. Am. Chem. Soc.* **2012**, *134*, 4963–4968. [CrossRef]
14. Satake, M.; Irie, R.; Hamamoto, Y.; Tachibana, K.; Holland, P.T.; Harwood, D.T.; Shi, F.; Beuzenberg, V.; Itoh, Y.; Hayashi, F.; et al. Polycyclic ether marine toxins from the dinoflagellate *Karenia brevisulcata*. *Heterocycles* **2018**, *96*, 2096–2105. [CrossRef]
15. Harwood, D.T.; Shi, F.; Satake, M.; Holland, P.T. A sensitive LC-MS/MS assay for brevisulcenal and brevisulcatic acid toxins produced by the dinoflagellate *Karenia brevisulcata*. *Toxicon* **2014**, *84*, 19–27. [CrossRef]
16. Satoh, T.; Sato, T.; Tamura, J. Development of a high-performance MALDI-TOF mass spectrometer utilizing a spiral ion trajectory. *J. Am. Soc. Mass Spectrom.* **2007**, *18*, 1318–1323. [CrossRef]
17. Satoh, T.; Sato, T.; Kubo, A.; Tamura, J. Tandem time-of-flight mass spectrometer with high precursor ion selectivity employing spiral ion trajectory and improvement offset parabolic reflectron. *J. Am. Soc. Mass Spectrom.* **2011**, *22*, 797–803. [CrossRef]
18. Murata, M.; Naoki, H.; Matsunaga, S.; Satake, M.; Yasumoto, T. Structure and partial stereochemical assignments for maitotoxin, the most toxic and largest natural non-biopolymer. *J. Am. Chem. Soc.* **1994**, *116*, 7098–7107. [CrossRef]
19. Murata, M.; Kumagai, M.; Lee, J.S.; Yasumoto, T. Isolation and structure of yessotoxin, a novel polyether compound implicated in diarrhetic shellfish poisoning. *Tetrahedron Lett.* **1987**, *28*, 5869–5872. [CrossRef]
20. Murata, M.; Gusovsky, F.; Sasaki, M.; Yokoyama, A.; Yasumoto, T.; Daly, J.W. Effect of maitotoxin analogues on calcium influx and phosphoinositide breakdown in cultured cells. *Toxicon* **1991**, *29*, 1085–1096. [CrossRef]
21. Terao, K.; Ito, E.; Oarada, M.; Murata, M.; Yasumoto, T. Histopathological studies on experimental marine toxins poisoning-5. The effects in mice of yessotoxin isolated from *Patinopecten yessoensis* and of a desulfated derivative. *Toxicon* **1990**, *28*, 1095–1104. [CrossRef]
22. Daiguji, M.; Satake, M.; Ramstad, H.; Aune, T.; Naoki, H.; Yasumoto, T. Structure and fluorometric HPLC determination of 1-desulfoyessotoxin, a new yessotoxin analog isolated from mussels from Norway. *Nat. Toxins* **1998**, *6*, 235–239. [CrossRef]

Article

Risk Assessment of Pectenotoxins in New Zealand Bivalve Molluscan Shellfish, 2009–2019

Michael J. Boundy [1,*], D Tim Harwood [1,2], Andreas Kiermeier [3], Cath McLeod [2], Jeane Nicolas [4] and Sarah Finch [5]

1   Cawthron Institute, 98 Halifax Street East, Nelson 7010, New Zealand; tim.harwood@cawthron.org.nz
2   New Zealand Food Safety Science and Research Centre, Massey University, Private Bag 11-222, Palmerston North 4442, New Zealand; C.McLeod@massey.ac.nz
3   Statistical Process Improvement Consulting and Training PTY Ltd., Gumeracha, SA 5233, Australia; andreas.kiermeier@gmail.com
4   Ministry for Primary Industries–Manatu Ahu Matua, P.O. Box 2526, Wellington 6140, New Zealand; Jeane.Nicolas@mpi.govt.nz
5   AgResearch Limited, Ruakura Research Centre, Private Bag 3123, Hamilton 3240, New Zealand; sarah.finch@agresearch.co.nz
*   Correspondence: Michael.Boundy@cawthron.org.nz

Received: 4 November 2020; Accepted: 4 December 2020; Published: 6 December 2020

Abstract: Pectenotoxins (PTXs) are produced by *Dinophysis* spp., along with okadaic acid, dinophysistoxin 1, and dinophysistoxin 2. The okadaic acid group toxins cause diarrhetic shellfish poisoning (DSP), so are therefore regulated. New Zealand currently includes pectenotoxins within the DSP regulations. To determine the impact of this decision, shellfish biotoxin data collected between 2009 and 2019 were examined. They showed that 85 samples exceeded the DSP regulatory limit (0.45%) and that excluding pectenotoxins would have reduced this by 10% to 76 samples. The incidence (1.3%) and maximum concentrations of pectenotoxins (0.079 mg/kg) were also found to be low, well below the current European Food Safety Authority (EFSA) safe limit of 0.12 mg/kg. Inclusion within the DSP regulations is scientifically flawed, as pectenotoxins and okadaic acid have a different mechanism of action, meaning that their toxicities are not additive, which is the fundamental principle of grouping toxins. Furthermore, evaluation of the available toxicity data suggests that pectenotoxins have very low oral toxicity, with recent studies showing no oral toxicity in mice dosed with the PTX analogue PTX2 at 5000 µg/kg. No known human illnesses have been reported due to exposure to pectenotoxins in shellfish, a fact which combined with the toxicity data indicates that they pose negligible risk to humans. Regulatory policies should be commensurate with the level of risk, thus deregulation of PTXs ought to be considered, a stance already adopted by some countries.

Keywords: Pectenotoxin; Exposure; Risk assessment; Diarrhetic shellfish poisoning

Key Contribution: The impact of including PTXs within DSP regulations was found to be low under NZ conditions. Despite this, the grouping of PTXs with the DSPs is scientifically flawed, so the PTX group should be removed from regulation in New Zealand. Furthermore, there is no evidence that PTXs pose any food safety risk to humans, so this toxin class should be deregulated.

1. Introduction

Pectenotoxins (PTXs) are produced by *Dinophysis* spp. [1], and during blooms of this microalgal species, filter feeding bivalve mollusks can accumulate the microalgae in their digestive glands and absorb lipophilic compounds they produce into the shellfish flesh. In addition to PTXs, *Dinophysis* spp. also produces okadaic acid group toxins; okadaic acid (OA), dinophysistoxin 1 (DTX1), and dinophysistoxin 2 (DTX2). Toxins from the OA group have been known to cause human illness since the late 1970s [2], inducing a syndrome called diarrhetic shellfish poisoning (DSP), which is dominated by the symptom of diarrhea. To minimize the incidence of this illness, regulatory limits have been set for OA group toxins found in shellfish. Historically, due to the co-production and co-occurrence of PTXs and okadaic acid group toxins by *Dinophysis* spp., PTXs have been included in DSP regulation, and this is still the case in New Zealand.

When monitoring methods moved away from the traditional mouse bioassay to analytical analysis, it was discovered that the PTXs consisted of a large array of 20 related analogues, although only PTX1 and PTX2 are included in the DSP regulation, with PTX1 not routinely monitored due to unavailability of suitable reference material. Shellfish samples that contain toxin concentrations above the maximum permissible level for DSP result in the closure of the shellfish harvesting area until the toxin levels have returned to safe concentrations. Little is known about the distribution of PTXs in New Zealand shellfish or the concentrations present in shellfish. The Ministry for Primary Industries and the New Zealand shellfish industry have tested for marine biotoxins in bivalve molluscan shellfish for many years, yielding a large set of data. The presence of PTXs in shellfish is typically monitored using liquid chromatograph-tandem mass spectrometry (LC-MS/MS) [3,4]. Using this approach, a range of PTXs have been reported, including the PTX analogue PTX2, and the non-regulated metabolites PTX2SA and 7-epi-PTX2SA, which are collectively reported as pectenotoxin 2 seco acids (PTX2SAs). When the method was first developed, PTX1, PTX11, and PTX6 were not routinely monitored due to instrument limitations [4]. However, with advancements in instrumentation, these three additional analogues are now acquired simultaneously by the LC-MS/MS method used for regulatory monitoring in New Zealand without impacting method performance. However, while these congeners have been added to the acquisition method, they are not included in the routine processing and quantitation due to the additional time and cost required.

In this study, to fill knowledge gaps surrounding PTXs in New Zealand shellfish, information gathered from 2009–2019 was used to collate prevalence data for OA, DTX1, and DTX2 (after hydrolysis), as well as PTX2 and its seco acids over the 10-year period. In addition, for selected bloom events, the concentrations of the PTX analogues, PTX1, PTX11, and PTX6, were obtained by manually reprocessing historical LC-MS/MS data acquired in order to retrospectively determine PTX profiles within New Zealand shellfish. Using these data, the impact of including PTXs in the DSP class of toxins was evaluated. To be able to conduct an exposure assessment for PTXs, the concentrations in shellfish must be combined with the quantity eaten by the consumer (meal sizes). Unfortunately, most consumption surveys are targeted to obtain data on consumption over time, which is best suited to chronic toxicity risk assessments. Because consumption surveys are often summarized as the "average amount of a food consumed over the survey period", it is usually impossible to discern the frequency and amount per serving. Knowing only the average amount consumed (e.g., 50 g/day) does not provide information on whether a consumer eats consistent portions daily throughout the week, or whether larger portions (e.g., 175 g/meal) are consumed on average a couple of times per week. In the Oct 2008–Oct 2009 New Zealand Adult Nutrition Survey [5], a 24-h recall of 4721 adults aged 15+, including 1040 Maori and 757 Pacific peoples, was used. It was not stated if people consumed more than one type of the seafood listed, so a total mollusk consumption could not be determined. The highest 97.5 percentile portion size across the shellfish species was 268 g (paua), followed by 256 g for mussels. While insufficient data are available to create a robust meal size distribution for risk modeling, an approximation can be made using simulations, such as a triangular distribution [6].

To conduct a risk assessment of the PTXs, information on not only exposure, but also toxicity of the compounds is required. There are many reports of intraperitoneal injection (i.p.) toxicity of PTXs in mice. However, information on the feeding method, strain, and sex of mice is not documented in most of the available publications, which makes the interpretation and accurate comparison of the data difficult. It is clear that PTX1, PTX2, PTX3, and PTX11 are of similar toxicity by i.p. administration, with lethal doses of between 219 and 411 µg/kg; PTX4 and PTX6 appear to be slightly less toxic, with lethal doses of

770 and 500 μg/kg, respectively, and PTX7, PTX8, PTX9, PTX2SA, and 7-epi-PTX2SA are of low toxicity, with no mouse deaths observed even at a dose rate of 5000 μg/kg [1,7–12]. In comparison to the i.p. route of administration, there have been few studies conducted to investigate the acute oral toxicity of PTXs. The first report was by Ishige et al. in 1988 [13], which stated that the lowest observed adverse effect level (LOAEL) was 250 μg/kg based on a single mouse dosed by gavage with PTX2 of unspecified purity. The effects observed in the study involved fluid accumulation in the intestine and damage to intestinal villi of the mouse. Using this figure, the EFSA CONTAM Panel derived an acute reference dose (ARfD) of 0.8 μg PTX2 equivalents/kg b.w., and derived a safe level of 0.12 mg/kg in shellfish flesh based on a 400 g large portion size [6]. Although focused on yessotoxins, a study in the 1990s [14] reported what appeared to be the oral acute toxicity of PTX2. In this study, the oral toxicity of PTX2 was reported to be similar to the toxicity by i.p. injection. In contrast, the study by Miles et al. [10] showed no signs of toxicity in any of the five mice dosed with PTX2 at a dose rate of 5000 μg/kg using well-characterized material. The acute oral toxicity of PTX2SA [10] and PTX11 [12] was found to be equally low, with no signs of toxicity observed in any of the five mice dosed with either compound at a dose rate of 5000 μg/kg. The severe diarrhea in mice attributed to PTX2 in the earlier study by Ishige may have been due to contamination of the sample with an okadaic acid derivative, which is co-extracted with PTX2 [10]. The question of the toxicity of PTXs is essential in conducting a risk assessment, and underpins whether they should be regulated. Furthermore, the validity of including PTXs with the OA group toxins is investigated. Various areas of the world handle the regulation of PTXs differently, so, in this study, we will review the available literature, which is often conflicting, and present a rationale for the interpretation of the data.

2. Results

2.1. Distribution of PTXs in New Zealand

2.1.1. Spatial Distribution of PTXs

PTX2 and *Dinophysis* spp. were both detected throughout the country with notably elevated concentrations and occurrence observed in Banks Peninsula, the Firth of Thames, and Port Underwood (Figure 1). Relative concentrations of PTX2, PTX2SAs, and DSP were similar across the different regions, with typically PTX2SAs > DSP > PTX2 (DSP toxins = OA, DTX1, and DTX2). However, in some bloom events, there were notably relatively less PTX2 and PTX2SAs compared to DSP toxins. These may be due to blooms of other species, such as *Prorocentrum* spp., which are known to produce OA group toxins but not PTXs. Benthic species, such as *Prorocentrum* spp., do not reliably get detected with routine phytoplankton monitoring. The locations of *Dinophysis* spp. detection were similarly consistent with the observations of PTX2, PTX2SAs, and DSP toxins. However, no phytoplankton samples from the West Coast have been tested where a PTX2/DSP bloom was detected due to the difficulty in obtaining samples caused by inaccessible terrain and weather. The concentrations of *Dinophysis* spp. cell counts did not correlate to detections of PTX2, PTX2SAs, and DSP toxins, with some higher cell counts not resulting in higher concentrations of PTX2, PTX2SAs, and DSP. This is likely due to differences in the production of toxins between algal species, and potentially non-producing strains.

2.1.2. Temporal Distribution of PTXs

The concentrations of PTX2, PTX2SAs (sum of PTX2SA and 7-epi-PTX2SA), and DSP toxins in New Zealand (independent of sample site) over the 2009–2019 period were plotted over time, together with the *Dinophysis* spp. cell concentrations (Figure 2).

Results were grouped by year in order to assess potential changes in occurrence over the 2009–2019 period. PTX2 results are summarized in Table 1.

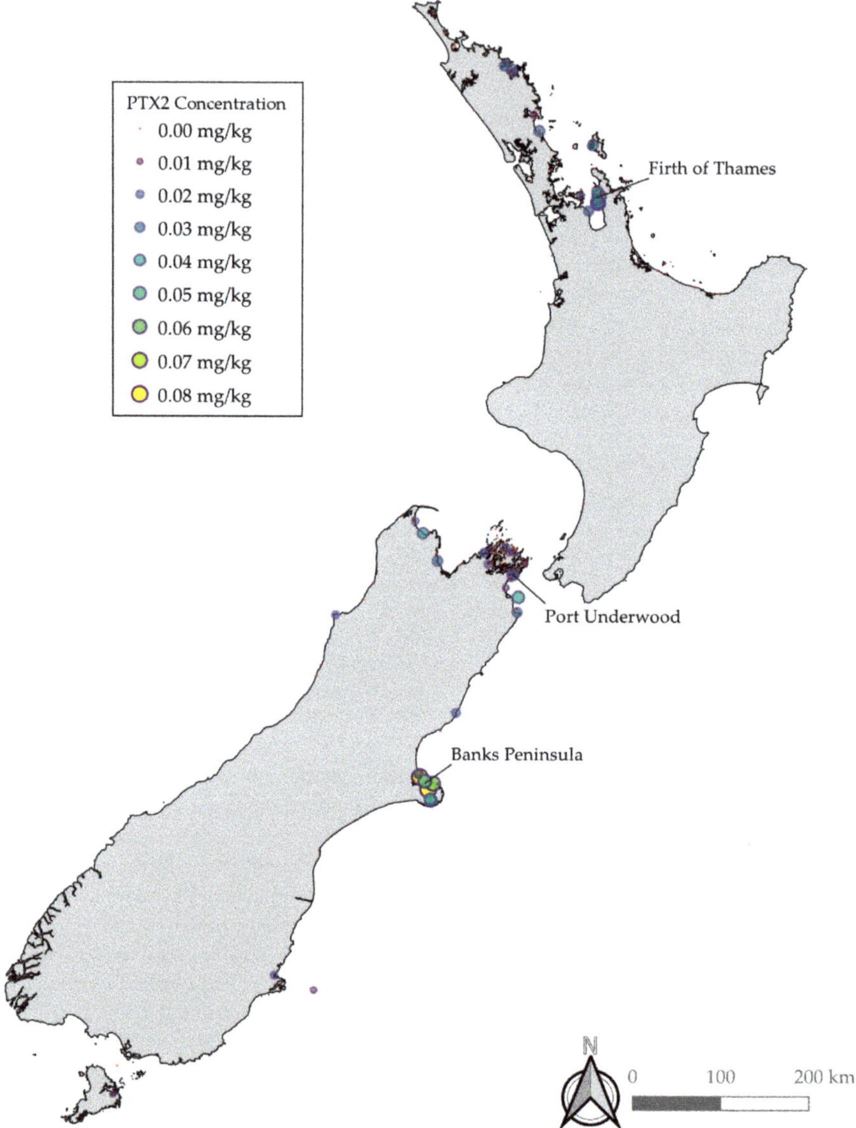

Figure 1. Maximum concentrations of PTX2 at sampling locations throughout New Zealand. Marker color and size are on a continuous scale.

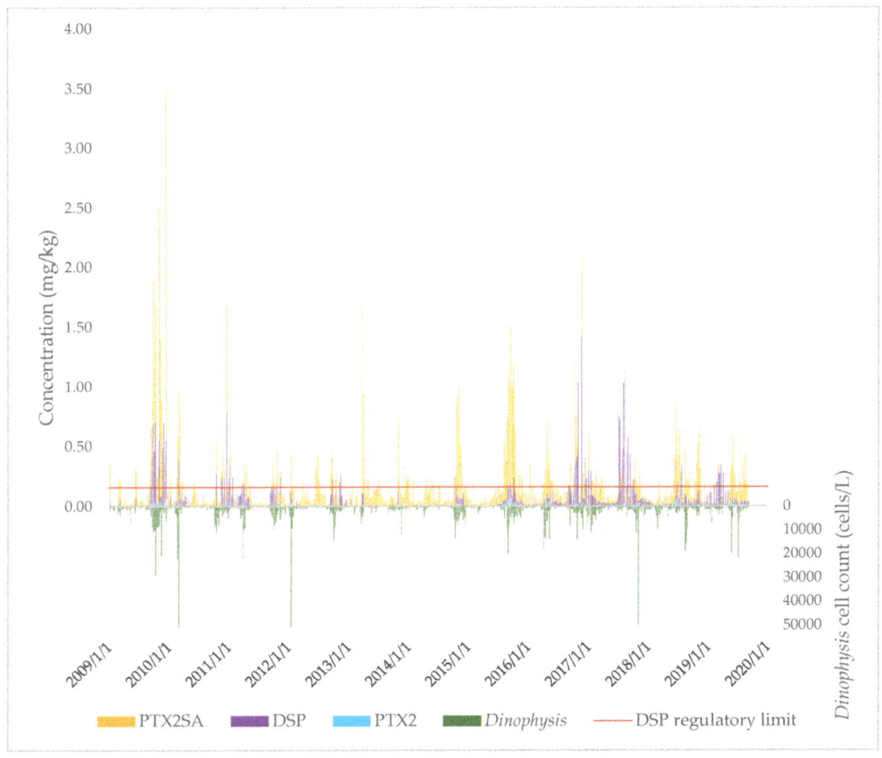

Figure 2. Concentrations of PTX2, PTX2SAs, DSP, and *Dinophysis* spp. throughout New Zealand over the 2009–2019 period.

Table 1. Summary of the number of samples analyzed, detections, and minimum, maximum, mean, median, and 97.5 percentile (PCTL) concentrations (mg/kg) of PTX2 in different years in New Zealand over the 2009–2019 period.

Year	No. Samples	Detections	% Detected	Min	Max	Mean	Median	97.5 PCTL
2009	1688	56	3.3%	0.010	0.063	0.019	0.015	0.048
2010	1618	14	0.9%	0.010	0.041	0.014	0.011	0.035
2011	1684	21	1.2%	0.010	0.043	0.016	0.014	0.038
2012	1647	13	0.8%	0.011	0.025	0.015	0.013	0.024
2013	1723	5	0.3%	0.010	0.021	0.017	0.019	0.021
2014	1776	10	0.6%	0.010	0.016	0.013	0.014	0.016
2015	1871	66	3.5%	0.010	0.059	0.021	0.017	0.053
2016	1836	21	1.1%	0.010	0.079	0.026	0.021	0.078
2017	1924	14	0.7%	0.010	0.027	0.017	0.017	0.026
2018	1857	12	0.6%	0.011	0.058	0.023	0.017	0.054
2019	1323	19	1.4%	0.010	0.024	0.014	0.012	0.023
Total	18947	251	1.3%	0.010	0.079	0.019	0.015	0.052

Both 2009 and 2015 showed elevated bloom occurrence, with 3.3–3.5% of samples having detectable PTX2 compared to the other years, where only 0.6–1.4% of the samples had detectable PTX2.

Results were grouped by month in order to assess potential seasonality, with PTX results summarized in Table 2. Detections of PTX2 were observed in all months of the year, with the largest number of detections in September–October, and maximum concentrations observed in November–December.

Table 2. Summary of the number of samples analyzed, detections, and minimum, maximum, mean, median, and 97.5 percentile (PCTL) concentrations (mg/kg) of PTX2 in different months of the year in New Zealand over the 2009–2019 period.

Month	No. Samples	Detections	% Detected	Min	Max	Mean	Median	97.5 PCTL
January	1615	10	0.6%	0.011	0.043	0.020	0.016	0.041
February	1617	30	1.9%	0.010	0.027	0.014	0.012	0.026
March	1679	10	0.6%	0.010	0.023	0.013	0.011	0.021
April	1594	11	0.7%	0.010	0.039	0.016	0.014	0.035
May	1634	10	0.6%	0.011	0.022	0.015	0.015	0.022
June	1574	15	1.0%	0.010	0.058	0.020	0.016	0.052
July	1594	9	0.6%	0.011	0.027	0.016	0.015	0.026
August	1563	21	1.3%	0.010	0.052	0.022	0.018	0.047
September	1514	47	3.1%	0.010	0.059	0.021	0.018	0.054
October	1593	50	3.1%	0.010	0.046	0.017	0.015	0.034
November	1542	28	1.8%	0.010	0.079	0.024	0.017	0.078
December	1428	10	0.7%	0.010	0.063	0.021	0.013	0.058
Total	18947	251	1.3%	0.010	0.079	0.019	0.015	0.052

2.1.3. Species Distribution of PTX2

Sample results were sorted by type of shellfish, and results for PTX2 are summarized in Table 3. The data available from the laboratory information management system (LIMS) database only identified species by a common name. The most commonly tested type of shellfish was green-lipped mussels (*Perna canaliculus*, 84%), followed by Pacific oyster (*Crassostrea gigas*, 6%), clams (unspecified, 5%), scallops (*Pecten novaezealandiae*, 2%), and dredge oyster (*Ostrea chilensis*, 1%). Small numbers of other shellfish species (< 1% each) were also analyzed. Blue mussels (*Mytilus edulis*) had the highest detection rate for any shellfish type (12.5%). This observation is likely impacted by sampling bias, as blue mussels are typically not analyzed as part of routine monitoring in New Zealand, and are instead taken from areas in response to a bloom event.

Table 3. Summary of the number of samples analyzed, detections, and minimum, maximum, mean, median, and 97.5 percentile (PCTL) concentrations (mg/kg) of PTX2 in different types of shellfish analyzed in New Zealand over the 2009–2019 period.

Organism [1]	Sites	No. Samples	Detections	% Detected	Min	Max	Mean	Median	97.5 PCTL
Green-lipped mussel	83	15947	186	1.2%	0.010	0.079	0.019	0.015	0.056
Pacific oyster	22	1141	40	3.5%	0.010	0.027	0.015	0.015	0.026
Clam	11	1042	6	0.6%	0.013	0.027	0.018	0.016	0.026
Scallop	20	298	4	1.3%	0.012	0.032	0.020	0.017	0.031
Dredge oyster	8	228	1	0.4%	0.043	0.043	0.043	0.043	0.043
Surf clam	6	97	5	5.2%	0.010	0.024	0.015	0.012	0.023
Blue mussel	12	56	7	12.5%	0.011	0.042	0.021	0.020	0.039
Queen scallop	2	52	2	3.8%	0.010	0.011	0.011	0.011	0.011
Tuatua	5	28	0						
Pipi	2	19	0						
Cockle	3	17	0						
Oyster	5	9	0						
Abalone	3	8	0						
Geoduck	3	5	0						
Total	144	18947	251	1.3%	0.010	0.079	0.019	0.015	0.052

[1] Organism as identified in the LIMS database.

2.1.4. Pectenotoxin Profiles

Samples from five bloom events were reprocessed to quantify PTX1, PTX11, and PTX6. Multiple reaction monitoring (MRM) transitions for these analogues are acquired using the LC-MS/MS method of analysis, although they are not routinely processed [4]. The three blooms with the highest observed concentration of PTX2 were selected for reprocessing, as well as the two highest concentration blooms from areas where Pacific oyster and scallops were most commonly sampled. There were no detections of PTX1, PTX11, or PTX6 above the 0.01 mg/kg reporting limit in any of the 389 reprocessed samples. Trace detections were observed for PTX1 and PTX11 in some samples, and PTX6 was not detected in any samples. As only trace detections were observed, profiles were assessed including all trace detections and including those below the quantitation and reporting limits. A bloom event in the Coromandel region in 2015 affected the largest number of sites, species, and samples, and provided the richest dataset of the bloom events observed in the 2009–2019 period. PTX profiles from this 2015 bloom event are shown for green-lipped mussels (n = 182), Pacific oysters (n = 10), and scallops (n = 3) in Figure 3.

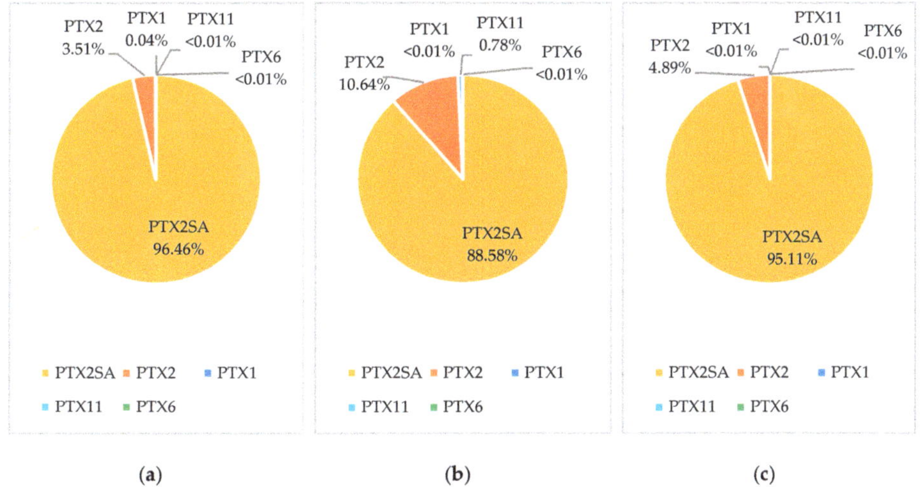

Figure 3. Pectenotoxin profiles based on the 97.5 percentile concentrations for PTX analogues in the Coromandel 2015 bloom for: (**a**) green-lipped mussels; (**b**) Pacific oyster; (**c**) scallops.

Green-lipped mussels showed a trace detection of PTX1, with an average of 1.1% of the concentration of PTX2. The PTX1 detections showed a similar accumulation and depuration trend to PTX2. PTX1 and PTX6 have previously been reported to form via metabolism in Japanese scallops (*M. yessoensis*) [7,15]. However, PTX1 is only observed at a very low abundance, in contrast to that observed in *M. yessoensis*. Pacific oysters showed a trace detection of PTX11, with an average of 13.1% of the concentration of PTX2. As there were only three Pacific oyster samples analyzed during the bloom that showed trace levels of PTX11, there was not enough information to observe an accumulation and depuration trend. Pacific oysters also showed a lower abundance of PTX2SAs, and this could be due to different binding of the compounds within the flesh, lower levels of PTX2 metabolism than in green-lipped mussels, or due to competing metabolism to form other congeners, such as PTX11. Scallops contained no detectable levels of either PTX1 or PTX11.

2.1.5. Impact of PTXs Contribution to DSP Levels

In New Zealand, PTXs are currently included in the DSP regulation. To compare the impact of the inclusion of PTXs in regulatory monitoring, the DSP (excluding PTX2) concentration and sum of DSP and PTX2 concentrations were calculated for each sample. As PTX2 is the only PTX group congener that is routinely monitored and is the most dominant congener, apart from the non-regulated seco acids, PTX2 was used as a surrogate for PTXs. These results were then compared against the current regulatory limit, with results shown in Figure 4A. As the OA group toxin concentration increases with samples where PTX2 is detected, the DSP + PTX2 concentration becomes closer to the DSP concentration, as the PTX2 concentration does not proportionately increase with the DSP concentrations (Figure 4B).

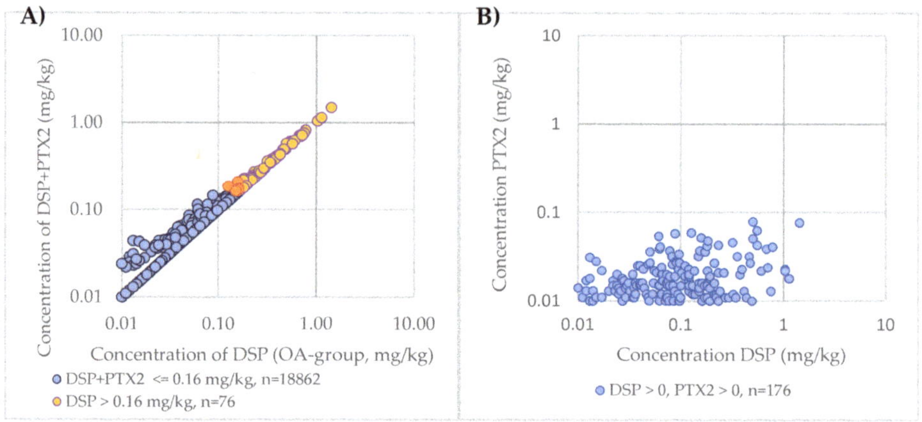

Figure 4. (**A**) Comparison of PTX2 contribution to DSP regulation in New Zealand over the 2009–2019 period on a logarithmic scale; (**B**) Comparison of PTX2 concentrations to DSP concentrations over the 2009–2019 period on a logarithmic scale.

Of the 18,947 samples analyzed, a total of 76 samples were above the regulatory limit for DSP (0.4%). An additional nine samples were considered above the DSP regulatory limit with the inclusion of PTX2. Of the nine samples in the 2009–2019 period that were pushed above the regulatory limit by including the PTX2 concentration, three of the OA group concentrations were at the regulatory limit (0.16 mg/kg), five were at 0.15 mg/kg, and one at 0.13 mg/kg; PTX2 concentrations for these samples were between 0.01 and 0.059. There were 75 samples that contained reportable levels of PTX2 and no reportable OA group toxins, with 90% of these samples analyzed prior to July 2015, when the limit of reporting for the OA group analogues was 0.05 mg/kg rather than 0.01 mg/kg due to use of a less sensitive instrument.

2.2. Risk Assessment

2.2.1. Deterministic Risk Assessment

Three portion sizes were used to assess exposure to PTX2: 100 g, the standard portion size [16]; 268 g, the highest 97.5 percentile portion size of shellfish species for New Zealand consumers; and 400 g, the large portion size adopted by EFSA for risk assessment [6]. The exposure for a consumer of a large (400 g) portion of shellfish meat contaminated with the maximum concentration of PTX2 observed from all samples over the 2009–2019 period is 0.53 µg PTX2/kg b.w. (Table 4), which is still less than the ARfD of 0.8 µg PTX2/kg b.w. proposed by EFSA [6]. A 60 kg person would have to consume approximately 608 g of shellfish at 0.079 mg PTX2/kg to reach this conservative ARfD.

The main components of the exposure assessment and risk characterization were the consumption amount of bivalve mollusks and the distribution of PTX2 concentrations in bivalve mollusks. A probabilistic estimate of dietary exposure to PTX2 was performed by a Monte Carlo simulation to generate the amount of PTX2 consumed in a sitting (adjusted by kg body weight). The focus of the risk assessment was on the acute exposure of the consumption of PTX2.

Table 4. Deterministic intake of PTX2 based on all samples

Parameter	Units	97.5 Percentile	Maximum
Concentration PTX2	mg PTX2/kg	0.01	0.079
Exposure by eating 100 g	μg PTX2/person	1.0	7.9
	μg PTX2/kg b.w.	0.02	0.13
Exposure by eating 268 g	μg PTX2/person	2.7	21.2
	μg PTX2/kg b.w.	0.04	0.35
Exposure by eating 400 g	μg PTX2/person	4.0	31.6
	μg PTX2/kg b.w.	0.07	0.53

2.2.2. Probabilistic Risk Assessment

The risk characterization is the comparison of the exposure distributions to the corresponding Health Based Guidance Value, which for PTX2 is the conservative ARfD of 0.8 μg/kg b.w. proposed by EFSA [6]. None of the 1,000,000 iterations in either of the models for the Monte Carlo simulations resulted in an exposure exceeding the ARfD (based on the maximum of 0.533 μg/kg b.w.). This is consistent with the absence of reported human illnesses due to exposure to PTXs.

3. Discussion

An examination of biotoxin data collected in New Zealand between 2009 and 2019 showed PTXs to be present throughout the country, in a range of shellfish species, with detections more frequent in September and October and maximum PTX2 concentrations observed in November (0.063 mg/kg) and December (0.079 mg/kg). However, the number of PTX2 detections was low, as demonstrated by the observation that only 3.3–3.5% of shellfish samples collected in the years with the highest number of detections (2009 and 2015) contained PTX2. The PTX profiles were examined in three shellfish species, which showed PTX2SA to be the dominant PTX compound (89–96%), followed by PTX2 (3.5–10.6%), PTX11 (0–0.78%), and PTX1 (0–0.04%); PTX6 was not detected in any of the shellfish samples.

Since New Zealand includes PTXs in the DSP regulation, in contrast to International Codex Standard 292-2008, the impact of PTX2 to DSP levels was investigated in this research. It was found to be minor. Over the 10 years of data examined, 76 shellfish samples were determined to be above the DSP regulatory limit (excluding PTX2) (0.4%), and only an additional nine samples (0.05%) were pushed over the regulatory limit by the inclusion of PTX2. When comparing the contribution of PTX2 to OA group toxins in shellfish where PTX2 was detected, there was a relatively higher contribution of PTX2 at lower concentrations of OA group toxins. This could be due to the metabolism of PTX2 to PTX2SA in New Zealand shellfish. As PTX2 and OA group toxins are accumulated by the shellfish, PTX2 is metabolized to PTX2SA over time, resulting in relatively lower PTX2 concentrations compared to OA group toxins as the bloom progresses and OA group toxin concentrations increases. From the deterministic risk assessment of PTX2, the highest concentration observed in shellfish over the 2009–2019 period would require a large 608 g portion size to be consumed in order to reach the conservative ARfD proposed by EFSA. With the probabilistic risk assessment of PTX2, there were no simulated cases exceeding this ARfD.

The grouping of related toxins for the assessment of human exposure is essential, as toxicity is generally not due to one individual compound, but rather a mixture of related structural analogues. Since the mouse bioassay has been proven to be inaccurate and is considered by many countries to be unethical for routine screening, this is now handled by instrumental chemical analysis of shellfish samples for all known analogues of the DSP toxin class. Since analogues will have different toxicities, to translate this into an estimate of overall toxicity, the relative toxicities of the individual components must be applied. To determine toxicity equivalence factors (TEFs), toxicity data is considered with the following order of importance: data from human cases (outbreaks) > oral LD_{50} in animals > i.p. LD_{50} in animals > mouse bioassay and in vitro data [17]. The fundamental principle for grouping toxins is that they must have a shared mechanism, hence, their toxicities are additive [18]. This requirement is met for

OA and the DTXs, as both are active on protein phosphatases. However, PTXs are inactive on protein phosphatases, and instead exert their effect by action on F-actin [19]. In our view, including the PTX group as part of the DSP regulation is therefore scientifically not justified. This position is consistent with the view expressed by numerous scientific opinions and FAO/WHO/IOC committees [6,16,18,20–22]. Despite these clear and numerous scientific opinions, some countries, including New Zealand, Canada, Chile, and the EU, currently include PTX2 in the DSP regulation, whereas other countries, including Australia, Japan, the United States of America, and Mexico, do not.

To provide an estimate of the acute risk of PTXs to human health, the most relevant parameter is the toxic dose by oral administration. The non-toxicity of PTX2 observed by Miles et al. [10] is at odds with the early study by Ishige et al. [13]. Another difference in the studies was that, in contrast to the early study, the one conducted in 2004 reported no diarrhea in mice dosed with PTX2. While diarrhea is a well-recognized symptom of the OA group toxins, whether PTXs induce diarrhea or not is a key point in assessing the validity of the Ishige et al. and Miles et al. toxicity assessments of PTX2. PTX1 has been shown to induce no diarrhea when injected into either suckling mice [23] or when administered by gavage [24]. Furthermore, using intestinal models, it has been shown that PTX1, unlike OA or the DTXs, caused no fluid accumulation in rabbit or mouse intestinal loops [24]. Since PTXs are co-extracted with the OA group toxins, and they are difficult to separate, it appears likely that the early report of PTX2 toxicity by gavage utilized material contaminated with an OA derivative, hence inducing diarrhea and giving an incorrect assessment of toxicity [10]. The other report of PTX2 inducing oral toxicity is the study by Ogino et al. [14], who found that the oral toxicity of PTX2 was similar to that generated by i.p. injection. However, the results reported are dubious, as no dose dependency was observed. The mortality recorded at a dose rate of 25 µg/kg (25%) was higher than that seen in mice dosed at both 100 µg/kg (0%) and 200 µg/kg (20%), while the mortality observed in mice dosed at 400 µg/kg (25%) was equal to that of the 25 µg/kg (25%) group, and lower than that recorded in mice dosed at 300 µg/kg (40%). It is difficult to account for this observed non-dose dependent mortality, but it should be noted that there can be a high incidence of gavage-associated deaths and that the administration technique can impact on the results [25–27]. The most robust study of PTX2 toxicity is therefore that of Miles et al. [10], which reported no signs of toxicity in mice dosed with 5000 µg/kg of well-characterized PTX2. PTX11, which has a similar i.p toxicity as PTX2, was equally non-toxic at an oral dose rate of 5000 µg/kg [12]. The major metabolic product of PTX2, PTX2SA, was also non-toxic orally at this dose rate [10].

Since EFSA based the ARfD on the oral toxicity reported in 1988, which we now believe to be incorrect, the ARfD should be reevaluated. The highest dose tested in the more recent toxicity studies was 5000 µg/kg, which induced no toxic effects and hence represents a no observed adverse effect level (NOAEL) rather than an LOAEL or LD_{50}, which may be considerably higher. If this figure of 5000 µg/kg is used to calculate an ARfD, taking into account a 10-fold safety factor for a possible toxicity difference in species and a 10-fold safety factor for possible toxicity variation within species, an ARfD of 50 µg/kg is generated, over 60-fold higher than the ARfD proposed by EFSA. For a 60 kg standard adult human, applying the 400 g large portion meal size proposed by EFSA gives a level of PTX2 equivalents that would be considered safe of 7.5 mg PTX2 equivalents/kg mollusk flesh. This is approximately 100-fold higher than the maximum observed concentration of PTX2 in New Zealand shellfish over the 2009–2019 period.

There is a total lack of toxicity in mice dosed with PTX2, PTX11, or PTX2SA orally at a dose rate of 5000 µg/kg, and a lack of toxicity observed in mice dosed with PTX7, PTX8, PTX9, PTX2SA, and 7-epi-PTX2SA by i.p. at a dose rate of 5000 µg/kg. Although no oral sub-chronic toxicity data are available, the difference between the i.p. and oral toxicity seen for PTX2 and PTX11 can be explained on the basis of a low or lack of absorption of these compounds in the gastrointestinal tract. Consistent with this assumption, after an oral administration of a mixture of PTX2 and PTX2SA, the majority of the toxins remained within the gastrointestinal tract and were excreted in the feces [28]. On this basis, the risk of cumulative toxicity is very low, and the acute toxicity data would give a good estimation of the overall risk posed by PTXs. These data show no oral toxicity, even at very high dose rates, which is consistent with the total lack of any evidence implicating PTXs in human illness, a fact recognized in various EFSA and WHO documents [6,16,29]. A review of all available data therefore suggests that PTXs do not pose a health risk. This view is also shared by FAO/WHO/IOC panels, who have regularly discussed PTXs, with the consensus being that there is no recommendation to regulate this toxin class [16,18,20,22,30].

This study has shown that the contribution of the PTXs to the DSP group toxin concentration is small and that the risk to human health posed by the occurrence of PTXs in shellfish in New Zealand is negligible, with the probabilistic risk assessment showing no simulated cases that exceed even the current ARfD. A comparison of the mechanisms of action for PTXs and DSP group toxin classes show them to be different, indicating that they cannot be co-located in the same toxin class. A review of the available pectenotoxin toxicity data indicates that the current ARfD that has been set by EFSA needs reviewing, and that an oral dose rate in animal studies of 5000 µg/kg of PTX2 showed no toxicity. Given the foregoing, it is clear that the risk posed by PTXs in shellfish is negligible, risk management controls should be commensurate/relatable to the level of risk posed, and therefore, consideration should be given to the omission of this group from the DSP toxin suite.

4. Materials and Methods

4.1. Exposure Data

Biotoxin testing performed on commercial and non-commercial samples in New Zealand uses liquid chromatography-tandem mass spectrometry (LC-MS/MS) [4]. Several changes have occurred with the implementation of this method of analysis over the years, with improvements to the technology resulting in improved performance (e.g., limit of detection). Three different tandem quadrupole mass spectrometry systems contributed to the data over the years, a Micromass Quattro Ultima (Manchester, UK) using a Phenomenex Luna C18 150 × 2 mm 5 µm column (Torrance, CA, USA), Micromass Quattro Premier XE (Manchester, UK) using a Phenomenex Luna C18(2) 50 × 1 mm 2.5 µm column (Torrance, CA, USA), and Waters Xevo TQ-S with a Waters Acquity BEH Shield RP18 50 × 2.1 mm 1.7 µm column (Milford, MA, USA). Routinely, a fixed limit of reporting is established, which is reliably able to be achieved by the instrumentation. During the 2009–2019 period, all of the PTX2 results were reported with a reporting limit of 0.01 mg/kg. DSP was calculated as a sum of OA, DTX1, and DTX2 after hydrolysis. Between 2009 and June 2015, the reporting limit for OA, DTX1, and DTX2 was 0.05 mg/kg. It was then reduced to 0.01 mg/kg.

Biotoxin testing and phytoplankton raw data for 2009–2019 were sourced from the Cawthron laboratory information management system (LIMS) database, excluding samples with null entries to either site code or results. For each result, data were exported, including identifiers, site code, site description, sample ID, sample type, sampled date, received date, analysis method, reported name, reported result, and unit. Results for PTX2, PTX2SAs (sum of PTX2SA and 7-epi-PTX2SA), total OA, total DTX1, and total DTX2 were extracted for each sample. DSP was calculated by adding the total OA and DTXs toxins following hydrolysis, i.e., excluding the PTX group. As PTX1, PTX11, and PTX6 were not processed and quantified as part of the monitoring program, no data were available for these congeners to be exported from the LIMS database. Samples from five bloom events were reprocessed to retrospectively quantify PTX1, PTX11, and PTX6, which are acquired in the LC-MS/MS. Raw data for the reprocessed batches (including trace results below the reporting limit) were exported directly from the TargetLynx processing software (Waters Corporation Milford, MA, USA).

For the biotoxin data, data from unclassified site locations, such as overseas product testing (n = 55), imported products (n = 12), and Chatham Island (n = 5) were removed. This yielded a total of 18,947 sample results, with sampling dates spanning 4 January 2009 to 2 September 2019. For phytoplankton data, data from unidentifiable sites (n = 1173) were removed. This yielded a total of 35,277 sample results, with sampling dates spanning 4 January 2009 to 9 September 2019.

Bloom events were classified for shellfish sites within New Zealand from 2009–2019 by first grouping the sites by their sampling zone. Where many samples with overlapping blooms were detected, the zones were separated into subzones by identifying natural barriers which isolate the different regions within the shellfish zones. Blooms were characterized by visually looking at accumulation/depuration patterns in the concentrations over time. Bloom events were assigned if any of the below conditions were observed in at least one sample within the event:

(a)   PTX2 was at or above reportable levels (0.01 mg/kg).
(b)   DSP toxins were at or above reportable levels (0.05 mg/kg until June 2015, 0.01 mg/kg after June 2015).
(c)   PTX2SAs (sum of PTX2SA and 7-epi-PTX2SA) was at or above 0.1 mg/kg (10-fold higher than the reporting limit of 0.01 mg/kg).

The bloom event was determined to start at the first detection of any of the above groups and end at the last detection of any of the above groups. In several cases, if a new bloom had started prior to the previous bloom depurating and the blooms were decided to be far enough apart to be considered as separate events, then the lowest concentration point was used to divide the two events. All samples within the zone or subzone were assigned as part of the bloom event over the time period established. Each bloom event was then reviewed, and any sites that were observed to not have had any toxin detections were excluded.

4.2. Risk Assessment

For the deterministic risk assessment, the exposure amount is calculated with the product of the concentration in the meal and the portion size. The exposure was then calculated by dividing the exposure amount by the body weight, assumed to be 60 kg for an adult for comparison against the ARfD. Both the 97.5 percentile and maximum concentrations of PTX2 were used for the exposure calculations with three portion sizes: 100 g, the standard portion size [16]; 268 g, the highest 97.5 percentile portion size of shellfish species for New Zealand consumers; and 400 g, the large portion size adopted by EFSA for risk assessment [6].

For the probabilistic risk assessment, an excel spreadsheet containing PTX2 and DSP data for New Zealand sites/zone and different bivalve species was loaded into the statistical software R 3.6.1. [31] for analysis and the risk characterization simulation. The mc2d package (version 0.1-18) for R was used in the development of the simulation and risk characterization [32].

A similar approach to EFSA [6] was taken for portion sizes, that is, a triangular distribution was used for the portion sizes because insufficient information was available to determine a distribution shape. This distribution was defined by the minimum value of 0 g, most likely value (mode) of 100 g, and maximum value of 400 g. The 400 g large portion is likely an overestimate and hence the likely exposure to PTX2 would also be overestimated.

Two approaches were undertaken to estimate distributions of PTX2 during bloom events only for the exposure modeling: Model 1: A binomial distribution with probabilities of a detection/non-detection that are equal to those in the bloom data set (i.e., 6.55% and 93.45%, respectively). Detects are generated from a log-normal distribution (parameters: meanlog = −4.098, sdlog = 0.445) that was the best fit to the detections, and this was left truncated at the limit of reporting of 0.01 mg/kg. Non-detects are assigned a PTX2 concentration of 0.01 mg/kg, resulting in a conservative, i.e., overestimate of risk. Model 2: Using the empirical distribution of PTX2 concentrations from the bloom data set. Non-detects are assigned a PTX2 concentration of 0.01 mg/kg.

Author Contributions: Conceptualization, D.T.H. and C.M.; Data curation, M.J.B. and A.K.; Formal analysis, M.J.B., A.K. and S.F.; Funding acquisition, D.T.H. and C.M.; Investigation, M.J.B., J.N. and S.F.; Methodology, M.J.B.; Project administration, M.J.B.; Supervision, D.T.H.; Visualization, M.J.B. and A.K.; Writing—original draft, M.J.B. and S.F.; Writing—review & editing, M.J.B., D.T.H., C.M., J.N. and S.F. All authors have read and agreed to the published version of the manuscript.

Funding: This research was funded by the New Zealand Ministry for Primary Industries, agreement number 405964. The manuscript preparation and article publishing charge was funded by the New Zealand Seafood Safety research platform, contract number CAWX1801.

Acknowledgments: Thanks to Brian Roughan for helping shape the proposal.

Conflicts of Interest: The authors declare no conflict of interest. The funders had a role in the review of the manuscript, and in the decision to publish the results.

References

1. Yasumoto, T.; Murata, M.; Oshima, Y.; Sano, M.; Matsumoto, G.; Clardy, J. Diarrhetic shellfish toxins. *Tetrahedron* **1985**, *41*, 1019–1025. [CrossRef]

2. Yasumoto, T.; Oshima, Y.; Yamaguchi, M. Occurence of a new type of shellfish poisoning in the Tohoku district. *Bull. Jpn. Soc. Sci. Fish.* **1978**, *44*, 1249–1255. [CrossRef]

3. Suzuki, T.; Beuzenberg, V.; MacKenzie, L.; Quilliam, M.A. Liquid chromatography–mass spectrometry of spiroketal stereoisomers of pectenotoxins and the analysis of novel pectenotoxin isomers in the toxic dinoflagellate Dinophysis acuta from New Zealand. *J. Chromatogr. A* **2003**, *992*, 141–150. [CrossRef]

4. McNabb, P.; Selwood, A.I.; Holland, P.T.; Aasen, J.; Aune, T.; Eaglesham, G.; Hess, P.; Igarishi, M.; Quilliam, M.; Slattery, D.; et al. Multiresidue Method for Determination of Algal Toxins in Shellfish: Single-Laboratory Validation and Interlaboratory Study. *J. Aoac Int.* **2005**, *88*, 761–772. [CrossRef] [PubMed]

5. University of Otago and Ministry of Health. *A Focus on Nutrition: Key Findings of the 2008/09 New Zealand Adult Nutrition Survey*; Ministry of Health: Wellington, New Zealand, 2011; ISBN 978-0-478-37348-6.

6. EFSA (The European Food Safety Authority). Scientific Opinion of the Panel on Contaminants in the Food Chain on a Request from the European Commission on Marine Biotoxins in Shellfish—Pectenotoxin Group. *Efsa J.* **2009**, *1109*, 1–47.

7. Yasumoto, T. Polyether toxins produced by dinoflagellates. In *Mycotoxins and Phycotoxins '88*; Natori, S., Hashimoto, K., Ueno, Y., Eds.; Elsevier: Amsterdam, The Netherlands, 1989; pp. 375–382.

8. Yoon, Y.M.; Kim, Y.C. [Acute toxicity of pectenotoxin-2 and its effects on hepatic metabolising enzyme system in mice]. *Korean J. Toxicol.* **1997**, *13*, 183–186. (In Korean)

9. Sasaki, K.; Wright, J.L.C.; Yasumoto, T. Identification and Characterization of Pectenotoxin (PTX) 4 and PTX7 as Spiroketal Stereoisomers of Two Previously Reported Pectenotoxins. *J. Org. Chem.* **1998**, *63*, 2475–2480. [CrossRef]

10. Miles, C.O.; Wilkins, A.L.; Munday, R.; Dines, M.H.; Hawkes, A.D.; Briggs, L.R.; Sandvik, M.; Jensen, D.J.; Cooney, J.M.; Holland, P.T.; et al. Isolation of pectenotoxin-2 from Dinophysis acuta and its conversion to pectenotoxin-2 seco acid, and preliminary assessment of their acute toxicities. *Toxicon* **2004**, *43*, 1–9. [CrossRef]

11. Miles, C.O.; Wilkins, A.L.; Munday, J.S.; Munday, R.; Hawkes, A.D.; Jensen, D.J.; Cooney, J.M.; Beuzenberg, V. Production of 7-epi-Pectenotoxin-2 Seco Acid and Assessment of Its Acute Toxicity to Mice. *J. Agric. Food Chem.* **2006**, *54*, 1530–1534. [CrossRef]

12. Suzuki, T.; Walter, J.A.; LeBlanc, P.; MacKinnon, S.; Miles, C.O.; Wilkins, A.L.; Munday, R.; Beuzenberg, V.; MacKenzie, A.L.; Jensen, D.J.; et al. Identification of pectenotoxin-11 as 34S-hydroxypectenotoxin-2, a new pectenotoxin analogue in the toxic dinoflagellate Dinophysis acuta from New Zealand. *Chem. Res. Toxicol.* **2006**, *19*, 310–318. [CrossRef]

13. Ishige, M.; Satoh, N.; Yasumoto, T. Pathological studies on mice administered with the causative agent of diarrhetic shellfish poisoning (okadaic acid and pectenotoxin-2). *Hokkaidoritsu Eisei Kenkyushoho* **1988**, *38*, 15–18.

14. Ogino, H.; Kumagai, M.; Yasumoto, T. Toxicologic evaluation of yessotoxin. *Nat. Toxins.* **1997**, *5*, 255–259. [CrossRef]

15. Suzuki, T.; Mitsuya, T.; Matsubara, H.; Yamasaki, M. Determination of pectenotoxin-2 after solid-phase extraction from seawater and from the dinoflagellate Dinophysis fortii by liquid chromatography with electrospray mass spectrometry and ultraviolet detection. Evidence of oxidation of pectenotoxin-2 to pectenotoxin-6 in scallops. *J. Chromatogr. A* **1998**, *815*, 155–160. [PubMed]

16. FAO/IOC/WHO. *Report of the Joint FAO/IOC/WHO ad hoc Expert Consultation on Biotoxins in Bivalve Molluscs*; Food and Agriculture Organization of the United Nations: Oslo, Norway, 2004; p. 31.

17. Botana, L.M.; Hess, P.; Munday, R.; Nathalie, A.; DeGrasse, S.L.; Feeley, M.; Suzuki, T.; Berg, M.V.D.; Fattori, V.; Gamarro, E.G.; et al. Derivation of toxicity equivalency factors for marine biotoxins associated with Bivalve Molluscs. *Trends Food Sci. Technol.* **2017**, *59*, 15–24. [CrossRef]
18. FAO; WHO. *Technical paper on Toxicity Equivalency Factors for Marine Biotoxins Associated with Bivalve Molluscs*; FAO: Rome, Italy, 2016; 108p, ISBN 978-92-5-109345-0.
19. Fladmark, K.E.; Serres, M.H.; Larsen, N.L.; Yasumoto, T.; Aune, T.; Døskeland, S.O. Sensitive detection of apoptogenic toxins in suspension cultures of rat and salmon hepatocytes. *Toxicon* **1998**, *36*, 1101–1114. [CrossRef]
20. FAO; WHO; IOC. *Report of the Working Group Meeting to Assess the Advice from the Joint FAO/WHO/IOC Ad Hoc Expert Consultation on Biotoxins in Bivalve Molluscs*; CX/FFP 06/28/6-Add.1; FAO: Beijing, China, 2006.
21. EFSA (The European Food Safety Authority). Scientific Opinion of the Panel on Contaminants in the Food Chain on a request from the European Commission on Marine Biotoxins in Shellfish—Summary on regulated marine biotoxins. *EFSA J.* **2009**, *130*, 1–23.
22. FAO; WHO. *Report of the Thirty-Second Session of the Codex Committee on Fish and Fishery Products*; FAO: Bali, Indonesia, 2013.
23. Terao, K.; Ito, E.; Ohkusu, M.; Yasumoto, T. A comparative Study of the Effects of DSP-toxins on mice and rats. In *Toxic Phytoplankton Blooms in the Sea: Proceedings of the Fifth International Conference on Toxic Marine Phytoplankton, Newport, RI, USA, 28 October–1 November 1991*; Theodore, Y.S., Smayda, J., Eds.; Elsevier: Amsterdam, The Netherlands, 1993; pp. 581–587.
24. Hamano, Y.; Kinoshita, Y.; Yasumoto, T. Enteropathogenicity of Diarrhetic Shellfish Toxins in Intestinal Models. *J. Food Hyg. Soc. Jpn.* **1986**, *27*, 375–379. [CrossRef]
25. Rao, G.N.; Peace, T.A.; Hoskins, D.E. Training could prevent deaths due to rodent gavage procedure. *Contemp. Top. Lab. Anim. Sci.* **2001**, *40*, 7–8.
26. Damsch, S.; Eichenbaum, G.; Tonelli, A.; Lammens, L.; van den Bulck, K.; Feyen, B.; Vandenberghe, J.; Megens, A.; Knight, E.; Kelley, M. Gavage-related reflux in rats: Identification, pathogenesis, and toxicological implications. *Toxicol. Pathol.* **2011**, *39*, 348–360. [CrossRef]
27. Munday, R. Toxicology of Seafood Toxins: A Critical Review. In *Seafood and Freshwater Toxins: Pharmacology, Physiology, and Detection*; Botana, L.M., Ed.; CRC Press: Boca Raton, FL, USA, 2014; pp. 197–290.
28. Burgess, V.A. *Toxicology Investigations with the Pectenotoxin-2 Seco Acids*; Griffith University: Queensland, Australia, 2003.
29. Munday, R. Toxicology of Seafood Toxins: Animal Studies and Mechanisms of Action. *Recent Adv. Anal. Mar. Toxins.* **2017**, *78*, 211.
30. FAO; WHO. *Report of the Thirty-Third Session of the Codex Committee on Fish and Fishery Products*; FAO: Bergen, Norway, 2014.
31. R Core Team. *A Language and Environment for Statistical Computing*; R Foundation for Statistical Computing; R Core Team: Vienna, Austria, 2019. Available online: https://www.R-project.org (accessed on 31 October 2019).
32. Pouillot, R.; Delignette-Muller, M.-L. Evaluating variability and uncertainty separately in microbial quantitative risk assessment using two R packages. *Int. J. Food Microbiol.* **2010**, *142*, 330–340. [CrossRef] [PubMed]

 *toxins*

Article

Dihydrodinophysistoxin-1 Produced by *Dinophysis norvegica* in the Gulf of Maine, USA and Its Accumulation in Shellfish

Jonathan R. Deeds [1,*], Whitney L. Stutts [1,†], Mary Dawn Celiz [1], Jill MacLeod [2], Amy E. Hamilton [2,‡], Bryant J. Lewis [2], David W. Miller [2], Kohl Kanwit [2], Juliette L. Smith [3], David M. Kulis [4], Pearse McCarron [5], Carlton D. Rauschenberg [6], Craig A. Burnell [6], Stephen D. Archer [6], Jerry Borchert [7] and Shelley K. Lankford [8]

[1] Office of Regulatory Science, United States Food and Drug Administration, Center for Food Safety and Applied Nutrition, College Park, MD 20740, USA; wlstutts@ncsu.edu (W.L.S.); MaryDawn.Celiz@fda.hhs.gov (M.D.C.)

[2] Maine Department of Marine Resources, West Boothbay Harbor, ME 05475, USA; Jill.MacLeod@maine.gov (J.M.); Amy.Hamilton@maryland.gov (A.E.H.); Bryant.J.Lewis@maine.gov (B.J.L.); David.W.Miller@maine.gov (D.W.M.); Kohl.Kanwit@maine.gov (K.K.)

[3] Virginia Institute of Marine Science, College of William and Mary, Gloucester Point, VA 23062, USA; jlsmith@vims.edu

[4] Department of Biology, Woods Hole Oceanographic Institute, Woods Hole, MA 02543, USA; dkulis@whoi.edu

[5] Biotoxin Metrology, National Research Council Canada, Halifax, NS B3H 3Z1, Canada; Pearse.McCarron@nrc-cnrc.gc.ca

[6] Bigelow Analytical Services, Bigelow Laboratory for Ocean Sciences, East Boothbay, ME 04544, USA; carlton.rauschenberg@gmail.com (C.D.R.); cburnell@bigelow.org (C.A.B.); sarcher@bigelow.org (S.D.A.)

[7] Washington State Department of Health, Olympia, WA 98504, USA; Jerry.Borchert@DOH.WA.GOV

[8] Washington State Department of Health Public Health Laboratories, Shoreline, WA 98155, USA; Shelley.Lankford@DOH.WA.GOV

* Correspondence: jonathan.deeds@fda.hhs.gov; Tel.: +1-(240)-402-1474

† Current Address: Molecular Education, Technology, and Research Innovation Center, North Carolina State University, Raleigh, NC 27695, USA.

‡ Current Address: Tidewater Ecosystem Assessment, Maryland Department of Natural Resources, Annapolis, MD 21401, USA.

Received: 25 July 2020; Accepted: 17 August 2020; Published: 20 August 2020

Abstract: Dihydrodinophysistoxin-1 (dihydro-DTX1, (M-H)$^-$ *m/z* 819.5), described previously from a marine sponge but never identified as to its biological source or described in shellfish, was detected in multiple species of commercial shellfish collected from the central coast of the Gulf of Maine, USA in 2016 and in 2018 during blooms of the dinoflagellate *Dinophysis norvegica*. Toxin screening by protein phosphatase inhibition (PPIA) first detected the presence of diarrhetic shellfish poisoning-like bioactivity; however, confirmatory analysis using liquid chromatography-tandem mass spectrometry (LC-MS/MS) failed to detect okadaic acid (OA, (M-H)$^-$ *m/z* 803.5), dinophysistoxin-1 (DTX1, (M-H)$^-$ *m/z* 817.5), or dinophysistoxin-2 (DTX2, (M-H)$^-$ *m/z* 803.5) in samples collected during the bloom. Bioactivity-guided fractionation followed by liquid chromatography-high resolution mass spectrometry (LC-HRMS) tentatively identified dihydro-DTX1 in the PPIA active fraction. LC-MS/MS measurements showed an absence of OA, DTX1, and DTX2, but confirmed the presence of dihydro-DTX1 in shellfish during blooms of *D. norvegica* in both years, with results correlating well with PPIA testing. Two laboratory cultures of *D. norvegica* isolated from the 2018 bloom were found to produce dihydro-DTX1 as the sole DSP toxin, confirming the source of this compound in shellfish. Estimated concentrations of dihydro-DTX1 were >0.16 ppm in multiple shellfish species (max. 1.1 ppm) during the blooms in 2016 and 2018. Assuming an equivalent potency and molar response to DTX1, the authority initiated precautionary shellfish harvesting closures in both years. To date, no illnesses have been associated with the presence of dihydro-DTX1 in shellfish in the Gulf of Maine region and studies are underway to determine the potency of this new toxin relative to the currently regulated DSP toxins in order to develop appropriate management guidance.

Keywords: diarrhetic shellfish poisoning; dihydro-DTX1; *Dinophysis norvegica*; Gulf of Maine USA

Key Contribution: This work describes the first record of dihydrodinophysistoxin-1, produced by *Dinophysis norvegica*, in shellfish in the Gulf of Maine USA. Also provided are data on accumulation of this compound in commercial shellfish species as well as preliminary methods of analysis.

---

1. Introduction

Shellfish harvesting closures due to Diarrhetic Shellfish Toxins (DSTs) in excess of the 0.16 ppm total okadaic acid equivalents (OA eq.) regulatory guidance level are a relatively recent occurrence in the United States (US). The first such closure occurred in the state of Texas (Gulf of Mexico region) in 2008 due to a bloom of *Dinophysis ovum* [1,2]. The first closure in the state of Washington (west coast Puget Sound region) occurred in 2011 due to a mixture of species, primarily *D. acuminata* [3,4]. On the east coast of the US, a large bloom of *D. acuminata* prompted a precautionary shellfish harvesting closure in the Potomac River bordering the states of Maryland and Virginia in 2002, but only trace concentrations of DSTs were found [5]. More recently, *D. acuminata* has been documented in increasing abundance in the Mid-Atlantic region with DSTs in excess of guidance levels in shellfish (non-commercial) occurring sporadically since 2011 [6,7], but to date, no additional shellfish harvesting closures have occurred in this region. In the east coast New England region, limited shellfish harvesting closures have occurred in the Nauset Marsh system in Massachusetts since 2015 due to DSTs from blooms of *D. acuminata* (M Brosnahan, Woods Hole Oceanographic Institute, personal communication).

Although *D. acuminata* and *D. ovum* have been responsible for the majority of diarrhetic shellfish poisoning (DSP)-related shellfish harvesting closures in the US to date, other potentially toxigenic species have been documented in lower abundance, namely *D. fortii* and *D. norvegica* [4,7]. In the central coast of the Gulf of Maine, *Dinophysis* spp. commonly reach peak abundances in the summer months, and when blooms occur, they are typically predominated by *D. norvegica* (Maine Department of Marine Resources, personal communication). From July 5 to August 29, 2016, a large monospecific bloom of *D. norvegica* occurred in the Penobscot and Frenchman Bay regions of the central coast of the Gulf of Maine, USA (Figure 1). Multiple samples with cell concentrations > 2000 cells L^{-1} were observed with a maximum concentration of 54,300 cells L^{-1} recorded on July 17. Shellfish collected from several sites throughout the bloom area were screened using a commercial protein phosphatase inhibition assay (PPIA), which indicated the presence of DSTs in excess of 0.16 ppm, prompting a ban on shellfish harvesting on July 20 (Figure 2). Subsequent confirmatory testing using liquid chromatography tandem mass spectrometry (LC-MS/MS) failed to detect OA, dinophysistoxin-1 (DTX1), or dinophysistoxin-2 (DTX2) from samples collected during the bloom. Testing in three additional laboratories using both methods provided the same results, positive by PPIA and negative by LC-MS/MS. Further exploratory testing using liquid chromatography-high resolution mass spectrometry (LC-HRMS) to screen for additional lipophilic shellfish toxins was also negative. Therefore, the harvesting ban was lifted on August 5 and the PPIA results were considered as false positives.

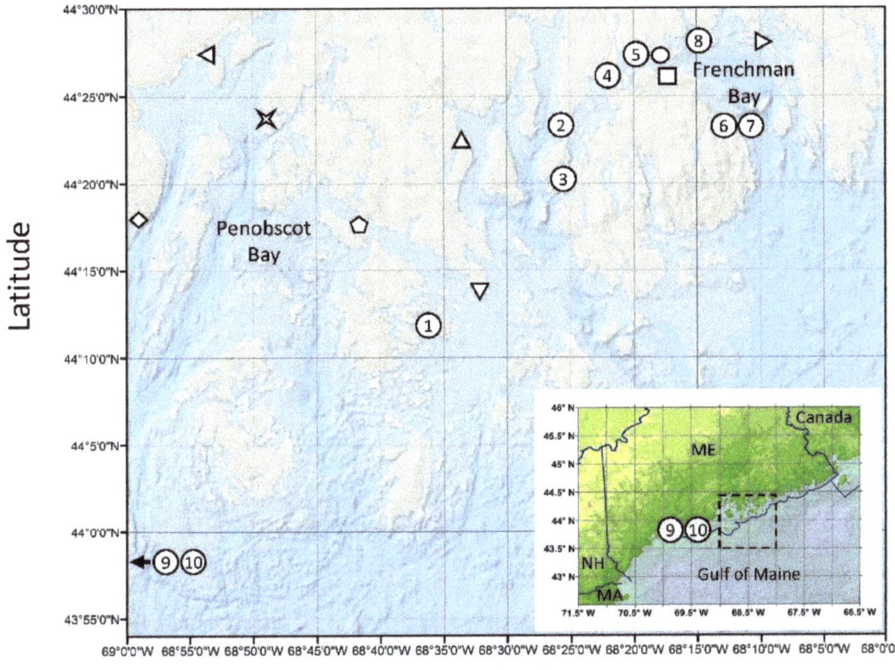

Figure 1. Map of sampling locations during the 2016 *Dinophysis norvegica* bloom on the central coast of the Gulf of Maine, USA. Symbols indicate sampling locations for both water and shellfish and correspond to data in Figure 2: [◊] Lincolnville, [◁] Searsport, [⋈] Dice Head, [□] Eggemoggin Reach, [△] Blue Hill Falls, [▽] Flye Point, [○] Lamoine State Park, [⬠] Salsbury Cove, [▷] Waukeag. Numbers indicate shellfish sampling only and correspond to data in Table 1: (1) Stinson Neck Causeway, (2) Oak Point, (3) Pretty Marsh Harbor, (4) Trenton Sea Plane Ramp, (5) Googins Ledge, (6,7) Bar Harbor, (8) Raccoon Cove, (9,10) Lumbo's Hole (control site outside of bloom area). Abbreviations: ME—Maine, NH—New Hampshire, MA—Massachusetts. Sampling site coordinates provided in Table S1.

In 2017, an investigation was initiated to determine the source of the DST-like bioactivity occurring in shellfish harvested during blooms of *D. norvegica* in Maine. Further testing using several commercial enzyme-linked immuno-sorbent assays (ELISAs) specific for DSTs provided the same results as those found by PPIA testing, confirming the presence of compound(s) with both structures and bioactivities similar to DSTs. Bioactivity-guided fractionation identified a single PPIA-positive fraction similar in retention time to DTX1. LC-MS/MS analyses of this semi-purified fraction tentatively identified the unknown DST as dihydro-DTX1. This same compound was identified in additional shellfish samples collected during the *D. norvegica* bloom in 2016 and again in 2018, as well as in a filtered plankton sample collected in 2018. Finally, this compound was confirmed to be the only DST produced in two cultured isolates of *D. norvegica* collected from the Gulf of Maine in 2018. Detailed here are the investigations leading to the discovery of this novel DST as well as a proposed strategy to manage this potential new DSP risk until an analytical standard can be produced and the potency of this new toxin in relation to the other known DSTs can be determined.

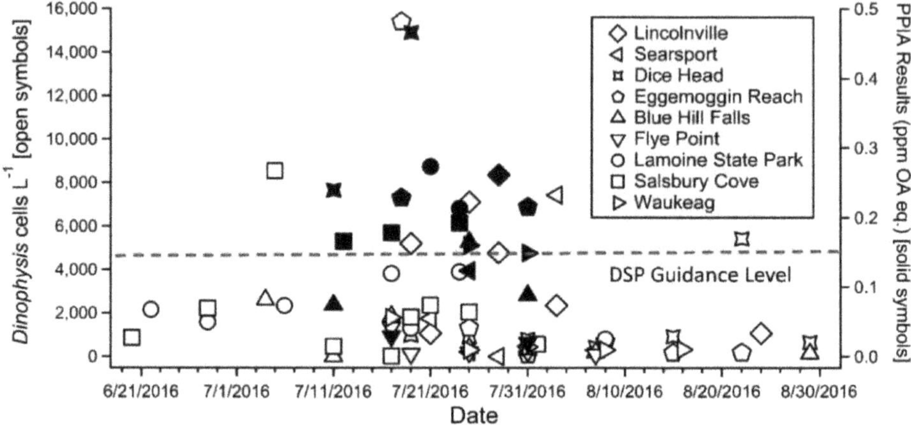

Figure 2. *Dinophysis* spp. cell counts (open symbols) and corresponding protein phosphatase inhibitory activity (closed symbols) in shellfish collected during the 2016 *Dinophysis norvegica* bloom in the Gulf of Maine. Dashed line indicates guidance level for DSP toxins in shellfish of 0.16 ppm okadaic acid equivalents (OA eq.). Symbols correspond to sampling sites in Figure 1. Sampling site coordinates provided in Table S1.

Table 1. Results from the analysis for mussel (*Mytilus edulis*) samples collected during the 2016 *Dinophysis norvegica* bloom in the Gulf of Maine. (PPIA) protein phosphatase inhibition assay (Okadaic Acid (DSP) PP2A kit, Abraxis Inc., USA), (ELISA) enzyme-linked immuno-sorbent assay (qualitative-NEOGEN Reveal 2.0 for DSP, quantitative-Bioo Scientific MaxSignal Okadaic Acid (DSP)), (LC-MS/MS) liquid chromatography-tandem mass spectrometry performed (1) according to the selected reaction monitoring (SRM) method approved by the National Shellfish Sanitation Program for analyzing clams for OA, DTX1, and DTX2 and (2) with the additional transitions for dihydro-DTX1. All units are in ppm.

Sample [a]	PPIA	ELISA		LC-MS/MS	
		Qualitative	Quantitative	OA, DTX1, DTX2	Dihydro-DTX1 [b]
1	0.20	Positive	0.24	ND	0.24
2	>0.35	Positive	0.67	ND	1.08
3	0.08	Negative	0.08	ND	0.06
4	0.10	Negative	0.07	ND	0.07
5	0.28	Positive	0.34	ND	0.79
6	0.17	Positive	0.16	ND	0.20
7	0.19	Negative	0.14	ND	0.21
8	0.14	Negative	0.15	ND	0.16
9	<0.06	Negative	<LOD	ND	ND
10	<0.06	Negative	0.06	0.04 DTX1	Trace

[a] Numbers correspond to sampling locations shown in Figure 1: (1) Stinson Neck Causeway, (2) Oak Point, (3) Pretty Marsh Harbor, (4) Trenton Sea Plane Ramp, (5) Googins Ledge, (6,7) Bar Harbor, (8) Raccoon Cove, (9,10) Lumbo's Hole (control site outside of bloom area). Sampling site coordinates provided in Table S1. [b] (M-H)⁻ m/z 819.5 fragmenting to m/z 255.2 (for quantification) and 151.1 (for confirmation), quantified using an external DTX1 standard. ND: Not detected. Trace: >LOD 0.2 ppb and <LOQ 0.6 ppb.

2. Results

2.1. Initial Testing of Shellfish Collected during the 2016 D. norvegica Bloom in the Gulf of Maine

The 2016 *D. norvegica* bloom in the Penobscot and Frenchman Bay regions of the central coast of the Gulf of Maine lasted for approximately two months (July 5th–August 29th) (Figures 1 and 2). During that time, numerous shellfish samples were found to display DST-like activity, based on PPIA screening, in excess of 0.16 ppm (Figure 2), but subsequent testing for OA, DTX1, and DTX2 by confirmatory LC-MS/MS analysis could not confirm the presence of these compounds.

In an attempt to resolve this discrepancy, ten initial samples of frozen homogenized mussels (*Mytilus edulis*) were sent to three additional laboratories for follow-up testing. Eight were samples collected during the 2016 *D. norvegica* bloom that had all previously tested positive for DST-like activity based on PPIA screening while an additional two samples, collected prior to the bloom and outside of the bloom area, were sent as negative controls. With the exception of one sample, all samples were found to be <LOD for OA, DTX1, and DTX2 by LC-MS/MS testing. All samples were also shown to be <LOD for the additional lipophilic shellfish toxins azaspiracids, pectenotoxins, and yessotoxins as tested using LC-HRMS analysis. The single sample found to contain a detectable concentration of DSTs by LC-MS/MS was actually one of the presumptive negative control samples collected on 6/27/16, prior to the *D. norvegica* bloom and from south of the bloom area (Sample 10 from Figure 1; 0.04 and 0.05 ppm DTX1 only as tested by two independent laboratories). All samples were re-tested at the FDA Center for Food Safety and Applied Nutrition (CFSAN) by PPIA and LC-MS/MS and results were found to be consistent with previous testing. Further testing using both a qualitative lateral flow ELISA (NEOGEN Reveal 2.0 for DSP) and a quantitative ELISA (Bioo Scientific MaxSignal Okadaic Acid (DSP)) showed results consistent with PPIA testing, suggesting the presence of DST-like compound(s) based on both bioactivity and structure, but a compound distinct from OA, DTX1, DTX2, or any esterified derivatives thereof (Table 1).

2.2. Bioactivity-Guided Fractionation

To first test the utility of the bioactivity-guided fractionation procedure, 500 µL of hydrolyzed extract of clam (*Mercenaria mercenaria*) homogenate that had previously been spiked with 0.16 ppm each of OA, DTX1, and DTX2 was fractionated and screened for PPIA activity. Three fractions were found to contain DST-like activity: the fractions collected from 23–24 min, 24–25 min, and 26–27 min. Subsequent testing of these fractions by LC-MS/MS confirmed them as containing OA, DTX2, and DTX1, respectively (Figure 3A). Next, 500 µL of hydrolyzed mussel extract that showed the highest PPIA activity from the initial set of 10 samples supplied by ME Department of Marine Resources (DMR) (Sample 2 from Table 1, >0.35 ppm OA eq. activity using the PPIA kit and 0.67 ppm OA eq. using the MaxSignal quantitative ELISA) was also fractionated. PPIA testing showed a prominent peak in bioactivity in the 26-27 min fraction, which was closest in elution order to DTX1 (Figure 3B). To determine the molecular ion of the compound responsible for the PPIA activity in fraction 26 of the mussel extract, full-scan experiments were performed by Q1 scanning in negative polarity (see Section 4.5.2). From an extracted ion chromatogram for *m/z* 800–840, the expected mass range for DST-like compounds, an abundant peak was observed at a retention time of 7.3 min, close to the known retention time of DTX1, using the chromatographic method described herein. The predominant ion observed in negative ionization mode for this chromatographic peak was *m/z* 819.5, two Da larger than DTX1.

A product ion scan was then performed to obtain MS/MS fragmentation information for *m/z* 819.5 and to select product ions for SRM method development. The top three most abundant product ions were the same as those for known DSTs, *m/z* 255.2, 151.1, and 131.1 (Figure S1).

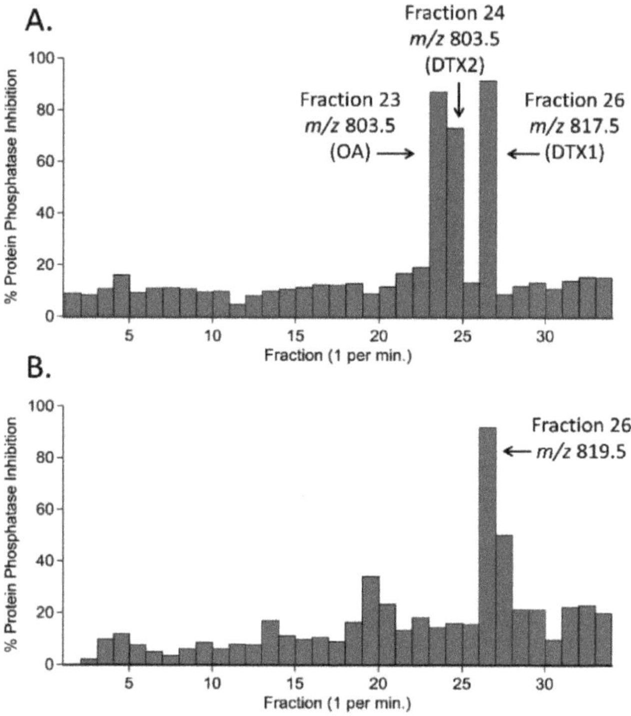

Figure 3. Results from the bioactivity-guided fractionation procedure for (**A**) control extract of clam (*Mercenaria mercenaria*) homogenate spiked with 0.16 ppm each of OA, DTX1, and DTX2, and (**B**) extract of mussel (*Mytilus edulis*) collected during 2016 *Dinophysis norvegica* bloom in the Gulf of Maine (corresponds to Sample 2 in Figure 1 and Table 1). LC-MS/MS confirmed OA, DTX2, and DTX1 in the spiked control sample, and the tentative toxin at (M-H)⁻ *m/z* 819.5 in the mussel extract.

2.3. Liquid Chromatography-High Resolution Mass Spectrometry (LC-HRMS) Measurements of the PPIA Active Fraction

A peak detected in the 26-27 min fraction showed a measured accurate mass of *m/z* 819.4908, which corresponds to the (M-H)⁻ of $C_{45}H_{71}O_{13}$ (Δm = 0.9376 ppm). This compound differed from the measured accurate mass for a DTX1 standard of *m/z* 817.4753 (Δm = 1.1020 ppm for $C_{45}H_{69}O_{13}$) by the addition of two hydrogen atoms, suggesting it was a dihydro derivative of DTX1. Furthermore, the LC-high-resolution MS/MS spectrum of this compound showed several product ions that were also consistent with the MS/MS spectrum of the DTX1 standard. These product ions were *m/z* 255.1240 as the most abundant ion, followed by *m/z* 113.0609 and *m/z* 151.0766. The corresponding product ions in the DTX1 standard were *m/z* 255.1242, *m/z* 113.0610, and *m/z* 151.0766. An additional product ion observed for the tentative dihydro-DTX1 was *m/z* 565.3014, corresponding to the chemical formula $C_{30}H_{45}O_{10}$ (Δm = −0.7591 ppm), and is 2 mass units higher than the corresponding product ion (*m/z* 563.2861; $C_{30}H_{43}O_{10}$, Δm = −0.1296 ppm) for DTX1, indicating that the two additional hydrogen atoms are contained in this segment of the molecule (Figure 4). Specifically, the double bond at C14,15 is potentially modified on the tentative dihydro-DTX1.

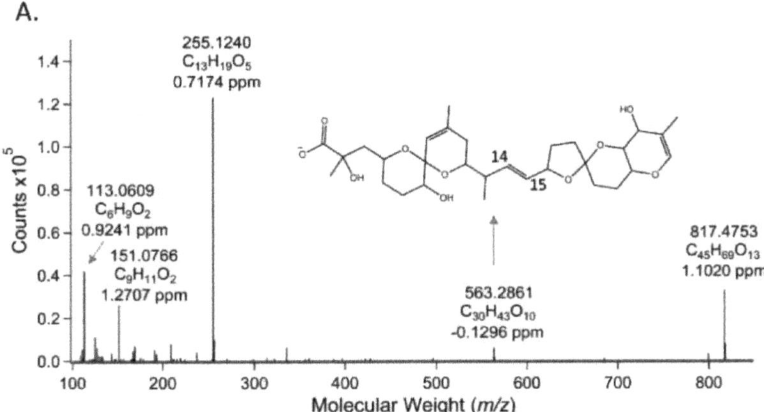

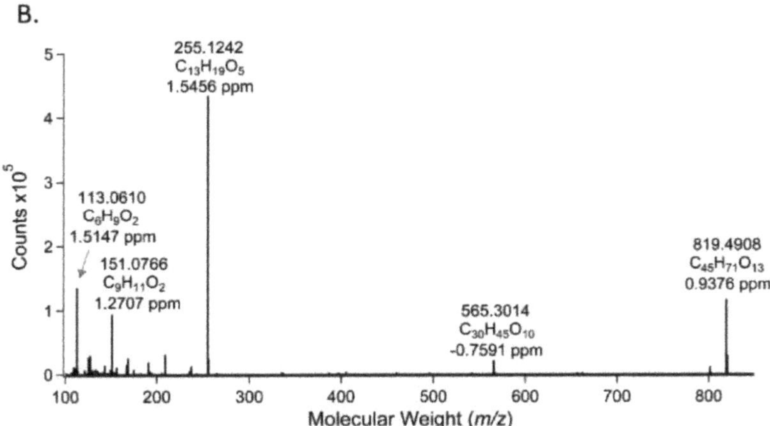

Figure 4. MS/MS spectra of (**A**) DTX1 certified reference material at (M-H)$^-$ *m/z* 817.5 and (**B**) suspect DST at (M-H)$^-$ *m/z* 819.5 from the 26-27-minute fraction from the PPIA bioactivity-guided fractionation procedure (depicted in Figure 3) for Sample 2 from Figure 1 and Table 1. Structure in panel A depicts (M-H)$^-$ *m/z* 563.2861 product ion showing the location of the 14-15 double bond. Structure of corresponding product ion in panel B with (M-H)$^-$ *m/z* 565.3014, from compound tentatively identified as dihydro-DTX1, has yet to be confirmed. Full structures of OA, DTX1, and DTX2, as well as proposed structures for the 255.2, 151.1, and 113.1 product ions used in the SRM analysis, are provided in Figure S2.

2.4. Selected Reaction Monitoring (SRM) Analysis for DSP Toxins and Dihydro-DTX1 in Water and Shellfish Samples and Comparison with PPIA

Using the data acquired from the previous analyses, minor modifications were made to the NSSP-approved LC-MS/MS method for the analysis of total (free plus esterified) DSP toxins in shellfish to include the preliminary detection of dihydro-DTX1 (detailed in Section 4.5.4 below). Using the modified LC-MS/MS method with data acquired in SRM scan mode, samples collected from the Gulf of Maine in 2016 and 2018 during blooms of *D. norvegica* were analyzed for the presence of these four compounds, and dihydro-DTX1 was confirmed to be the only DST present in shellfish as well as water (Figure 5).

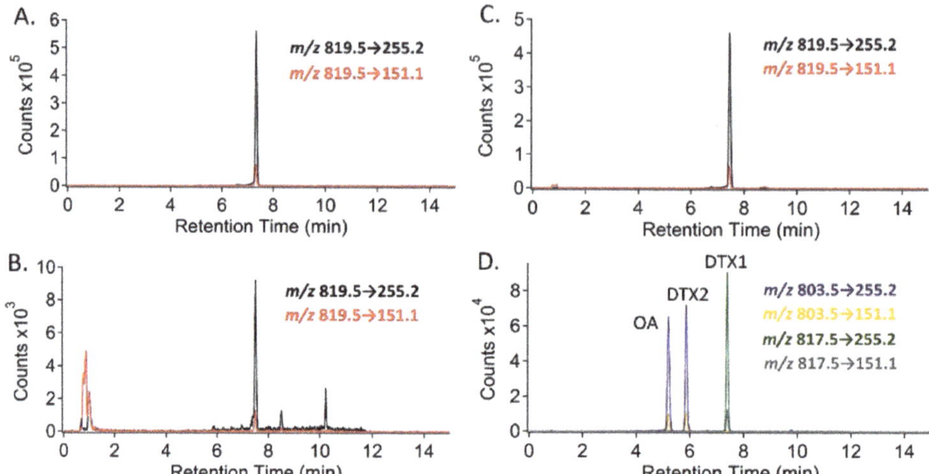

Figure 5. Extracted ion chromatograms from LC-MS/MS SRM analysis for dihydro-DTX1 ((M-H)⁻ *m/z* 819.5), fragmenting to *m/z* 255.2 (for quantitation) (black trace) and 151.1 (for confirmation) (red trace), for (**A**) 26–27 min fraction from bioactivity guided fractionation procedure on mussel (*M. edulis*) extract from 2016 *D. norvegica* bloom in the Gulf of Maine, (**B**) filtered water sample collected during 2018 *D. norvegica* bloom in the Gulf of Maine, and (**C**) representative mussel (*M. edulis*) sample collected during 2018 *D. norvegica* bloom in the Gulf of Maine. OA ((M-H)⁻ *m/z* 803.5), DTX1 ((M-H)⁻ *m/z* 817.5), and DTX2 ((M-H)⁻ *m/z* 803.5), all fragmenting to *m/z* 255.2 and 151.1, were monitored for and were not detected (not shown). Panel (**D**) shows extracted ion chromatograms for a 12.8 ng/mL standard mix of OA, DTX1, and DTX2 run in the same analytical batch as panel C for comparison of relative retention times using the current chromatography for the approved LC-MS/MS method.

Next, the original 10 shellfish samples analyzed by multiple labs in 2016 were re-analyzed using the modified LC-MS/MS method and samples 1–8, collected during the 2016 *D. norvegica* bloom, were all confirmed to contain dihydro-DTX1 at concentrations that matched well with previous results generated by PPIA and ELISA (Table 1). In addition, shellfish samples from multiple species, including mussels (*Mytilus edulis*), oysters (*Crassostrea virginica*), and clams (*Spisula solidissima* and *Mya arenaria*), collected in 2016 and 2018 (N = 48 total) were analyzed both by PPIA and by the modified LC-MS/MS method and results were compared using linear regression and correlation analyses for any sample >LOD of the PPIA kit (N = 42). In the absence of an analytical standard for dihydro-DTX1, a DTX1 standard was used for external calibration, assuming an equivalent molar response [8]. The linear regression analysis found a slope of 1.13 ± 0.07, with an r² of 0.86. Correlation analysis of results from these same samples found the two analyses to be significantly correlated (*p* < 0.0001) with a Pearson r score of 0.9260 (Figure 6).

Lastly, in order to test for species-specific differences in the bioaccumulation of dihydro-DTX1 in various commercial shellfish species from the Gulf of Maine, three species of shellfish: mussels (*M. edulis*), clams (*S. solidissima*), and oysters (*C. virginica*), were collected approximately weekly between May 30th and June 18th from a single location (Blue Hill Falls, symbol (△) from Figure 1) during the 2018 *D. norvegica* bloom and analyzed by LC-MS/MS. All three species were found to bioaccumulate dihydro-DTX1 with the rank order mussels > clams > oysters. Even though clams and oysters were found to accumulate less toxin overall compared to mussels, all three species would have exceeded regulatory guidance levels on at least one occasion during the course of the bloom, assuming equivalent potency and molar response to DTX1 (Figure 7).

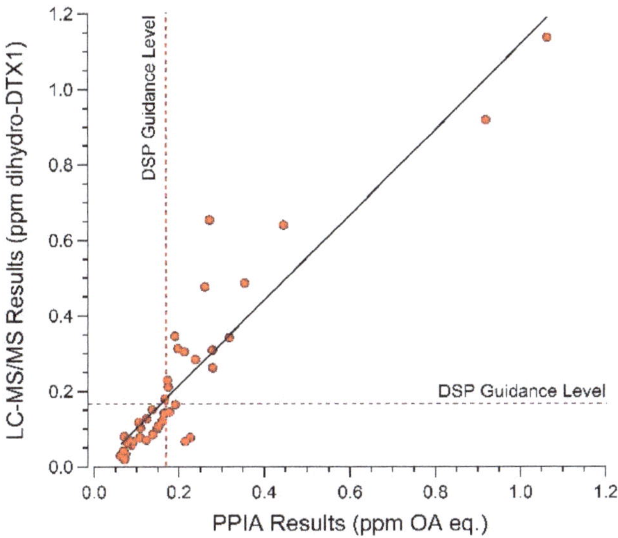

Figure 6. Comparison of protein phosphatase inhibition, as measured using a commercial PPIA kit, and dihydro-DTX1 concentration, as measured by LC-MS/MS SRM analysis, for mussels (*Mytilus edulis*), oysters (*Crassostrea virginica*), and clams (*Spisula solidissima,* and *Mya arenaria*) (N = 42 total) collected in 2016 and 2018 from the central coast of the Gulf of Maine during blooms of *Dinophysis norvegica*. Dashed line indicates DSP guidance level of 0.16 ppm OA eq. Solid line indicates best fit linear regression for the two data sets (slope 1.13 ± 0.07, r^2 0.86).

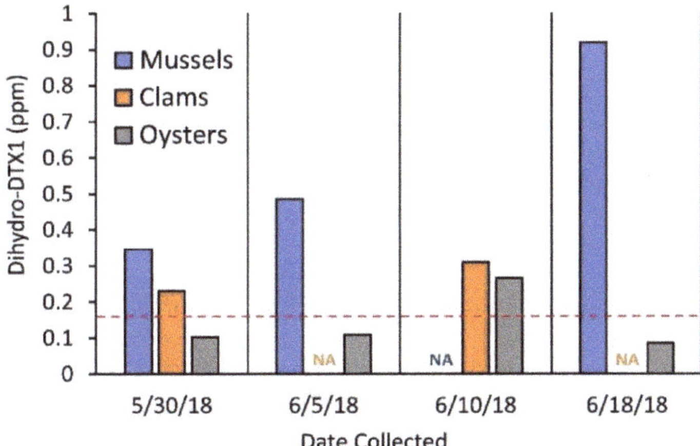

Figure 7. Dihydro-DTX1 accumulation in mussels (*Mytilus edulis*), clams (*Spisula solidissima*), and oysters (*Crassostrea virginica*) collected between May 30th and June 18th from Blue Hill Falls, Maine during the 2018 Gulf of Maine *Dinophysis norvegica* bloom. Each bar represents a single composite shellfish sample (≥12 individuals each). Dashed line indicates DSP guidance level of 0.16 ppm OA eq.

2.5. Production of Dihydro-DTX1 in a Culture of Gulf of Maine D. norvegica

Isolates of *D. norvegica* from the Gulf of Maine contained dihydro-DTX1 and pectenotoxin 2 (PTX2) in both the intracellular and extracellular fractions of the cultures (Table 2). OA, DTX1, and DTX2 were not detected in the extracts of *D. norvegica* cultures.

Table 2. Concentrations of dihydro-DTX1 and PTX2 in methanolic extracts from the intracellular and extracellular fractions of two cultures of *D. norvegica* isolated from the Gulf of Maine in 2018, as measured by LC-MS/MS. OA, DTX1, and DTX2 were also analyzed, but were not detected.

Isolate	dihydro-DTX1 (ppb)						PTX2 (ppb)	
	Intracellular			Extracellular			Intracellular	Extracellular
	Free	Esterified	Total	Free	Esterified	Total		
DNBH-FB4	1	60.6	61.5	1.3	5.3	6.5	36.3	3.9
DNBH-B3F	0.5	43.1	43.6	1.4	7.1	8.5	26.5	2.5

Once subjected to alkaline hydrolysis, the toxin concentrations of dihydro-DTX1 in the intracellular fraction increased 60–80 fold, indicating the extensive presence of esterified toxin derivatives (Table 2). More specifically, the esterified toxins made up 98% of the total intracellular dihydro-DTX1 pool (free + esterified dihydro-DTX1) in both isolates. Extracellular dihydro-DTX1 derivatives were also present in the medium and dominated over the parent toxin with 80–83% present as esterified toxins. Overall, the amount of the parent congener, dihydro-DTX1, was similar inside and outside the cells; however, once esterified dihydro-DTX1 toxins were included in the analysis, significantly more toxin was located within the cells.

3. Discussion

Multiple cases of severe vomiting and diarrhea associated with the consumption of mussels and scallops, with a failure to detect any known pathogenic microorganisms, occurring in Japan in 1976–1977 led to the first description of the syndrome known as Diarrhetic Shellfish Poisoning (DSP) [9]. The source organism for this toxic syndrome, in Japan, was determined to be the dinoflagellate *Dinophysis fortii* and the unknown toxin was named dinophysistoxin [10]. Structural elucidation of this toxin, termed dinophysistoxin-1 (DTX1), determined it to be a novel polyether derivative of a C38 fatty acid that was closely related to the similar toxins okadaic acid (35-desmethyl-DTX1) and acanthifolicin (9,10-episulfide-35-desmethyl-DTX1) that had previously been identified during bioassay-guided searches for novel antitumor agents from marine sponges [11–13]. OA was later determined to also be produced by dinoflagellates, such as *Prorocentrum lima* and *D. acuminata* and to be the primary DSP toxin found in shellfish in Europe [14–16]. Another dinophysistoxin (DTX2, 31-desmethyl-35-methyl-OA) was later found in shellfish from Ireland [17]. To date, these three toxins, along with their 7-*O*-acyl shellfish-derived metabolites, have been the primary toxins found in shellfish associated with DSP events worldwide.

Since the original descriptions of OA, DTX1, and DTX2, a limited number of additional DST-derivatives have been described, although they have never been found to be present in shellfish in sufficient quantities to cause DSP. 2-Deoxy-OA and 7-deoxy-OA were both isolated as minor constituents from the dinoflagellate *P. lima* [18,19], 19-epi-OA was isolated as a minor constituent from *P. belezianum* [20], 19-epi-DTX1 and 19-epi-DTX2 were identified as minor impurities during the production of DSP-certified reference materials [21], while 14,15-dihydro-OA was isolated as a minor constituent from the marine sponge *Halichondria okadai* during the original isolation of OA [22]. Using a similar bioassay-guided isolation approach as was used to first identify OA from *H. okadai*, with the goal of identifying novel biologically active compounds from marine organisms, 14,15-dihydrodinophysistoxin-1, along with roughly equivalent amounts of OA and DTX1, was isolated from marine sponges (*Phakellia* sp.) collected in waters from the central coast of Maine in 1985 and 1986 [18,23]. Although this compound was first reported over 20 years ago, its biological source was never determined, and it has not appeared in the scientific literature again to date.

The first record of *D. norvegica* blooming in North America occurred in Bedford Basin, Nova Scotia on the Atlantic coast of Canada in 1990 [24]. Maximum cell densities were 4.5×10^5 cells L^{-1} and DST(s) were reported to be present both in net tow plankton samples as well as in experimentally exposed scallops (*Placopecten magellanicus*) based on a commercial ELISA. In that same year, the first cases of DSP were reported in North America from mussels harvested from Mahone Bay, Nova Scotia, Canada [25,26]. During that event, *D. norvegica* was reported in plankton samples collected from Mahone Bay, but no DSP toxins were detected [25] (methods not described). Shellfish from the DSP event were positive for DST-like activity by mouse bioassay and DTX1 was confirmed by a combination of methods including LC-MS and proton NMR spectroscopy [26]. *P. lima*, cultured from water samples collected during the same time period, was subsequently shown to produce both OA and DTX1 and the event was attributed to *P. lima* [25]. At that time, the methods required to culture *Dinophysis* spp. had not yet been established, so the contribution of *Dinophysis* to this DSP event could not be determined.

Not long after the first DSP event in eastern Canada, several unexplained incidents of shellfish-related gastroenteritis in Maine, USA prompted a study to assess the prevalence of potential DST-producing organisms and the occurrence of DST-like toxicity in mussels along the entire state coast [27]. In that study, DST-like activity in mussel hepatopancreas was detected using a PPIA test only in the central coastal region in the vicinity of Eastern Bay and Frenchman Bay, the same region as our current study. The most prevalent potentially DST-producing species found in this region at the time was *D. norvegica*, but analysis of net tow samples were negative for protein phosphatase inhibitory (PPI) activity. Both mussels and plankton were also negative for OA and DTX1 by LC-MS/MS testing, but the authors noted that the bioactivity detected in mussels by PPIA was likely below the detection limit for LC-MS/MS. Although lower in number, *P. lima* was documented in the epiphytic community and testing of a concentrated sample of this material detected DST-like activity using the PPIA test with the presence of DTX1 confirmed by LC-MS/MS; therefore, the authors concluded that *P. lima* was the source of the PPI activity in mussels during that study. Follow-on studies performed between 2001 and 2003 looking at the seasonal distribution of potential DST-producing dinoflagellates and DSP-like bioactivity in plankton and shellfish in several northeastern states found a weak but significant correlation between PPI activity and the presence of *P. lima* in the epibiota [28,29]. The authors also noted the presence of both *D. acuminata* and *D. norvegica* throughout the study region, with *D. norvegica* being the more prevalent of the two species in the coastal Gulf of Maine, but found PPI activity associated with a bloom of *D. norvegica* only once [29]. Overall, toxins in shellfish were found only rarely and only at low levels and therefore the authors concluded that the threat of DSP in the region was minimal despite the presence of several potential DSP-producing species [29].

We show here that *D. norvegica* in the central coast of the Gulf of Maine, both in culture and in-situ, produces a DST-like toxin distinct from OA, DTX1, or DTX2, the only DSTs associated with DSP worldwide to date. The toxin has been tentatively identified as a dihydro-derivative of DTX1 and is likely 14,15 dihydro-DTX1 as was described from marine sponges collected from this same region over 30 years ago, although confirmation of this will require further structural elucidation studies. The presence of this toxin, as quantified against a DTX1 standard by LC-MS/MS analysis, the current reference method for DST testing under the National Shellfish Sanitation Program in the US [30], correlated well with PPIA testing, indicating that this toxin alone can explain the DST-like activity detected in this region in association with blooms of *D. norvegica* in 2016 and 2018. This also suggests that dihydro-DTX1 binds equivalently to the PP2A enzyme used in the commercial PPIA kit compared to OA, the standard utilized in the kit. During the original description of 14,15-dihydro-DTX1, this compound was reported to be equally potent as compared to OA and DTX1 using a cytotoxicity assay against L-1210 leukemia cells [23], but in other studies looking at the relative inhibitory potencies of various OA-derivatives, including 14,15-dihydro-OA, this compound was reported to have a higher dissociation constant (lower affinity) for both PP1 and PP2A, isolated from rabbit skeletal muscle, compared to OA and DTX1 [22]. This was proposed to be due to the conversion of the double bond between C-14 and C-15 to a single bond, which the authors hypothesized to stabilize the circular confirmation known to be important for the binding of these compounds to protein phosphatases [22]. Determination of the relative potency of the dihydro-DTX1 produced by *D. norvegica* compared to the other known DSP-causing DSTs will require further testing once a better characterized, analytically pure preparation of this compound is available. In the meantime, it appears that PPIA screening with confirmatory LC-MS/MS testing utilizing *m/z* 819.5 with product ions at *m/z* 255.2, for quantitation, and 151.1 and/or 113.1 for confirmation, in negative ion mode with acidic chromatography according to the current NSSP method is a viable means of screening for this toxin in shellfish until a formal analyte extension study can be performed. As with all marine biotoxins, chemical analytical methods such as HPLC or LC-MS/MS depend greatly on the availability of accurate toxicity equivalency factors (TEFs) [31]. The determination of a TEF for the dihydro-DTX1 compound is a priority so that the risk this toxin poses to human consumers can be properly assessed, especially considering that this toxin has not been conclusively linked to any DSP-like illnesses to date. This work is currently in progress.

The central coast of the Gulf of Maine is not the only region in the US where *D. norvegica* occurs, although it does appear to be the only region in the US that we are aware of where it is the predominant species of *Dinophysis* present. *D. norvegica* has been documented in lower abundance compared to species such as *D. acuminata* in the mid-Atlantic region [7], the Pacific Northwest [4], and along the central coast of California [32]. Historically, *D. norvegica* was shown to reach high abundances on the Atlantic coast of Canada, particularly Nova Scotia [24], and still occurs occasionally in high abundance along with *D. acuminata* in that region today (Nancy Lewis, personal communication). Worldwide, *D. norvegica* occurs in many cold-water coastal environments, such as Norway, where it commonly co-occurs with other *Dinophysis* species such as *D. acuta* and *D. acuminata* [33], and the Baltic Sea [34], but it is often considered a minor contributor to DSP events [35,36]. Studies looking at toxins in pooled cells of *D. norvegica* picked from environmental samples in Norway [37] and Japan [38] found PTX in 10/10 samples, but found low levels of OA, as measured by targeted LC-MS/MS, in only 1/10 samples, with no DTX1 or DTX2 detected. It remains to be determined if *D. norvegica* in any of these additional locations also produces the dihydro-DTX1 compound and what potential risk this poses to shellfish consumers.

4. Materials and Methods

4.1. Phytoplankton and Shellfish Sampling

Individual phytoplankton samples were each comprised of 10 liters of surface seawater collected from shore and gravity filtered through a 20 μm sieve. Once the seawater had fully drained, the cylinder was inverted and placed onto a funnel attached to a 50 mL centrifuge tube. Phytoplankton collected on the sieve were rinsed into the collection tube using freshly filtered seawater. Sample tubes were suspended by a rack in a cooler held between 0–10 °C during transport. Once transported to the lab, samples were held at 0–4 °C and analyzed within 24 h. If samples could not be analyzed within this timeframe, they were fixed with 80 μL of Lugol's iodine and stored at 0–4 °C until analysis.

Swift M10 compound microscopes (Swift Optical Instruments, Schertz, TX, USA) were used at 100X magnification for phytoplankton enumeration. Each sample tube was gently inverted three times and 1 mL of the sample aliquoted onto a Sedgwick Rafter gridded slide for enumeration. Using light microscopy, 0.2 mL (200 grids) were analyzed, after which the sample was discarded, the slide rinsed with deionized water, and a second sample was loaded and analyzed. The genus *Dinophysis* was enumerated using morphological characteristics. The total counts of individual cells in the two separate aliquots were averaged with numbers reported in cells liter^{-1}.

Each bivalve sample was comprised of 12–15 specimens for a given species that were composited for analysis. Samples were transported live in a cooler held at a temperature of 0–10 °C. Upon delivery to the lab, samples were stored at 0–4 °C and were processed within 24 h of collection. For processing, each sample was rinsed, drained, shucked, and the tissues were homogenized by a blender. Homogenized samples were stored at 0–4 °C and extracted within 24 h. Samples not scheduled for analysis within 24 h were frozen at −20 °C until extraction. Remaining homogenates were stored at −20 °C and after 2 months, archived in long-term storage at −80 °C. Additional aliquots for further chemical analysis as part of this study were taken from these archived samples.

4.2. Standards and Reagents

Certified reference materials (CRMs) for OA, DTX1, and DTX2 were purchased from the National Research Council Canada (Halifax, NS, Canada). All reagents were purchased from Sigma-Aldrich (St. Louis, MO, USA) and were of analytical grade or better. All solvents were purchased from Fisher Scientific (Pittsburgh, PA, USA) and were LC-MS grade.

4.3. Commercial Test Kits

Three commercial test kits for the determination of DSP toxins in shellfish were utilized in the study: 1. Okadaic Acid (DSP) PP2A kit (PN 520025, Abraxis, Inc. Warminster, PA, USA), 2. Reveal 2.0 for DSP (NEOGEN Corporation, Lansing, MI, USA), and 3. MaxSignal Okadaic Acid (DSP) ELISA Test Kit (Bioo Scientific Corp, Austin, TX, USA). The Okadaic Acid (DSP) PP2A kit is an *in-vitro* test based on the inhibition of protein phosphatase 2A, a known biological target for DSTs. The Reveal 2.0 for DSP kit is a qualitative antibody-based lateral flow test, while the MaxSignal Okadaic Acid (DSP) ELISA kit is a quantitative micro-well plate antibody-based test. For all kits, the manufacturer's instructions were followed, with the exception of the Okadaic Acid (DSP) PP2A kit where 2 g of shellfish homogenate was extracted in a total of 20 mL solvent (1:10 dilution) as opposed to the current manufacturer's instructions of 5 g homogenate extracted in 25 mL solvent (1:5 dilution). To compensate for the different extraction dilutions, the total volume of the dilution in buffer after sample hydrolysis was changed from 20 mL to 10 mL (Section D, Step 12 of manufacturer's instructions). This adjustment corresponds to the standard extraction for the validated LC-MS/MS method for DSTs and allows the same extract to be used for both analyses.

4.4. Bioactivity-Guided Fractionation

Purification of DSP-like compound(s) from Maine shellfish was achieved through bioactivity guided fractionation following a modification of the method described in [39]. First, a homogenized composite sample of 12 mussels (*Mytilus edulis*), collected in 2016 from the Gulf of Maine during the *D. norvegica* bloom and previously shown to contain >0.35 ppm OA equivalent activity using the PPIA kit and 0.67 ppm OA eq. using the MaxSignal quantitative ELISA, both described in the "Commercial Test Kit" section above, was extracted following the standard procedure used in the LC-MS/MS analysis described below (2 g homogenate extracted twice with 9 mL each of MeOH with the final volume adjusted with MeOH to 20 mL). A 2 mL sub-sample was hydrolyzed with 250 μL of 2.5M NaOH followed by heating in a water bath at 76 °C for 40 min. After cooling to room temperature, the sample was neutralized with 250 μL of 2.5M HCl. A sub-sample of hydrolyzed extract (500 μL) was injected onto an Agilent 1200 Series HPLC system equipped with a G1363A 900 μL extended loop injection kit and a 1260 Series model G1364C fraction collector (Agilent Technologies, Waldbronn Germany). Separations were achieved using a 4.6 × 150 mm, 5 μm particle size, Zorbax Eclipse XDB-C18 column (Agilent Technologies, Waldbronn Germany) at a flow rate of 500 μL/min using the following elution profile: 70% A:30% B hold for 2 min, increasing linearly to 100%B over 20 min, hold at 100%B for 5 min, decrease to 70%A:30% B over 3 min, hold at 70% A:30% B for 5 min. Mobile phase A consisted of 100% water with 0.1% TFA and mobile phase B consisted of 100% acetonitrile with 0.1% TFA following [39]. Fractions were collected every minute over the entire 35 min run time. All fractions were tested for protein phosphatase inhibitory activity using the commercial kit described in Section 4.3. As a positive control, a homogenate of clam (*Mercenaria mercenaria*), previously shown to be <LOD for DSP toxins by LC-MS/MS, was spiked with 0.16 ppm each of methanolic reference solutions of OA, DTX1, and DTX2, then extracted, fractionated, and analyzed for PPIA activity as described above.

4.5. Liquid Chromatography-Mass Spectrometry

4.5.1. Lipophilic Toxin Screening by Liquid Chromatography-High Resolution Mass Spectrometry (LC-HRMS) Analysis

Initial lipophilic toxin screening using LC-HRMS analyses, performed at NRC Canada, of the first set of 10 samples collected by ME DMR in 2016 for DSTs as well as azaspiracids, pectenotoxins, and yessotoxins was performed according to [40] and [41].

4.5.2. Q1 Scanning and MS/MS Analysis of the Unknown DST-Like Compound

Initial LC-MS experiments on the semi-purified extract from the 26–27 min fraction obtained from the PPIA bioactivity-guided fractionation of Sample 2 (from Figure 1 and Table 1) were performed using an Acquity Ultra-Performance Liquid Chromatography system (Waters Corporation, Manchester, UK) coupled to a Sciex QTrap 5500 mass spectrometer equipped with a Turbo V ionization source (SCIEX, Framingham, MA, USA). The column used for separations was a Waters BEH C18 (1.7 μm, 1.0 mm × 150 mm) (Waters Corp., Milford, MA). The aqueous mobile phase (A) consisted of 2 mM ammonium formate and 50 mM formic acid in water. The organic mobile phase (B) consisted of 2 mM ammonium formate and 50 mM formic acid in 95% acetonitrile/5% water. For initial method development studies, gradient conditions started at 50% B, were maintained for two min at 50% B, and were then linearly increased to 70% B in four min, followed by 100% B in two min, held at 100% B for five min, then decreased to 50% B in 0.5 min and lastly, were held at 50% B for 4.5 min. The total run time was 18 min at a flow rate of 0.12 mL/min. The column temperature was maintained at 40 °C while the autosampler temperature was 10 °C. The injection volume was 5 μL.

The electrospray ionization (ESI) source parameters were as follows: source temperature 550 °C, ion spray voltage −4500 V, curtain gas 25 psi, gas 1 and 2 both 40 psi. Full scan, negative ionization mode data were collected using a mass range from 500 to 900 Da and a scan rate of 200 Da s^{-1}. Product ion scans for m/z 819.50 were collected using a mass range from 100 to 900 Da and a scan rate of 1000 Da s^{-1}. The declustering and entrance potentials were −110 V and −10 V, respectively, and for product ion scans the collision energy was −70 V and collision cell exit potential was −15 V. Analyst®chromatography software (ver. 1.6.2, SCIEX, Framingham, MA, USA) was used for data visualization and analysis.

4.5.3. LC-HRMS Analysis of Dihydrodinophysistoxin-1

LC-HRMS measurements of the semi-purified extract from the 26–27 min fraction of Sample 2 (from Figure 1 and Table 1) was performed using a Nexera LC system (Shimadzu, Columbia, MD, USA) coupled with a Q Exactive Hybrid Quadrupole-Orbitrap mass spectrometer (Thermo Scientific, Waltham, MA, USA). The column and mobile phases were the same as described in Section 4.5.2., with the LC run time shortened to 15 min. Gradient conditions starting at 50% B were maintained for two min, then linearly increased to 70% B in four min, followed by 99% B in two min, held at 99% B for two min, then decreased to 50% B in 0.5 min and held at this level for 4.5 min to re-equilibrate. All other chromatography and autosampler settings were the same as described in Section 4.5.2.

Analytes were ionized using ESI in negative mode with source conditions as follows: spray voltage −3000 V, capillary temperature 320 °C, sheath gas 5 arbitrary units (au), and aux gas 0 au. A targeted-single ion monitoring (SIM)/data dependent (dd)-MS^2 experiment was performed on the sample. The instrument was set to monitor and perform MS/MS on m/z 819.49002. The parameters for SIM were a mass resolution setting of 70,000, automatic gain control (AGC) of 2×10^5, maximum injection time of 200 ms, and an isolation window of 2 m/z. The parameters for dd-MS^2 were as follows:

mass resolution setting of 35,000, AGC 2×10^5, maximum injection time 100 ms, normalized collision energy 35. The dd settings to initiate MS/MS was a minimum AGC of 8×10^3.

4.5.4. LC-MS/MS Selected Reaction Monitoring (SRM) Analysis for OA, DTX1, DTX2, and Dihydro-DTX1

LC-MS/MS testing by SRM was performed using an Acquity Ultra-Performance Liquid Chromatography system coupled to a Sciex QTrap 5500 mass spectrometer equipped with a Turbo V ionization source. The protocol "LC-MS/MS Method for the Detection of DSP Toxins in Shellfish" [42] that was adopted in 2017 by the Interstate Shellfish Sanitation Conference (ISSC) for use in the US National Shellfish Sanitation Program (NSSP) [30] was followed with minor modifications to include the measurement of dihydro-DTX1 as detailed below. All samples were subjected to alkaline hydrolysis, following the referenced protocol, to measure total (free plus esterified) toxins. The column, mobile phase, gradient conditions, and ESI source parameters were the same as those used for LC-HRMS measurements (Section 4.5.2) and in the ISSC protocol. Data acquisition was in negative ionization mode using SRM. The SRM parameters for dihydro-DTX1 were as follows: Q1 m/z 819.5, Q3 m/z 255.2 and 151.1, dwell time 100 ms, declustering potential -110 V, entrance potential -10 V, collision energy -70 V, and collision cell exit potential -15 V. The peak area of the SRM transition m/z 819.5$\rightarrow$255.2 was used for quantitation, while the m/z 819.5$\rightarrow$151.1 SRM transition was used for confirmation. In the absence of a standard, quantitation of dihydro-DTX1 was performed using the calibration curve of DTX1. Analyst®chromatography software (ver. 1.6.2, SCIEX, Framingham, MA, USA) was used for peak area integration and quantitation.

4.6. Analysis of Gulf of Maine Shellfish for Dihydro-DTX1 by LC-MS/MS SRM and Comparison with PPIA

To compare the determination of dihydro-DTX1 by LC-MS/MS SRM (as described in Section 4.5.4) to PPIA (as described in Section 4.3), 48 shellfish samples collected by ME DMR during blooms of *D. norvegica* in 2016 and 2018, each comprising of ≥12 composited individuals and representing mussels (*Mytilus edulis*), oysters (*Crassostrea virginica*), and clams (*Spisula solidissima* and *Mya arenaria*), were analyzed using both methods. For any samples found to be greater than the working range of the PPIA kit (>0.35 ppm OA eq.), samples were diluted using the kit-provided dilution buffer and re-analyzed. All samples >LOD for the PPIA kit (0.063 ppm) (N = 42) were compared to results determined by LC-MS/MS SRM, quantified using an external DTX1 standard curve, using both linear regression and correlation analysis with the program GraphPad Prism (ver. 5.01 for Windows, GraphPad Prism Software, San Diego, CA, USA).

During the 2018 *D. norvegica* bloom in the Gulf of Maine, three species of shellfish: mussels (*Mytilus edulis*), clams (*Spisula solidissima*), and oysters (*Crassostrea virginica*), were collected approximately weekly (≥12 composited individuals per sample) between May 30th and June 18th from a single location (Blue Hill Falls, Figure 1) and analyzed by LC-MS/MS SRM (as described in Section 4.5.4) to look for species-specific differences in the accumulation of dihydro-DTX1.

4.7. Testing of a Gulf of Maine Dinophysis norvegica Culture for DST Production

Two new cultured clonal isolates of *D. norvegica* (DNBH-FB4 and DNBH-B3F) were established in culture in May 2018 from surface water collected from Blue Hill Falls, Maine, following the single-cell isolation methods described by [43]. At the time of water collection, the salinity was 30 psu and the water temperature was 12 °C. During isolation and maintenance of the culture, *D. norvegica* were fed *Mesodinium rubrum*, which had been previously raised on *Teleaulax amphioxeia* isolated from Japan [44] following the protocols of [45] as modified by [46]. The dinoflagellate, ciliate, and cryptophyte cultures were grown in modified f/6-Si medium [47] and a salinity of 30 at 15 °C in dim light (40 μmol photons·m-2·sec-1) under a 14 h:10 h light:dark photocycle.

To assess the toxigenicity of DNBH-FB4 and DNBH-B3F, cells were inoculated into fresh medium, salinity 30 psu, at 400 cells·mL^{-1}, fed *M. rubrum* at a 1:10 ratio of prey to predator, and monitored every 3 days for the complete consumption of *M. rubrum* by examining 1 mL subsamples in a Sedgewick-Rafter counting cell at 100x using an Olympus CX31 light microscope (Olympus America, Waltham, MA, USA). Three days after all ciliate prey were consumed, (i.e., during late exponential growth of the dinoflagellate) the culture was harvested for toxin analysis.

The harvested cultures were gently separated into cells (intracellular toxins) and medium (extracellular toxins) using a 10 μm sieve and the components were extracted and analyzed for toxins separately. Cells and medium were bath-sonicated at room temperature for 15 min (Branson 5800 Ultrasonic Cleaner, 5800) and loaded onto an Oasis HLB 60 mg cartridge (Waters Corporation, Millford, MA, USA) that was previously equilibrated with 3 mL of MeOH and 3 mL of GenPure water. The cartridge was then washed with 6 mL of GenPure water, blown dry, and eluted with 1 mL of 100% MeOH into a glass 1.5 mL high recovery LC vial and stored at −20 °C until analyzed. A portion of the sample underwent alkaline hydrolysis to enable the quantitation of total DSP toxins (free plus esterified) following [48]. Extracts, original and alkaline hydrolyzed, were analyzed using an Acquity liquid chromatography system coupled with a Xevo mass spectrometer with electrospray ionization (Waters, Milford, MA, USA) following the DSP and PTX2 analytical methods described by [49].

Dihydro-DTX1 was detected using selected reaction monitoring (SRM) in negative ion mode with the transitions m/z 819.5→255.1, 819.5→819.5, 819.5→151.1, and 819.5→113.1. Quantitation was performed using the former SRM transition; concentrations were calculated using an external DTX1 standard curve with MassLynx 4.1 software (Waters Corporation, Millford, MA, USA). Matrix spikes, final concentration of 12.5 ppb DTX1, were conducted to confirm separation from DTX1. Toxin data are presented as toxin concentration per mL of culture, and free vs. esterified, with the latter being calculated through the subtraction of free toxins from total toxins.

Supplementary Materials: The following are available online at http://www.mdpi.com/2072-6651/12/9/533/s1, Table S1: Sampling site coordinates for shellfish and phytoplankton samples collected during the 2016 *Dinophysis norvegica* bloom depicted in Figures 1 and 2. Figure S1: MS/MS spectrum of m/z 819.5 peak detected in the 26–27 min fraction from the bioactivity guided fractionation extract of mussel (*Mytilus edulis*) collected during 2016 *Dinophysis norvegica* bloom in the central coast of the Gulf of Maine, USA; Figure S2: Structure of precursor ions and proposed product ion structures for okadaic acid (OA), dinophysistoxin 1 (DTX1), and dinophysistoxin 2 (DTX2).

Author Contributions: Conceptualization, J.R.D., W.L.S., J.M., and J.L.S.; Investigation, J.R.D., W.L.S., M.D.C., J.M., A.E.H., J.L.S., D.M.K., P.M., C.D.R., C.A.B., and S.K.L.; Methodology, J.R.D., W.L.S., M.D.C., J.L.S., D.M.K., and P.M.; Resources, J.M., A.E.H., B.J.L., and D.W.M.; Supervision, K.K., S.D.A., J.B., and S.K.L.; Writing—original draft, J.R.D., W.L.S., M.D.C., A.E.H. and J.L.S.; Writing—review and editing, J.M., D.M.K., P.M., S.D.A., and S.K.L. All authors have read and agreed to the published version of the manuscript.

Funding: Partial support for this research was received from the National Oceanic and Atmospheric Administration, National Centers for Coastal Ocean Science Competitive Research, Ecology and Oceanography of Harmful Algal Blooms Program under awards NA17NOS4780184 and NA19NOS4780182 to Juliette Smith (VIMS) and Jonathan Deeds (US FDA), and Prevention, Control, and Mitigation of Harmful Algal Blooms program award NA17NOS4780179 to Stephen Archer. This paper is ECOHAB publication number EC0956.

Acknowledgments: The authors would like to express their gratitude to Marta Sanderson and Han Gao (VIMS) for their assistance in culture maintenance and toxin analysis, Satoshi Nagai (National Research Institute of Fisheries Science, Japan) for his sharing of *Mesodinium rubrum* and *Teleaulax amphioxeia* cultures, Stephen Conrad (CFSAN) for assistance with figure preparation, and Claude Mallet (Waters Inc.) and Sabrina Giddings (Biotoxin Metrology, NRC Canada) for their technical assistance.

Conflicts of Interest: The authors declare no conflict of interest

References

1. Campbell, L.; Olson, R.J.; Sosik, H.M.; Abraham, A.; Henrichs, D.W.; Hyatt, C.J.; Buskey, E.J. First harmful *Dinophysis* (Dinophyceae, Dinophysiales) bloom in the U.S. revealed by automated imaging flow cytometry. *J. Phycol.* **2010**, *46*, 66–75. [CrossRef]

2. Deeds, J.R.; Wiles, K.; Heideman, G.B., VI; White, K.D.; Abraham, A. First US report of shellfish harvesting closures due to confirmed okadaic acid in Texas Gulf coast oysters. *Toxicon* **2010**, *55*, 1138–1146. [CrossRef]

3. Lloyd, J.K.; Duchin, J.S.; Borchert, J.; Flores Quintana, H.; Robertson, A. Diarrhetic shellfish poisoning, Washington, USA, 2011. *Emerg. Infec. Dis.* **2013**, *19*, 1314–1316. [CrossRef]

4. Trainer, V.L.; Moore, L.; Bill, B.D.; Adams, N.G.; Harrington, N.; Borchert, J.; da Silva, D.A.M.; Eberhart, B.L. Diarrhetic shellfish toxins and other lipophilic toxins of human health concern in Washington State. *Mar. Drugs.* **2013**, *11*, 1815–1835. [CrossRef]

5. Tango, P.; Butler, W.; Lacouture, R.; Goshorn, D.; Magnien, R.; Michael, B.; Hall, S.; Browhawn, K.; Wittman, R.; Beatty, W. An unprecedented bloom of *Dinophysis acuminata* in Chesapeake Bay. In *Harmful Algae 2002*; Steidinger, K.A., Landsberg, J.H., Tomas, C.R., Vargo, G.A., Eds.; Florida Fish and Wildlife Conservation Commission, Florida Institute of Oceanography, and Intergovernmental Oceanographic Commission of UNESCO: St. Petersburg, FL, USA, 2004; pp. 358–360.

6. Hattenrath-Lehmann, T.K.; Marcoval, M.A.; Berry, D.L.; Fire, S.; Wang, Z.; Morton, S.L.; Gobler, C.J. The emergence of *Dinophysis acuminata* blooms and DSP toxins in shellfish in New York waters. *Harmful Algae* **2013**, *26*, 33–44. [CrossRef]

7. Wolny, J.L.; Egerton, T.A.; Handy, S.M.; Stutts, W.L.; Smith, J.L.; Whereat, E.B.; Bachvaroff, T.R.; Henrichs, D.W.; Campbell, L.; Deeds, J.D. Characterization of *Dinophysis* spp. (Dinophyceae, Dinophysiales) from the mid-Atlantic region of the US. *J. Phycol.* **2020**, *56*, 404–424. [CrossRef]

8. Zendong, Z.; Sibat, M.; Herrenknecht, C.; Hess, P.; McCarron, P. Relative molar response of lipophilic marine algal toxins in liquid chromatography electrospray ionization mass spectrometry. *Rapid. Comm. Mass Spec.* **2017**, *31*, 1453–1461. [CrossRef]

9. Yasumoto, T.; Oshima, Y.; Yamaguchi, H. Occurrence of a new type of shellfish poisoning in the Tohoku district. *Bull. Jpn. Soc. Sci. Fish.* **1978**, *44*, 1249–1255. [CrossRef]

10. Yasumoto, T.; Oshima, Y.; Sugawara, W.; Fukuyo, Y.; Oguri, H.; Igarashi, T.; Fujita, N. Identification of Dinophysis firtii as the causative organism of diarrhetic shellfish poisoning. *Bull. Jpn. Soc. Sci. Fish* **1980**, *46*, 1405–1411. [CrossRef]

11. Schmitz, F.J.; Prasad, R.S.; Gopichand, Y.; Hossain, M.B.; van der Helm, D. Acanthifolicin, a new episulfide-containing polyether carboxylic acid from extracts of the marine sponge *Pandaros acanthifolium*. *J. Am. Chem. Soc.* **1981**, *103*, 2467–2469. [CrossRef]

12. Tachibami, K.; Scheuer, P.J.; Tsukitani, Y.; Kikuchi, H.; Van Engen, D.; Clardy, J.; Gopichand, Y.; Schmitz, F.J. Okadaic acid, a cytotoxic polyether from two marine sponges of the genus *Halichondria*. *J. Am. Chem. Soc.* **1981**, *103*, 2469–2471. [CrossRef]

13. Murata, M.; Shimatani, M.; Sugitani, H.; Oshima, Y.; Yasumoto, T. Isolation and structural elucidation of the causative toxin of diarrhetic shellfish poisoning. *Bull. Jpn. Soc. Sci. Fish.* **1982**, *48*, 549–552. [CrossRef]

14. Murakami, Y.; Oshima, Y.; Yasumoto, T. Identification of Okadaic acid as a toxic component of a marine dinoflagellate *Prorocentrum lima. Bull. Jpn. Soc. Sci. Fish* **1982**, *48*, 69–72. [CrossRef]

15. Yasumoto, T.; Murata, M.; Oshima, Y.; Sano, M. Diarrhetic shellfish toxins. *Tetrahedron* **1985**, *41*, 1019–1025. [CrossRef]

16. Kumagi, M.; Yanagi, T.; Murata, M.; Yasumoto, T.; Kat, M.; Lassus, P.; Rodriguez-Vasquez, J.A. Okadaic acid as the causative toxin of diarrhetic shellfish poisoning in Europe. *Agric. Biol. Chem.* **1986**, *50*, 2853–2857.

17. Hu, T.; Doyle, J.; Jackson, D.; Marr, J.; Nixon, E.; Pleasance, S.; Quilliam, M.A.; Walter, J.A.; Wright, J.L.C. Isolation of a new diarrhetic shellfish poison from Irish mussels. *J. Chem. Soc. Chem. Commun.* **1992**, *1*, 39–41. [CrossRef]

18. Schmitz, F.J.; Yasumoto, T. The 1990 United States—Japan seminar on bioorganic marine chemistry, meeting report. *J. Nat. Prod.* **1991**, *54*, 1469–1490. [CrossRef]

19. Holmes, M.J.; Lee, F.C.; Khoo, H.W.; Ming Teo, S.L. Production of 7-deoxy-okadaic acid by a New Caledonian strain of *Prorocentrum lima* (Dinophyceae). *J. Phycol.* **2001**, *37*, 280–288. [CrossRef]

20. Cruz, P.G.; Hernández Daranas, A.; Fernández, J.J.; Norte, M. 19-epi-okadaic acid, A novel protein phosphatase inhibitor with enhanced selectivity. *Org. Lett.* **2007**, *9*, 3045–3048. [CrossRef]

21. Beach, D.G.; Crain, S.; Lewis, N.; LaBlanc, P.; Hardstaff, W.R.; Perez, R.A.; Giddings, S.D.; Martinez-Farina, C.E.; Stefanova, R.; Burtan, I.W.; et al. Development of certified reference materials for diarrhetic shellfish poisoning toxins. Part 1: Calibration solutions. *J. AOAC Int.* **2016**, *99*, 1151–1162. [CrossRef]

22. Takai, A.; Murata, M.; Torigoe, K.; Isobe, M.; Mieskes, G.; Yasumoto, T. Inhibitory effect of okadaic acid derivatives on protein phosphatases. *Biochem. J.* **1992**, *284*, 539–544. [CrossRef] [PubMed]

23. Sakai, R.; Rinehart, K.L. A new polyether acid from a cold water marine sponge, a *Phykellia* species. *J. Nat. Prod.* **1995**, *58*, 773–777. [CrossRef] [PubMed]

24. Rao, D.V.S.; Pan, Y.; Zitko, V.; Bugden, G.; Mackeigan, K. Diarrhetic shellfish poisoning (DSP) associated with a subsurface bloom of *Dinophysis norvegica* in Bedford Basin, eastern Canada. *Mar. Ecol. Prog. Ser.* **1993**, *97*, 117–126.

25. Marr, J.C.; Jackson, A.E.; McLachlan, J.L. Occurrence of *Prorocentrum lima*, a DSP toxin-producing species from the Atlantic coast of Canada. *J. Appl. Phycol.* **1992**, *4*, 17–24. [CrossRef]

26. Quilliam, M.A.; Gilgan, M.W.; Pleasance, S.; de Freitas, A.S.W.; Douglas, D.; Fritz, L.; Hu, T.; Marr, J.C.; Smyth, C.; Wright, J.L.C. Confirmation of an incident of diarrhatic shellfish poisoning in eastern Canada. In *Toxic Phytoplankton Blooms in the Sea*; Smayda, T.J., Shimizu, Y., Eds.; Elsevier Science Publishers: Amsterdam, The Netherlands, 1993; pp. 547–552.

27. Morton, S.A.; Leighfield, T.A.; Haynes, B.L.; Petitpain, D.L.; Busman, M.A.; Moeller, P.D.R.; Bean, L.; McGowan, J.; Hurst, J.W., Jr.; Van Dolah, F.M. Evidence of diarrhetic shellfish poisoning along the coast of Maine. *J. Shellfish Res.* **1999**, *18*, 681–686.

28. Maranda, L.; Corwin, S.; Hargraves, P.E. *Prorocentrum lima* (Dinophyceae) in northeastern USA coastal waters: I. Abundance and distribution. *Harmful Algae* **2007**, *6*, 623–631. [CrossRef]

29. Maranda, L.; Corwin, S.; Dover, S.; Morton, S.L. *Prorocentrum lima* (Dinophyceae) in northeastern USA coastal waters: II. Toxin loads in the epibiota and in shellfish. *Harmful Algae* **2007**, *6*, 632–641. [CrossRef]

30. NSSP (National Shellfish Sanitation Program). Guide for the Control of Molluscan Shellfish: 2017 Revision. Available online: http://www.fda.gov/Food/GuidanceRegulation/FederalStateFoodPrograms/ucm2006754.htm (accessed on 8 June 2020).

31. Botana, L.M.; Hess, P.; Munday, R.; Nathalie, A.; Degrasse, S.L.; Feeley, M.; Suzuki, T.; van der Berg, M.; Fattori, V.; Gamarro, E.G.; et al. Derivation of toxicity equivalency factors for marine biotoxins associated with bivalve molluscs. *Trends Food Sci. Tech.* **2017**, *59*, 15–24. [CrossRef]

32. Schultz, D.; Campbell, L.; Kudela, R.M. Trends in *Dinophysis* adundance and diarrhetic shellfish toxin levels in California mussels (*Mytilus californianus*) from Monterey Bay, California. *Harmful Algae* **2019**, *88*, 1–12. [CrossRef]

33. Séchet, V.; Safran, P.; Hovgaard, P.; Yasumoto, T. Causative species of diarrhetic shellfish poisoning (DSP) in Norway. *Mar. Biol.* **1990**, *105*, 269–274. [CrossRef]

34. Carpenter, E.J.; Janson, S.; Boje, R.; Pollehne, P.; Chang, J. The dinoflagellate *Dinophysis norvegica*: Biological and ecological observations in the Baltic Sea. *Eur. J. Phycol.* **1995**, *30*, 1–9. [CrossRef]

35. Reguera, B.; Velo-Suárez, L.; Raine, R.; Park, M.G. Harmful *Dinophysis* species: A review. *Harmful Algae* **2012**, *14*, 87–106. [CrossRef]

36. Reguera, B.; Riobó, P.; Rodriguez, F.; Díaz, P.A.; Pizarro, G.; Paz, B.; Franco, J.M.; Blanco, J. *Dinophysis* toxins: Causative organisms, distribution, and fate in shellfish. *Mar. Drugs* **2014**, *12*, 394–461. [CrossRef] [PubMed]

37. Miles, C.O.; Wilkins, A.L.; Samdall, I.A.; Sandvik, M.; Petersen, D.; Quilliam, M.A.; Naustvoll, L.J.; Rundberget, T.; Torgersen, T.; Hovgaard, P.; et al. A novel pectenotoxin, PTX-12, in *Dinophysis* Spp. and shellfish in Norway. *Chem. Res. Toxicol.* **2004**, *17*, 1423–1433. [CrossRef] [PubMed]

38. Suzuki, T.; Miyazono, A.; Baba, K.; Sugawara, R.; Kamiyama, T. LC-MS/MS analysis of okadaic acid analogs and other lipophilic toxins in single-cell isolates of several *Dinophysis* species collected in Hokkaido, Japan. *Harmful Algae* **2009**, *8*, 233–238. [CrossRef]

39. Holmes, C.F.B. Liquid chromatography-linked protein phosphatase bioassay; a highly sensitive marine bioscreen for okadaic acid and related diarrhetic shellfish toxins. *Toxicon* **1991**, *29*, 469–477. [CrossRef]

40. McCarron, P.; Wright, E.; Quilliam, M.A. Liquid chromatography/mass spectrometry of domoic acid and lipophilic shellfish toxins with selected reaction monitoring and optional confirmation by library searching of product ion spectra. *J. AOAC Int.* **2014**, *97*, 316–324. [CrossRef]

41. Blay, P.; Hui, J.P.M.; Chang, J.M.; Melanson, J.E. Screening for multiple classes of marine biotoxins by liquid chromatography-high resolution mass spectrometry. *Anal. Bioanal. Chem.* **2011**, *400*, 577–585. [CrossRef]

42. Liquid Chromatography Tandem Mass Spectrometry (LC-MS/MS) Method for the Determination of Diarrhetic Shellfish Poisoning (DSP) Toxins in Shellfish. Available online: http://www.issc.org/Data/Sites/1/media/00-2017biennialmeeting/--taskforcei2017/17-103-supporting-documentation.pdf (accessed on 8 June 2020).

43. Tong, M.; Kulis, D.M.; Fux, E.; Smith, J.L.; Hess, P.; Zhou, Q.; Anderson, D.M. The effects of growth phase and light intensity on toxin production by *Dinophysis acuminata* from the northeastern United States. *Harmful Algae* **2011**, *10*, 254–264. [CrossRef]

44. Nishitani, G.O.H.; Nagai, S.; Sakiyama, S.; Kamiyama, T. Successful cultivation of the toxic dinoflagellate *Dinophysis caudata* (Dinophyceae). *Plankton Benthos Res.* **2008**, *3*, 78–85. [CrossRef]
45. Park, M.G.; Kim, S.; Kim, H.S.; Myung, G.; Kang, Y.G.; Yih, W. First successful culture of the marine dinoflagellate *Dinophysis acuminata*. *Aquat. Microb. Ecol.* **2006**, *45*, 101–106. [CrossRef]
46. Hackett, J.D.; Tong, M.; Kulis, D.M.; Fux, E.; Hess, P.; Bire, R.; Anderson, D.M. DSP toxin production *de novo* in cultures of *Dinophysis acuminata* (Dinophyceae) from North America. *Harmful Algae* **2009**, *8*, 873–879. [CrossRef]
47. Anderson, D.M.; Kulis, D.M.; Doucette, G.J.; Gallagher, J.C.; Balech, E. Biogeography of toxic dinoflagellates in the genus *Alexandrium* from the northeastern United States and Canada. *Mar. Biol.* **1994**, *120*, 467–478. [CrossRef]
48. Hattenrath-Lehmann, T.K.; Lusty, M.W.; Wallace, R.B.; Haynes, B.; Wang, Z.; Broadwater, M.; Deeds, J.R.; Morton, S.L.; Hastback, W.; Porter, L.; et al. Evaluation of Rapid, Early Warning Approaches to Track Shellfish Toxins Associated with *Dinophysis* and *Alexandrium* Blooms. *Marine Drugs* **2018**, *16*, 28. [CrossRef] [PubMed]
49. Onofrio, M.D.; Mallet, C.R.; Place, A.R.; Smith, J.L. A screening tool for the direct analysis of marine and freshwater phycotoxins in organic SPATT extracts from the Chesapeake Bay. *Toxins* **2020**, *12*, 322. [CrossRef]

MDPI

St. Alban-Anlage 66

4052 Basel

Switzerland

Tel. +41 61 683 77 34

Fax +41 61 302 89 18

www.mdpi.com

Toxins Editorial Office

E-mail: toxins@mdpi.com

www.mdpi.com/journal/toxins